Sea Surface Salinity Remote Sensing

Sea Surface Salinity Remote Sensing

Special Issue Editors

Emmanuel Philippe Dinnat
Xiaobin Yin

MDPI • Basel • Beijing • Wuhan • Barcelona • Belgrade

Special Issue Editors
Emmanuel Philippe Dinnat
NASA/GSFC
USA

Xiaobin Yin
Beijing Piesat Information Technology Co. Ltd
China

Editorial Office
MDPI
St. Alban-Anlage 66
4052 Basel, Switzerland

This is a reprint of articles from the Special Issue published online in the open access journal *Remote Sensing* (ISSN 2072-4292) from 2018 to 2019 (available at: https://www.mdpi.com/journal/remotesensing/special_issues/sea_salinity)

For citation purposes, cite each article independently as indicated on the article page online and as indicated below:

LastName, A.A.; LastName, B.B.; LastName, C.C. Article Title. *Journal Name* **Year**, *Article Number*, Page Range.

ISBN 978-3-03921-076-3 (Pbk)
ISBN 978-3-03921-077-0 (PDF)

Contents

About the Special Issue Editors

Emmanuel Philippe Dinnat, (Senior Research Scientist) is with the Center of Excellence in Earth Systems Modeling & Observations (CEESMO) at Chapman University, Orange, CA, USA, and the Cryospheric Sciences Laboratory at the NASA Goddard Space Flight Center (GSFC), in Greenbelt, MD, USA. Dr. Dinnat was born in Sarcelles, France, in 1975. He received a Master of Advanced Studies degree in instrumental methods in astrophysics and spatial applications (1999), and a Ph.D. degree in computer science, telecommunications, and electronics (2003) from the University Pierre and Marie Curie in Paris, France. From 2003 to 2005, he was a research fellow at the European Space Agency (ESA), and the European Space Research and Technical Centre (ESTEC) in Noordwijk, The Netherlands. He joined the NASA GSFC in 2005 to work on the Aquarius instrument under the NASA Postdoctoral Program, with the National Research Council (NRC) and then with the Oak Ridge Associated Universities (ORAU). Between 2007 and 2010, he was an Assistant Research Scientist with the Goddard Earth Sciences and Technology Center/University of Maryland Baltimore County (GEST/UMBC). Dr. Dinnat's research interests include combined active and passive microwave remote sensing, atmospheric radiative transfer, scattering from rough surfaces, the remote sensing of sea surface salinity, temperature, wind, and numerical simulations. He is working on calibration, validation and algorithm improvements for Soil Moisture and Ocean Salinity (SMOS), Aquarius/SAC-D and Soil Moisture Active Passive (SMAP) missions. His latest research focuses on high latitude oceanography and the interactions between the cryosphere and oceans, as well as microwave radiometer intercalibration. Dr. Dinnat received the GSFC's Hydrospheric and Biospheric Sciences Laboratory peer review awards for outstanding Post-Doc/Research Associate and outstanding publication in 2006 and 2008, respectively. He received two NASA group achievement awards as a member of the Aquarius commissioning and calibration/validation teams in 2012 and 2013, respectively. Dr. Dinnat is a member of the American Geophysical Union (AGU) and a senior member of the Institute of Electrical and Electronics Engineers (IEEE).

Xiaobin Yin (Research Professor) received B.Sc. and Ph.D. degrees in ocean science and physical oceanography from the Ocean University of China, Qingdao, China, in 2003 and 2007, respectively. He has studied the remote sensing of sea surface salinity and sea surface wind with multifrequency microwave radiometers. From 2009 to 2015, he was involved in the validation of the SMOS L-band radiometric measurements and the improvement of the European Space Agency Level 2 ocean salinity retrieval algorithms with the Laboratoire d'Oceanographie et du Climat—Experimentation et Approches Numeriques, Paris, France. From 2009 to 2015, he was a research professor at the National Space Science Center and was involved in the pre-research of the SSS space mission of China. He is currently a research professor at Beijing Piesat Information Technology Co., Ltd., Beijing, China, a high-tech enterprise that specializes in the study and application of satellite technology. He is also adjunct professor at the Ocean University of China and Hohai University.

remote sensing

Editorial

Editorial for the Special Issue "Sea Surface Salinity Remote Sensing"

Emmanuel Dinnat [1,2,*] and Xiaobin Yin [3]

1 NASA Goddard Space Flight Center, Greenbelt, MD 20771, USA
2 CEESMO, Chapman University, Orange, CA 92866, USA
3 Beijing PIESAT Information Technology Co., Ltd., Beijing 100195, China; yinxiaobin@piesat.cn
* Correspondence: Emmanuel.dinnat@nasa.gov; Tel.: +1-301-614-6871

Received: 27 May 2019; Accepted: 30 May 2019; Published: 31 May 2019

Abstract: This Special Issue gathers papers reporting research on various aspects of remote sensing of sea surface salinity (SSS) and the use of satellites SSS in oceanography. It includes contributions presenting improvements in empirical or theoretical radiative transfer models; mitigation techniques of external interference such as radio frequency interferences (RFI) and land contamination; comparisons and validation of remote sensing products with in situ observations; retrieval techniques for improved coastal SSS monitoring, high latitude SSS monitoring and assessment of ocean interactions with the cryosphere; and data fusion techniques combining SSS with sea surface temperature (SST). New instrument technology for the future of SSS remote sensing is also presented.

Keywords: sea surface salinity; ocean surface roughness; microwave radiometry; remote sensing; forward model; retrieval algorithm

1. Introduction

Sea surface salinity (SSS) is an essential climate variable [1]. It is a key component of the water cycle, as a tracer of precipitation and evaporation, river outflow and ice melt/freeze. It is a key driver of oceanic circulation through its role in ocean density. It is also a critical parameter for understanding the variability of ocean carbon fluxes, providing information on water masses and of their chemical properties.

In situ SSS observation coverage is lacking due to limited accessibility and the rough environment of most oceanic regions, which limits long term deployment and maintenance of in situ instruments. The lack of long term observations with good spatial coverage is exacerbated in the high latitude oceans such as the Arctic and the Southern Oceans. Since the year 2000, the number of in situ observations and their spatial and temporal coverage have increased substantially with the deployment of the Argo network of free drifting profiling floats [2]. There are now approximately 4000 Argo floats deployed globally that measure conductivity, temperature and depth/pressure (CTD), from which vertical profiles of salinity and temperature are retrieved from 2000 m deep to 5-10 m below the surface. Despite the continuous deployment of Argo floats and other sensors in the global oceans, spatial and temporal coverage remains sparse in many regions, and combining satellite and in situ measurements is necessary to derive an overview of the complete system.

SSS in the open ocean has been monitored from space since 2010 by the European Space Agency's Soil Moisture and Ocean Salinity (SMOS) [3] mission, the National Air and Space Administration (NASA) and Comisión Nacional de Actividades Espaciales's Aquarius/SAC-D mission [4,5], and more recently by NASA's Soil Moisture Active Passive (SMAP) mission [6,7]. This special issue gathers contributions on research related to various aspects of remote sensing of sea surface salinity and the combined use of satellite SSS with other observations.

The topics covered by the special issue include improvements in empirical or theoretical radiative transfer models, mitigation techniques of external interference such as RFI, Sun, and land contamination, comparisons and validation of remote sensing products with in situ observations, retrieval techniques for improved coastal SSS monitoring, high latitude SSS and ocean interactions with the cryosphere, and data fusion techniques combining SSS with sea surface temperature (SST). New instrument technology for SSS remote sensing is also presented with the future Water Cycle Observation Mission (WCOM) and the microwave imager combined active/passive (MICAP) payload. The next section reports a short summary of the varied contributions to the special issue.

2. Overview of Contributions

The contributions reported in this special issue include retrieval algorithm improvements and validation of SSS from Aquarius, SMAP and SMOS missions, applications of space-borne SSS combined with in-situ and model SSS to oceanographic studies, and an introduction to future SSS missions.

2.1. Algorithm Improvements and Validation of SSS

The Aquarius Mission formally ended on 31 December 2017, a little more than 2 years after the spacecraft failure in June 2015. The last major milestone was the release of the final version of the salinity product Version 5. Le Vine et al. [8] provide an overview of the improvements in the algorithm of the final version and discuss the overall performance of the product. They also provide a discussion of the remaining issues to be addressed, such as the unknown origin of the SST-dependent bias in SSS, the seasonal variations in calibration and the observed biases over the low- (cold sky) and high- (land) end of the brightness temperature range. They conclude that the understanding developed with Aquarius is being transferred to SMAP, which shows an accuracy approaching that of Aquarius. Meissner et al. [9] present the changes in the retrieval algorithm for Aquarius Version 5 and for SMAP Version 3 SSS products, in particular they highlight the changes in corrections for the absorption by atmospheric oxygen, the SST dependence of the surface roughness and the reflected galaxy. The authors find that the difference between Aquarius Version 5 and Argo is within 0.1 psu, with an estimated global RMS uncertainty for monthly 100 km averages of 0.128 psu. Further evaluations of the Aquarius product are presented by Kao et al. [10]. From a triple point analysis using Aquarius, in situ observations and SSS from the Hybrid Coordinate Ocean Model (HYCOM) products, Kao et al. show that Aquarius' mission requirements (0.2 psu root mean square error) are exceeded, with the root mean square errors of Aquarius Level-2 and Level-3 products being estimated at 0.17 psu and 0.13 psu, respectively. The authors advise that caution should be exercised when using Aquarius salinity data in areas with high RFI and heavy rainfall, close to the coastlines where leakage of land signals may significantly affect the quality of the SSS retrievals, and in high-latitude oceans where the L-band radiometer has poor sensitivity to SSS.

A few contributions focus on SSS retrievals in challenging regions, such as enclosed seas, coastal waters and cold oceans in the high latitudes. Olmedo et al. [11] present a methodology for improving SMOS SSS retrievals in enclosed seas that usually suffer from large contaminations due to land emission or RFI. The method combines debiased non-Bayesian retrieval, data interpolating empirical orthogonal functions (DINEOF) and multifractal fusion (using SST products). The authors report improved SSS retrieval over the North Atlantic Ocean and the Mediterranean Sea. The error in SMOS SSS in the Mediterranean Sea is practically halved and the improved product reproduces the dynamic and fronts as reported by in situ data and high resolution SST products. Using a debiased non-Bayesian retrieval approach with an enhanced time-dependent bias correction, Olmedo et al [12] produce an improved SMOS SSS product for the Arctic and subarctic regions. With the new time-dependent calibration designed to mitigate seasonal bias, the standard deviation of the SSS difference with Argo measurements is between 0.25 and 0.35 psu and the major features of the inter-annual SSS variations observed by thermosalinographs are also captured by SMOS. The study's results also suggest that the use of remotely-sensed SSS may help better constrain models in regions lacking in situ observations. The Gulf of Mexico is another challenging region for SSS retrievals. Vazquez-Cuervo et al. [13] present an evaluation of three SMAP and one SMOS SSS products using continuous surface buoys and observations of opportunity from the World Ocean Database in order to mitigate the insufficient temporal and spatial coverage of the Argo network in such a variable region. While they find that the four products are remarkably consistent on seasonal time scales and that they reproduce dominant salinity features, the SMAP product at 70 km spatial resolution is lacking in terms of data availability in the nearshore region and it performs poorly within 100 km of the coast relative to other products. The other products (JPL SMAP, REMSS 40 km SMAP, and SMOS) show performances similar to each other, with root mean square differences (RMSD) between 0.5 and 1.5 psu further away than 100 km from the coast and increasing biases and RMSD at 100 km or less from the coast. REMSS 40 km generally shows the lowest RMSD in the nearshore region but has lower data availability. SMOS shows generally larger biases and RMSD, but has the lowest RMSD at some buoy locations. A few other contributions highlighting the use of satellite SSS in coastal or high latitude oceans are discussed in the next section.

Significant differences have been reported between the various satellite SSS products of SMOS, Aquarius and SMAP, as well as between satellite and in situ SSS. Dinnat et al. [14] present an overview of satellite SSS products and their differences, and an analysis of the impact of the retrieval parameters on the differences. All satellite products exhibit SSS errors with a strong dependence on the SST, but this dependence varies significantly with the sensor and version of the product. The authors show that these differences are first and foremost due to differences in the dielectric constant model, then to the atmospheric corrections, and to a lesser extent to the ancillary product used for the SST.

2.2. Oceanography with Space-Borne SSS

The South Asian Summer Monsoon (SASM) is a critical source of freshwater and its onset and intensity impacts drought and flood events in South Asia, agricultural yields, water resources, energy production, population health and the economy. Yuan et al. [15] assess the relationship between the SSS anomaly (SSSA) and SASM using observational and statistical evidence. They show that a positive SSSA leads to a thinning of the barrier layer and to a decrease in SST anomaly. A time delay of SSSA changes between the northern and southern hemisphere intensifies cross-equatorial SST gradients and currents, contributing to the onset of the summer monsoon.

In the Arctic Ocean, SSS is impacted by freshwater flows from river discharge, sea ice formation and melt, precipitations and evaporation, and oceanic transport from the North Pacific and Atlantic oceans. However, in situ SSS observations in the Arctic Ocean are sparse and insufficient to depict the spatial and temporal variability and to address the question of how climate variability affects the Arctic Ocean waters. Tang et al. [16] provide an assessment of SMAP SSS north of 50°N and its use in monitoring freshwater fluxes in the Arctic from rivers and transport from other basins. They find that SMAP SSS has a good correlation with in situ data and a RMSD 1.2 psu. In contrast, the HYCOM SSS has a smaller RMSD but suffers from significant systematic biases. They show that SMAP SSS in the Kara Sea has the potential to provide an assessment of variable interannual runoff where the model assimilates climatological runoffs without interannual changes. Despite its limited accuracy in the Arctic, SMAP observed large changes in SSS at the Arctic Ocean gateways. This confirms that satellite SSS could be used in the Arctic monitoring system as a proxy of the upper ocean layer freshwater exchanges with subarctic oceans.

Grodsky et al. [17] present another study of satellite SSS in cold coastal waters. They investigate the circulation dynamic in the Gulf of Maine and its interactions with the northwestern Atlantic Shelf using the SMAP data and in situ observations from moorings, ship-based thermosalinographs and gliders. Despite the presence of an SST-dependent bias that is amplified in winter/early spring and of a land contamination-induced bias, monthly SSS anomalies show important water intrusions in the Gulf of Maine in the winters of 2016/17 and 2017/18. The water intrusion patterns are generally consistent with measurements from SMOS, and the timing and general magnitude of a near-surface freshening in the Gulf of Maine in 2016 is confirmed by measurements by a couple of buoys. The authors conclude with an analysis of regional currents and wind anomalies to present a possible scenario for the water intrusions. They identify stronger than usual geostrophic currents along the Scotian Shelf Current

(SSC), likely due to a southwestward wind anomaly over the NW Atlantic reflecting a weakening of prevailing alongshore westerly winds on the shelf near southeastern Nova Scotia. This suggests that the strong increase in the SSC, which transports the fresh Scotian Shelf water southwestward down the coast, is the cause of the observed fresh anomalies.

Vazquez-Cuervo and Gomez-Valdes [18] compare SMAP SSS and in situ measurements from the California Cooperative Oceanic Fisheries Investigations (CalCOFI) to assess the 2015-2016 freshening event in the Southern California Current System. They report biases between SMAP SSS and CalCOFI of less than 0.1 PSU in periods of low stratification, increasing to greater than 0.4 during periods of high stratification. SMAP observed the southward propagation of freshening due to the Northeast Pacific heat wave, extending down the Baja California Coast. SMAP SSS was used to identify changes in salinity associated with a coastal upwelling system. The authors conclude that SMAP SSS can be used to monitor the freshening in a coastal region associated with a major warming event.

Castellanos et al. [19] analyze the seasonal variation of the surface currents and transports in three of the world's most energetic regions in the tropical and South Atlantic: the North Brazil Current Retroflection, the Brazil-Malvinas Confluence, and the Agulhas Current Retroflection. They use SSS from SMOS and Argo, and the SSS and surface velocity from the HYCOM model. After deriving the monthly functional relationship between SSS and surface velocity using the high resolution model, they used the SMOS SSS fields to characterize the flow in three retroflection regions. Consistent patterns of seasonal variability for the water and salt transports associated with the retroflections are obtained.

2.3. Future SSS Missions

Two papers report on future missions for SSS remote sensing. Zhang et al. [20] present a sensitivity study to assess the configuration of the microwave imager combined active/passive (MICAP) payload, which is designed to simultaneously retrieve SSS, SST and wind speed (WS) by means of multi-frequency radiometry at 1.4 GHz, 6.9 GHz, 18.7 GHz and 23.8 GHz, combined with a 1.26 GHz scatterometer. The simulations indicate that errors in retrieved SSS, SST, and WS achieve the accuracy requirements without the need for the 23.8 GHz channel, and that the L-band scatterometer is essential to contain the error. The Water Cycle Observation Mission (WCOM) is an Earth science mission focused on the observation of the water cycle global climate change intensity through three different payloads. WCOM's main payload is the interferometric microwave imager (IMI). IMI is a tri-frequency, one-dimensional aperture synthesis microwave radiometer operating at the L-, S-, and C-bands designed to perform measurements of soil moisture and ocean salinity. Li et al. [21] present simulations from the end-to-end simulator of WCOM/IMI that use antenna patterns measured using a prototype of IMI. They find general good agreement between original and retrieved SSS, with a root mean square error of 0.26 psu for a single measurement in open sea, but strong contamination is observed in coastal areas. They also find a large spatial error that could be alleviated through an increase in the number of antenna elements.

3. Conclusions

The contributions reported in this special issue highlight the thriving research on SSS remote sensing and its applications to a better understanding of the oceans and their interactions with the rest of the hydrological system. Two satellite missions are still ongoing to this date, and their calibration and retrieval algorithm continue to benefit from sustained research, leading to improved and enhanced products. Future missions to measure SSS are also being planned and should enable continuity in the regular and global observation of this essential climate variable. They should also provide expanded capabilities by using multiple frequencies.

Author Contributions: The two authors contributed equally to all aspects of this editorial.

Acknowledgments: The Guest Editors would like to thank the authors who contributed to this Special Issue and the reviewers who dedicated their time and provided the authors with valuable and constructive recommendations. They would also like to thank the editorial team of Remote Sensing for their support.

Conflicts of Interest: The authors declare no conflict of interest.

References

1. GCOS. The Global Observing System For Climate Implementation Needs. *World Meteorol. Organ.* **2016**, *200*, 316.
2. Gould, J.; Roemmich, D.; Wijffels, S.; Freeland, H.; Ignaszewsky, M.; Jianping, X.; Pouliquen, S.; Desaubies, Y.; Send, U.; Radhakrishnan, K.; et al. Argo profiling floats bring new era of in situ ocean observations. *Eos Trans. Am. Geophys. Union* **2004**, *85*, 185. [CrossRef]
3. Kerr, Y.H.; Waldteufel, P.; Wigneron, J.-P.; Delwart, S.; Cabot, F.; Boutin, J.; Escorihuela, M.-J.; Font, J.; Reul, N.; Gruhier, C.; et al. The SMOS Mission: New Tool for Monitoring Key Elements of the Global Water Cycle. *Proc. IEEE* **2010**, *98*, 666–687. [CrossRef]
4. Lagerloef, G.; Colomb, F.R.; Le Vine, D.; Wentz, F.; Yueh, S.; Ruf, C.; Lilly, J.; Gunn, J.; Chao, Y.; deCharon, A.; et al. The Aquarius/SAC-D Mission: Designed to Meet the Salinity Remote-Sensing Challenge. *Oceanography* **2008**, *21*, 68–81. [CrossRef]
5. Le Vine, D.M.; Dinnat, E.P.; Meissner, T.; Yueh, S.H.; Wentz, F.J.; Torrusio, S.E.; Lagerloef, G. Status of Aquarius/SAC-D and Aquarius Salinity Retrievals. *IEEE J. Sel. Top. Appl. Earth Obs. Remote Sens.* **2015**, *8*, 5401–5415. [CrossRef]
6. Entekhabi, D.; Njoku, E.G.; O'Neill, P.E.; Kellogg, K.H.; Crow, W.T.; Edelstein, W.N.; Entin, J.K.; Goodman, S.D.; Jackson, T.J.; Johnson, J.; et al. The soil moisture active passive (SMAP) mission. *Proc. IEEE* **2010**, *98*, 704–716. [CrossRef]
7. Meissner, T.; Wentz, F.; LeVine, D.; Dinnat, E.; Lagerloef, G. Ocean products from the SMAP radiometer: surface salinity and wind speeds. In Proceedings of the Microwave Radiometry and Remote Sensing of the Environment (MicroRad), Espoo, Finland, 11–14 April 2016.
8. Le Vine, D.M.; Dinnat, E.P.; Meissner, T.; Wentz, F.J.; Kao, H.Y.; Lagerloef, G.; Lee, T. Status of Aquarius and salinity continuity. *Remote Sens.* **2018**, *10*, 1585. [CrossRef]
9. Meissner, T.; Wentz, F.J.; Le Vine, D.M. The salinity retrieval algorithms for the NASA Aquarius version 5 and SMAP version 3 releases. *Remote Sens.* **2018**, *10*, 1121. [CrossRef]
10. Kao, H.Y.; Lagerloef, G.S.E.; Lee, T.; Melnichenko, O.; Meissner, T.; Hacker, P. Assessment of aquarius sea surface salinity. *Remote Sens.* **2018**, *10*, 1341. [CrossRef]
11. Olmedo, E.; Taupier-Letage, I.; Turiel, A.; Alvera-Azcárate, A. Improving SMOS sea surface salinity in the Western Mediterranean sea through multivariate and multifractal analysis. *Remote Sens.* **2018**, *10*, 485. [CrossRef]
12. Olmedo, E.; Gabarró, C.; González-Gambau, V.; Martínez, J.; Ballabrera-Poy, J.; Turiel, A.; Portabella, M.; Fournier, S.; Lee, T. Seven Years of SMOS sea surface salinity at high latitudes: Variability in Arctic and Sub-Arctic Regions. *Remote Sens.* **2018**, *10*, 1772. [CrossRef]
13. Vazquez-Cuervo, J.; Fournier, S.; Dzwonkowski, B.; Reager, J. Intercomparison of In-Situ and Remote Sensing Salinity Products in the Gulf of Mexico, a River-Influenced System. *Remote Sens.* **2018**, *10*, 1590. [CrossRef]
14. Dinnat, E.P.; Le Vine, D.M.; Boutin, J.; Meissner, T.; Lagerloef, G. Remote sensing of sea surface salinity: Comparison of satellite and in situ observations and impact of retrieval parameters. *Remote Sens.* **2019**, *11*, 750. [CrossRef]
15. Yuan, X.; Salama, M.S.; Su, Z. An observational perspective of sea surface salinity in the Southwestern Indian Ocean and its role in the South Asia Summer Monsoon. *Remote Sens.* **2018**, *10*, 1930. [CrossRef]
16. Tang, W.; Yueh, S.; Yang, D.; Fore, A.; Hayashi, A.; Lee, T.; Fournier, S.; Holt, B. The potential and challenges of using Soil Moisture Active Passive (SMAP) sea surface salinity to monitor Arctic Ocean freshwater changes. *Remote Sens.* **2018**, *10*, 869. [CrossRef]
17. Grodsky, S.A.; Vandemark, D.; Feng, H. Assessing coastal SMAP surface salinity accuracy and its application to monitoring Gulf of Maine circulation dynamics. *Remote Sens.* **2018**, *10*, 1232. [CrossRef]
18. Vazquez-Cuervo, J.; Gomez-Valdes, J. SMAP and CalCOFI observe freshening during the 2014-2016 Northeast Pacific Warm Anomaly. *Remote Sens.* **2018**, *10*, 1716. [CrossRef]
19. Castellanos, P.; Olmedo, E.; Pelegrí, J.L.; Turiel, A.; Campos, E.J.D. Seasonal Variability of Retroflection Structures and Transports in the Atlantic Ocean as Inferred from Satellite-Derived Salinity Maps. *Remote Sens.* **2019**, *11*, 802. [CrossRef]

20. Zhang, L.; Wang, Z.; Yin, X. Comparison of the retrieval of sea surface salinity using different instrument configurations of MICAP. *Remote Sens.* **2018**, *10*, 550. [CrossRef]
21. Li, Y.; Liu, H.; Zhang, A. End-to-End Simulation of WCOM IMI Sea Surface Salinity Retrieval. *Remote Sens.* **2019**, *11*, 217. [CrossRef]

remote sensing

MDPI

Article

Status of Aquarius and Salinity Continuity

David M. Le Vine [1,*], Emmanuel P. Dinnat [1,2], Thomas Meissner [3], Frank J. Wentz [3], Hsun-Ying Kao [4], Gary Lagerloef [4] and Tong Lee [5]

1 NASA/Goddard Space Flight Center, Greenbelt, MD 20771, USA; emmanuel.dinnat@nasa.gov
2 Chapman University, Orange, CA 92866, USA
3 Remote Sensing Systems, 444 Tenth Street, Suite 200, Santa Rosa, CA 95401, USA; meissner@remss.com (T.M.); frank.wentz@remss.com (F.J.W.)
4 Earth and Space Research, 2101 Fourth Ave, Suite 1310, Seattle, WA 98121, USA; Hkao@esr.org (H.-Y.K.); lager@esr.org (G.L.)
5 Jet Propulsion Laboratory, 4800 Oak Grove Drive, Pasadena, CA 91109, USA; tlee@jpl.nasa.gov
* Correspondence: david.m.levine@nasa.gov; Tel.: +1-301-614-5640

Received: 5 September 2018; Accepted: 18 September 2018; Published: 2 October 2018

Abstract: Aquarius is an L-band radar/radiometer instrument combination that has been designed to measure ocean salinity. It was launched on 10 June 2011 as part of the Aquarius/SAC-D observatory. The observatory is a partnership between the United States National Aeronautics and Space Agency (NASA), which provided Aquarius, and the Argentinian space agency, Comisión Nacional de Actividades Espaciales (CONAE), which provided the spacecraft bus, Satelite de Aplicaciones Cientificas (SAC-D). The observatory was lost four years later on 7 June 2015 when a failure in the power distribution network resulted in the loss of control of the spacecraft. The Aquarius Mission formally ended on 31 December 2017. The last major milestone was the release of the final version of the salinity retrieval (Version 5). Version 5 meets the mission requirements for accuracy, and reflects the continuing progress and understanding developed by the science team over the lifetime of the mission. Further progress is possible, and several issues remained unresolved at the end of the mission that are relevant to future salinity retrievals. The understanding developed with Aquarius is being transferred to radiometer observations over the ocean from NASA's Soil Moisture Active Passive (SMAP) satellite, and salinity from SMAP with accuracy approaching that of Aquarius are already being produced.

Keywords: ocean salinity; microwave remote sensing; remote sensing

1. Introduction

Aquarius is an L-band radar/radiometer instrument combination that has been designed to measure ocean salinity. It was launched 10 June 2011 and lost on 7 June 2015, almost four years to the day later when a failure in the power distribution network resulted in loss of control of the spacecraft. Mission operations ended soon afterward, and scientific operations (e.g., final algorithm) ended 31 December 2017. Aquarius was unique in that it was the only satellite dedicated solely to remote sensing of sea surface salinity (SSS). This manuscript reports an overview of the status of the Aquarius research on mapping the global surface salinity field as of the end of the scientific part of the mission on 31 December 2017. This overview begins with a short history and description of the mission (Section 2: Background). This is followed by a brief description of the salinity retrieval algorithm, Version 5.0, as it existed at the end of the mission (Section 3). A more detailed description of the algorithm can be found in Meissner, Wentz and Le Vine [1], and in this special issue [2]. Section 4 describes several of the important research issues that remain at end of the mission. This includes small residual changes in the temporal signal, a dependence on sea surface temperature (SST), and tuning of the calibration to improve the performance at the cold and warm ends. These "features" are hidden in Version 5 because they are either corrected empirically (temporal variations and SST-dependence) or are not manifested over the ocean (e.g., warm end occurs over land/ice). These features are pointed out here because further progress in improving the remote sensing algorithm requires a more fundamental understanding of the root cause of these issues. Finally (Section 5), the paper concludes with an introduction to the work that is continuing on the remote sensing of salinity from space using the observations of the SMAP L-band radiometer over the ocean.

2. Background

The Aquarius instrument is an L-band active/passive (radar/radiometer) combination designed to map surface salinity over the ocean [3]. It was part of the Aquarius/SAC-D observatory, which was a partnership between the United States National Aeronautics and Space Agency (NASA) and the Argentine space agency, Comisión Nacional de Actividades Espaciales (CONAE). Figure 1 is an artist drawing of the observatory. Aquarius consisted of three radiometers (the feed horns can be seen in the figure) arranged to image in pushbroom fashion looking toward the left with the spacecraft oriented as shown and flying perpendicular to the page (e.g., see Figure 3 in Le Vine et al. [3]). A NASA team from the Goddard Space Flight Center (radiometer) and the Jet Propulsion Laboratory (radar) built Aquarius. CONAE provided the spacecraft bus, SAC-D, and CONAE and its partners (Italy, France, and Canada) provided several other instruments [3,4] that are indicated on the Earth-viewing side of the bus.

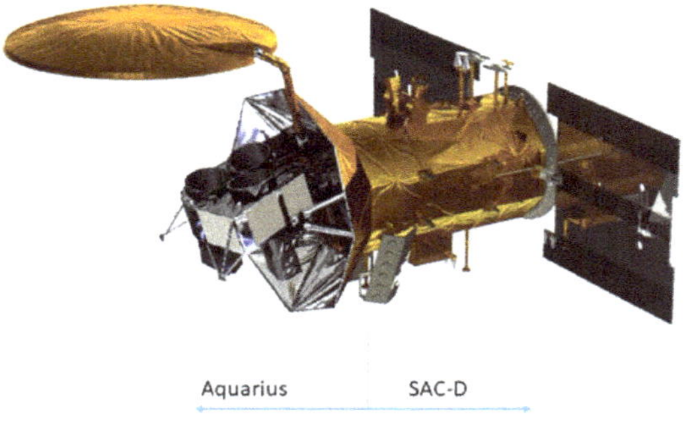

Figure 1. The Aquarius/SAC-D observatory. Aquarius is the feed-reflector assembly to the left. SAC-D is the observatory bus to the right, plus several instruments.

Aquarius was unique among the recent L-band missions, (i.e., also the European Space Agency Soil Moisture and Ocean Salinity mission, SMOS, and the NASA Soil Moisture Active Passive, SMAP, mission) in that it focused exclusively on the measurement of sea surface salinity. For example, the radar was included to help correct for the effect of ocean surface roughness (waves), which is a major source of uncertainty in the salinity retrieval algorithm [5]. The radiometers were polarimetric (measured the third Stokes parameter) to provide an in situ measurement of Faraday rotation [6,7] that can be significant at L-band [8]. Even the orbit was chosen to optimize the retrieval of salinity. The orbit was sun-synchronous with a ground track close to the day–night terminator (18:00 equatorial crossing, ascending) so that the radiometer beams could look to the nighttime side to avoid the reflection of L-band radiation from the Sun into the antenna main beam (sun glint). In addition, the orbit was an exact repeat with a seven-day cycle that enabled Aquarius to view the same footprints each cycle to facilitate averaging to reduce noise in the measurement. The three radiometers provided sufficient coverage so that Aquarius mapped the globe completely in each seven-day cycle with overlap at higher latitude, again to facilitate averaging to meet the accuracy goal of 0.2 psu (global root mean square (RMS), monthly, and at 150-km spatial resolution [3]). The radar used the same feeds as the radiometer, but there was only one radar that was cycled among three feeds. The goal was to have almost simultaneous passive/active looks at the same spot on the surface.

Significant attention was paid to the thermal control of the instrument and included both active and passive elements. As part of the passive control, Aquarius carried a large thermal shield (the large umbrella-like shade separating Aquarius from the SAC/D spacecraft bus seen in Figure 1) to mitigate the effects of the Sun. There also was active thermal control to keep the critical parts of the radiometer at a stable temperature [3]. The requirement was for a variation of the radiometer front end of less than 0.1 °C per orbit, and the performance achieved on orbit was better than this [9].

Two other features new on Aquarius were the internal calibration and provisions for detection/mitigation of radio frequency interference (RFI). The internal calibration included reference diodes, and was the product of several years of research [10,11]. RFI was known to be a potential problem based on experience with airborne instruments [12,13], and provisions were made to address this problem. In particular, rapid sampling (10 ms per sample) and an algorithm to detect and remove pulses were implemented, because RFI at L-band from air traffic control radar was known to be a problem [3,14].

Aquarius released its first salinity map in September 2011 almost one month after it was turned on (Figure 5 in Le Vine et al. [9]), and functioned almost flawlessly and within specification until the observatory was lost on 7 June 2015. The problem was a failure in the power distribution network in the spacecraft. A component used in the power switches began to fail early in the mission. Since the switches were redundant, there was no early impact. Available information suggested a random failure mode, and that the probability of a failure in both primary and backup switch was small. However, in June 2015, power to the attitude control system was lost. Both primary and backup units had failed (the primary unit failed much earlier). Without attitude control, contact with the spacecraft was lost, and the retrieval of data was not possible. Operations ended soon thereafter, and the science team focused on consolidating its research for the release of a final version of the salinity retrieval. Version 5.0, the final Aquarius salinity product, was released in November 2017, and the Aquarius mission formally ended on 31 December 2017.

3. Results: Aquarius Version 5.0

This section provides a brief overview of the status of the final version, Version 5.0, of the Aquarius Project salinity retrieval algorithm. This includes the changes made to the algorithm, an assessment of its performance, and a list developed by the algorithm team at the end of the mission of issues that remained to be resolved.

3.1. Changes in the Retrieval Algorithm

The approach to calibration and the retrieval of salinity has not changed fundamentally since it was outlined in the pre-launch documentation [2,15,16]. However, changes have been made based on the actual data and the performance of the hardware on orbit. Each improvement led to better understanding of the calibration and algorithm, which in turn led to further improvements and further understanding. When significant changes were made to the retrieval, the data was reprocessed, and a new version of the salinity product was released to the public. In total, there were five versions (data releases). The changes are documented in a series of appendices to the pre-launch algorithm theoretic basis document (ATBD), which were issued with each new version of the algorithm to describe the changes incorporated in that version [16]. The changes made in the development of Version 5.0 are:

- The ancillary sea surface temperature (SST) field was changed from the National Oceanographic and Atmospheric Administration (NOAA) Optimally Interpolated (OI) SST to the SST field from the Canadian Meteorological Center (CMC);
- The reference sea surface salinity (SSS) field used in the sensor calibration and in the derivation of expected antenna temperature, TA_expected, in the forward algorithm was changed from SSS

obtained from the Hybrid Coordinate Ocean Model (HYCOM) to the analyzed monthly Scripps Argo SSS;

- The model for the celestial radiation reflected from the surface into the radiometer antenna was changed to values derived from fore and aft observations of the SMAP radiometer. The advantage of this approach is that it includes the effects of surface roughness;
- The empirical symmetrization correction that corrects asc/dsc differences [1] was re-derived to reflect the improvements that resulted from the improved model for the reflected celestial radiation (above).
- The model for absorption by atmospheric oxygen was changed from Meissner et al. [17] to the original model [18].
- The surface roughness correction was updated (i.e., compared to that described in Meissner, Wentz and Riciardulli [19]):
    - The SST dependence was adjusted.
    - The dependence on significant wave height (SWH) was omitted.
    - In addition, the correction table for dependence on wind speed and radar backscatter was updated, and as a consequence, the initial guess for the SSS field used to derive the final wind speed (i.e., "HHH" wind speed) was also updated.
- Observations at vertical and horizontal polarization are given equal weight in the retrieval of salinity (i.e., in the maximum likelihood estimate used in the last step in the retrieval);
- The L2 files include instantaneous rain rates based on the NOAA rain product, CMORPH (Climate Prediction Center Morphing). They are used to filter data for rain in the calibration and also for validating the Aquarius salinity versus in situ measurements.

Since Version 5.0 is the final version of the salinity retrieval for the Aquarius Project, it was decided to rewrite the ATBD to reflect the algorithm in place for Version 5.0 [1]. The data itself and a complete set of documents describing the algorithm can be found at the NASA Physical Oceanography Distributive Active Archive Center (PO.DAAC): https://podaac.jpl.nasa.gov/aquarius. Also, a description of the algorithm in its Version 3 form can be found in Boutin et al. [20]. The NASA PO.DAAC website also contains a historical record of the documents issued for each version (including evaluation metrics, users guide, and addenda to the ATBD).

3.2. Evaluation of the Version 5.0 Salinity

The release of each Aquarius salinity product was accompanied by an evaluation of the product. The assessment is made by comparing the retrieved salinity with in situ measurements (mostly Argo floats [21]). In the matchup, Argo data closest to the surface is selected and matched to the nearest Aqaurius beam center within a search radius of ±3.5 days and 75 km. Then, 11 Aquarius observations (the data product is one salinity value every 1.44 s) are averaged centered on this point (see Kao et al. [22] for additional details). Figure 2 shows the RMS difference between the retrieved and in situ salinity averaged over the entire mission. The three bars represent the three Aquarius beams (beam 1 is the innermost beam), and results are shown for versions 2, 3, 4, and 5. The continuous progress and improvement in the algorithm since the early phase of the mission is evident in the gradual decrease in the level of error. The results shown for Version 5 (V5) are for the entire mission, and consist of individual matchups as described above, with no additional spatial or temporal averaging.

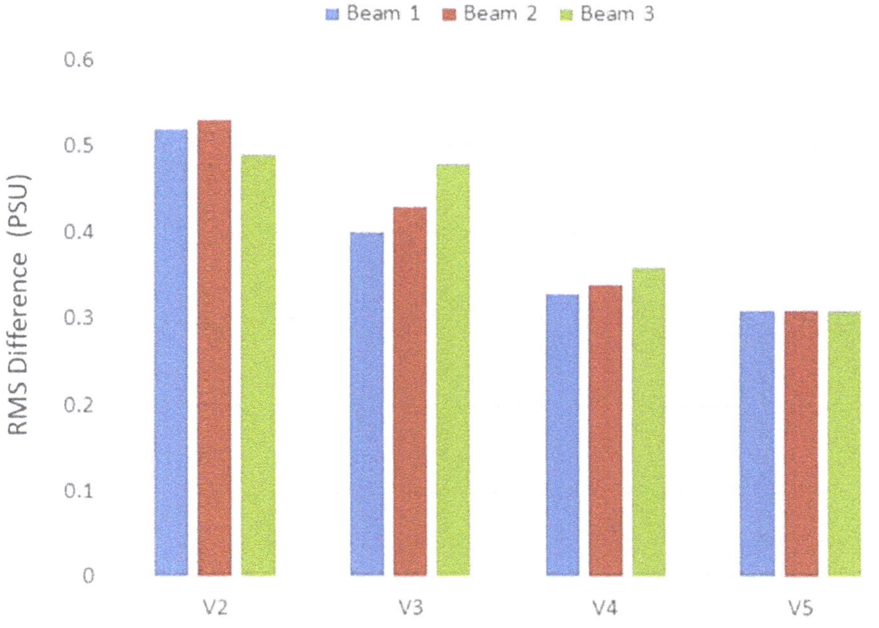

Figure 2. The root mean square (RMS) difference between Aquarius-retrieved sea surface salinity (SSS) and in situ ocean observations. The three bars in each group represent the three Aquarius beams. The groups are for different versions of the retrieval algorithm from Version 2 to the final Version 5. The data are from Figure 3 in Kao et al. [22].

The statistics shown in Figure 2 include "error" in both the Aquarius retrieval and the error of the in situ measurements. The latter includes sampling error due to the relatively sparse in situ measurements and because they are not truly measurements at the surface (i.e., within the 1–2 cm where the microwave signal originates). In general, the discrepancy on the measurement depth between the satellite and in situ is negligible, but it could be a factor in the case of large precipitation and low mixing of the ocean upper layer [20]. One way to try to isolate the error in the Aquarius retrieval from the other sources is to use a triple-point analysis (e.g., see Appendix B in Kao et al. [22]). An example is shown in Figure 3 for beam 2 (the Aquarius middle beam). The data is the same as used in Figure 2. The dashed line represents the Aquarius mission requirement of an accuracy of 0.2 psu for the global RMS error on a monthly basis and a spatial resolution of 150 km. This triple-point analysis employs data from Aquarius, Argo [21], and the HYCOM [23] salinity field. The version of HYCOM used by the Aquarius project is the HYCOM + NCODA Global 1/12° Analysis (GLBA0.08, see https://hycom.org/data/glba0pt08). These data are not strictly independent. For example, HYCOM assimilates Argo data, so its SSS are not strictly independent from Argo measurements. However, HYCOM SSS are also affected by model ocean dynamics, evaporation–precipitation forcing, and a relaxation of HYCOM SSS toward a seasonal climatology (to prevent model drift). These factors tend to create some level of independence between HYCOM and Argo. Aquarius uses Argo during calibration (but not during retrieval). On the other hand, the data represent single comparisons, whereas the Aquarius requirement is for a monthly map. Since Aquarius maps the globe in seven days, one can expect four or more measurements in one month, and consequently, the monthly RMS is likely lower than the 0.17 psu shown in Figure 3 for Version 5.0. Calculations suggest the number is on the order of 0.13 psu (Table 4 in Kao et al. [22]).

Figure 4 is an example of the assessment of Level 3 (L3) data for V5 (L2 data is available in swath coordinates every 1.44 s, and Level 3 data is averaged into latitude and longitude bins of 1 degree × 1 degree resolution.) In Figure 4, the data are averaged for the entire mission [22]. The bars in each group represent different versions of the Aquarius salinity retrieval (V3, V4, and V5), and the groups show the data averaged spatially with increasing size, from the native one-degree to 10-degree bin sizes. Figure 4 illustrates two points: (a) the improvement in the retrieval with each new version, and (b) the impact of spatial averaging on reducing noise. It is evident from Figure 4 that V5.0 Level 3 exceeds the goal of 0.2 psu at 1° x1° and, as expected, improves with further spatial averaging.

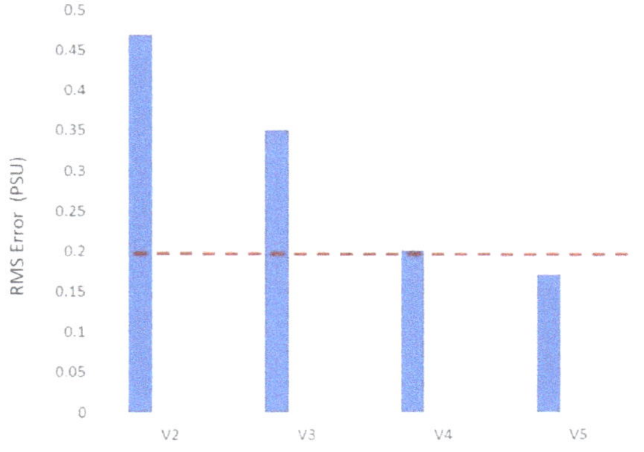

Figure 3. Aquarius SSS retrieval RMS error estimated with a triple point analysis using Aquarius, Hybrid Coordinate Ocean Model (HYCOM), and Argo data. The results shown are for the Aquarius middle beam. The dashed line represents the Aquarius mission requirement for monthly maps at 150-km spatial resolution. The data are from Table 1 in Kao et al. [22]. The version of HYCOM used by the Aquarius project is the HYCOM + NCODA Global 1/12° Analysis (GLBA0.08).

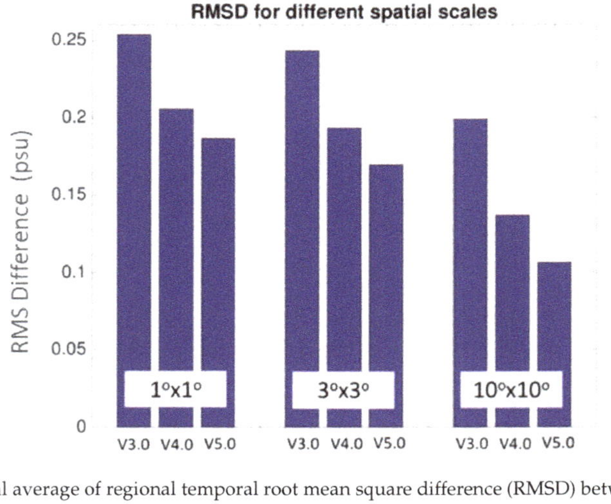

Figure 4. Global average of regional temporal root mean square difference (RMSD) between Aquarius Level 3 monthly SSS and monthly Argo gridded maps as a function of the spatial scale for the entire mission. The three bars represent Version 3, 4, and 5 retrievals. The Argo data used in the comparison are from the gridded product from the Scripps Institute of Oceanography: http://apdrc.soest.ucsd.edu/Gridded_fields.html. The data are from Section 10 of Kao et al. [22], which also contains more detail.

The root mean square difference (RMSD) with respect to gridded Argo maps not only contains the errors of the Aquarius Level 3 SSS, but also the sampling and mapping errors of the Argo maps. The latter was found to be quite significant on 1° × 1° and 3° × 3° scales [24]; for example, the global average RMSD between two gridded Argo products is 0.1 psu on a 1° × 1° scale.

3.3. Work Remaining to Improve the Salinity Product

Aquarius Version 5.0 meets and exceeds the mission requirements. However, more can be done to improve the product. Among the issues identified by the Aquarius Cal/Val and Ocean Salinity Science Teams that will need to be addressed are:

- Determining the physical reason for an SST-dependent bias (which is empirically removed in Version 5; See Section 4.3 below);
- Refining the model for the dielectric constant of sea water (i.e., functional dependence on SST and SSS), which varies among the two models currently in use, Klein and Swift [25] and Meissner and Wentz [26], and also the model being developed at the George Washington University [27];
- Identifying and correcting a remaining small annual cycle (not due to changes in salinity);
- Merging Aquarius, SMOS, and SMAP salinity maps into a single product;
- Improving the theory for the effect of surface roughness on emission and the correction for the reflection of signals such as the galactic background;
- Improving the level of missed detection in the RFI algorithm;
- Improving the performance in cold water;
- Addressing regional biases (e.g., North Pacific and southern Indian Ocean);
- Improving calibration over the full range of expected targets (i.e., cold sky, ocean, and land).

The advantage of having a very stable instrument is that each improvement in the retrieval algorithm opens the door to additional possibilities. The manifestation of this can be seen in the steady improvement from Version 2 to Version 5 that is evident in Figures 2–4. Examples are presented below to illustrate how this process unfolded for Aquarius for three of the issues listed above (SST-dependent bias, residual annual cycle, and calibration over the full range), and to describe the status of these issues as it exists in V5.

4. Discussion: Remaining Issues

4.1. Background

It helps for understanding the issues remaining in V5 to review, very briefly, two aspects of the salinity retrieval process: calibration (counts to TA) and the algorithm that converts the calibrated antenna temperature, TA, at the spacecraft to salinity at the surface.

4.1.1. Calibration

The internal hardware aspect of calibration (e.g., linearity correction, correction for temperature dependence, loss model, etc.) is based on pre-launch measurements and described in a summary document prepared for Version 5 [15]. The radiometers include internal reference sources (noise diodes) that are sampled periodically [11,15,28] and used to compute radiometer gain and offset. Although the values of the reference loads are measured pre-launch, an on-orbit check is required. In an ideal case, this should be a one-time check; however, (see "gain drift" below) the Aquarius radiometers were very good, but not perfect. In addition, largely because the antenna cannot be measured well enough on the ground, an on-orbit calibration of radiometer bias is necessary. In the case of Aquarius, the global average antenna temperature (which relies on ancillary salinity and temperature fields) is used as a reference for these external elements of the calibration, as will be explained in more detail below.

4.1.2. Retrieval of Salinity

Figure 5 is a schematic outlining the major steps in the transformation from the calibrated antenna temperature, TA, to the retrieval of salinity. Details can be found in the Aquarius ATBD, both the end of mission ATBD [1] and also the pre-launch ATBD and associated addenda [16]. The addenda were written to update the pre-launch ATBD to make it consistent with the revised processing each time a new version of the data was released. The end of mission ATBD is a rewrite of the ATBD to reflect the process in place for Version 5, which is the final Aquarius data set.

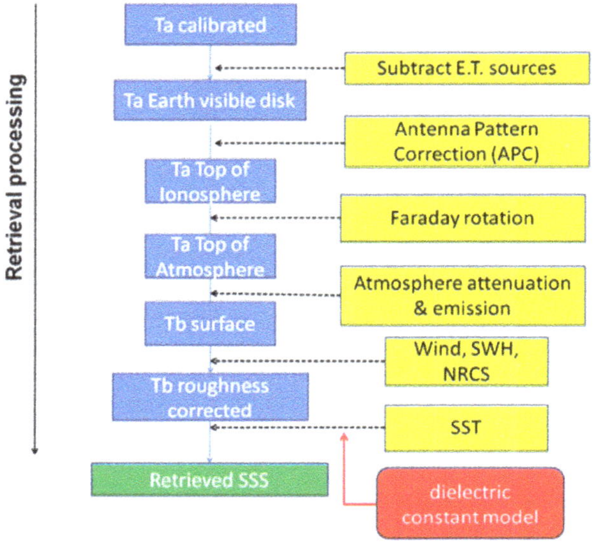

Figure 5. Flow diagram illustrating the steps in the Aquarius salinity retrieval algorithm. The process starts at the top with calibrated antenna temperature, TA, from the radiometer and ends with a measurement of SSS at the surface.

The retrieval of salinity consists of the transfer of TA from the spacecraft to the surface, and then the calculation of surface salinity. Figure 5 illustrates the major steps in going from TA to the brightness temperature at the surface. These are in order (yellow boxes): off-Earth sources (e.g., the Sun) are subtracted; a correction is made to account for antenna imperfections; a correction is made for Faraday rotation (which changes the polarization) and for attenuation and emission from the atmosphere; then, the brightness temperature is corrected for surface roughness (e.g., waves that are parameterized as a function of wind and normalized radar cross-section); and finally, the SST is used with a model for the dielectric constant of sea water to retrieve SSS. The path shown in Figure 5 is used in two ways: to compute salinity from calibrated TA, and in reverse to compute TA from the known values of salinity. The latter is called "expected TA" in the Aquarius literature, and is used in the calibration of the radiometer. In particular, the upward version of the algorithm evaluated over reference scenes is used to predict the value of antenna temperature that is expected at the radiometer output. This is compared with the actual value as part of calibration. In essence, calibration consists of matching what is measured and "expected" TA (called TA_expected). This was done once after the radiometer was first turned on in orbit to adjust the radiometer bias and set the value of the reference noise diode. In addition, this was done periodically with the radiometer looking at cold sky to check for stability [29]. It was also done while looking at the ocean and using a salinity reference model (e.g., the ocean model HYCOM or a set of in situ observations such as from the Argo floats) to fine-tune gain and offset (see "drift and bias" below).

4.2. Example Issue: Drift and Wiggles

After the initial on-orbit calibration to establish the value for the reference noise diode and bias [15], the radiometer performance was monitored by comparing the global average TA (over the ocean) with the TA_expected using surface salinity from HYCOM. In Version 4, some regions with known issues (e.g., RFI and ice/land contamination) were eliminated to form a more reliable calibration subset. In Version 5, areas with rain were added to this list, and in addition, the reference salinity was changed to the Scripps optimally interpolated Argo salinity field [30].

Figure 6a shows the time history of the difference, TA—TA_expected, as it looked at the end of the mission [29]. The grey shows the difference per orbit, and the green curve is a seven-day average (sliding window). Red is an exponential fit to the data. The vertical axis is the difference (Kelvin), and the abscissa is time (months) since the beginning of data collection (Aquarius was turned on in August 2011) until the end (June 2015). The total change is small, but this must be considered in the context of the requirements for measurement of salinity (a sensitivity of about 0.5 K/psu and accuracy goal of 0.2 psu). There was great concern at the beginning of the mission with the (relatively) rapid change (i.e., the decrease evident in Figure 6a during the first three months of the mission). However, the exponential nature of the change soon became clear, and based on the leveling of the change, it was assumed that some manner of outgassing was the cause. It was decided to model this feature with an exponential and remove it as part of calibration. The correction was made by adjusting the temperature of the reference noise diode [31].

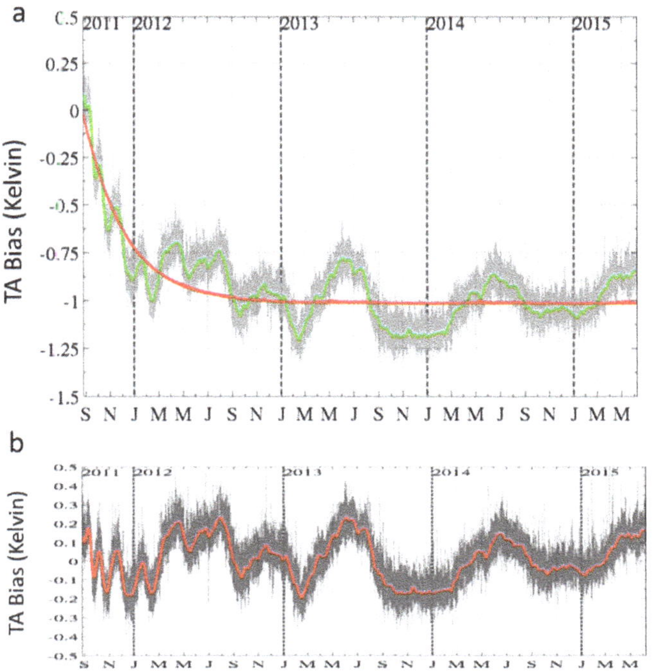

Figure 6. The difference between observed and expected antenna temperature (TA bias) for beam 2 and horizontal polarization (**a**) before and (**b**) after removing an exponential fit for the temporal drift. Grey is the difference per orbit. Green (**a**) and red (**b**) are the average of the difference over a seven-day period (i.e., global average); red (**a**) is an exponential fit to the temporal drift (based on figures in Dinnat et al. [29]).

Figure 6b shows the same data after correction for the exponential drift (the scale is the same as in Figure 6a to make the comparison easier). The irregular "oscillations" around zero were given the name, "wiggles", by the Aquarius team for want of a better name. They remained a mystery for some time. However, around the time that V4 was being prepared, it was discovered that a potential source for this behavior was an instability in the radiometer backend voltage to the frequency converter [32]. A model for the behavior was developed based entirely on the radiometer hardware, and the model was used to make a correction for this feature in Version 5 (it was first employed in Version 4).

Figure 7 shows the corrected difference (beam 1, horizontal polarization). Figure 7a is the same as Figures 6b and 7b is the residual, TA—TA_expected, after correction for the backend instability. Again, grey is the difference per orbit, and green is a seven-day average. The instrument-only correction reduced the "wiggles", but did not remove them. This correction also dramatically changed their character. There is a very clear annual cycle in the residual after this correction. The grey and green curves are data collected in the nominal mode: viewing ocean. The red curve in Figure 7c shows the difference calculated for data collected when looking at cold sky and superimposed on the data in Figure 7b. The crosses "+" indicate the actual data points. It is evident from Figure 7c that this behavior is not scene-dependent, and therefore, it is most likely not associated with an error in the radiative transport model of the ocean used to compute the grey and green curves. The shape of the curve and, in particular, the small hump between the larger peaks is suggestive of the solar beta angle (Figures 7 and 8 in Dinnat and Le Vine [33]), which suggests a potential correlation with instrument temperature. The blue curve in Figure 7d shows the temperature of the reference noise diode shared by the two polarizations (i.e., correlated noise diode [3]) centered on zero (i.e., mean value over mission lifetime subtracted). An effort was made to find a connection between temperature variations and the residual in Figure 7. For example, temperatures were varied in the radiometer loss model [15], and although changes were observed in TA, it was not possible to reproduce changes of the order of magnitude that was seen in the data.

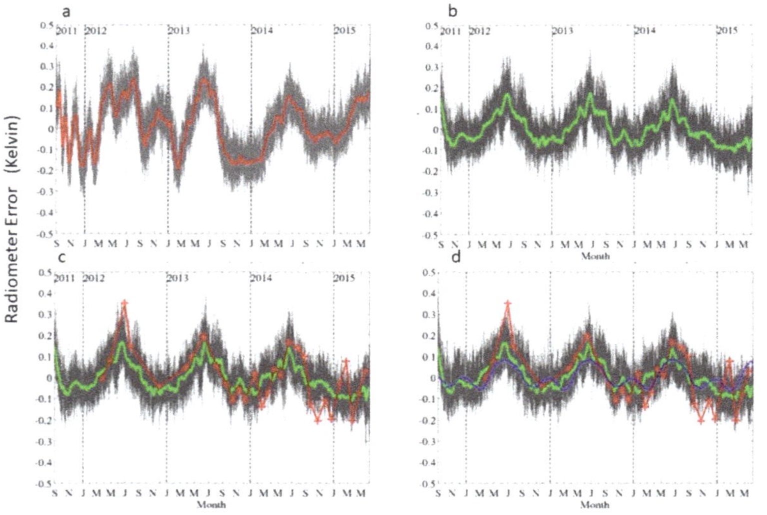

Figure 7. Radiometer residual error, TA—TA_exp, after removing the exponential drift (**a**) and then correcting for the radiometer backend instability (**b–d**). The data is for the Aquarius outer beam and horizontal polarization. Grey is the difference per orbit, and green is a seven-day average. Red in (**c**,**d**) shows the residual when looking at cold sky superimposed on the data in (**b**) (with a shift of 0.05 to align the curves). Crosses "+" indicate the data points. Blue in (**d**) shows the temperature of correlated noise diode centered on zero.

The source of the residual shown in Figure 7 remains unknown. It is corrected empirically in Version 5. This was done by calculating the difference, TA—TA_expected, and averaging using a seven-day sliding window (i.e., seven days centered on the current time). This average difference is treated like a bias and subtracted from TA. This removes the artifact from the retrieval of salinity; however, the annual signature strongly suggests a physical cause which, if identified, will lead to better understanding and likely a better salinity product.

4.3. Example Issue: SST Dependence

During the research leading to release of Aquarius Version 3.0, regional biases were observed in the error of the salinity retrieval. The bias had a strong zonal behavior, with salty biases noticed in mid-high latitudes, and fresh bias in the tropics and subtropics (see Section 6 of Addendum III of [17,34]). The zonal character of those biases suggested a correlation with SST, which was confirmed, and Version 3 was released with both a nominal version and an option for the removal of an SST-dependent bias (e.g., see Addendum III in the Aquarius ATBD [16]).

Figure 8 shows an example of the SST dependence. The vertical axis (ordinate) is the difference between the retrieved salinity and the value reported by Argo floats, and the abscissa is SST. The grey are individual differences, and the red curve is the median value. The matchup is with a smoothed map of Argo values formed by binning the near-surface Argo observations. Notice the fresh bias at very warm temperatures, which is typical of the tropics and low latitude, and the salty bias, which is typical at cooler temperatures and higher latitude. The behavior at very cold temperatures is different and an issue of its own (see below). The data in Figure 8 are from Version 3 of the algorithm. This choice was made because in versions 4 and 5, empirical adjustments have been made that hide the extent of the dependence on SST.

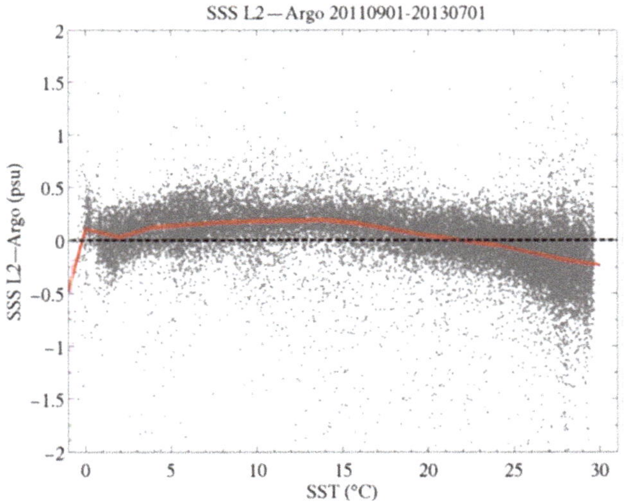

Figure 8. Salinity error (difference between retrieved SSS and in situ salinity from Argo) as a function of sea surface temperature, SST. Grey are the data and red is the median value. (Author's figure from Le Vine et al. [4]).

The reason for the SST dependence of the error is not known, but it is likely associated with the residual errors in one of the geophysical model functions used in the SSS retrieval algorithm. The most probable candidates are the model function for the dielectric constant of sea water (which has a strong dependence on SST) and the model for atmospheric absorption and emission, which has a latitudinal dependence. A temperature dependence was also included in the empirical model that is employed to correct for the effect of surface roughness on emissivity (e.g., Section 2 of Addendum III of the Aquarius ATBD [16]), which is somewhat arbitrary and not necessarily physically correct.

In the research leading to the release of Version 5.0, the model used in previous versions for attenuation by oxygen was identified as a contributor to this bias. The problem appeared to be with a modification made to the Liebe model for the effect of oxygen in the calculation of atmospheric absorption. In Version 5, this modification was abandoned in favor of the original model [18].

The return to the original Liebe model in Version 5 improved the performance of the algorithm at high latitude and decreased the SST dependence. The remaining SST bias is removed empirically in Version 5 by empirically adjusting the temperature dependence in the correction for roughness [1]. In Version 5, an additional temperature dependence is added to the correction of emissivity for roughness (e.g., Equation (13) in Appendix V of the end of mission ATBD [1]) and adjusted empirically). As a consequence, Version 5.0 is able to remove the SST-dependent bias to within ±0.1 psu, even in cold water.

This empirical adjustment of the roughness correction is an effective solution, but it does not give insight into the physical source of the SST dependence. In particular, it clearly depends on the model function that is used for the dielectric constant of sea water. This is illustrated in Figure 9, which shows the dependence of the salinity error on SST when different model functions are used in the retrieval from Aquarius brightness temperature to SSS. The solid black curve is the Klein–Swift model function [25], which is used by SMOS. The dashed blue curve is the Meissner–Wentz model function [26,35], which is used in the retrievals by the Aquarius Project, and the dashed red curve is the model function reported by Zhou et al. [27], and based on their measurements at 1.413 GHz [36,37]. The models by Klein and Swift and Meissner and Wentz have been used in the SSS retrievals by various teams using SMOS, SMAP, and Aquarius brightness temperature. The model by Zhou et al. is more recent, and has not yet been used in SSS products.

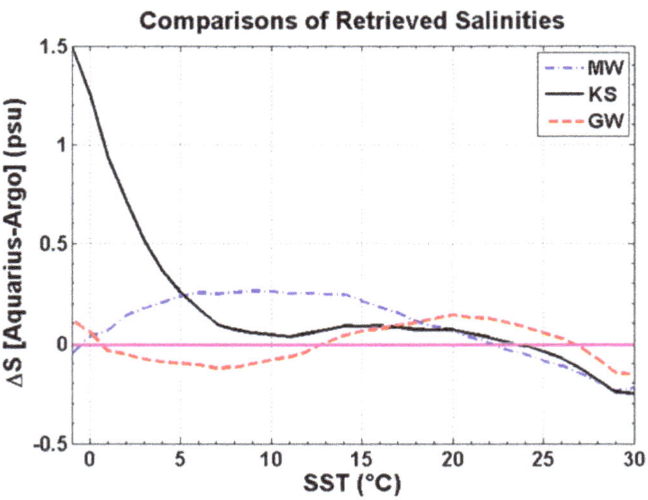

Figure 9. Salinity error as a function of sea surface temperature, SST, when using three different model functions for the dielectric constant of sea water in the retrieval. The three model functions are: Klein–Swift [25], Meissner–Wentz [26], and George Washington University [36]. (Author's figure from Zhou et al. [36]).

Figure 9 was obtained by keeping the retrieval algorithm fixed and replacing the dielectric constant with one of the three model functions. The example in Figure 9 was produced using Version 3 of salinity retrieval, because changes were made in V4 and V5 that empirically corrected for the

SST-dependent bias, and are hard to remove. It is evident that the dependence of error on SST is different for the three model functions, especially at very low temperatures. It is also clear from Figure 8 that there is an issue with the retrieval at temperatures near 0 °C (although the data are sparse at these temperatures), and Figure 9 shows a very different and inconsistent dependence on the model function that is used in the retrieval at low temperatures. This is an especially important issue because of the importance of cold water with changing climate (i.e., understanding the effect of melting ice and the associated in flux of fresh water on ocean circulation) and because cold water is also a region of decreased sensitivity to changes in salinity (e.g., Figure 2 in Le Vine, Lagerloef and Torrusio [38]).

As mentioned above, the SST dependence is removed in Aquarius Version 5 by correcting the model for atmospheric absorption, which reduced the bias, and then empirically removing the residual bias. The empirical fix is good at removing the SST dependence; however, there are likely physical causes for this dependence, such as the model function for the dielectric constant of sea water, which when more fully understood will lead to improved retrievals in the future and lessen the need for empirical adjustments.

4.4. Example Issue: Whole Range Calibration

4.4.1. Introduction

The goal of Aquarius is to measure sea surface salinity in the open ocean, and the calibration of the instrument has been tuned to achieve this goal. There are two calibration scenes that are sufficiently large and uniform, and sufficiently well-known to be appropriate for this purpose: cold sky and the ocean itself. Both have been used in the calibration of Aquarius, but after the initial removal of a bias, the calibration uses only the ocean (e.g., see discussion of drift above). After the initial post-launch check for bias, the cold sky was used only to look for temporal changes as a check on stability and adjust the model for the antenna patterns [29].

In the ideal case, the Aquarius calibration would use two points (e.g., cold sky and ocean) and be checked at a third (e.g., land). Even better, the calibration would use cold sky and a non-ocean scene such as land or ice, and then be verified over the ocean. Unfortunately, this is not practical, because scenes at the warm end (e.g., 250 K which is typical of land and ice at L-band) are generally not known with sufficient accuracy over a large enough area to be compatible with the Aquarius footprint (3-dB radius on the order of 100 km [3]). Furthermore, because the brightness temperature of the open ocean at L-band has a very limited dynamic range, it is possible to have a calibration that is accurate over the ocean, but not sufficiently accurate at the warm extremes for applications over land such as the retrieval of soil moisture This is illustrated in Figure 10, which shows a plot of the measured Aquarius antenna temperature, TA_meas, against the expected value computed given the appropriate surface truth (salinity and water temperature for ocean). The data are for the horizontal polarization and the middle beam, and the dashed line is the Aquarius calibration curve. The insets show expanded views over the ocean and at cold sky. It is clear that the calibrated operating curve for the radiometer (dashed line) is well fitted to the ocean data, but that the radiometer is too cold at the cold end (inset for cold sky, magenta) and is too warm at the warm end (red dots near 250 K). The data at the warm end at 250 K are from measurements over the United States Department of Agriculture (USDA) Little River research watershed.

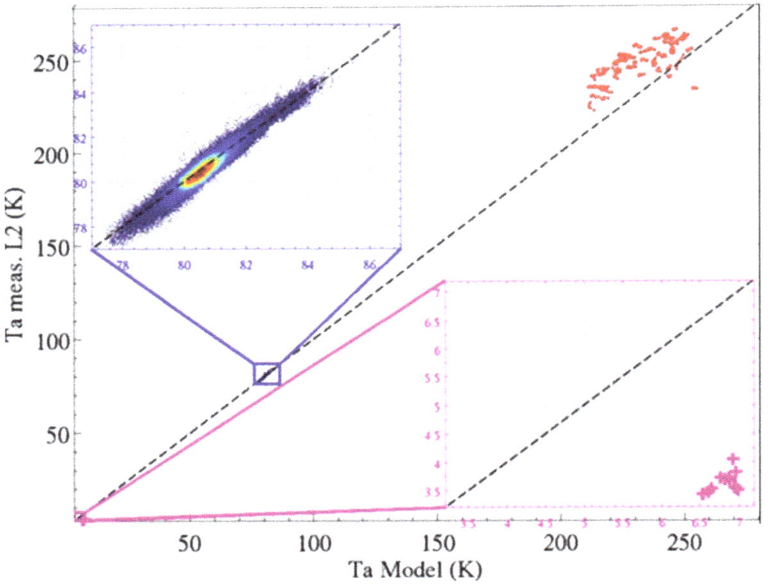

Figure 10. Aquarius measured antenna temperature, Ta_measured, plotted against the antenna temperature predicted with the forward model, Ta_model, for three regimes: cold sky (magenta insert), ocean (blue insert), and land (Little River watershed, red dots). The data is for horizontal polarization, and the middle beam and the measurements are from V2.0 of the algorithm. (Author's figure from Le Vine et al. [4]).

4.4.2. Whole Range Calibration V5.WR

An initial attempt has been made in Version 5 to provide a better calibration for the full range of applications by including the cold sky in a two-point (cold sky and ocean) calibration. Essentially, this amounts to drawing a straight line between these two references, and extending it to warm brightness temperatures [39]. This calibration is provided as an extension of Version 5, and called Version 5.WR [40,41].

The development of Version 5.WR begins with V5, and adds an additional step in which the cold sky and ocean observations (V5) are used to tune the calibration. The change is a linear transformation from the original V5 antenna temperatures to the new, V5.WR antenna temperatures: TA_V5.WR = a TA_V5.0 + b. The large difference in antenna temperature between cold sky (4 K) and ocean (80 K) helps determine the slope as a function of the target TA in a way that is not possible with only ocean observations, which have a small dynamic range (e.g., see the insets in Figure 10). The changes have minimal effect on the data over ocean or the retrieved salinity [40], because the ocean global average TA is kept unchanged by design.

The coefficients (a, b) are given in Table 1. They were derived from a linear regression between two points as follows:

- At the cold end, using the difference between the mean of the TA measured by Aquarius for the 30 cold sky calibrations and the mean of the corresponding TA_expected for the cold sky look computed from radiative transport theory;
- Over the ocean, using the mean of the TA measured by Aquarius globally for the year 2012 (filtered for RFI, and with a water fraction of $\geq$99.9%) and the mean of the corresponding TA_expected.

Table 1. Values for the coefficients (a, b).

	V-Pol		H-Pol	
	a	*b*	*a*	*b*
Beam 1	1.003350568406014	−3.592857280825468	1.007405352181248	−6.595668007611077
Beam 2	1.008337848581688	−9.655610862597533	1.003498234086013	−2.948722716952730
Beam 3	1.017695610594212	−2.233911609523380	1.001311195384693	−1.075651639210478

A full evaluation of the whole range calibration, V5.WR, is not complete, largely because of the lack of reference sites of sufficient size, but a check has been made over the USDA Little Washita and Little River watersheds, which are instrumented for monitoring soil moisture, as shown in Figure 11. Due to the large footprint and fixed orbit of Aquarius, it was not possible to use the same scene for all of the beams. The improvement compared to earlier versions of the algorithm without the whole range calibration is evident by comparing it with the example in Figure 10. The improvement at the cold end is significant. The change over land when compared with V5 is an improvement by as much as 2.2 K (H-pol and outer beam), and generally on the order of 1 K. These changes are small compared to the uncertainty in the model due to uncertainties in the surface truth (soil moisture) and effect of vegetation, and the small size of the scene compared to the footprint of the radiometer. Further validation of the results at the warm end is needed. Due to this, and to avoid confusion with V5, the version V5.WR has not been listed in the public database; however, it is available to the public upon request from the PO.DAAC.

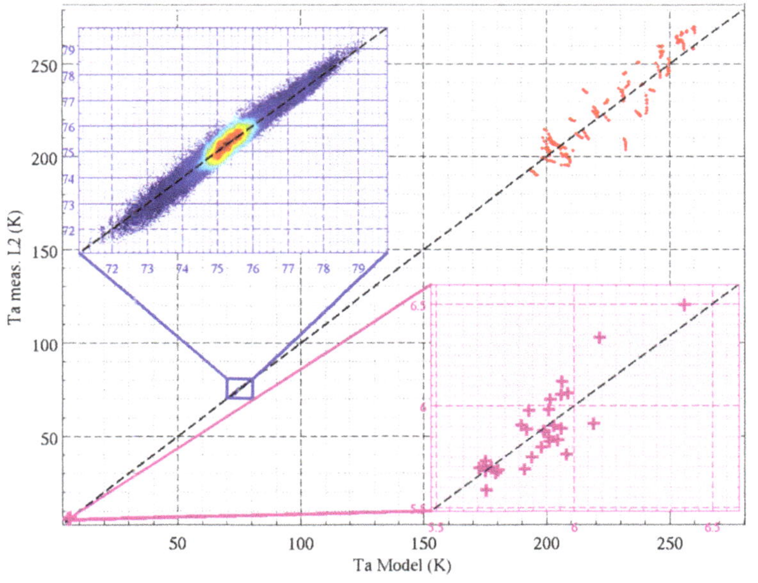

Figure 11. Aquarius measured antenna temperature, Ta_measured, plotted against the antenna temperature predicted with the forward model, Ta_model, for three regimes: cold sky (magenta insert), ocean (blue insert), and Little Washita watershed (red dots). The data are for the outer Aquarius beam and horizontal polarizations, and Version 5.WR, the final release of the algorithm with the full range calibration. Insets are close-ups of ocean and Sky data.

5. Conclusions: Future of SSS Remote Sensing

Although the Aquarius mission has ended, research on the remote sensing of sea surface salinity from space by NASA continues by shifting the effort of retrieving SSS to SMAP. This work actually started several years ago within the Aquarius project. Soon after SMAP was launched, a subset of the Aquarius science team submitted a proposal to NASA to look at the feasibility of adapting the Aquarius salinity retrieval algorithm to retrieve salinity using the SMAP radiometer observations over ocean. Retrieving salinity from SMAP presented several obvious challenges. (1) The SMAP radar failed soon after launch, which means that a model function and external source for wind speed are needed for the roughness correction. (2) The loss and temperature dependence of the SMAP antenna system are not well-known (compared to Aquarius), which requires addition empirical adjustments that are not necessary with Aquarius. The objective of this work was to transfer the algorithm and understanding gained with Aquarius to SMAP data to see if a scientifically meaningful SSS product could be obtained.

The work with SMAP data began during the Aquarius project, and initial results have been reported [42–44]. Figure 12 is an example. Figure 12a shows the mean salinity field from SMAP data reported by remote sensing systems [44] from algorithm Version 2 (based on Version 3 of SMAP Level 1B brightness temperatures) for the two years 2015–2016. The dominant features of the global surface salinity field are clear (the Atlantic Ocean is saltier than the Pacific; dipole-like structure with salty mid-latitudes separated by the fresher tropical convergence zone; salty Arabian Sea; fresher Indian Ocean). However, there is clearly work to be done to improve the overall performance of the retrieval. This is illustrated in Figure 12b, which shows the salinity error map (the difference between the retrieved SSS and in situ measurements from Argo). A notable feature is the strong zonal dependence with salty bias (positive) in the low latitudes and fresh bias (negative) at mid and high latitudes. Plotting the error as a function of temperature demonstrates an SST dependence that is just the opposite of that present in Aquarius V5. (The SMAP V2 salinity retrieval uses the same model function as Aquarius V4, which suggests that this difference is due to something that is specifically associated with SMAP such as the external wind speed that goes into the roughness correction, or the loss model for the reflector model, or both.)

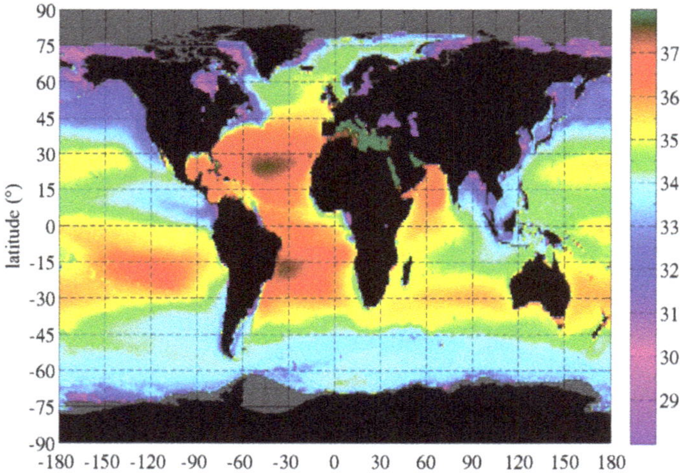

(a)

Figure 12. *Cont.*

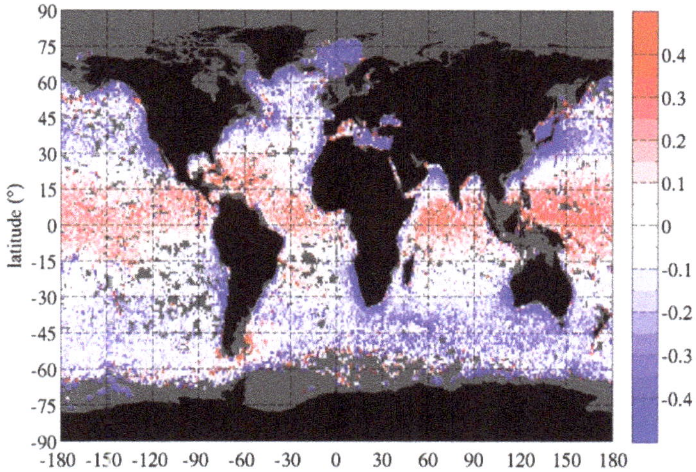

(b)

Figure 12. Salinity from the Soil Moisture Active Passive (SMAP) satellite: (**a**) global maps of SSS, remote sensing systems, Version 2, two-year average, 2015–2016; (**b**) salinity error, difference between retrieved SSS and in situ from Argo.

Now that the Aquarius Mission has ended, plans are in place to build upon this structure to form a NASA salinity continuity project using SMAP data. A new ocean salinity science team is being formed to work on this data. It will include research at the Jet Propulsion Laboratory and Goddard Space Flight Center, and a nominal algorithm with processing the responsibility of remote sensing systems (which was responsible for the Aquarius algorithm). Data will be available at the PO.DAAC, and there will also be an independent website (https://salinity.oceanscience.org) with information for the public and links to other data products, similar to what was available for Aquarius.

An important part of the work on Aquarius was the assessment of the data product. This meant matching in situ measurements with the retrievals from space, and taking into account the error of the pointwise in situ measurements to represent the salinity values on Aquarius' measurement scales. It is planned to continue this work and cooperate with the effort that is currently underway at the European Space Agency (ESA) to develop a matchup database to support salinity retrievals of SMOS [45]. Hopefully, a common database for this purpose can be developed that helps in the evaluation and eventual merging of the two sources of salinity observations.

Finally, work is underway to prepare for the future. Aquarius is gone, and SMOS is aging. SMAP is already nearly four years old. At least two missions with L-band radiometers are under development in China [46–48], and one of them, the Water Cycle Observation Mission (WCOM), is scheduled for launch in the near future [47]. Studies are also underway within NASA, at the Jet Propulsion Laboratory (JPL) and the Goddard Space Flight Center (GSFC), to define the NASA L-band mission of the future. Designs for broadband sensors including lower frequencies such as P-band and with and without radar have been reported [49,50]. However, this is preliminary design, and the optimum choice is far from clear [32]. Nor is it clear that a do-it-all mission that can meet the future needs of both the soil moisture community (high spatial resolution and short revisit time) and ocean salinity community (higher sensitivity to address cold water and better spatial resolution to address coastal regions) can be met with one instrument.

Author Contributions: This work reported here was conducted as part of the Aquarius Project. All authors were members of the Project and/or the associated Science Team and are listed because they contributed substantial to aspects of the manuscript (as indicated by the references cited). The lead author was responsible for most of the writing. Editing was a team effort.

Funding: Some of this research was funded by the National Aeronautics and Space Administration grant number NNX14AR31G.

Acknowledgments: Aquarius was a team effort. The authors wish to acknowledge the contributions of the entire Aquarius team whose dedication made Aquarius a success and this report possible.

Conflicts of Interest: The authors declare no conflict of interest.

References

1. Meissner, T.; Wentz, F.; Le Vine, D.M. Aquarius Salinity Retrieval Algorithm: End of Mission Algorithm Theoretical Basis Document (ATBD). *RSS Tech. Rep.* **2017**, *120117*.
2. Meissner, T.; Wentz, F.J.; Le Vine, D.M. The Salinity Retrieval Algorithm for the NASA Aquarius Version 5 and SMAP Version 3 Releases. *Remote Sens.* **2018**, *10*, 1121. [CrossRef]
3. Le Vine, D.M.; Lagerloef, G.S.E.; Colomb, F.R.; Yueh, S.H.; Pellerano, F.A. Aquarius: An instrument to monitor sea surface salinity from space. *IEEE Trans. Geosci. Remote Sens.* **2007**, *45*, 2040–2050. [CrossRef]
4. Le Vine, D.M.; Dinnat, E.P.; Meissner, T.; Yueh, S.H.; Wentz, F.J.; Torrusio, S.E.; Lagerloef, G. Status of Aquarius/SAC-D and Aquarius Salinity Retrievals. *IEEE J. Sel. Top. Appl. Earth Obs. Remote Sens.* **2015**, *8*, 5401–5415. [CrossRef]
5. Yueh, S.H.; West, R.; Wilson, W.J.; Li, F.K.; Njoku, E.G.; Rahmat-Samii, Y. Error sources and feasibility for microwave remote sensing of ocean surface salinity. *IEEE Trans. Geosci. Remote Sens.* **2001**, *39*, 1049–1060. [CrossRef]
6. Yueh, S.H. Estimates of Faraday rotation with passive microwave polarimetry for microwave remote sensing of Earth surfaces. *IEEE Trans. Geosci. Remote Sens.* **2000**, *38*, 2434–2438. [CrossRef]
7. Le Vine, D.M.; Abraham, S.; Utku, C.; Dinnat, E.P. Aquarius third stokes parameter measurements: Initial results. *IEEE Geosci. Remote Sens. Lett.* **2013**, *10*, 520–524. [CrossRef]
8. Le Vine, D.M.; Abraham, S. The effect of the ionosphere on remote sensing of sea surface salinity from space: Absorption and emission at L band. *IEEE Trans. Geosci. Remote Sens.* **2002**, *40*, 771–782. [CrossRef]
9. Le Vine, D.M.; Lagerloef, G.S.E.; Ruf, C.; Wentz, F.; Yueh, S.; Piepmeier, J.; Lindstrom, E.; Dinnat, E. Aquarius: The instrument and initial results. In Proceedings of the 12th Specialist Meeting on Microwave Radiometry and Remote Sensing of the Environment (MicroRad), Rome, Italy, 5–9 March 2012.
10. Wilson, W.J.; Tanner, A.; Pellerano, F.; Horgan, K. Ultrastable radiometers for future sea surface salinity missions. *JPL Rep.* **2005**, *D-31794*.
11. Pellerano, F.A.; Piepmeier, J.; Triesky, M.; Horgan, K.; Forgione, J.; Caldwell, J.; Wilson, W.J.; Yueh, S.; Spencer, M.; McWatters, D.; et al. The {A}quarius Ocean Salinity Mission High Stability {L-band} Radiometer. In Proceedings of the IEEE International Conference on Geoscience and Remote Sensing Symposium, Denver, CO, USA, 31 July–4 August 2006; pp. 1681–1684.
12. Le Vine, D.M. ESTAR experience with RFI at L-band and implications for future passive microwave remote sensing from space. In Proceedings of the IEEE International Geoscience and Remote Sensing Symposium, Toronto, ON, Canada, 24–28 June 2002; pp. 847–849.
13. Skou, N.; Misra, S.; Balling, J.E.; Kristensen, S.S.; Sobjaerg, S.S. L-band RFI as experienced during airborne campaigns in preparation for SMOS. *IEEE Trans. Geosci. Remote Sens.* **2010**, *48*, 1398–1407. [CrossRef]
14. Piepmeier, J.R.; Pellerano, F.A.; Freedman, A. Aquarius L-Band Microwave Radiometer: 3 Years of Radiometric Performance and Systematic Effects. *IEEE J. Sel. Topics Appl. Earth Obs. Remote Sens.* **2006**, *8*, 5416–5423. [CrossRef]
15. de Matthaeis, P.; Peng, J.; Piepmeier, J.; Le Vine, D. Overview of Aquarius Radiometer Post-Launch Measurement Counts to Antenna Temperature Processing for Product Version 5. *AQ-014-PS-0029*, 2018.
16. Aquarius Project, "ATBD History. Available online: https://podaac.jpl.nasa.gov/aquarius,December31,2017 (accessed on 1 October 2018).
17. Meissner, T.; Wentz, F.J.; Scott, J.; Vazquez-Cuervo, J. Sensitivity of Ocean Surface Salinity Measurements From Spaceborne L-Band Radiometers to Ancillary Sea Surface Temperature. *IEEE Trans. Geosci. Remote Sens.* **2016**, *54*, 7105–7111. [CrossRef]

18. Liebe, H.J.; Rosenkranz, P.W.; Hufford, G.A. Atmospheric {60-GHz} oxygen spectrum: New laboratory measurements and line parameters. *J. Quant. Spectrosc. Radiat. Transf.* **1992**, *48*, 629–643. [CrossRef]
19. Meissner, T.; Wentz, F.J.; Ricciardulli, L. The emission and scattering of L-band microwave radiation from rough ocean surfaces and wind speed measurements from the Aquarius sensor. *J. Geophys. Res. C Ocean.* **2014**, *119*, 6499–6522. [CrossRef]
20. Boutin, J.; Chao, Y.; Asher, W.E.; Delcroix, T.; Drucker, R.; Drushka, K.; Kolodziejczyk, N.; Lee, T.; Reul, N.; Reverdin, G.; et al. Satellite and In Situ Salinity: Understanding Near-Surface Stratification and Subfootprint Variability. *Bull. Amer. Meteor. Soc.* **2016**, *97*, 1391–1407. [CrossRef]
21. Roemmich, D.; Johnson, G.; Riser, S.; Davis, R.; Gilson, J.; Owens, W.B.; Garzoli, S.; Schmid, C.; Ignaszewski, M. The Argo Program: Observing the Global Oceans with Profiling Floats. *Oceanography* **2009**, *22*, 34–43. [CrossRef]
22. Kao, H.-Y.; Lagerloef, G.; Lee, T.; Melnichenko, O.; Hacker, P. Aquarius Salinity Validation Analysis. AQ-014-PS-0016. 28 February 2018.
23. Chassignet, E.P.; Hurlburt, H.E.; Smedstad, O.M.; Halliwell, G.R.; Hogan, P.J.; Wallcraft, A.J.; Baraille, R.; Bleck, R. The HYCOM (HYbrid Coordinate Ocean Model) data assimilative system. *J. Mar. Syst.* **2007**, *65*, 60–83. [CrossRef]
24. Lee, T. Consistency of Aquarius sea surface salinity with Argo products on various spatial and temporal scales. *Geophys. Res. Lett.* **2016**, *43*, 3857–3864. [CrossRef]
25. Klein, L.A.; Swift, C.T. An improved model for the dielectric constant of sea water at microwave frequencies. *IEEE J. Ocean. Eng.* **1977**, *AP-25*, 104–111. [CrossRef]
26. Meissner, T.; Wentz, F.J. The Emissivity of the Ocean Surface Between 6 and 90 GHz Over a Large Range of Wind Speeds and Earth Incidence Angles. *IEEE Trans. Geosci. Remote Sens.* **2012**, *50*, 3004–3026. [CrossRef]
27. Zhou, Y.; Lang, R.H.; Dinnat, E.P.; Le Vine, D.M. L-Band Model Function of the Dielectric Constant of Seawater. *IEEE Trans. Geosci. Remote Sens.* **2017**, *55*, 6964–6974. [CrossRef]
28. Wilson, W.J.; Yueh, S.H.; Dinardo, S.J.; Li, F.K. High-Stability L-Band Radiometer Measurements of Saltwater. *IEEE Trans. Geosci. Remote Sens.* **2004**, *42*, 1829–1835. [CrossRef]
29. Dinnat, E.P.; Le Vine, D.M.; Piepmeier, J.R.; Brown, S.T.; Hong, L. Aquarius L-band Radiometers Calibration Using Cold Sky Observations. *IEEE J. Sel. Top. Appl. Earth Obs. Remote Sens.* **2015**, *8*, 5433–5449. [CrossRef]
30. Scripps Institution of Oceanography. Global gridded NetCDF Argo only dataset produced by optimal interpolation Available online: http://apdrc.soest.ucsd.edu/Gridded_fields.html (accessed on 1 October 2018).
31. Piepmeier, J.R.; Hong, L.; Pellerano, F.A. Aquarius L-Band Microwave Radiometer: 3 Years of Radiometric Performance and Systematic Effects. *IEEE J. Sel. Top. Appl. Earth Obs. Remote Sens.* **2015**, *8*, 5416–5423. [CrossRef]
32. Misra, S. Enabling the Next Generation of Salinity, Sea Surface Temperature and Wind Meaurements from Space: Instrument Challenges. In Proceedings of the Global Ocean Salinity and Water Cycle Workshop, Woods Hole, MA, USA, 22–26 May 2017.
33. Dinnat, E.P.; Le Vine, D.M. Impact of sun glint on salinity remote sensing: An example with the aquarius radiometer. *IEEE Trans. Geosci. Remote Sens.* **2008**, *46*, 3137–3150. [CrossRef]
34. Dinnat, E.P.; Boutin, J.; Yin, X.; Le Vine, D.; Waldteufel, P.; Vergely, J.-L. Comparison of SMOS and Aquarius sea surface salinity and analysis of possible causes for the differences. In Proceedings of the 2014 XXXIth URSI General Assembly and Scientific Symposium (URSI GASS), Beijing, China, 16–23 August 2014.
35. Meissner, T.; Wentz, F.J. The complex dielectric constant of pure and sea water from microwave satellite observations. *IEEE Trans. Geosci. Remote Sens.* **2004**, *42*, 1836–1849. [CrossRef]
36. Lang, R.; Zhou, Y.; Dinnat, E.; Le Vine, D. The Dielectric Constant Model Function and Implications for Remote Sensing of Salinity. In Proceedings of the IEEE International Geoscience and Remote Sensing Symposium (IGARSS), Fort Worth, TX, USA, 23–28 July 2017; pp. 3572–3574.
37. Lang, R.; Zhou, Y.; Utku, C.; Le Vine, D. Accurate measurements of the dielectric constant of seawater at L band. *Radio Sci.* **2016**, *51*, 2–24. [CrossRef]
38. Le Vine, D.M.; Lagerloef, G.S.E.; Torrusio, S.E. Aquarius and remote sensing of sea surface salinity from space. *Proc. IEEE* **2010**, *98*, 688–703. [CrossRef]

39. Dinnat, E.P.; Le Vine, D.M.; Bindlish, R.; Piepmeier, J.R.; Brown, S.T. Aquarius whole range calibration: Celestial Sky, ocean, and land targets. In Proceedings of the 13th Specialist Meeting on Microwave Radiometry and Remote Sensing of the Environment (MicroRad), Pasadena, CA, USA, 24–27 March 2014; pp. 192–196.
40. Le Vine, D.M.; Dinnat, E.P. Whole Range Calibration: Version 5.WR. *AQ-014-PS-0030*; 28 February 2018. Available online: https://podaac.jpl.nasa.gov/aquarius (accessed on 1 October 2018).
41. Dinnat, E.P.; Le Vine, D.M.; Hong, L. Aquaruius Final Release Product and Full Range Calibration of L-Band Radiometer. In Proceedings of the IEEE International Geoscience and Remote Sensing Symposium (IGARSS2018), Valencia, Spain, 22–27 July 2018.
42. Dinnat, E.; Le Vine, D.; Soldo, Y.; de Matthaeis, P. Theoretical algorithm for the retrieval of sea surface salinity from SMAP observations at L-band. In Proceedings of the 15th Specialist Meeting on Microwave Radiometry and Remote Sensing of the Environment, Cambridge, MA, USA, 27–30 March 2018.
43. Meissner, T.; Wentz, F.; Ricciardulli, L.; Mears, C.; Manaster, A. Ocean Surface Salinity and Wind Speed from the SMAP L-Band Radiometer. In *Microwave Radiometry and Remote Sensing of the Earth's Surface and Atmosphere*; VSP: Rancho Cordova, CA, USA, 2018.
44. Meissner, T.; Wentz, F. *Remote Sensing Systems SMAP Ocean Surface Salinities Level 2C, Version 2.0 Validated Release*; Remote Sensing Systems: Santa Rosa, CA, USA, 2016.
45. Sabia, R. SMOS Pilot-Mission Exploitation Platform (PI-MEP): A Hub for Validation and Expolitation of ESA SMOS Sea Surface Salinity Data. In Proceedings of the Ocean Sciences Meeting, Portland, OR, USA, 11–16 February 2018.
46. Dong, X.; Shi, J.; Zhang, S.; Liu, H.; Wang, Z.; Zhu, D.; Zuo, L.; Chen, C.; Chen, W. Prelminary design of water cycle observation mission (WCOM). In Proceedings of the IEEE International Geoscience and Remote Sensing Symposium (IGARSS), Beijing, China, 10–15 July 2016; pp. 3434–3437.
47. Shi, J.; Dong, X.; Zhao, T.; Du, Y.; Liu, H.; Wang, Z.; Zhu, D.; Ji, D.; Xiong, C.; Jiang, L. The Water Cycle Observation Mission (WCOM): Overview. In Proceedings of the IEEE International Geoscience and Remote Sensing Symposium (IGARSS), Beijing, China, 10–15 July 2016; pp. 3430–3433.
48. Xu, X.; Yun, R.; Dong, X.; Zhu, D.; Yin, X.; Liu, H. Data Pre-Processing of MICAP (Microwave Imager Combined Active and Passive) Scatterometer. In Proceedings of the IEEE International Geoscience and Remote Sensing Symposium (IGARSS), Beijing, China, 10–15 July 2016; pp. 4776–4779.
49. Dinnat, E.; de Amici, G.; Le Vine, D.; Piepmeier, J. Next generation spaceborne instrument for monitoring ocean salinity with application to the coastal zone and cryosphere. In Proceedings of the 15th Specialist Meeting on Microwave Radiometry and Remote Sensing of the Environment, Cambridge, MA, USA, 27–30 March 2018.
50. Brown, S. A Next Generation Spaceborne Ocean State Observatory: Surface Salinity, Temperature and Ocean Winds from Equator to Pole. In Proceedings of the Globasl Ocean Salinity and the Water Cycle Workshop, Woods Hole, MA, USA, 22–26 May 2017.

remote sensing

MDPI

Article

The Salinity Retrieval Algorithms for the NASA Aquarius Version 5 and SMAP Version 3 Releases

Thomas Meissner [1,*], Frank J. Wentz [1] and David M. Le Vine [2]

1 Remote Sensing Systems, 444 Tenth Street, Suite 200, Santa Rosa, CA 95401, USA; frank.wentz@remss.com
2 NASA Goddard Space Flight Center, Greenbelt, MD 20771, USA; david.m.levine@nasa.gov
* Correspondence: meissner@remss.com; Tel.: +1-545-2904 (ext. 22)

Received: 14 June 2018; Accepted: 30 June 2018; Published: 15 July 2018

Abstract: The Aquarius end-of-mission (Version 5) salinity data set was released in December 2017. This article gives a comprehensive overview of the main steps of the Level 2 salinity retrieval algorithm. In particular, we will discuss the corrections for wind induced surface roughness, atmospheric oxygen absorption, reflected galactic radiation and side-lobe intrusion from land surfaces. Most of these corrections have undergone major updates from previous versions, which has helped mitigating temporal and zonal biases. Our article also discusses the ocean target calibration for Aquarius Version 5. We show how formal error estimates for the Aquarius retrievals can be obtained by perturbing the input to the algorithm. The performance of the Aquarius Version 5 salinity retrievals is evaluated against salinity measurements from the ARGO network and the HYCOM model. When stratified as function of sea surface temperature or sea surface wind speed, the difference between Aquarius Version 5 and ARGO is within ± 0.1 psu. The estimated global RMS uncertainty for monthly 100 km averages is 0.128 psu for the Aquarius Version 5 retrievals. Finally, we show how the Aquarius Version 5 salinity retrieval algorithm is adapted to retrieve salinity from the Soil-Moisture Active Passion (SMAP) mission.

Keywords: sea surface salinity; remote sensing; aquarius; SMAP; retrieval algorithm; calibration; validation

1. Introduction

NASA's Aquarius mission [1] measured ocean surface salinity from late August 2011 until early June 2015. The end-of mission (Version 5) data release [2] in December 2017 provides an important legacy data set of ocean salinity for the oceanographic research community. It is the goal of this publication to highlight the most important aspects of the Aquarius Version 5 salinity retrieval algorithm. Many technical details of the algorithm are given in the Algorithm Theoretical Basis Document (ATBD) [3], to which the interested reader is referred. Here, we will focus on the parts of the Version 5 algorithm that have not yet been published. Another goal of our paper is to demonstrate the improvements in the Version 5 algorithm over prior releases [4]. Finally, we will show how the Aquarius Version 5 salinity retrieval algorithm can be adapted to the SMAP (Soil Moisture Active Passive) Mission [5,6]. SMAP has been measuring ocean surface salinity [7] since 1 April 2015 and thus is continuing the Aquarius legacy data.

Our paper is organized as follows: Section 2 gives a schematic overview of the steps in the salinity retrieval algorithm. It also contains a brief discussion of the necessary ancillary inputs and the concept of the expected antenna temperature, which is computed by a reference salinity field. We will then discuss various components of the algorithm: surface roughness correction (Section 3), atmospheric correction (Section 4), reflected galaxy correction (Section 5) and correction for side-lobe intrusion from land surfaces (Section 6). The ocean target calibration and calibration drift correction is the subject of Section 7. In Section 8 we discuss the various error sources and how formal estimates for the retrieval uncertainties can be obtained. Section 9 briefly summarizes the validation of the Aquarius Version 5 retrievals versus in-situ measurements from the ARGO network and exhibits how the Version 5 results have improved over prior versions. Finally, in Section 10 we discuss how the Aquarius Version 5 algorithm can be adapted to retrieving salinity with SMAP. Section 11 summarizes our main findings.

2. Major Steps of the Salinity Retrieval Algorithm

2.1. Basic Algorithm Flow

The basic inputs to the Aquarius Version 5 salinity retrieval algorithm are the antenna temperature (TA) measurements from the Aquarius radiometer, which have been filtered for radio frequency interference (RFI) [8,9], and radar backscatter (σ_0) measurements from the Aquarius scatterometer. The output is sea-surface salinity (SSS) and many intermediate variables required for the salinity calculation. The algorithm (Figure 1) consists of a number of steps that are intended to remove unwanted signals: radiation from galaxy, sun, moon and the cosmic microwave background (CMB), antenna cross-polarization contamination, Faraday rotation in the Earth's ionosphere, attenuation in the Earth's atmosphere and radiation from land surfaces. The result of these successive corrections is the brightness temperature (TB) that is emitted from the ocean surface. Finally, the effect of the wind roughening of the ocean surface is removed leading to the brightness temperature (T_{B0}) that is emitted from a flat ocean surface. A maximum likelihood estimator (MLE) is then used to estimate salinity from T_{B0}. The MLE minimizes the sum of square (SOS) difference between the Aquarius measurements of T_{B0} of the two polarizations (vertical V and horizontal H) and the values of the radiative transfer model calculation for T_{B0}. This model calculation which is based on the dielectric constant of sea-water [10–12]. Figure 1 shows a schematic flow diagram of the salinity retrieval algorithm.

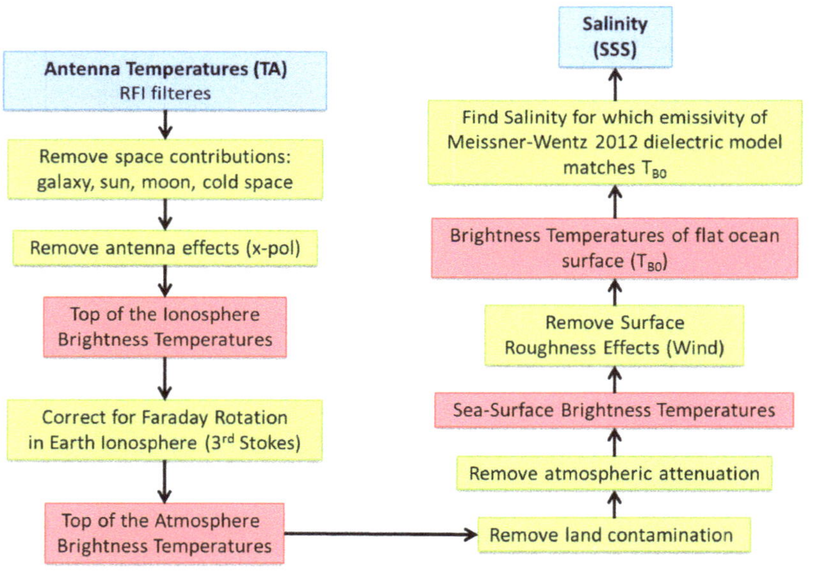

Figure 1. Schematic flow diagram of the Aquarius Version 5 salinity retrieval algorithm.

2.2. Ancillary Inputs

The salinity retrieval algorithm also requires a number of ancillary input fields. The most important ones are discussed in this section:

2.2.1. Sea Surface Temperature (SST)

SST is a crucial input to the dielectric constant model [10–12], which enters the computation of the emissivity of a flat ocean surface. An evaluation was made of the performance of the SSS retrieval algorithm with various ancillary SST fields [13]. The result of the analysis was that the best performance is obtained with the daily GHRSST (Gridded High-Resolution SST) Level 4 field from the Canadian Meteorological Center (CMC). It is gridded at 0.2° resolution and available from the PO.DAAC web site (https://podaac.jpl.nasa.gov/dataset/CMC0.2deg-CMC-L4GLOB-v2.0). Version 5.0 uses this ancillary SST field. The CMC field is linearly interpolated in space and time to the boresight location of the Aquarius observation.

2.2.2. Atmospheric Profiles

The atmospheric profiles for pressure, temperature, relative humidity, and cloud water mixing ratio are used in the correction for atmospheric absorption. They are obtained from the NCEP (national Centers for Environmental Prediction) GDAS (General Data Assimilation System) 1-deg 6-hourly fields are used, which are available from http://nomads.ncep.noaa.gov/. All fields are linearly interpolated in space and time to the boresight location of the Aquarius observation.

2.2.3. Wind Speed Background Field

The NCEP GDAS 1-deg, 6-h scalar wind speed field is used as background field in the Aquarius wind speed retrievals [12]. It is available from http://nomads.ncep.noaa.gov/. It is linearly interpolated in space and time to the boresight location of the Aquarius observation.

2.2.4. Wind Direction

The NCEP GDAS 1-deg, 6-h scalar wind direction field is used in the surface roughness correction and in the Aquarius wind speed retrievals [12]. It is available from http://nomads.ncep.noaa.gov/. It is linearly interpolated in space and time to the boresight location of the Aquarius observation.

2.2.5. Land Mask

We use the static 1-km land/water mask from the OCEAN DISCIPLINE PROCESSING SYSTEM (ODPS). It is based on World Vector Shoreline (WVS) database and World Data Bank and was provided as courtesy of Fred Patt, Goddard Space Flight Center, frederick.s.patt@nasa.gov. From the land mask two values of the land fraction are computed. The first one, called f_{land}, is defined as the fraction of land area within an antenna 3 dB footprint, where the land area consist of the sum of 1-km land pixels that fall within the footprint. A second land fraction, called g_{land}, is the fractional land area weighted by the antenna gain pattern, and the integration is taken over the whole Earth field of view. For details see ([3], page 34 + 35).

2.2.6. Rain Rate and Rain Flagging

The instantaneous rain rate (IRR) is used as a quality control indicator in the ocean target calibration (Section 7) and in the validation (Section 9) when matching Aquarius observations with in-situ salinity measurements from ARGO drifters. Aquarius measures ocean salinity within a few centimeters of the surface, whereas ARGO measures salinity at 5 m depth. In order to avoid mismatches between the two measurements resulting from salinity stratification within the upper ocean layer under precipitation [14], it is necessary to have information on rain rate, which is used to flag observations with rain. Our IRR, provided by the Aquarius Rain Accumulation (RA) product [15], is used as an ancillary data set that aids users of the V5.0 Aquarius Level 2 data to better understand the salinity stratification changes due to rain. The product uses as input the surface rain rates from the NOAA CMORPH (CPC-Climate Prediction Center-Morphing technique) global precipitation data set [16]. The average instantaneous rain rate is calculated using a structure of 13 CMORPH pixels around the center of the Aquarius IFOV, where the weight associated with each pixel is based on the antenna gain. We want to emphasize that an Aquarius observation that has been rain flagged according to this procedure is not to be regarded as degraded or bad. The surface rain freshening is a real signal that is picked up by the satellite but not by ARGO floats or models. The rain flag solely serves the purpose to indicate not to use this observation when validating Aquarius salinity measurements against ARGO drifters.

2.3. Forward Model and Expected TA

The forward model calculates the Aquarius TA, also called *expected TA*, for a given Earth scene by reversing the steps of the retrieval algorithm. The expected TA serves as an important diagnostic tool in the assessment of the geophysical model that is used in the salinity retrieval and also in the sensor calibration. The computation of TA expected requires an external reference salinity field as crucial input in the computation of T_{B0} for a flat ocean surface. The reference salinity field for Aquarius Version 5 is the monthly 1-degree gridded interpolated ARGO SSS field provided by Scripps (http://www.argo.ucsd.edu/Gridded_fields.html). It is linearly interpolated in space and time to the boresight location of the Aquarius observation. At high latitudes (above 65 N/S) and very close to land where no ARGO data exist V5.0 uses the HYCOM salinity (www.hycom.org) as reference field, after linearly interpolating in space and time to the Aquarius observation. It should be noted that only the global average of the expected TA enter the salinity retrieval algorithm through the ocean target calibration (Section 7). Because of that, the Aquarius salinity retrieval algorithm uses effectively only the global 7-day average of the Scripps ARGO field.

3. Surface Roughness Correction

The correction of the wind induced surface roughness and the model for the wind induced emissivity in the Aquarius salinity retrieval algorithm is based on the results of [12]. In Version 5 a few changes were implemented. The purpose for these changes was the mitigation of biases that have been found in earlier releases and that could be traced back to the surface emissivity model. The major change from [12] concerns the SST dependence of the wind emissivity ΔE_W. The functional form $\rho(T_S)$ of this SST dependence in Version 5, which is shown in Figure 2, is given as follows:

$$\begin{aligned} \Delta E_W &= \delta(W, \varphi_r) \cdot \rho(T_S) \\ \rho(T_S) &= \left[\frac{E_0(T_S)}{E_0(T_{ref})} + \rho'(T_S) \right], \end{aligned} \tag{1}$$

where T_S denotes SST and $T_{ref} = 20$ °C. The harmonic expansion $\delta(W, \varphi_r)$ of the wind emissivity, which depends on the wind speed W and wind direction, φ_r relative to the azimuthal look, is given in Equation (7) of [12]. The 1st term in the expression for $\rho(T_S)$ is the SST dependence that was used in [12]. It is proportional to the flat surface emissivity E_0. As explained in [11] it can be understood within the geometrics optics approach for the rough ocean surface by the fact that the wind roughened surface mixes the vertical and horizontal polarizations of the specular surface and the mixing increases with increasing emissivity of the specular surface. The additional term $\rho'(T_S)$ is the empirically derived change that was made in the Version 5 algorithm.

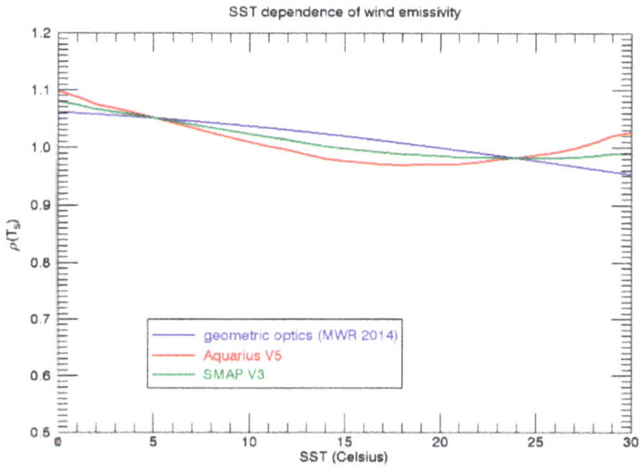

Figure 2. SST dependence of the wind induced emissivity for Aquarius horn 2 H-pol. The blue line in the SST dependence from [12], which is predicted by the geometric optics model for the wind induced surface emission and which was used in earlier Aquarius releases. The red line is the SST dependence used in the Aquarius Version 5 release. The green line is the SST dependence used in the SMAP Version 3 release (Section 10).

4. Atmospheric Absoprtion Correction

4.1. *Atmospheric Absoprtion and Correction Algorithm*

The radiation that is emitted from the ocean surface is attenuated by the Earth's atmosphere. The brightness temperature at the top of the atmosphere (TOA) is given by [3,11,17]:

$$T_{B,TOA} = T_{BU} + \tau \cdot (E \cdot T_S + T_{B\Omega}). \tag{2}$$

T_{BU} is the brightness temperature of the upwelling atmospheric radiation, τ is the atmospheric transmittance, E is the ocean surface emissivity, T_S is the sea surface temperature and $T_{B\Omega}$ is the downwelling sky radiation that is scattered off the ocean surface in the direction of the observation. At L-band frequencies, it is a very good approximation to write [3]:

$$T_{B\Omega} = R \cdot (T_{BD} + \tau \cdot T_{B\,\cos}), \tag{3}$$

where T_{BD} is the downwelling atmospheric radiation that is incident on the ocean surface, $T_{B\,\cos}$ is the cosmic microwave background radiation and $R = 1 - E$ is the ocean surface reflectivity.

The values for τ, T_{BU} and T_{BD} are given as integrals over the vertical profiles atmospheric absorption coefficients [3,11,17]. At L-band frequencies the only significant sources of atmospheric attenuation are due absorption by oxygen, water vapor and cloud liquid water. The calculation of the atmospheric absorption coefficients requires the atmospheric profiles for pressure, temperature, relative humidity, and cloud water mixing ratio as ancillary input (see Section 2.2.2).

The atmospheric absorption correction in the salinity retrieval algorithm (Figure 1) inputs the TOA brightness temperatures $T_{B,TOA}$ and outputs ocean surface emissivities E and the surface brightness temperatures $T_{B,SUR} = E \cdot T_S$ by inverting Equation (2) using Equation (3).

4.2. *Oxygen Absorption Model*

The oxygen absorption is the largest constituent among the three sources of atmospheric attenuation at L-band. Its radiometric contribution to the total TOA TB amounts to a few Kelvin.

At L-band frequencies, which are far from the oxygen absorption lines, the oxygen absorption is caused almost entirely by non-resonant continuum absorption, which is difficult to determine experimentally.

Prior to Version 5, all Aquarius salinity releases had used the oxygen absorption model of Wentz and Meissner [18], which has been extensively used in retrieval algorithms at higher frequencies. The salinity retrievals of these prior versions showed spurious seasonal salty biases at high N and S latitudes, which were largest the N Pacific during April and May. These biases have been tracked back to the oxygen absorption model. It has been found that these biases can be significantly reduced by reverting to the oxygen absorption model of Liebe et al. [19], which is based on laboratory measurements. The Version 5 algorithm uses the oxygen absorption model of [19]. The difference between the two oxygen absorption models is the dependence of the oxygen absorption coefficient of air temperature T_{air}. This temperature dependence is given by the expression $(300K/T_{air})^{\alpha}$. The numerical value of the exponent α is 1.5 in [18] and it is 0.8 in [19]. This means that the change of the absorption coefficient as function of air temperature when going from warm to cold temperatures has effectively been reduced in Version 5 by about 50% compared to prior versions. Figure 3 depicts the radiometric impact of the change in the O_2 absorption by showing the difference in the correction (TOA TB minus surface TB) between the two absorption models as function of time and latitude.

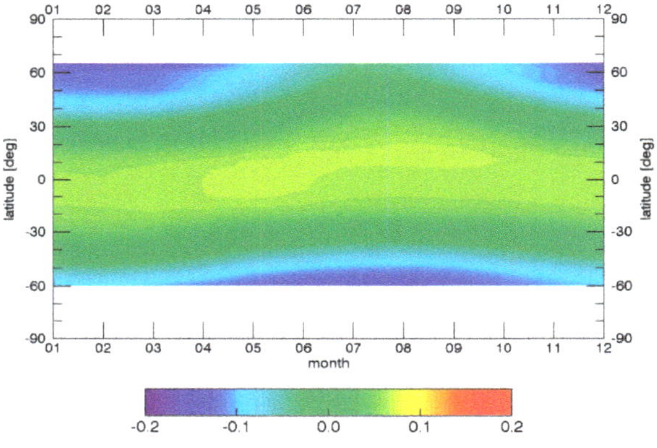

Figure 3. Effect of atmospheric absorption: Top of the atmosphere TB minus surface TB for Aquarius H-pol horn 2 (in Kelvin). The figure shows the difference between the O_2 absorption model by Liebe et al. [19], which is used in the Version 5 algorithm and the Wentz Meissner O_2 absorption model [18], which had been used in prior versions.

5. Reflected Galaxy Correction

Another large source of error that needs to be corrected in the salinity retrieval algorithm is the reflected radiation from the galaxy [20]. It can be large (5 Kelvin) and is difficult to deal with as it requires an accurate knowledge of the location and strengths of the galactic radiation, an analytic model for the wind induced roughness of the ocean surface and the antenna gain pattern.

5.1. Geometric Optics Model

For a flat ocean the contribution of the reflected galactic radiation to the antenna temperature $T_{A,gal,ref}$ is given by integrating radiation from the galactic sources and reflected at the ocean surface over the antenna gain pattern. Location and strength of the galactic sources at L-band are taken from the galactic map [20,21], which was derived from radio-astronomy observations. In actuality, bistatic scattering from a rough ocean will result in galactic radiation entering the mainlobe of the antenna from many different directions. In effect, a rough ocean surface tends to add additional spatial smoothing to $T_{A,gal,ref}$. Modeling of this effect is based on the geometric optics (GO) approach, in which the rough ocean surface is modeled as a collection of tilted facets, with each facet acting as an in-dependent specular reflector. The formulation for this model is given by [22] and the technical details of its application to calculating the TB of reflected galactic radiation at a rough ocean surface are spelled out in [3]. A crucial input to the GO model is the distribution of the slopes of the tilted facets of the rough ocean surface, which depends on wind speed W. The Aquarius algorithm uses the slope variance from [11]. At L-band frequencies this value represents about a 50% reduction in the slope variance from the classic Cox and Munk experiment [23], which measured the ocean sun glitter distribution.

The accuracy of the GO model can be estimated by looking at the zonal differences between the morning (descending) and the evening (ascending) swaths over the same ocean (Figure 4) and comparing them with the size of the reflected galactic correction (not shown). Ideally, the differences between ascending and descending swaths should vanish when averaged over weekly to monthly time scales. From Figure 4, we conclude that the GO model removes the reflected galactic radiation correctly to about 90%. The remaining 10% shows up as spurious signal in the Aquarius salinity retrievals.

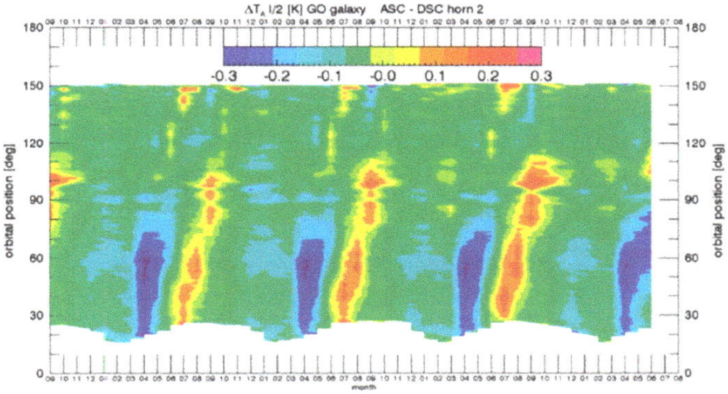

Figure 4. Difference of ascending minus descending the TA measured—expected (in Kelvin) for Aquarius horn 2 using the reflected galaxy from the GO model. The figure shows the value of the average $\frac{1}{2}(T_{A,V} + T_{A.H})$ as function of time (SEP 2011–JUN 2015) and latitude.

5.2. SMAP Fore—Aft Analysis

Observations from SMAP provide an opportunity to improve the correction for the reflected galactic signal. SMAP performs a full 360° scan in less than 5 s and thus observes each location in forward (fore) and backward (aft) direction within a couple minutes. The (relatively) strong reflected radiation emanating from the plane of the galaxy can appear in both the forward and the backward look but usually not at the same time. Radiation from directions other than the plane of the galaxy are generally quite small [20]. If all other signals that depend on look direction (Faraday rotation, wind direction, solar and lunar radiation) have been accurately removed [7], then taking the difference between fore and aft measured TA produces the reflected galactic radiation:

$$T_{A,gal,ref}\left(\phi_{fore}\right) - T_{A,gal,ref}\left(\phi_{aft}\right) = T_A\left(\phi_{fore}\right) - T_A\left(\phi_{aft}\right). \quad (4)$$

Here, ϕ denotes the azimuthal look angle. This equation can be used to derive an empirical galactic correction separate for the SMAP fore and aft looks. For example, looking for cases where the signal from the aft look is small (<2 K) and assuming that the model (theory) for the SMAP aft look reflected galactic radiation is correct if it is smaller than 2 K, then the empirical correction for the fore look can then be obtained from (4) as:

$$T_{A,gal,ref,emp}\left(\phi_{fore}\right) = T_{A,meas}\left(\phi_{fore}\right) - T_{A,meas}\left(\phi_{aft}\right) + T_{A,gal,ref,model}\left(\phi_{aft}\right). \quad (5)$$

Likewise, assuming that the computed SMAP fore look galaxy model is correct if it is smaller than 2 K, then the empirical galaxy model for the aft look can then be obtained from (4) as:

$$T_{A,gal,ref,emp}\left(\phi_{aft}\right) = T_{A,meas}\left(\phi_{aft}\right) - T_{A,meas}\left(\phi_{fore}\right) + T_{A,gal,ref,model}\left(\phi_{fore}\right). \quad (6)$$

When performing the analysis, observations were discarded for which the reflected solar radiation is not negligible. Reflected solar radiation differs between fore and aft looks and currently the correction for reflected solar radiation in the SMAP algorithm is not accurate enough to correct for the difference. It is possible to find observations for all times and orbit positions for which both the reflected solar radiation is negligible and either the TA galaxy of the fore or the aft look are less than 2 K. Therefore, it is possible to derive empirical galactic corrections with the SMAP sensor for both look directions using Equations (5) and (6). Separate derivations are performed for different wind speed regimes using 5 m/s intervals.

The largest part of the SMAP fore—aft results can be reproduced using a tilted facet calculation as explained in Section 5.1 but adding 2 m/s to the wind speed when calculating the RMS slope variance. The effective increase in slope variance increases the surface roughness at L-band frequencies and this increase brings the slope variance from the value in Section 5.1 closer to the Cox-Munk value [23].

Based on the prescription to add 2 m/s to the wind speed when deriving the reflected galactic correction from the geometric optics calculation (Section 5.1) a revised correction for the reflected galaxy for Aquarius can be derived, which takes the results from the SMAP fore—aft analysis into account [3].

Figure 5 shows the significant improvement in the biases between ascending (evening) and descending (morning) swaths when using the SMAP fore—aft result compared with the original geometric optics calculation (Figure 4). However, Figure 5 also shows that even with this improvement in the galactic reflected model based on SMAP, some residual ascending—descending biases still remain, which would cause unacceptably large inaccuracies in the retrieved Aquarius salinity. Mitigating these residual ascending—descending biases is the goal of the empirical zonal symmetrization correction, which is the subject of the following section.

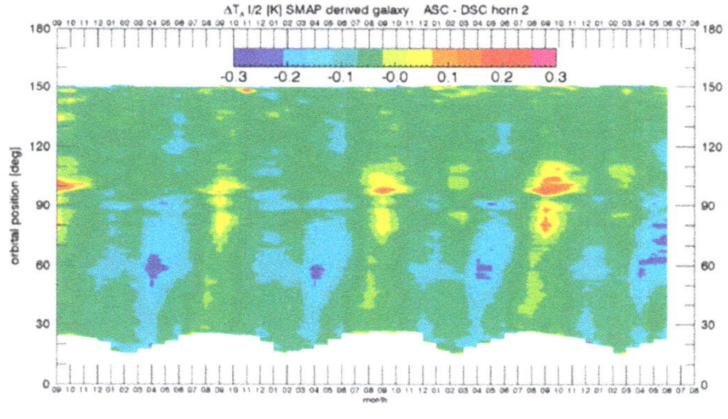

Figure 5. Difference of ascending minus descending the TA measured—expected (in Kelvin) for Aquarius horn 2 using the reflected galaxy from the SMAP fore—aft analysis. The figure shows the value of the average $\frac{1}{2}(T_{A,V} + T_{A.H})$ as function of time (SEP 2011–JUN 2015) and latitude.

5.3. Emprirical Zonal Symmetrization

There are several possible reasons for the remaining inaccuracies in the reflected galaxy correction:

1. The value of the variance of the slope distribution is not completely correct, even after effectively increasing the roughness by adding 2 m/s to the wind speed based on the SMAP fore—aft results.
2. Errors in the antenna gain patterns used to derive the tables of the GO model.
3. Other ocean roughness effects, which cause reflection of galactic radiation but cannot be modeled with an ensemble of tilted facets (e.g., Bragg scattering at short waves, breaking waves and/or foam, and net directional roughness features on a large scale).

The galactic tables themselves, which were derived from radio astronomy measurements [20,21]. For example, there is a small polarized component and Cassiopeia A is very strong and variable.

Such effects are very difficult or impossible to model. We have therefore decided to derive and use an empirical correction for the reflected galactic radiation, which is added to the GO calculation. The danger in doing this is that other geophysical issue (i.e., not associated with reflected radiation from the galaxy) could be masked. But, it was decided to accept this risk for V5.0.

This empirical correction is based on symmetrizing the ascending and the descending Aquarius swaths. The basic assumptions are:

A. There are no zonal ascending—descending biases in ocean salinity on weekly or larger time scales.
B. The residual zonal ascending—descending biases that are observed are all due to the inadequacies (either over or under correction) in the GO model calculation for the reflected galactic radiation.
C. The size of the residual ascending—descending biases is proportional to the strength of the reflected galactic radiation.

Assumption A is based on current understanding of the structure of the salinity field for which there no known physical processes that would cause such a difference. Assumption B results from analyses of the salinity fields and known limitation of the GO model. Assumption C is based on theory for scattering from rough surfaces and the assumption that the source of any difference is reflected galactic radiation and the fact that the source and surface are independent. It is expected to hold in some mean sense over the footprint.

A symmetrization of the ascending and descending Aquarius swaths can be done on the basis of a zonal average. According to Assumption C above the symmetrization weights will be determined by the strength of the reflected galactic radiation. We describe the symmetrization procedure for the 1st Stokes parameter, which is the sum of the brightness temperatures at the ocean surface and will be denoted by T_B. In the equations below, $\langle\ldots\rangle$ denotes the zonal average and the variable z denotes the orbital angle (z-angle). If z lies in the ascending swath, then $-z$ (or $360^\circ - z$) lies in the descending swath and vice versa. $T_B(z)$ is first Stokes parameter as measured by Aquarius at the surface at z. $T_{A,gal,ref}(z)$ is the value of the reflected galactic radiation received by Aquarius as computed based on the SMAP fore—aft results (Section 5.2). The symmetrization term, $\Delta(z)$, which is the basis of the empirical correction, is given as:

$$\begin{aligned} \Delta(z) &= [p \cdot \langle T_B(z) \rangle + q \cdot \langle T_B(-z) \rangle] - \langle T_B(z) \rangle \\ p &= \frac{\langle T_{A,gal,ref}(-z) \rangle}{\langle T_{A,gal,ref}(z) \rangle + \langle T_{A,gal,ref}(-z) \rangle} \\ q &= \frac{\langle T_{A,gal,ref}(z) \rangle}{\langle T_{A,gal,ref}(z) \rangle + \langle T_{A,gal,ref}(-z) \rangle}. \end{aligned} \tag{7}$$

The probabilistic channel weights p and q add up to 1: $p + q = 1$. The symmetrized surface brightness temperature, called T_B', is given by:

$$T_B'(z) = T_B(z) + \Delta(z). \tag{8}$$

It is not difficult to see that this symmetrization has the following features:

1. Assume that z lies in the ascending swath and therefore $-z$ lies in the descending swath. If there is no reflected galactic radiation in the ascending swath, i.e., $\left\langle T_{A,gal.ref}(z)\right\rangle = 0$, then $p = 1$ and $q = 0$. That means that the symmetrization term and thus the whole empirical correction $\Delta(z)$ vanishes, and therefore: $T'_B(z) = T_B(z)$.
2. If, on the other hand, there is no reflected galactic radiation in the descending swath, i.e., $\left\langle T_{A,gal.ref}(-z)\right\rangle = 0$, then $p = 0$ and $q = 1$. That implies $\Delta(z) = \langle T_B(-z)\rangle - \langle T_B(z)\rangle$ and thus $\langle T'_B(z)\rangle = \langle T_B(-z)\rangle$.
3. The zonal average of T'_B is symmetric: $\langle T'_B(z)\rangle = \langle T'_B(-z)\rangle$.
4. If the reflected galactic radiation is the same in ascending and descending swaths $\left\langle T_{A,gal.ref}(z)\right\rangle = \left\langle T_{A,gal.ref}(-z)\right\rangle$, then $p = q = \frac{1}{2}$ and thus the global average (sum of ascending and descending swaths) does not change after adding the symmetrization term: $\langle T_B{}'(z)\rangle + \langle T'_B(-z)\rangle = \langle T_B(z)\rangle + \langle T_B(-z)\rangle$.
5. If the zonal T_B averages are already symmetric $\langle T_B(z)\rangle = \langle T_B(-z)\rangle$, then the symmetrization term and thus the whole empirical correction $\Delta(z)$ vanishes, and therefore: $T'_B(z) = T_B(z)$. That means that our method will not introduce any additional ascending—descending biases that were not already there.

Figure 6 shows the size and pattern of the empirically derived symmetrization Δ in relation to the value of $T_{A,gal,ref}$ from the GO in time—z-angle space. For the GO computation we have assumed an average wind speed of 7.5 m/s. Sizeable contributions for Δ are observed in the vicinity of the galactic pattern that is obtained from the GO model. The magnitude of the peak values of Δ is about 0.2 K compared to about 3 K in $T_{A,gal,ref}$ from the GO model.

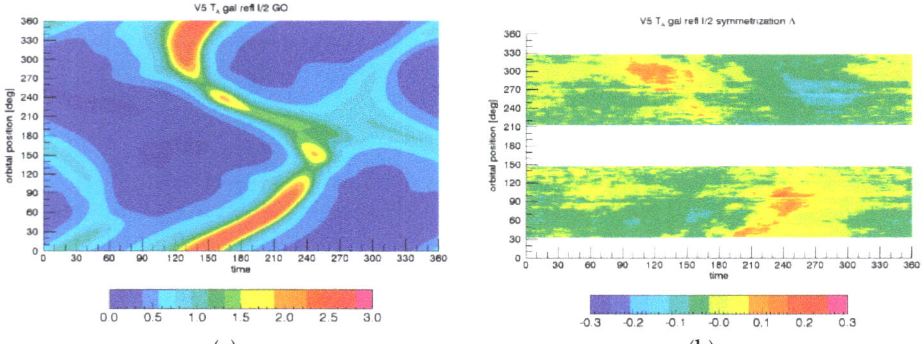

(a) (b)

Figure 6. (**a**) Reflected galactic radiation computed from the GO model for an average wind speed of 7.5 m/s. (**b**) Value of the empirical symmetrization Δ derived in this section. The figures show the values of the averages $\frac{1}{2}(T_{A,V} + T_{A.H})$ (in Kelvin) as function of time (day of year) and orbital position (z-angle) for Aquarius horn 2.

An important feature of this symmetrization procedure is the fact that it is derived from Aquarius measurements only and does not rely on or need any auxiliary salinity reference fields such as ARGO or HYCOM.

It is assumed that the galactic radiation itself is unpolarized and polarization occurs only through the reflection at the ocean surface. Ignoring Faraday rotation of the galactic radiation in the empirical correction term, its 2nd (Q) and 3rd Stokes (U) components are:

$$\Delta_Q \approx \frac{R_V - R_H}{R_V + R_H} \cdot \Delta_I \approx \frac{T_{A,gal,ref,Q}}{T_{A,gal,ref,I}} \cdot \Delta_I, \quad \Delta_U \approx 0, \tag{9}$$

where $R_{V,H}$ are the reflectivity for V and H polarization of an ideal (i.e., flat) surface.

Figure 7 shows the final ascending—descending biases after including the empirical zonal symmetrization. It is evident that all residual zonal biases have been effectively removed. That means, the empirical zonal symmetrization procedure is working as designed.

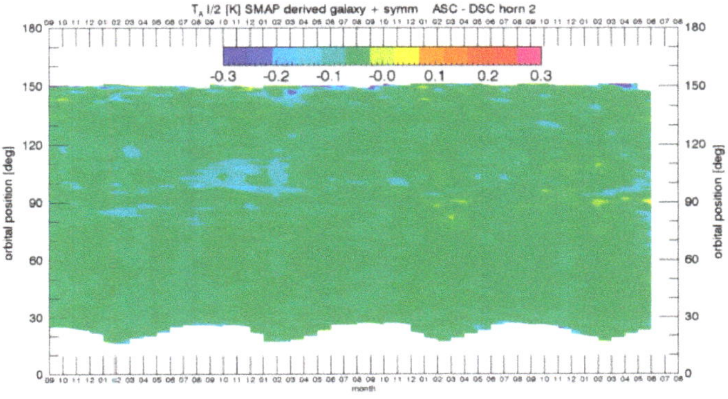

Figure 7. Same as Figure 5 but after applying the empirical symmetrization correction.

6. Correction for Sidelobe Intrusion from Land Surfaces

The Aquarius salinity retrievals degrade quickly as the footprint approaches land closer than 400 km. This land-contamination error occurs because the land is much warmer than the ocean. A correction for land entering the antenna sidelobes when the Aquarius observation gets close to land can be derived from simulated Aquarius brightness temperatures [3].

The land contamination is most conveniently dealt with the TOA (top of the atmosphere) (Figure 1). The error due to land contamination is given as:

$$\Delta T_{B,TOA} = \hat{T}_{B,TOA} - \overline{T}_{B,TOA,ml}. \tag{10}$$

The 1st term on the right hand side of (10) is the *observed* (i.e., measured) signal, which is computed from simulated TOA Earth brightness temperatures containing representative ocean and land scenes and integrating over the antenna gain patterns of the 3 Aquarius feedhorns. The 2nd term of the right hand side of (10) is the *true* TOA TB coming from the antenna main beam. Using the simulated Aquarius TB a table of the $\Delta T_{B,TOA}$ is computed one time off-line before the algorithm is run. This table is stratified according to the spacecraft nadir longitude (2881 elements in 0.125° increments), the spacecraft position in orbit (i.e., z-angle) (2881 elements in 0.125° increments), month (12 elements), polarization (V-pol, H-pol), and horn (inner, middle, and outer). When running the Aquarius salinity retrieval algorithm the value of $\Delta T_{B,TOA}$ is found by linearly interpolating the table to the exact spacecraft position. The interpolated value of $\Delta T_{B,TOA}$ is then subtracted from the actual TOA TB and then the retrieval process proceeds as outlined in Figure 1. The 3rd dimension of month is required to account for seasonal variability in the land brightness temperatures, although this effect is fairly minor. In order to calculate the true (i.e., theoretical) TB of the land surface we use a static monthly climatology of soil moisture and land surface temperatures that has been computed from NCEP fields (http://nomads.ncep.noaa.gov/.)

Figure 8 demonstrates the improvement in the salinity retrievals that is achieved by including the land correction. It shows bias (a) and standard deviation (b) between Aquarius and HYCOM salinity as function of the gain weighted land fraction g_{land} (Section 2.2.5).

Even with the sidelobe correction for intrusion from land surfaces, the salinity retrieval algorithm degrades if the observation is too close to the coast. Figure 8a suggests that observations should be flagged if the value of g_{land} exceeds 1%. We want to mention that the salinity retrieval algorithm does not attempt to correct for or flag freshwater plumes or discharges of rivers into the ocean. In contrast to the spurious signal caused by contamination from land surfaces, the river discharge is a real signal that is picked up by the satellite. It might not be picked up correctly by the models (HYCOM) or by the ARGO drifters, as the drifters cannot be deployed close to the continental shelf. Therefore, the Aquarius salinity measurements can provide valuable information in these cases.

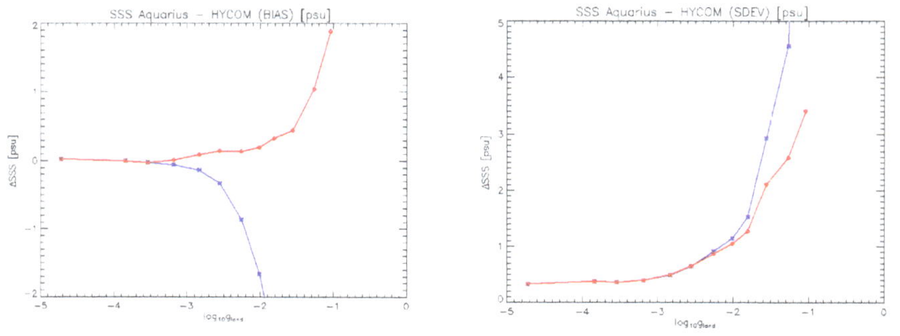

(a) (b)

Figure 8. Comparison between Aquarius horn 2 and HYCOM salinity as function of the logarithmic gain weighted land fraction g_{land}. Blue: Without correction for land intrusion. Red: With correction for land intrusion. (**a**) Bias. (**b**) Standard deviation.

7. Ocean Target Calibration and Calibration Drift Correction

The Aquarius radiometer uses a reference load and noise diode injection as internal calibration targets [24] when converting radiometer counts to TA. The calibration is performed at each Level 2 report interval, which lasts 1.44 s. Unfortunately, the accuracy of the pre-launch values for the noise diode injection temperatures T_{ND} are not sufficient for retrieving salinity. Moreover, immediately after launch it became evident that the values of T_{ND} were drifting (probably outgassing) by several Kelvin, and also other components of the instrument change over time. It is therefore necessary to have a stable and well-known calibration target over long time scales (weekly—monthly) for determining the value of T_{ND} and its time dependence. The global average of the ocean was adopted for this purpose.

The ocean target calibration uses a 7-day (103 orbit) running TA average, denoted by $\langle \ldots \rangle_{OT}$ over the global ocean as calibration target. The goal of the calibration is to tune the calibration parameters, essentially gain and offset, so that for each Aquarius channel:

$$\langle T_{A,cal} - T_{A,\exp} \rangle_{OT} = 0. \quad (11)$$

$T_{A,cal}$ are the Aquarius TA measurements after the calibration adjustments are performed. $T_{A,\exp}$ are the expected TA and calculated as described in Section 2.3. The crucial input in this calculation is the reference salinity field, which is taken from Scripps ARGO (Section 2.3). That means essentially that the ocean target calibration forces the retrieved Aquarius salinity to the Scripps ARGO reference salinity on a global 7-day average. Note that only the global 7-day average of the Scripps ARGO field is used in the ocean target calibration. A consequence of this constraint is that it is not possible to make an independent prediction of the global salinity average from Aquarius, as its value is forced to agree with the one from Scripps ARGO. The average (11) is calculated for each orbit. The 7-day averaging window is chosen, because the Aquarius orbit repeat cycle is 7 days.

It is crucial that both the measured and expected TA in the average $\langle \ldots \rangle_{OT}$ of the ocean target calibration, which determine the instrument calibration parameters, are of the highest possible quality and as much as possible free of errors. Therefore, a number of additional quality control checks, more rigorous than for the regular SSS product, are applied when calculating this average [3].

Equation (11) was used during the first week of Aquarius operations to adjust the prelaunch calibration of radiometers (i.e., gain and bias). The gain was tuned by adjusting the values of the reference noise diodes $T_{ND,0}$. As the mission progressed, it was apparent that the Aquarius calibration changes over time and therefore it is necessary to update the values for the noise diode T_{ND} from these initial values. Two major drift components have been identified and analyzed, which have different time scales and different origins.

The first one is a transient gain drift, most likely caused by out-gassing of the noise diodes. It is slowly varying and decreasing with time with a total change over the mission on the order of one Kelvin. It varies by channel (horn and polarization). It is well modeled as an exponential time dependence of T_{ND}:

$$T_{ND}(n_{orbit}) = T_{ND,0} \cdot (1 - c) \qquad c = A - B \cdot \exp(-D \cdot n_{orbit}) \tag{12}$$

where n_{orbit} denotes the orbit number. The coefficients A, B and D in this exponential fit (12) are determined once using the time series of Aquarius measurements over the entire mission using the ocean target calibration condition (11). The updated values for T_{ND} from (12) are used when converting radiometer counts to TA.

The second type calibration drift manifests itself in pseudo-periodic oscillations in $T_{A,meas} - T_{A,\exp}$, which appear to be superimposed on the exponential drift and occur on the time scales of weeks—months. Its magnitude is in the order a few tenth of a Kelvin and it varies with channel. The oscillations are termed "pseudo" because the calibration anomaly is not periodic in nature and only has a rough appearance of periodicity. One of the root causes for this oscillation was determined to be a locking issue in the backend Voltage to Frequency Converter (VFC), which impacts all counts of the radiometer including the reference calibration load. A correction for the pseudo-periodic oscillations ("wiggles") has been developed and implemented that uses a hardware-based model that only requires inputs from the Aquarius radiometer and does not depend on the ocean or an external salinity reference field [25]. However, this correction scheme does not remove the "wiggles" completely and it is necessary to remove the residual to achieve the very high level of accuracy necessary for retrieving ocean salinity. An analysis of cold space maneuvers showed similar oscillations in several channels [26]. This indicates that these residual oscillations are not predominantly caused by the errors in the geophysical model used in the salinity retrievals but are likely an instrument issue whose root cause is currently not known. Therefore, it is warranted to remove them empirically in the instrument calibration process. It was decided for the Version 5 to treat this residual as an offset and to use the ocean target calibration (11) to remove them In order to remove the residuals, an offset correction is performed at each orbit to obtain the final calibrated TA:

$$T_{A,cal} = \widetilde{T}_A - \left\langle \widetilde{T}_A - T_{A,\exp} \right\rangle_{OT}. \tag{13}$$

Here, $\widetilde{T}_A$ denotes the antenna temperature that is obtained *after* the exponential drift correction has been applied and the average is over 7 days. It is estimated that the residuals are removed to a level of about 0.01 K.

8. Error Sources and Formal Uncertainty Estimation

8.1. Methodology

Each Level 2 (L2) and Level 3 (L3) Aquarius Version 5 salinity retrieval [22] is associated with a formal uncertainty value [3,27,28]. In this section we present a method for formally assessing random and systematic uncertainties in the Aquarius salinity retrievals. The method is based on performing multiple retrievals by perturbing the various inputs to the retrieval algorithm and calculating the sensitivity of the Aquarius salinity to these inputs. Together with an error model for the uncertainties in the input parameters it is possible to calculate the uncertainty in the retrieved salinity.

The basic approach to formally assess an uncertainty of the Aquarius salinity retrieval $S(x_i, \ldots)$ to a parameter x_i, for example wind speed, is to calculate the sensitivity of S to x_i. This is done by running the Version 5 Aquarius L2 algorithm after perturbing the input x_{i0} by a small amount $\pm\Delta x_i$. Assuming that we have an uncertainty estimate $\Delta x_i(x_{i0})$ for the input parameter x_{i0}, then the corresponding uncertainty in S is given by:

$$\Delta S_i(x_{i0}) \approx \frac{\partial S}{\partial x_i}(x_{i0}) \cdot \Delta x_i \approx \frac{S(x_{i0} + \Delta x_i) - S(x_{i0} - \Delta x_i)}{2 \cdot \Delta x_i} \cdot \Delta x_i. \quad (14)$$

The assessment of the uncertainty in S consists therefore in two parts:

1. The computational/algorithm part, i.e., running each retrieval algorithm with the perturbed parameter values.
2. Obtaining a realistic error model for all the uncertainties that are involved. This part is done offline and its results are fed into the perturbed retrievals.

When running the algorithm for a perturbed variable, all the other variables are left unperturbed. Performing the uncertainty estimation this way takes into account that a given uncertainty in one of the input parameter can translate to very different uncertainties in the retrieved salinity depending on the environmental scene. For example, the same error in the input wind speed that is used in the surface roughness correction or in the reflected galactic radiation will result in a much larger uncertainty in salinity in cold water where the sensitivity of the T_B to salinity is low than it would in warm water where the sensitivity is higher. The SST of the scene is a major driver in the size of the salinity uncertainty.

It is necessary to separate the uncertainty Δx_i in each input x_{i0} into a random and a systematic component. As a general guideline:

1. Uncertainties that fluctuate on larger time and spatial scales (1 month, >100 km) are treated as systematic uncertainties.
2. Uncertainties that fluctuate on shorter time and length scales are treated as random uncertainties.

The distinction between random and systematic uncertainties becomes important when propagating the uncertainties from the 1.44 s measurement (L2) to the monthly or weekly L3 averages [2]. Whereas the random components are suppressed by a factor $1/\sqrt{N}$ when averaged over N samples, the systematic components are simply the average of the individual uncertainties.

8.2. Error Sources

This section briefly discusses the major error sources of the Aquarius Version 5 salinity retrieval algorithm and the quantitative assessment of their uncertainty. Further details can be found in [3,27,28].

8.2.1. NEDT

The radiometer noise (NEDT) is approximately the standard deviation of the noise in each 10 ms sample. The effective noise for salinity retrieval is the noise in the basic 1.44 s Aquarius data block used in processing. This effective NEDT is computed as the standard deviation of the RFI filtered antenna temperatures (TF) in each 1.44 s cycle divided by the square root of the number of measurements within that data block. This error (effective NEDT) is treated as random. We compute the effective NEDT and the resulting error in the salinity for all 3 channels: V-pol, H-pol and the 3rd Stokes parameter. It is assumed that these 3 components are independent and that the resulting errors in the salinity can be added as root sum squares.

8.2.2. Wind Speed

The uncertainty in the Aquarius HHH wind speed, which is used in the surface roughness correction [3,12] can be estimated by comparing the Aquarius HHH wind speed with other wind speed sources, for example WindSat [29]. This uncertainty has both a random and a systematic component.

8.2.3. Wind Direction

For the uncertainty in the auxiliary NCEP wind direction field (Section 2.2.4) we assume 10° and treat it as random error. This value is suggested by comparing the NCEP wind direction with measurements from buoys [30] or satellites [31], for example QuikSCAT or WindSat.

8.2.4. SST

An estimate of the uncertainty in the ancillary SST input from CMC (Section 2.2.1) can be obtained by comparing the CMC SST field with other SST sources, for example from WindSat [29].

8.2.5. Reflected Galaxy

The estimated uncertainty in the correction for the reflected galactic radiation are treated as systematic and based on the bias of $T_{A,meas} - T_{A,\exp}$ as a function of $T_{A,gal,ref}$ and the Aquarius HHH wind speed. It is assumed that $T_{A,meas} - T_{A,\exp}$ characterizes the degradation of the salinity retrievals.

8.2.6. Land Contamination

The estimated uncertainty due to intrusion of radiation from land and sea ice surfaces into the sidelobes of the Aquarius antenna is treated as systematic and its estimation is based on the RMS of $T_{A\ meas} - T_{A,\exp}$ as a function of the gain-weighted land fraction g_{land}.

8.2.7. Undetected RFI

The signal from the Aquarius instrument is filtered for radio-frequency interference (RFI). The RFI filtering is performed during the conversion from radiometer counts to TA based on detecting and removing outliers in the time series of the 10 ms samples (short accumulations) before they get averaged into the 1.44 s intervals, which are used in the Level 2 salinity retrievals [8,9]. Unfortunately, there are cases which are missed by this RFI detection. This happens mainly with low level RFI coming from the antenna sidelobes. The uncertainty from undetected RFI can be estimated from the SSS differences between ascending and descending Aquarius swaths. It is treated as systematic uncertainty. For the formal uncertainty estimate we first create a static 3-year map of the difference between ascending (PM) and descending (AM) Aquarius SSS summing over all three horns (Figure 9). The next step is to create a mask of areas where undetected RFI is likely present. This can be done by creating peak hold maps of RFI filtered (TF)—unfiltered (TA) antenna temperatures, separately for ascending and descending swaths (Figure 10a), mask cells where this difference exceeds a threshold (0.2 K) and then extend this mask by a certain amount (±4°) in order to account for the fact that the undetected RFI can also affect adjacent footprints (Figure 10b).

The final step is to look for the overlap in the ascending—descending maps (Figure 9) and the mask (Figure 10b). Because undetected RFI always results in a low salinity value, we create maps for the ascending swaths where $SSS_{asc} - SSS_{dsc} < 0$ and for the descending swaths where if

$SSS_{asc} - SSS_{dsc} > 0$. This results in the two maps of Figure 11, which show the uncertainty (difference in SSS). This method aims to avoid that differences in the ascending—descending SSS maps are getting falsely identified as RFI. For example, the ascending—descending differences close to the Antarctica in Figure 9 are not present anymore in the final RFI uncertainty maps of Figure 11. These biases are likely caused by sea-ice contamination and not by RFI and thus they do not appear in the RFI peak-hold map (Figure 10a) or in the RFI mask map (Figure 10b).

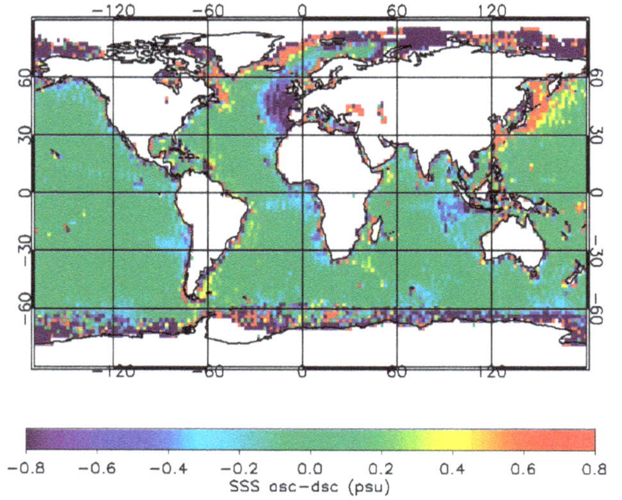

Figure 9. Map of the salinity difference between ascending (asc) and descending (dsc) Aquarius swaths for SEP 2011–AUG 2014.

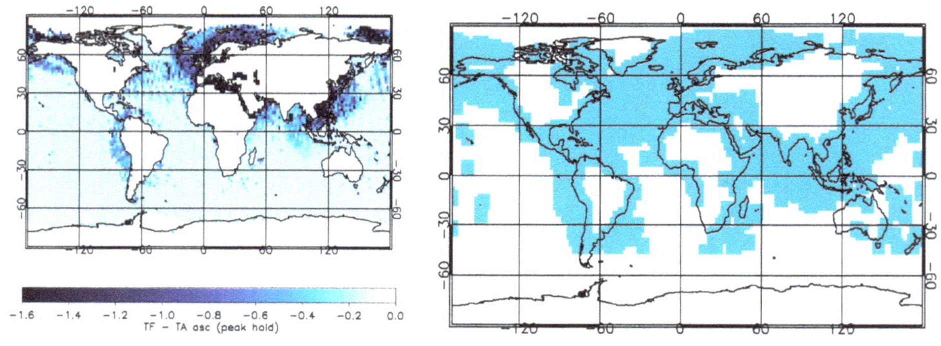

(**a**) (**b**)

Figure 10. (**a**) RFI peak hold map: The map shows the largest monthly average value of the difference between RFI-filtered (*TF*) and unfiltered (*TA*) antenna temperatures over the time period SEP 2011–AUG 2014 of the ascending Aquarius swath. In producing the figure we have averaged over all 3 horns. (**b**) RFI mask for the ascending Aquarius swath: The map is obtained from the peak-hold map in (**a**) by taking all cells for which $|TF - TA| > 0.2$ K and then extending this area by $\pm 4°$. This mask indicates the geographical area where undetected RFI might be present.

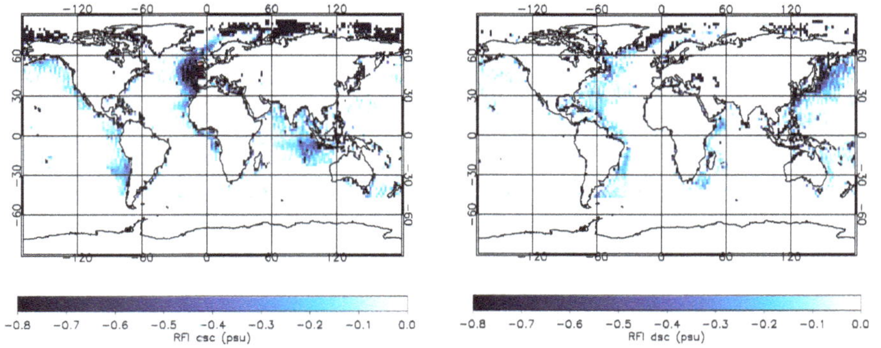

(a) (b)

Figure 11. Estimated uncertainty in the retrieved Aquarius salinity due to undetected RFI for the ascending swath (**a**) and the descending swath (**b**). In producing the figures we have averaged over all 3 horns.

Unlike all other uncertainties, the uncertainty estimate due to RFI is done directly on the salinity level. The uncertainty maps are static, i.e., we assume the same values for the whole Aquarius mission and we average over all three horns, i.e., our uncertainty estimates due to RFI are not horn specific.

8.3. Error Allocations at Level 2 and Level 3

Figure 12 shows the contributions of the various components of the error model (Section 8.2) to the total formal uncertainty estimate for (a) the Aquarius L2 product (1.44 s) and (b) the monthly 1° L3 salinity product, respectively. The dominant contributions at the 1.44 s are the NEDT and the random and systematic uncertainties in the wind speed that is used in the surface roughness correction algorithm. At the monthly 1° Level 3 product all the random uncertainties including the NEDT are reduced to low levels. The dominant uncertainty contribution at the monthly L3 product is the systematic uncertainty in the HHH wind speed.

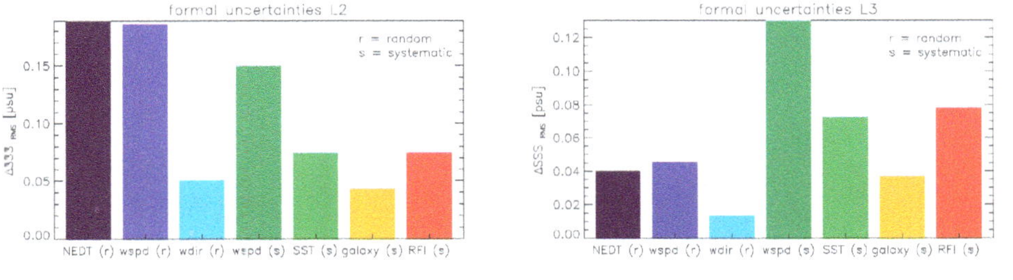

(a) (b)

Figure 12. Contribution of the various uncertainties to the total estimated uncertainty for open ocean scenes of: (**a**) the Aquarius Level 2 salinity that is observed at the 1.44 s cycle; (**b**) the 1° Aquarius Level 3 salinity maps.

9. Validation and Improvements from Previous Releases

The ultimate assessment of the performance of the Aquarius Version 5 salinity algorithm is done by comparing the Aquarius Version 5 salinity retrievals with other salinity measurements and sources such as ARGO or HYCOM.

A comprehensive validation analysis for the Aquarius Version 5 products have been performed in [32,33]. The root mean square error (RMSE) of the Aquarius Version 5 salinity have been estimated from a triple point analysis using individual match-ups between Aquarius, ARGO floats and HYCOM. The estimated RMSE for Aquarius is 0.17 psu for the Level 2 product (1.44 s) and 0.128 psu for monthly 100 km averages. For computing this value, observations in very cold water (SST < 5 °C) where the sensitivity is low were excluded. These values are significantly better than the Aquarius prime mission requirement, which allocated and RMSE of 0.2 psu for monthly 100 km averages.

It is also worthwhile to look how the improvements in the Version 5 algorithm compared with previous releases and match the steps in the algorithm updates these performance improvements. A good way to do this is to analyze time—latitude Hovmoeller diagrams (Figure 13) and maps of global differences (Figure 14) between Aquarius and Scripps ARGO (Section 2.3) salinity fields. For computing these figures the Aquarius observations in rain were filtered out in order to exclude mismatches between Aquarius and the ARGO due to stratification on the upper ocean layer that can occur in precipitating scenes (see Section 2.2.6). Both figures show clear improvement in Version 5 for most of the temporal and zonal biases that were observed in the Version 4 and earlier releases. The reduction of the salty biases at mid-high latitudes, which were particularly strong in the NW and N Pacific during April—June and in the S Atlantic and S Indian Ocean during October - December are mainly due to the change in the atmospheric oxygen absorption model (Section 4.2) and partly also to the changes in the SST dependence of the wind induced emissivity (Section 3). The small improvements in the fresh biases at high S latitudes during the summer months can be traced back to the changes in the reflected galaxy correction (Section 5).

Another important achievement in the performance of the Version 5 salinity retrievals is evident from Figure 15, which shows bias and standard deviation of- the difference between the Aquarius Level 2 salinity and Scripps ARGO as function of SST for both Version 4 and Version 5. In Version 5 the difference between Aquarius Version 5 and ARGO is within ± 0.1 psu, even in very cold water. That applies to all 3 Aquarius horns. That means that with the algorithm updates in Version 5 the SST dependent biases that have been observed in earlier releases [4] have been essentially eliminated. We note that the difference between Aquarius Version 5 and ARGO is also within ± 0.1 psu if the stratification is done as a function of wind speed.

Despite the significant improvements in Aquarius Version 5 from prior releases, Figure 14b still indicates noticeable residual biases in some areas, for example salty biases in the S Atlantic and S Indian Ocean. The cause of these residual biases is currently not fully understood. So, there is still room for some improvements of the Aquarius salinity algorithm in possible future releases. We also want to be clear that it might well be possible to use a different geophysical model function than the one that we have used for the Aquarius Version 5 release and presented here, which might result in the same or even better performance. Such a change in the geophysical model could, for example include a different model for the dielectric constant of sea water.

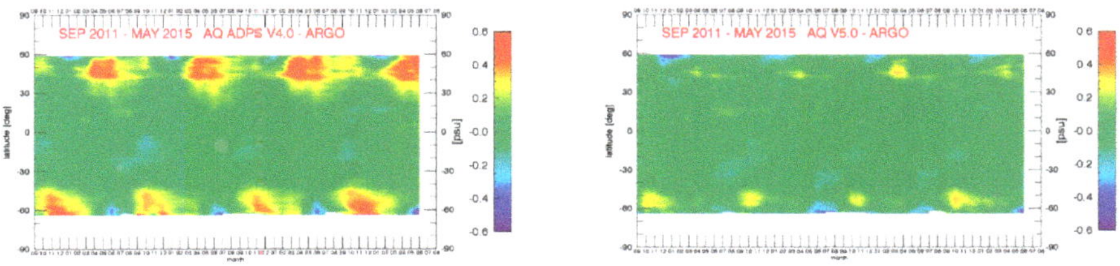

(a) (b)

Figure 13. Hovmoeller plot of the difference between Aquarius and Scripps ARGO salinity. Panel (**a**) is for the Version 4 release. Panel (**b**) is for the Version 5 release. The x-axis in both plots is the time (month) over the time period September 2011–May 2015. The y-axis is the latitude. For the comparison Aquarius observations containing rain have been discarded (see Section 2.2.6).

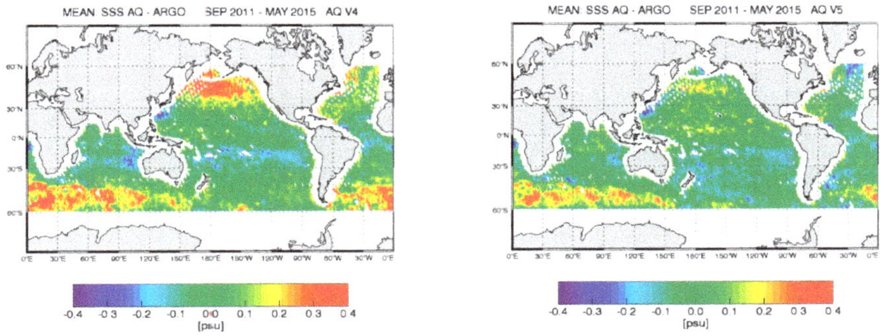

(a) (b)

Figure 14. Average difference between Aquarius and Scripps ARGO over the time period September 2011–May 2015. Panel (**a**) is for the Version 4 release. Panel (**b**) is for the Version 5 release. For the comparison observations containing rain have been discarded (see Section 2.2.6).

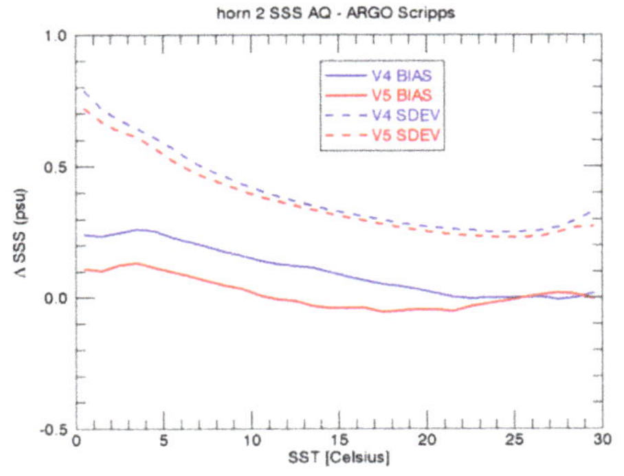

Figure 15. Difference between L2 Aquarius horn 2 and Scripps ARGO salinity as function of SST. The blue curves are for the Version 4 release. The red curves are for the Version 5 release. Full lines show the biases and dashed lines show the standard deviations. For the comparison observations containing rain have been discarded (see Section 2.2.6).

10. Adaption to Version 3 SMAP Salinity Retrievals

The last section in this paper deals with the adaption of the Aquarius Version 5 salinity retrieval algorithm to SMAP. Specifically, we are considering the NASA RSS (Remote Sensing Systems) SMAP Version 3 salinity release, which is scheduled for late summer 2018. For most of the part, adapting the Aquarius salinity retrieval algorithm (Figure 1) is rather straightforward and amounts to deriving or interpolating the corrections that were derived for Aquarius to the SMAP orbit and pointing. There are two important exceptions, which we discuss in the following.

10.1. SMAP Emissive Reflector

The emissivity of the Aquarius antenna was negligibly small for all practical purposes. However, the SMAP mesh reflector has an emissivity of about 1%, which is large enough so that a correction needs to be applied in the salinity retrieval [7]. If T_A is the antenna temperature before the radiation hits the reflector whose physical temperature is denoted by T_{refl} and whose emissivity is ε_{refl}, then after antenna temperature T'_A after the reflection, which enters the receiver, is given by:

$$T'_A = \left(1 - \varepsilon_{refl}\right) \cdot T_A + \varepsilon_{refl} \cdot T_{refl} = T_A + \varepsilon_{refl} \cdot \left(T_{refl} - T_A\right). \tag{15}$$

In order to perform the correction, i.e., determine the value of T_A from the measured T'_A according to Equation (15), it is necessary to know the values of both the reflector emissivity ε_{refl} and its physical temperature T_{refl}.

The value of the reflector emissivity ε_{refl} can be determined by performing a linear regression of the SMAP $\Delta T_A = T_{A,meas} - T_{A,\text{exp}}$ against $T_{refl} - T_{A,meas}$ before performing any correction for the emissive reflector. The slope of this regression is ε_{refl}. We have determined values of $\varepsilon_{refl} = 0.01012$ for both V-pol and H-pol. It is worth to point out that these values for the reflector emissivity are about 4 times larger than the values that were determined pre-launch.

Unfortunately, there are no direct measurements of the physical temperature T_{refl} of the SMAP mesh antenna. Only the results of thermal model for the SMAP reflector, which was developed and run by the Jet Propulsion Laboratory (JPL) thermal modeling team is available (Figure 16a). The values of this JPL thermal model are used and included in the SMAP L1B files [34]. Our analysis has revealed that the JPL thermal model is not accurate enough to retrieve ocean salinity from SMAP without adjustments. This can be seen from the Hovmoeller diagram in Figure 17, which shows the bias of $T_{A,meas} - T_{A,\text{exp}}$ as function to time (day of year) and orbital position (z-angle) if the JPL thermal model is used in the emissive reflector correction. In the computation of $T_{A,\text{exp}}$ we have used Scripps ARGO as reference salinity (Section 2.3). The zonal and temporal biases get large when the spacecraft goes in and out of solar eclipse during the summer months and it is also large during the winter months. In those instances, rapid cooling or heating of the SMAP reflector occurs. Apparently, the thermal models can overestimate or underestimate the rate of these thermal changes. The observed zonal and temporal biases in Figure 17 are largely independent of the SMAP look direction. They differ significantly between ascending (lower half of the diagram) and descending (upper half of the diagram) swaths, because the thermal heating and cooling of the SMAP antenna is not symmetric between the two swaths. That makes us believe that they are indeed caused by inaccuracies in the JPL thermal model rather than by other sources. For example errors in the correction for galaxy or sun intrusion would strongly depend on look direction. On the other hand, errors in dielectric model or surface roughness correction are expected to be largely the same in the ascending and descending swaths. It was decided for SMAP salinity retrievals to make an empirical adjustment to the JPL thermal model, whose purpose is to minimize the zonal and temporal biases in $\Delta T_A = T_{A,meas} - T_{A,\text{exp}}$ when the correction for the emissive reflector is performed with this empirical adjusted model. This can be done by taking the values for the ΔT_A biases from Figure 17 and computing the corresponding biases of ΔT_{refl} using (15). The result for the empirical adjusted thermal model in the SMAP Version 3 salinity release is shown in Figure 16b. We use the same thermal model adjustments for V-pol and H-pol.

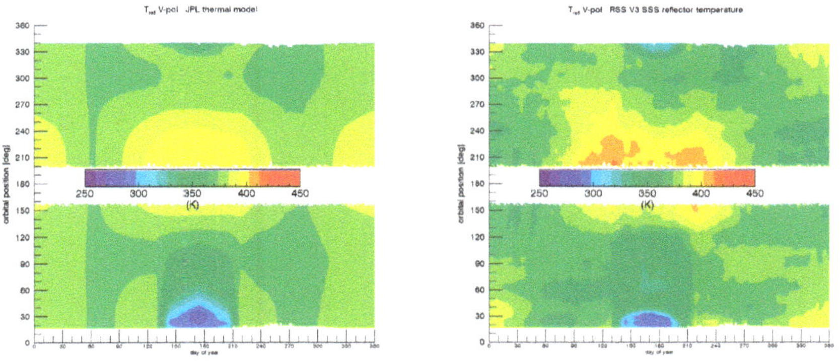

(a) (b)

Figure 16. Physical temperature of the reflector. (**a**) JPL thermal model that is used in the SMAP L1B files [34]. (**b**) Empirical adjustment in the RSS SMAP Version 3 salinity release.

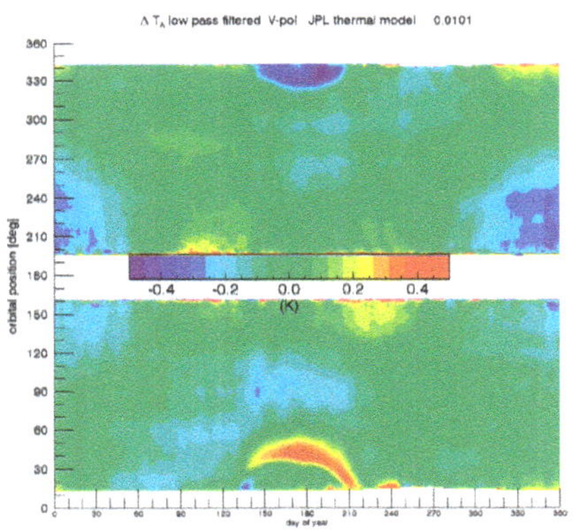

Figure 17. Hovmoeller diagram of SMAP $\Delta T_A = T_{A,meas} - T_{A,\exp}$ over the open ocean using the JPL thermal model for the SMAP mesh antenna. The x-axis is time (day of year) and the y-axis is orbital position (z-angle). For the computation of $T_{A,\exp}$ we have used Scripps ARGO as reference salinity (Section 2.3). The computation of this diagram is based on 2 years of SMAP data (September 2015–August 2017). A simple spatial and temporal low-pass filter was applied by performing a running average in both dimensions.

A final note in our approach of the empirical determination of both ε_{refl} and T_{refl} for SMAP concerns the fact that we have tried to avoid folding a potential error in one quantity into the other. When determining the value of ε_{refl} from the linear regression, we have used only cases where we can regard the JPL thermal model as accurate, i.e., where the biases in Figure 17 are small (less than 0.1 K).

10.2. SMAP Surface Roughness Correction

The second major difference between Aquarius and SMAP that needs to be taken into account when transferring the Aquarius Version 5 salinity retrieval algorithm to SMAP is the surface roughness correction. The crucial ancillary input to the surface roughness correction is the surface wind speed. The Aquarius salinity retrievals use the Aquarius HHH wind speed, which is obtained from the Aquarius HH-pol L-band scatterometer and H-pol radiometer observations [3,12]. For the SMAP salinity retrievals scatterometer observations are not available, as the SMAP radar failed in July 2015.

It is necessary to use an external ancillary wind speed in the SMAP salinity retrieval algorithm, as the two SMAP radiometer channels (V-pol and H-pol) do not carry sufficient information to perform the surface roughness correction without ancillary wind speed input. For the SMAP Version 3 it was decided to use the CCMP (Cross Calibrated Multi-Platform) Version 2.0 wind fields [35–37] as ancillary input for both wind speed and wind direction. CCMP is a gridded (0.25°) Level 4 wind vector product that combines various satellite wind measurements from the RSS Version 7 ocean suite with a background field from a numerical weather prediction model using a variational assimilation method (VAM). A near-real time (NRT) version of CCMP V2.0 is produced at RSS, whose latency is short enough to be used as ancillary input in the Version 3 SMAP salinity retrievals. This NRT V2.0 CCMP wind product currently ingests RSS V7.0 WindSat, GMI, SSMIS and AMSR2 wind speed observations and uses NCEP GDAS 0.25° wind speed and direction as background field.

The crucial point for developing the geophysical model of the wind induced emissivity for SMAP Version 3 is that the CCMP ancillary field is slightly different from the Aquarius HHH wind speed that has been used in the Aquarius Version 5 algorithm. That means that there are small biases in the order of a few tenth of m/s between these two ancillary wind fields and these biases depend on wind speed and also on SST. Because of the high level of accuracy that is required for retrieving salinity, these biases need to be taken into account when deriving the wind induced emissivity model function for SMAP Version 3 using the method outlined in [12]. As a consequence of the slightly different ancillary wind speed inputs to the Aquarius Version 5 and SMAP Version 3 salinity retrieval algorithms, the geophysical model functions for the wind emissivities also slightly differ. This is most important for the wind speed dependence 0th harmonic coefficient of the wind induced emissivity, i.e., the isotropic part. This is shown in Figure 18 for both SMAP polarizations. Small differences are observable at very low and at very high wind speeds. This coincides with the instances where small differences between Aquarius HHH and CCMP wind speeds exist. In addition, we have also found slight differences in the SST dependence $\rho(T_S)$ (c.f. Section 3) of the wind induced emissivity, which is shown in Figure 2 for the h-pol. The term $\rho'(T_S)$ in Equation (1), which is empirically determined and which is the deviation from the theoretical value predicted by the geometric optics model, is half as large in SMAP Version 3 than it was in Aquarius Version 5. Consequently, the value of $\rho(T_S)$ in SMAP Version 3 lies between the theoretical value of the geometrics optics model, which is given by $E_0(T_S)/E_0\left(T_{ref}\right)$, and the value of Aquarius Version 5.

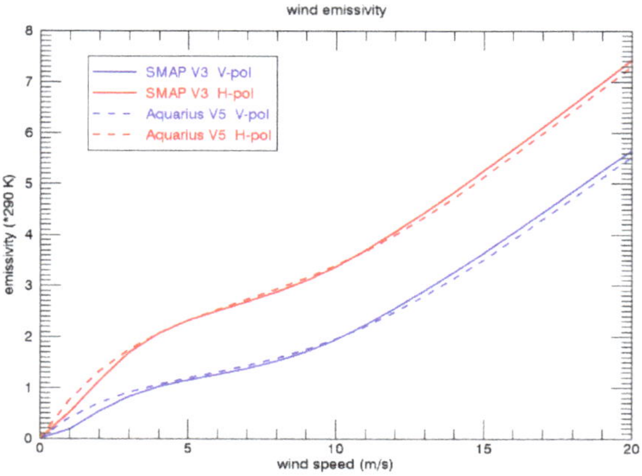

Figure 18. Isotropic (wind-direction independent) part of the wind induced emissivity that is used in the Aquarius Version 5 after interpolating to the SMAP Earth Incidence Angle (dashed lines) and the SMAP Version 3 (full lines) releases. Blue: V-pol. Red: H-pol. The figure shows the 0th harmonic of the wind induced excess emissivity [3,12] multiplied by 290 K.

11. Summary and Conclusions

To summarize our main results:

Our paper gives an overview of the steps in the salinity retrieval algorithm for Aquarius Version 5. We have highlighted and elaborated on the issues that have not been previously published or that are new and that help improving the performance of Version 5 from previous releases. The most important components are the corrections for the absorption by atmospheric oxygen, the SST dependence of the surface roughness correction and the reflected galaxy correction. The Aquarius Version 5 reflected galaxy correction requires an empirical zonal symmetrization to remove spurious biases between the ascending and descending Aquarius orbit segments. The ocean target calibration effectively constraints the global average of the Aquarius salinity to the value from ARGO. This is necessary to determine the main calibration parameters of the internal radiometer calibration references (noise-diode injection temperatures) and to accurately correct temporal drifts in the calibration system. A comparison with ground truth observations from ARGO floats shows that most of the temporal and zonal biases that have been identified in previous Aquarius releases have been removed or reduced to a low level in the Version 5 release. The accuracy of Aquarius Version 5 is significantly better than the prime mission requirement of 0.2 psu. The Aquarius Version 5 salinity retrievals are accompanied by estimates for random and systematic uncertainties. The basis for these uncertainty estimates in the salinity retrievals are realistic estimates of the uncertainties in the input parameters to the algorithm.

The Aquarius Version 5 salinity retrieval algorithm can be adapted to SMAP, which is straightforward for most parts. Important differences between Aquarius and SMAP that need to be taken into account are the correction for the emissive SMAP antenna and the wind induced emissivity.

Our analysis has also shown that there is still room and necessity of further improvements in the salinity retrieval algorithms for both instruments. This will be part of future releases. Direct comparisons between the retrieved Aquarius Version 5 and SMAP Version 3 salinities will be an important tool in analyzing remaining biases and in deriving future algorithm improvements.

Author Contributions: T.M. and F.J.W. designed the basics of salinity retrieval algorithms for Aquarius with input from the Aquarius Cal/Val team and adapted the algorithm to SMAP. T.M. implemented the changes that were developed for Aquarius Version 5 and for SMAP Version 3, performed an assessment of the algorithm performance and developed the formal uncertainty estimation. F.J.W. and D.L.M.V. developed the algorithm basics as expressed in the Aquarius ATBD. F.J.W. developed the galaxy correction based on SMAP fore—aft looks and also the sidelobe correction for land intrusion.

Funding: The work at RSS has been funded by NASA contracts NNG04HZ29C, NNH15CM44C and 80HQTR18C0015. The funding covers the costs for open access publication.

Acknowledgments: We would like to thank the members of the Aquarius and SMAP Cal/Val and science teams for providing valuable input and suggestions over the course of this work.

Conflicts of Interest: The authors declare no conflict of interest. The funding sponsors had no role in the design of the study; in the collection, analyses, or interpretation of data; in the writing of the manuscript, and in the decision to publish the results.

References

1. Le Vine, D.; Lagerloef, G.; Colomb, F.; Yueh, S.; Pellerano, F. Aquarius: An instrument to monitor sea surface salinity from space. *IEEE Trans. Geosci. Remote Sens.* **2007**, *45*, 2040–2050. [CrossRef]
2. Aquarius Official Release Level 2 Sea Surface Salinity & Wind Speed Data V5.0. Available online: https://podaac.jpl.nasa.gov/dataset/AQUARIUS_L2_SSS_V5 (accessed on 10 July 2018).
3. Meissner, T.; Wentz, F.; Le Vine, D. Aquarius Salinity Retrieval Algorithm Theoretical Basis Document (ATBD), End of Mission Version; RSS Technical Report 120117. 1 December 2017. Available online: http://podaac-ftp.jpl.nasa.gov/allData/aquarius/docs/v5/AQ-014-PS-0017_Aquarius_ATBD-EndOfMission.pdf (accessed on 10 July 2018).
4. Le Vine, D.; Dinnat, E.; Meissner, T.; Yueh, S.; Wentz, F.; Torrusio, S.; Lagerloef, G. Status of Aquarius/SAC-D and Aquarius salinity retrievals. *IEEE J. Sel. Top. Appl. Earth Obs. Remote Sens.* **2015**, *8*, 5401–5415. [CrossRef]

5. Entekhabi, D.; Joku, E.G.; O'Neill, P.E.; Kellogg, K.H.; Crow, W.T.; Edelstein, W.N.; Entin, J.K.; Goodman, S.D.; Jackson, T.J.; Johnson, J.; et al. The Soil Moisture Active Passive (SMAP) mission. *Proc. IEEE* **2010**, *98*, 704–716. [CrossRef]
6. Peng, J.; Misra, S.; Piepmeier, J.R.; Dinnat, E.P.; Hudson, D.; Le Vine, D.M.; De Amici, G.; Mohammed, P.N.; Bindlish, R.; Yueh, S.H.; et al. Soil Moisture Active/Passive L-Band microwave radiometer postlaunch calibration. *IEEE Trans. Geosci. Remote Sens.* **2017**, *55*, 5339–5354. [CrossRef]
7. RSS SMAP Level 2C Sea Surface Salinity V2.0 Validated Dataset. Available online: https://podaac.jpl.nasa.gov/dataset/SMAP_RSS_L2_SSS_V2 (accessed on 10 July 2018).
8. Misra, S.; Ruf, C. Detection of radio-frequency interference for the Aquarius radiometer. *IEEE Trans. Geosci. Remote Sens.* **2008**, *46*, 3123–3128. [CrossRef]
9. Le Vine, D.; de Matthaeis, P.; Ruf, C.; Chen, D. Aquarius RFI Detection and Mitigation Algorithm: Assessment and Examples. *IEEE Geosci. Remote Sens.* **2014**, *52*, 4574–4584. [CrossRef]
10. Meissner, T.; Wentz, F. The complex dielectric constant of pure and sea water from microwave satellite observations. *IEEE Trans. Geosci. Remote Sens.* **2004**, *42*, 1836–1849. [CrossRef]
11. Meissner, T.; Wentz, F. The emissivity of the ocean surface between 6 and 90 GHz over a large range of wind speeds and Earth incidence angles. *IEEE Trans. Geosci. Remote Sens.* **2012**, *50*, 3004–3026. [CrossRef]
12. Meissner, T.; Wentz, F.; Ricciardulli, L. The emission and scattering of L-band microwave radiation from rough ocean surfaces and wind speed measurements from Aquarius. *J. Geophys. Res. Oceans* **2014**, *119*. [CrossRef]
13. Meissner, T.; Wentz, F.; Scott, J.; Vasquez-Cuervo, J. Sensitivity of ocean surface salinity measurements from spaceborne L-Band radiometers to ancillary sea surface temperature. *IEEE Trans. Geosci. Remote Sens.* **2016**, *54*, 7105–7111. [CrossRef]
14. Boutin, J.; Chao, Y.; Asher, W.E.; Delcroix, T.; Drucker, R.; Drushka, K.; Kolodziejczyk, N.; Lee, T.; Reul, N.; Reverdin, G.; et al. Satellite and in situ salinity: Understanding near-surface stratification and subfootprint variability. *Bull. Am. Meteorol. Soc.* **2016**, *97*, 1391–1407. [CrossRef]
15. Santos-Garcia, A.; Jacob, M.; Jones, W.L.; Asher, W.; Hejazin, Y.; Ebrahimi, H.; Rabolli, M. Investigation of rain effects on Aquarius sea surface salinity measurements. *J. Geophys. Res. Oceans* **2014**, *119*, 7605–7624. [CrossRef]
16. Joyce, R.; Janowiak, J.; Arkin, P.; Xie, P. CMORPH: A method that produces global precipitation estimates from passive microwave and infrared data at high spatial and temporal resolution. *J. Hydrometeorol.* **2004**, *5*, 487–503. [CrossRef]
17. Wentz, F.; Meissner, T. Algorithm Theoretical Basis Document (ATBD), Version 2, AMSR Ocean Algorithm, RSS Tech. Report 121599A-1. Available online: http://images.remss.com/papers/rsstech/2000_121599A-1_Wentz_AMSR_Ocean_Algorithm_ATBD_Version2.pdf (accessed on 10 July 2018).
18. Wentz, F.; Meissner, T. Atmospheric absorption model for dry air and water vapor at microwave frequencies below 100 GHz derived from spaceborne radiometer observations. *Radio Sci.* **2016**, *51*, 381–391. [CrossRef]
19. Liebe, H.; Rosenkranz, P.; Hufford, G. Atmospheric 60-GHz oxygen spectrum: New laboratory measurements and line parameters. *J. Quant. Spectrosc. Radiat. Transf.* **1992**, *48*, 629–643. [CrossRef]
20. Le Vine, D.; Abraham, S. Galactic noise and passive microwave remote sensing from space at L-band. *IEEE Trans. Geosci. Remote Sens.* **2004**, *42*, 119–129. [CrossRef]
21. Dinnat, E.; Le Vine, D.; Abraham, S.; Floury, N. Map of Sky Background Brightness Temperature at L-Band. Available online: https://podaac-tools.jpl.nasa.gov/drive/files/allData/aquarius/L3/mapped/galaxy/2018 (accessed on 7 July 2018).
22. Wentz, F. The forward scattering of microwave solar radiation from a water surface. *Radio Sci.* **1978**, *13*, 131–138. [CrossRef]
23. Cox, C.; Munk, W. Measurement of the roughness of the sea surface from photographs of the sun's glitter. *J. Opt. Soc. Am.* **1954**, *44*, 838–850. [CrossRef]
24. Piepmeier, J. Calibration of passive microwave polarimeters that use hybrid coupler-based correlators. *IEEE Trans. Geosci. Remote Sens.* **2004**, *43*, 391–400. [CrossRef]
25. Misra, S.; Brown, S. Enabling the extraction of climate-scale temporal salinity variations from Aquarius: An instrument based long-term radiometer drift correction. *IEEE Trans. Geosci. Remote Sens.* **2017**, *55*, 2913–2923. [CrossRef]

26. Dinnat, E.; Le Vine, D. Cold sky calibration (CSC) biases and time Series with hardware-only wiggle correction. Presented at the Aquarius Cal/Val Meeting, Santa Rosa, CA, USA, 9–11 January 2017. Available online: https://aquarius.umaine.edu/cgi/meetings.htm (accessed on 7 July 2018).
27. Meissner, T. Assessment of Uncertainties in Aquarius Salinity Retrievals, RSS Technical Report 061015. 10 June 2015. Available online: http://podaac-ftp.jpl.nasa.gov/allData/aquarius/docs/v4/AQ-014-PS-0017_AquariusATBD_uncertainties_Addendum5_DatasetVersion4.0.pdf (accessed on 7 July 2018).
28. Meissner, T.; Wentz, F.; Le Vine, D.; Lee, T. Estimate of uncertainties in the Aquarius salinity retrievals. In Proceedings of the 2015 IEEE International Geoscience and Remote Sensing Symposium (IGARSS), Milan, Italy, 26–31 July 2015; pp. 5324–5327. Available online: http://images.remss.com/papers/rssconf/Meissner_igarss_2015_Milan_AquariusErrors.pdf (accessed on 7 July 2018). [CrossRef]
29. Wentz, F.; Ricciardulli, L.; Gentemann, C.; Meissner, T.; Hilburn, K.; Scott, J. Remote Sensing Systems Coriolis WindSat Environmental Suite on 0.25 deg Grid, Version 7.0.1, Remote Sensing Systems. Santa Rosa, CA, 2013. Available online: www.remss.com/missions/windsat (accessed on 7 July 2018).
30. Yu, T.-W.; Gerald, V.M. Evaluation of NCEP operational model forecasts of surface wind and pressure fields over the oceans. In Proceedings of the 20th Conference on Weather Analysis and Forecasting/16th Conference on Numerical Weather Prediction. 2004. Available online: http://polar.ncep.noaa.gov/mmab/papers/tn233/mmab233.pdf (accessed on 7 July 2018).
31. Ricciardulli, L.; Meissner, T.; Wentz, F. Towards a climate data record of satellite ocean vector winds. In Proceedings of the 2012 IEEE International Geoscience and Remote Sensing Symposium (IGARSS), Munich, Germany, 22–27 July 2012; pp. 2067–2069. Available online: http://images.remss.com/papers/rssconf/ricciardulli_igarss_2012_munich.pdf (accessed on 7 July 2018). [CrossRef]
32. Aquarius Salinity Validation Analysis (Data Version 5.0). Available online: https://podaac.jpl.nasa.gov/dataset/AQUARIUS_L2_SSS_V5?ids=SpatialCoverage:TemporalResolution&values=Global:Weekly (accessed on 7 July 2018).
33. Kao, H.-Y.; Lagerloef, G.; Lee, T.; Melnichenko, O.; Meissner, T.; Hacker, P. Assessment of Aquarius Sea Surface Salinity Data using Aquarius Validation Data System (AVDS) and other statistical methods. *Remote Sens.* **2018**, submitted.
34. Piepmeier, J.; Mohammed, P.; Peng, J.; Kim, E.; De Amici, G.; Ruf, C. *SMAP L1B Radiometer Half-Orbit Time-Ordered Brightness Temperatures*; Version 3; CRID 13080; NSIDC: Boulder, CO, USA, 2016. [CrossRef]
35. Atlas, R.; Hoffman, R.; Ardizzone, J.; Leidner, S.M.; Jusem, J.; Smith, D.; Gombos, D. A cross-calibrated, multiplatform ocean surface wind velocity product for meteorological and oceanographic applications. *Bull. Am. Meteorol. Soc* **2011**, *92*, 157–174. [CrossRef]
36. Wentz, F.; Scott, J.; Hoffman, R.; Leidner, M.; Atlas, R.; Ardizzone, J. *Remote Sensing Systems Cross-Calibrated Multi-Platform (CCMP) 6-Hourly Ocean Vector Wind Analysis Product on 0.25 deg Grid*; Version 2.0; Remote Sensing Systems: Santa Rosa, CA, USA, 2015; Available online: http://www.remss.com/measurements/ccmp (accessed on 7 July 2018).
37. Ricciardulli, L.; National Center for Atmospheric Research Staff (Eds.) The Climate Data Guide: CCMP: Cross-Calibrated Multi-Platform Wind Vector Analysis. 2017. Available online: https://climatedataguide.ucar.edu/climate-data/ccmp-cross-calibrated-multi-platform-wind-vector-analysis (accessed on 7 July 2018).

remote sensing

MDPI

Article

Assessment of Aquarius Sea Surface Salinity

Hsun-Ying Kao [1,*], Gary S. E. Lagerloef [1], Tong Lee [2], Oleg Melnichenko [3], Thomas Meissner [4] and Peter Hacker [3]

1 Earth & Space Research, 2101 Fourth Ave., Suite 1310, Seattle, WA 98121-2350, USA; lager@esr.org
2 Jet Propulsion Laboratory, California Institute of Technology, Pasadena, CA 91109, USA; tlee@jpl.nasa.gov
3 International Pacific Research Center, School of Ocean and Earth Science and Technology, University of Hawaii, Honolulu, HI 96822, USA; oleg@hawaii.edu(O.M.); phacker@hawaii.edu (P.H.)
4 Remote Sensing Systems, 444 Tenth Street, Suite 200, Santa Rosa, CA 95401, USA; meissner@remss.com
* Correspondence: hkao@esr.org; Tel.: +1-206-726-0501 (ext. 140)

Received: 28 June 2018; Accepted: 18 August 2018; Published: 22 August 2018

Abstract: Aquarius was the first NASA satellite to observe the sea surface salinity (SSS) over the global ocean. The mission successfully collected data from 25 August 2011 to 7 June 2015. The Aquarius project released its final version (Version-5) of the SSS data product in December 2017. The purpose of this paper is to summarize the validation results from the Aquarius Validation Data System (AVDS) and other statistical methods, and to provide a general view of the Aquarius SSS quality to the users. The results demonstrate that Aquarius has met the mission target measurement accuracy requirement of 0.2 psu on monthly averages on 150 km scale. From the triple point analysis using Aquarius, in situ field and Hybrid Coordinate Ocean Model (HYCOM) products, the root mean square errors of Aquarius Level-2 and Level-3 data are estimated to be 0.17 psu and 0.13 psu, respectively. It is important that caution should be exercised when using Aquarius salinity data in areas with high radio frequency interference (RFI) and heavy rainfall, close to the coast lines where leakage of land signals may significantly affect the quality of the SSS data, and at high-latitude oceans where the L-band radiometer has poor sensitivity to SSS.

Keywords: Aquarius satellite; sea surface salinity; Aquarius Validation Data System (AVDS)

1. Introduction

Aquarius/Satélite de Applicaciones Científicas (SAC)-D was a collaboration between NASA and Argentina's space agency, Comision Nacional de Actividades Espaciales (CONAE) [1,2]. NASA's Aquarius was the primary instrument on the SAC-D spacecraft. The Aquarius mission was developed to study the connections between ocean circulation and the global water cycle by measuring sea surface Salinity (SSS). The Aquarius satellite has successfully collected global high resolution SSS data from 25 August 2011 until SAC-D spacecraft ceased operating because of an on-board power failure on 7 June 2015. In all, Aquarius collected 45 complete months of data (September 2011–May 2015), exceeding its 36-month science requirement by 9 months. One specific goal of Aquarius was to monitor the seasonal and inter-annual variations of the large-scale features of the SSS. Numerous scientific results have been published using Aquarius SSS, taking advantage of its unprecedented spatial and temporal resolution. Some research reveals the SSS signals that were not captured by the in situ data. For example, Aquarius SSS data captures many fine scale ocean structures, including the salinity fronts in the tropical Pacific [3], tropical instability waves in both the Pacific and Atlantic [4], the haline wake over the Amazon plume after the passage of hurricanes [5], and salinity anomalies and fluxes associated with the ocean eddy field [6].

This manuscript reports the Aquarius SSS measurement uncertainty characteristics, including residual errors in the final version (V5.0) of the Aquarius data. This version of data was released by the Aquarius Project when the Aquarius mission ended in December 2017. Most of the results are also documented in the Aquarius validation [7]. Here we further consolidate the information and make it more practical to the general data users. We evaluate the Aquarius Level-2 and Level-3 SSS using co-located in situ salinity measurements and gridded maps based on in-situ measurements, respectively. It should be noted that the matchup statistics between Aquarius level-2 SSS and in situ observations not only include Aquarius SSS uncertainty, but also the differences in sampling (e.g., spatial scales) between Aquarius data (averaged over the Aquarius footprint) and the point-wise in-situ measurements. Likewise, the differences between Aquarius Level-3 SSS with the gridded in-situ data also contain the sampling and mapping errors of the gridded maps of the in-situ measurements. These points will be re-iterated when presenting the results of the comparison (Section 3). For further discussions about the uncertainties in the Aquarius salinity retrieval algorithm, readers can refer to [8]. Random and systematic uncertainties have been included in the Level-3 monthly data distributed via PO.DAAC, as well.

Here we use 45 months of data observed during the whole mission period from September 2011 to May 2015 for validation analysis. The rain filters are applied to the SSS used in this paper for both Level-2 and Level-3 data when the instantaneous rain rate is larger than 0.25 mm/h. The heavy rain events, which cause larger SSS biases in Aquarius observations, are removed with the rain masks. If the users are interested in the Aquarius SSS under strong precipitation, the data without rain masks should be used. Otherwise, data with rain masks are advised to be used for general studies.

The Aquarius/SAC-D mission and sensor design, sampling pattern, salinity remote sensing principles, and pre-launch error analysis are described in [1,9]. The Aquarius/SAC-D satellite was positioned on a polar sun-synchronous orbit crossing the equator at 6 pm (ascending) and 6 am (descending) local time with a repeat cycle of one week. The Aquarius instrument consisted of three passive microwave radiometers that were "looking" along three beams at different angles relative to the sea surface. The beams formed three elliptical footprints on the sea surface (76 × 94 km, 84 × 120 km, and 96 × 156 km) aligned across a ~390-km-wide swath. The emission from the sea surface, measured as an equivalent brightness temperature, was converted to SSS subject to corrections for various geophysical effects. Individual observations along each orbit/beam consisted of a sequence of data points sampled at a 1.44-s (~10 km) interval. Each individual observation represented the average salinity in the upper 1–2 cm layer and over a ~100 km footprint [1,9].

The sensor calibration was done with a forward model to estimate the antenna temperature at the satellite, then differencing that estimate from the measured antenna temperature on a global average [10]. The forward model included the surface emission, geophysical corrections, antenna pattern correction, etc. This helped remove the quasi-monthly, non-monotonic variations, also called the "wiggles", seen in the earlier version (V1.3) of Aquarius data [11].

The surface emission for the forward model is derived from ancillary sea surface temperature (SST) and SSS. The source of the ancillary SST field is from the Canadian Meteorological Center (CMC) [12]. More details for the reference SST can be found in [13]. The ancillary SSS data have been derived from the US Navy Hybrid Coordinate Ocean Model (HYCOM) daily averaged data-assimilative analysis [14]. The operational data are produced by the U.S. Naval Oceanographic Office (NAVO), and the digital output is distributed by Florida State University. The analyzed monthly Scripps Argo SSS has been used in the sensor calibration and in the derivation of expected brightness temperature (Aquarius antenna temperature, TA) (i.e., forward algorithm).

The objectives in this paper include (1) quantify the differences between Aquarius and in situ data on different spatial and temporal resolutions; (2) retrieve the root-mean-square errors of Aquarius SSS data from triple-point analysis; and (3) summarize cautions for the general users when addressing the results using Aquarius data.

2. Materials and Methods

2.1. Aquarius Data

The Aquarius V5.0 data sets are described and discoverable via the PO.DAAC data portal (https://podaac.jpl.nasa.gov/datasetlist?ids=Collections&values=Aquarius). The Aquarius project produces three data sets: Level-1A (raw data), Level-2 (science data in swath coordinates and matching ancillary data), and Level-3 (gridded 1-degree daily, weekly and monthly salinity and wind speed maps, as well as sea water density and spice). This paper evaluates both Level-2 and Level-3 salinity data.

The Level-3 maps are generated from Level-2 salinity data without any added adjustment for climatology, reference model output or in situ data. The smoothing interpolation applies a bi-linear fit within specified search radius [15]. The standard Aquarius Level-3 data produced by the Aquarius Data Processing System (ADPS) use the criterion for land fraction set as 0.01 (severe), which means values are excluded for a grid point with the fraction of land area larger than 1%. Therefore, more salinity information near the coastal regions is including compared to the salinity maps using a criterion for land fraction set as 0.001 (moderate, fraction of land area >0.1%). As a result, the standard deviations of the salinity biases are higher in the ADPS Level-3 data due to land contamination. When using the ADPS Level-3 mapped data, users should be careful when analyzing the salinity data near the coasts. Likewise, measurement sensitivity of L-band brightness temperature to salinity reduces from the tropics (relatively high sea surface temperature, SST) to high-latitude oceans (relatively low SST). As a result, L-band salinity data such as those from Aquarius are more prone to errors in the high latitudes than in the tropics. This information is documented in the algorithm theoretical basis document (ATBD) [13,16].

Salinity measurements are on the practical salinity scale (PSS-78), technically a dimensionless number, but practical salinity units (psu) are used in this paper.

2.2. Argo Data

2.2.1. Argo Profiles

Argo float measurements shallower than 6-m depth and flagged as good from each Argo profile are used for the analysis. Argo profile data are from the US Global Ocean Data Assimilation Experiment (GODAE) and are available at ftp://usgodae.org/pub/outgoing/argo. Typically, Argo floats rise to the surface once every 10 days and remain at the surface for a few hours. The data are collected randomly at any time of day.

2.2.2. Gridded Argo Maps

Two Argo monthly 1°-gridded salinity products are used for comparison: One from the Scripps Institution of Oceanography (SIO) (http://www.argo.ucsd.edu/Gridded_fields.html) and the other one from the Asia Pacific Data Research Center (APDRC) of the University of Hawaii (UH) (http://apdrc.soest.hawaii.edu/projects/Argo/data/gridded/On_standard_levels/index-1.html). Both Argo gridded products are used for Aquarius validation analysis, and both of them show the same features of the differences with Aquarius data. It is important to note that Aquarius measurements represent salinity in the top centimeter of the ocean. Near-surface salinity stratification in the upper few meters can cause differences between Aquarius and Argo SSS, especially under rain bands [17].

2.3. Aquarius Validation Data System (AVDS)

The Aquarius Validation Data System (AVDS) compares the Aquarius Level 2 samples and Level 3 gridded maps with near-surface in situ salinity data, including those from Argo floats and the global tropical moored buoy array from Pacific Marine Environmental Laboratory (PMEL, http://www.pmel.noaa.gov/tao/data_deliv/). The shallowest sampling depths of the Argo data are generally 3–5 m below the surface. The shallowest sampling depth of the tropical buoy array is 1 m. Under most conditions (e.g., moderate to high winds) the surface ocean mixed layer extends much deeper, and the buoy provides an accurate estimate of the 1–2 cm surface layer that emits the microwave signal seen by the satellite. However, at river outflow regions and under persistently rainy conditions (especially under low winds when vertical mixing is small), there are often vertical gradients between the surface and the buoy measurement depth.

For each in situ observation, we search for the closest point of approach (CPA) from the Aquarius Level-2 (swath) data. The time window is ±3.5 days to gather all in situ data within the 7-day orbit repeat cycle. The search radius is 75 km between the in situ location and the bore sight position of the Aquarius footprint. The Aquarius data are averaged over 11 samples (~100 km) centered on the match-up point (i.e., the CPA). The development and ongoing operations of the AVDS, which include the collection, processing, and quality analysis of surface in situ ocean salinity and temperature for the calibration and validation of satellite salinity measurements, and the delivery of these data to the broader community. The AVDS is a facility developed at Earth and Space Research (ESR) to gather useable surface ocean validation data, run validation processes, and serve the data via the Internet to the broader scientific community. Details can be found at ftp://podaac-ftp.jpl.nasa.gov/allData/aquarius/docs/v5/AQ-014-PS-0028_V5_AVDS_Tech_Memo.pdf.

3. Results

3.1. Quantification of the Time-Mean, Seasonal, and Non-Seasonal Differences from In-Situ Measurements

Figure 1a shows the Aquarius retrieved salinity at the in situ matchup points for all 45 months of observation from September 2011 to May 2015. The results are obtained from the AVDS. All the individual in situ salinity data at the same matchup points for the 45 months are shown in Figure 1b. The matchup processes are described in Section 2.3. The criterion for land fraction is set to 0.001, so the validation information is missing near the coasts due to the land contamination. The correspondence is visibly quite clear, with Aquarius Level-2 data resolving the salient large-scale ocean features. The SSS values in the open ocean generally range from 32 to 37 psu. Overall, Aquarius is able to accurately capture the SSS signature over the globe. High salinity is seen in the subtropical gyres, and the salinity maximum is located in the North Atlantic. Low salinity is observed in high latitudes, under the Intertropical Convergence Zone (ITCZ), and around major river outflows (including the Bay of Bengal). Figure 1c shows the Aquarius and in situ differences. Few anomalous values are observed near the islands. Positive biases up to 0.5 psu locally appear at high latitude (50 degrees poleward); they may be related to the galaxy reflection term that is not correctly adjusted for wind [16]. In the open ocean, only small differences (<0.2 psu) are generally present.

In the Northern Hemisphere, the negative biases in the eastern Atlantic and in the western Pacific are thought to be related to low-level radio frequency interference (RFI) from adjacent land that is not adequately detected by the standard RFI filter algorithm. This causes a positive brightness temperature bias and, thus, a negative salinity bias.

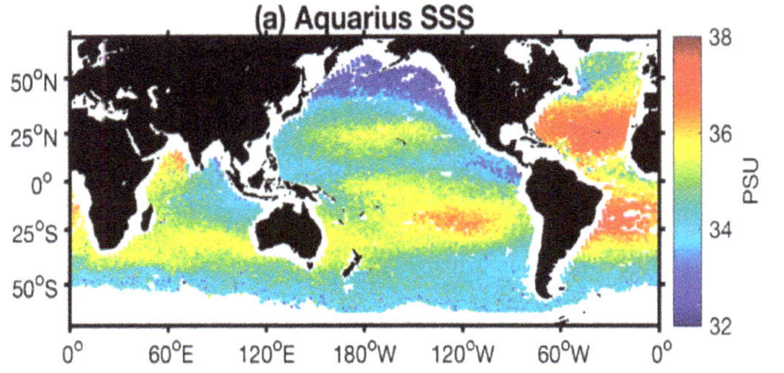

Figure 1. *Cont.*

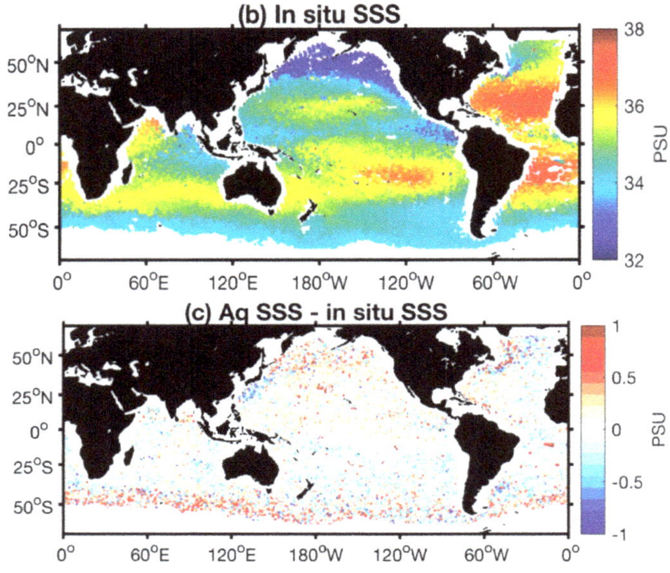

Figure 1. (**a**) Aquarius Level-2 and (**b**) in situ co-located salinity data gathered from September 2011 to May 2015. (**c**) The sea surface salinity (SSS) differences defined as the Aquarius minus the co-located in situ salinity.

Figure 2 shows the time series of daily median of global differences between Aquarius SSS from each of the three beams and co-located in situ data from the AVDS analysis. The values of the standard deviation (STD), the square root of the variance, are also labeled in the figures. All three beams show small differences with little variations for the global median. The remaining differences may be related to the uncertainties of salinity observations, such as near-surface stratification or the sub-footprint variations [17]. The important conclusion from Figure 2 is that Aquarius measurements exhibit no spurious trends and drifts.

The time-mean bias error statistics for the Level-3 SSS rain-flagged product are presented in Figure 3. The error statistics were computed by comparing Argo float measurements from each profile for a given week with SSS values at the same locations obtained by interpolation of the corresponding Level-3 SSS maps. The root-mean-square deviation (RMSD) is defined as RMSD = $\text{sqrt}(\text{bias}^2 + \text{STD}^2)$. The geographical distribution of the time-mean (static) Aquarius minus Argo differences is shown in Figure 3a. Large positive biases (up to 0.2 psu locally) are observed in the sub-polar North Pacific and in the Southern Ocean poleward of about 40°S. Large negative biases (up to −0.2 psu) are observed in the subtropical South Pacific and along the continental boundaries. These regions of positive and negative biases tend to cancel each other in the global average, producing nearly zero global bias (Figure 2). The regional biases shown in Figure 3a are similar to those shown in Figure 1c.

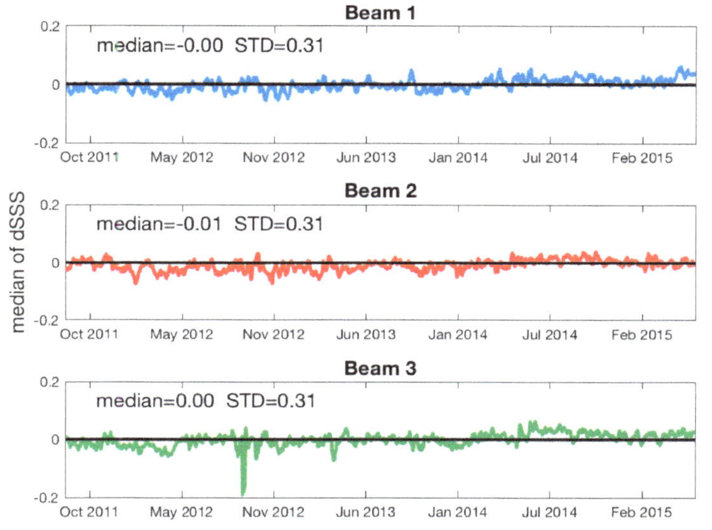

Figure 2. Time series of daily median of global Aquarius—in situ SSS in three beams. 7-day average is applied to smooth the data. The median and the standard deviation (STD) labeled in the figures are based on the average of differences for the whole mission period from September 2011 to May 2015.

The RMSD between the weekly Level-3 analysis and concurrent Argo float data is smaller than 0.29 psu for nearly all weeks over the nearly 4-year period of comparison (not shown). The mean RMSD over the whole mission period (September 2011 through May 2015) is 0.247. The geographical distribution of the RMSD for the weekly Level-3 product is shown in Figure 3b. The RMSD is computed in 8°-longitude by 8°-latitude bins to ensure an adequate number of collocations (>100) in each bin. Over most of the ocean, the RMSD between weekly SSS maps and collocated in situ data do not exceed 0.2 psu. Figure 3b also demonstrates that the largest RMSD, exceeding 0.2 psu, are found in the regions of strong variability in SSS, such as along the North Pacific and North Atlantic ITCZ, the North Pacific sub-polar front, the Gulfstream, and near outflows of major rivers such as the Amazon in the tropical North Atlantic. In this regard, the observed relatively large RMSD between the Aquarius and Argo float data in some areas are not necessarily due to errors in Aquarius measurements only, but RMSD may include the disparity between time and space scales captured by two different observational platforms [17–19] and the difference in measurement depth between Aquarius (ocean surface) and Argo (~5 m depth) [17]. Larger bias and RMSD is also observed in the high latitudes. As mentioned in the introduction, this is partly due to lower measurement sensitivity of L-band brightness temperature to salinity in high-latitude oceans (relatively low SST).

To examine temporal variability in the bias fields, Figure 4 shows the latitude-time distribution of the zonally averaged differences between the weekly Aquarius SSS maps and the corresponding Argo data. The zonally averaged biases are calculated weekly by averaging these statistics over 5°-latitude bins. The latitude-time distribution shows significant positive biases at high latitudes and negative biases in the subtropics. Besides the residual static bias, there is a clear seasonal cycle in the bias distribution. To emphasize the time-varying part, the 3-year average (September 2011 to August 2014) in each zonal bin is subtracted from the time series and is shown in Figure 4b. The peak-to-peak amplitude of the anomalous annual cycle can reach 0.2 psu locally. Whether this is significant or not depends on the amplitude of the "true" annual cycle in SSS and the signal-to-noise ratio.

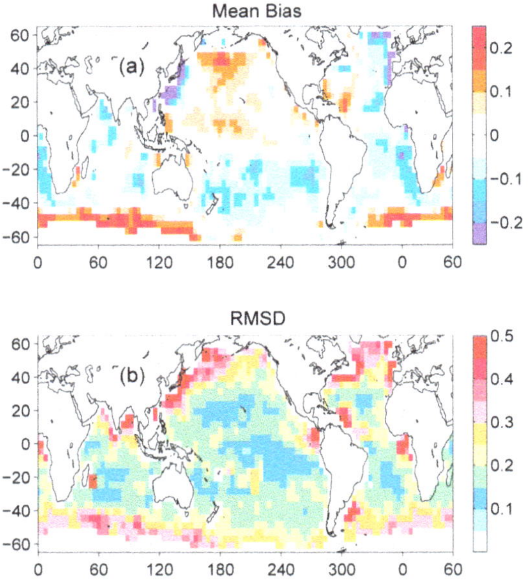

Figure 3. Geographical distribution of (**a**) mean spatial bias (psu), and (**b**) root-mean-square deviation (RMSD) (psu) between the Aquarius weekly Level-3 SSS product and Argo float observations. The error statistics were computed by comparing Argo float measurements for a given week with SSS values at the same locations obtained by interpolation of the corresponding Level-3 SSS maps. The geographical distributions are computed in 8°-longitude by 8°-latitude bins.

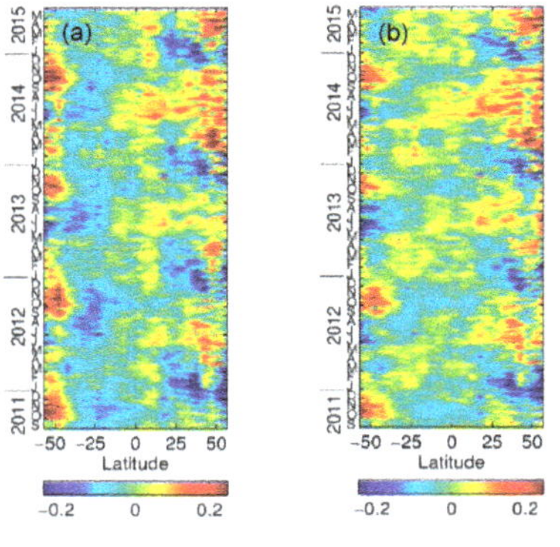

Figure 4. (**a**) Latitude-time distribution of the zonally averaged differences (psu) between the weekly Level-3 SSS maps and the corresponding Argo data. The error statistics were computed by comparing Argo float measurements for a given week with SSS values at the same locations obtained by interpolation of the corresponding Level-3 SSS maps. The zonally averaged biases were computed by averaging these statistics over 5°-latitude bins. (**b**) The same as in (**a**), but with the 3-year mean (September 2011 to August 2014) subtracted.

In Figure 5, the geographical distribution of the seasonal signal in Aquarius SSS is assessed against the seasonal cycle in the Argo-SIO gridded SSS. Because the Argo-derived product is very smooth, to match the spatial scales, the weekly Level-3 SSS maps from Aquarius were smoothed with a 2D running Hanning window of half-width of 4°, generally consistent with the smoothness properties of the Argo-derived salinity fields. The annual and semi-annual variations of the differences between Aquarius and Argo data are calculated from the harmonic analysis.

The amplitudes of the annual harmonic in the bias fields are presented in Figure 5a. Over most of the ocean, the amplitude of the annual signal in the Aquarius and Argo SSS differences is smaller than 0.05 psu. There are a few areas, however, where the amplitudes can reach 0.2 psu and more: The western boundary areas in the North Pacific and North Atlantic, the eastern boundary region of the North Atlantic, a zonal band going along the Southern Ocean, the Bay of Bengal and the Arabian Sea in the Indian Ocean, as well as a relatively small area in the eastern equatorial Pacific. Compared to the annual cycle in the Argo gridded SSS data (Figure 5b), the difference from the annual signal in Aquarius SSS appears to be minor, except for the areas where the vertical salinity gradient is known to be large. Although the relatively sparse sampling of Argo measurements in these regions may contribute to the differences in seasonal variability between Aquarius and Argo data, Aquarius data users are advised to exercise caution when analyzing seasonal variability in these areas considering the fundamental differences in how these observations are taken.

The amplitudes of the semi-annual harmonic in the bias fields are presented in Figure 5c. For comparison, the amplitudes of the semi-annual cycle in the Argo-derived SSS fields are presented in Figure 5d. Although generally small compared to the annual cycle biases, the semi-annual harmonic in the time-varying bias can be important compared to the Argo-derived variability regionally. Significantly affected areas are in the subtropical South Pacific, particularly along a quasi-zonal band stretching across the basin from about 30°S in the east to close to the equator in the west, and along the Southern Ocean.

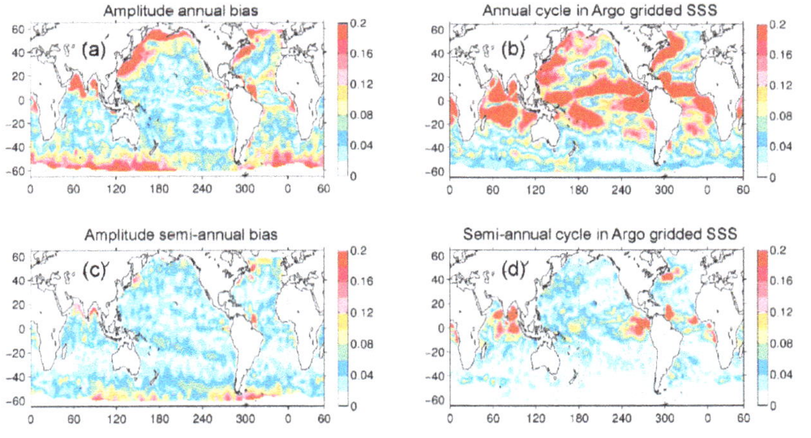

Figure 5. (**a**) Amplitude of the annual cycle in the Aquarius minus Argo bias; (**b**) Amplitude of the annual cycle in SSS from the Argo-derived gridded SSS fields produced by the Scripps Institution of Oceanography (SIO); (**c**) Amplitude of the semi-annual cycle in the Aquarius minus Argo bias; (**d**) Amplitude of the semi-annual cycle in SSS from the Argo-derived gridded SSS fields produced by the SIO.

It is worth emphasizing here that although an attempt has been made to quantify seasonal biases in Aquarius SSS, the results should rather be viewed as qualitative, serving as both a guidance and a note of caution, especially in light of potentially inadequate sampling of the Argo measurements in regions with strong intraseasonal variability.

3.2. Evaluation of Aquarius Level-3 SSS Data at Different Spatial Scales

Figure 6a shows the global average value of the regional temporal RMSD between Aquarius and Argo-SIO SSS for $1^\circ \times 1^\circ$, $3^\circ \times 3^\circ$, and $10^\circ \times 10^\circ$ scales. The RMSD decreases with larger smoothing scale. The RMSD are 0.19, 0.17, and 0.11 psu, respectively. The RMSD values include the contributions of differences in time-mean values and differences for temporal anomalies (deviation from time mean) between Aquarius and Argo SSS. Many applications do not concern time-mean biases but focus on variability (e.g., temporal and spatial changes). Therefore, we also assess the STD of the differences between Aquarius and Argo SSS. The time-mean differences between Aquarius and Argo products do not contribute to the STD values as they do to the RMSD values.

Figure 6b shows the global averages of regional temporal STD values of Aquarius SSS with respect to Argo-SIO SSS for various spatial scales. The difference between Figure 6a,b is that the latter excluded the time-mean biases of Aquarius SSS. The global average STD values for the $1^\circ \times 1^\circ$, $3^\circ \times 3^\circ$, and $10^\circ \times 10^\circ$ scales are 0.15, 0.13, and 0.07 psu, respectively. Note that the RMSD and STD values with respect to Argo-SIO not only contain the errors of the Aquarius SSS, but the errors of the Argo-gridded product as well. Vinogradova and Ponte [18] Compared the Argo-SIO and Argo-APDRC gridded products and found a global average STD value of 0.10, 0.09, and 0.04 psu on the $1^\circ \times 1^\circ$, $3^\circ \times 3^\circ$, and $10^\circ \times 10^\circ$ scales for the differences between the monthly Argo-SIO and Argo-APDRC products (Figure 1 in [20]). These values reflect the sampling error of individual Argo float data to represent the averages on different spatial scales as well as the mapping error of the Argo gridded products.

Figure 6c,d shows the global average of STD values for composite seasonal anomalies and non-seasonal anomalies (referenced to the composite seasonal cycle) of the SSS differences between Aquarius and Argo-SIO. The non-seasonal anomalies are associated with the differences between Aquarius and Argo-SIO in representing non-seasonal signals associated with phenomena such as tropical instability waves [4,21], Madden-Julian Oscillation [22,23], El Niño-Southern Oscillation [24], and Indian Ocean Dipole [25]. It is noteworthy that the global average STD value for non-seasonal SSS anomalies is less than 0.1 psu for $1^\circ \times 1^\circ$ scale and 0.04 psu for $10^\circ \times 10^\circ$ scale. Again, these values include the uncertainties of the Argo gridded product on these spatial and temporal scales.

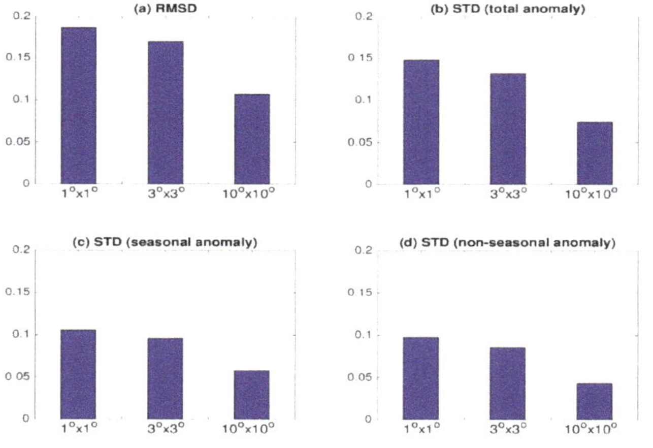

Figure 6. (**a**) Global average of regional temporal RMSD values between Aquarius and Argo-SIO SSS on various spatial scales as indicated by legends along the horizontal axis. Global average of regional temporal STD values for the differences between Aquarius and Argo-SIO SSS on various spatial scales for (**b**) total anomalies; (**c**) composite seasonal anomalies; and (**d**) non-seasonal anomalies. All RMSD and STD values are in the unit of psu.

The RMSD and STD values of Aquarius-Argo SSS differences are not only due to the uncertainty of Aquarius SSS, but also the uncertainty of the Argo gridded datasets as well. This is because Argo float distributions are too sparse to produce robust monthly mean values at 1° × 1° scale in regions with strong spatiotemporal variability such as tropical rain bands (e.g., under the ITCZ), near river plumes, western boundary currents, and the Antarctic Circumpolar Current. High-resolution in-situ thermosalinograph (TSG) data indeed show significant variations of SSS within the scales of Aquarius footprint [17]. Therefore, the Argo gridded data in these regions can be significantly affected by the sampling errors of the Argo array. In fact, Lee [20] found that the STD values between two of the Argo-SIO and Argo-APDRC products are as large as or even larger than the STD values between Aquarius and either of the Argo product in some of these regions.

This is also reflected in Figure 7 by comparing the STD maps for Aquarius-Argo salinity for various resolutions of the Aquarius data as well as the STD for the difference between Argo-SIO and Argo-APDRC data. This issue also exists for the comparison of STD on 3° × 3° scale (Figure 7c,d). However, for the 10° × 10° scale, Argo data have sufficient sampling to represent the large scale monthly mean values, so the difference between the two Argo products is much smaller (Figure 7e,f). The message for the comparison shown in Figure 7 is that RMSD and STD values of the difference between Aquarius and Argo gridded data can be used as an indication of Aquarius data uncertainty. However, these values also contain the sampling uncertainty associated with the data products generated from the individual Argo profiles.

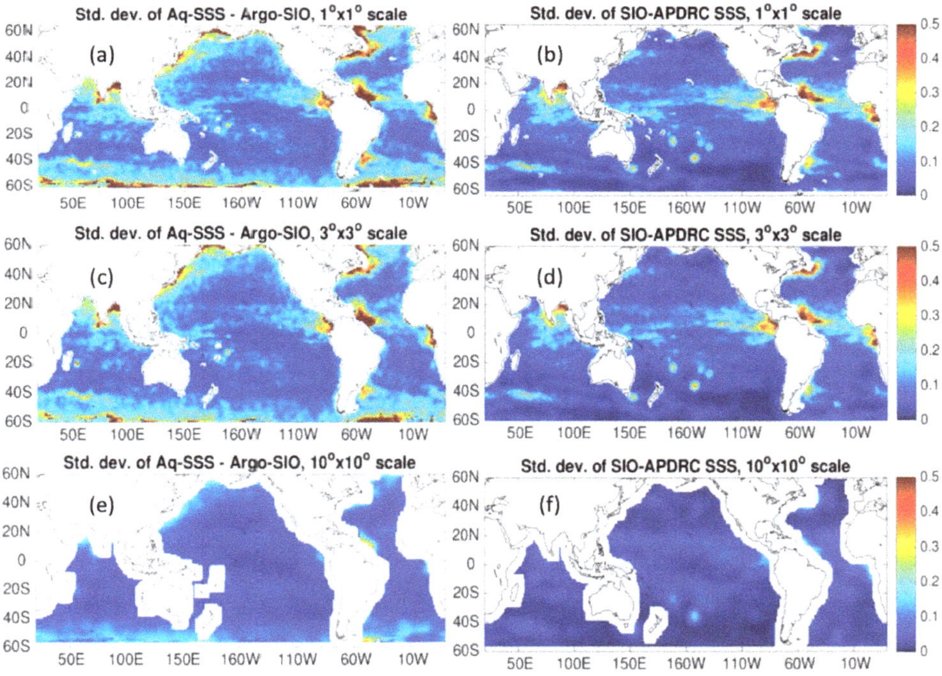

Figure 7. The STD of the difference between Aquarius and Argo-SIO SSS on (**a**) 1° × 1°, (**c**) 3° × 3° and (**e**) 10° × 10° scale. The STD of the difference between Argo-SIO and Argo-APDRC are shown in (**b**), (**d**), and (**f**).

3.3. Triple-Point Analysis of Aquarius, In Situ, and HYCOM

As discussed earlier, the differences between Aquarius and in-situ measurements contain uncertainties of the Aquarius data, as well as the effect of sampling differences between Aquarius and in situ measurements. It is therefore of interest to estimate the uncertainty of the in situ data in representing the truth on the spatiotemporal scales of the Aquarius measurements, as well as the uncertainty of the Aquarius measurements that take into account the representativeness error of the in-situ measurements on Aquarius' measurement scales. One such approach is the so-called triple collocation analysis or triple-point analysis (Appendix A). Here, we apply the triple-point approach to assess the Aquarius root mean square error (RMSE) by taking into consideration the representativeness error of the in situ measurements on Aquarius measurement scales. From the bias and standard deviation (STD) differences between the Aquarius data and in situ observations, the root mean square deviation (RMSD) is obtained as the square-root of the (bias^2 + STD^2). The RMSD per se includes both the Aquarius SSS error and the effect due to the sampling differences between Aquarius and in situ measurement, whereas our goal here is to isolate the Aquarius RMSE. The detailed formulas used to calculate the RMSE are documented in Appendix A.

The triple-point analysis requires three independent datasets. Here we use the HYCOM operational ocean analysis (a data-assimilation product), in addition to the Aquarius SSS and in-situ measurements. See Appendix A in the Aquarius user guide for more details about HYCOM data. Aquarius data are independent of HYCOM as the Aquarius satellite information is not used for HYCOM assimilation. Note that HYCOM assimilates Argo measurements, which account for the majority of in-situ measurements over the global ocean as a whole. In this sense, HYCOM is not fully independent from in situ measurements. However, HYCOM salinity is not only constrained by Argo measurements, but also by evaporation-precipitation forcing, ocean dynamics, and a relaxation of model SSS to a seasonal climatology to prevent model drift. These factors help maintain some level of independence of HYCOM SSS from in-situ measurements. Nevertheless, the results of the triple-point analysis is to some extent affected by the assumption that the three datasets are independent.

Figure 8 gives the matchup statistics for Aquarius-in situ, HYCOM-in situ, and Aquarius-HYCOM (at the in situ locations), for each of the three beams. Note that all three data sets are biased $\leq |0.01|$ relative to each other, on average, at all Aquarius beam locations. The root-mean-square deviation (RMSD) is defined as RMSD = sqrt(bias^2 + STD^2). Figure 8 shows HYCOM-in situ RMSD ~0.23 psu overall. The RMSD for Aquarius-in situ and Aquarius-HYCOM are 0.31 psu and 0.27 psu, respectively. The co-located statistics allow us to estimate the root-mean-square error (RMSE) of each of the three measurements. The results of Aquarius co-located matchup points are given in Table 1. The Aquarius, HYCOM and in situ RMSE are approximately 0.17 psu, 0.08 psu, and 0.14 psu, respectively. Recall that these Aquarius matchup statistics are for individual match-ups, with no averaging.

Table 1. Estimated Root Mean Square Error (RMSE) in the unit of psu for each data type based on the triple point analysis of co-located point measurements.

V5.0	Beam 1	Beam 2	Beam 3
Aquarius RMSE	0.17	0.17	0.16
HYCOM RMSE	0.09	0.08	0.09
In situ RMSE	0.13	0.14	0.15

For the Level-3 SSS error analysis, the three data sets for the triple-point collocation are (1) Aquarius ADPS Level-3 data; (2) similarly gridded HYCOM SSS monthly $1^\circ \times 1^\circ$ maps; and (3) the in situ data set (un-gridded). Next, we find the RMSD of three data pairs: (1) Aquarius-in situ; (2) HYCOM-in situ; and (3) Aquarius-HYCOM. The process finds all the in situ data points within the mapped $1^\circ \times 1^\circ$ boxes for each month, averages those, differences that from the gridded monthly value for that grid-box, and then computes the RMSD of all the matched $1^\circ \times 1^\circ$ grid-boxes over the globe for that month. Aquarius-HYCOM is simply the RMSD between the respective monthly 1° × 1° maps. The RMSD accumulations also ensure that only the 1° × 1° grid boxes containing in situ samples are counted, to ensure common sampling. We also note that the standard Level-3 gridding masks and flags are applied; thus, cold regions (SST < 5°C) and regions higher than the threshold for land contamination are omitted (See Table 1 in AQ-014-PS-0018_AquariusLevel2specification_DatasetVersion5.0 for the description of data quality flags and masks).

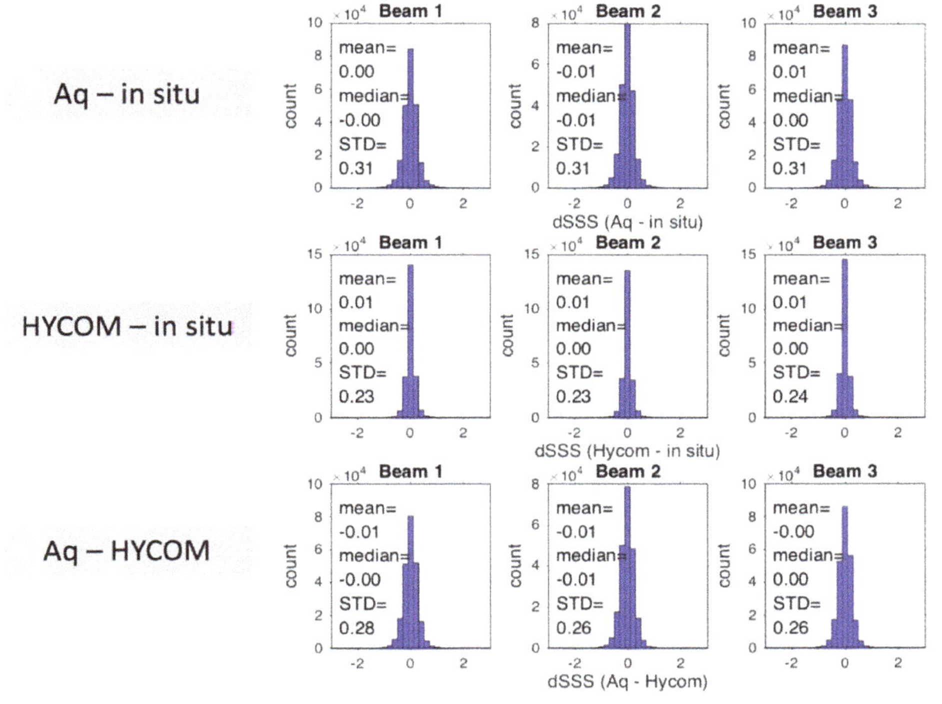

Figure 8. Co-located difference histograms for each beam of Aquarius L2 data. (**top**) Aquarius—in situ, (**middle**) Hybrid Coordinate Ocean Model (HYCOM)—in situ, (**bottom**) Aquarius—HYCOM.

The triple-point analyses giving estimated RMSE of each measurement system (Aquarius, HYCOM, in situ) are presented in Table 2. Note that the largest RMSE belongs to the in situ data. These are a combination of in situ measurement and representativeness errors. The latter include spatial and temporal variations of the in situ observations within the 1 × 1 grid box during the month, plus the salinity differences between the in situ sampling depths and the surface. However, the results from such an analysis are sometimes met by skepticism because of the assumptions such as independent errors among the three products and usage of model-based ground truth (HYCOM).

The Aquarius monthly RMSE estimates are <0.2 psu for all months of the mission, and the average over all months is 0.128 psu. Given that 0.20 psu is the mission accuracy requirement for monthly average maps, this calculation verifies that the Aquarius data exceed the mission requirement 0.2 psu by a substantial margin.

Table 2. Triple-point analysis: Monthly Root Mean Square Error (RMSE) differences for Aquarius, Hybrid Coordinate Ocean Model (HYCOM), and in situ fields. Each panel represents one year (September–August) beginning September 2011. At the bottom of the fourth panel, green highlight, are the average RMSE over the 45 months. Note that the Aquarius mean value is 0.128 psu.

	Aquarius	HYCOM	In situ
	RMSE	RMSE	RMSE
Sep-11	0.105	0.142	0.118
Oct-11	0.129	0.134	0.114
Nov-11	0.137	0.106	0.139
Dec-11	0.111	0.121	0.134
Jan-12	0.135	0.128	0.120
Feb-12	0.132	0.137	0.125
Mar-12	0.130	0.131	0.120
Apr-12	0.133	0.103	0.115
May-12	0.135	0.096	0.131
Jun-12	0.134	0.118	0.113
Jul-12	0.115	0.120	0.106
Aug-12	0.115	0.138	0.109

	Aquarius	HYCOM	In situ
	RMSE	RMSE	RMSE
Sep-12	0.112	0.158	0.126
Oct-12	0.127	0.144	0.149
Nov-12	0.116	0.159	0.184
Dec-12	0.132	0.139	0.113
Jan-13	0.124	0.135	0.093
Feb-13	0.121	0.118	0.096
Mar-13	0.123	0.078	0.098
Apr-13	0.128	0.074	0.091
May-13	0.117	0.072	0.121
Jun-13	0.150	0.054	0.156
Jul-13	0.128	0.063	0.133
Aug-13	0.129	0.076	0.114

	Aquarius	HYCOM	In situ
	RMSE	RMSE	RMSE
Sep-13	0.110	0.076	0.129
Oct-13	0.142	0.058	0.080
Nov-13	0.124	0.078	0.102
Dec-13	0.126	0.089	0.121
Jan-14	0.125	0.075	0.125
Feb-14	0.135	0.092	0.212
Mar-14	0.148	0.096	0.165
Apr-14	0.137	0.103	0.206
May-14	0.103	0.089	0.204
Jun-14	0.144	0.063	0.178
Jul-14	0.125	0.071	0.198
Aug-14	0.106	0.078	0.212

	Aquarius	HYCOM	In situ
	RMSE	RMSE	RMSE
Sep-14	0.121	0.079	0.183
Oct-14	0.135	0.084	0.162
Nov-14	0.134	0.074	0.193
Dec-14	0.150	0.074	0.213
Jan-15	0.144	0.056	0.178
Feb-15	0.131	0.079	0.171
Mar-15	0.139	0.063	0.240
Apr-15	0.136	0.069	0.173
May-15	0.127	0.066	0.206
Mean RMSE Values	0.128	0.097	0.146

3.4. Contrasting Ascending and Descending Passes

We now examine the differences of SSS between the ascending (northward, 6p.m.) and descending (southward, 6a.m.) orbits; the mission long average is shown in Figure 9. In principle, the ascending and descending maps are expected to be nearly identical. However, a large blue patch in the eastern Atlantic, and red zones in the western Atlantic and Asia-Pacific are observed. These ascending and descending differences are related to the RFI. The RFI asymmetry between ascending and descending tracks is the results of the opposite viewing angle (toward or away from the land emitting sources) between the two sides of the orbit. The antenna faces eastward on ascending passes and westward on descending passes.

In Figure 9, the RFI influenced areas, including both sides of the Atlantic Ocean, eastern Indian Ocean, and the northwestern Pacific are masked out for the large ascending minus descending differences. Users should note that regions with excessive ascending and descending differences due to RFI are masked when mapping. This shows gaps in the maps of only ascending or only descending passes. The ascending masked regions differ from the descending ones, and do not overlap. See AQ-014-PS-0006_ProposalForFlags&Masks_DatasetVersion3.0, which is included in the Data Version V5.0 documentation. There are visible ascending and descending differences at high southern latitudes. One can see a zonal variation between light blue and light red blobs around 60°S. It is not clear at this point what the reasons are for these ascending minus descending differences.

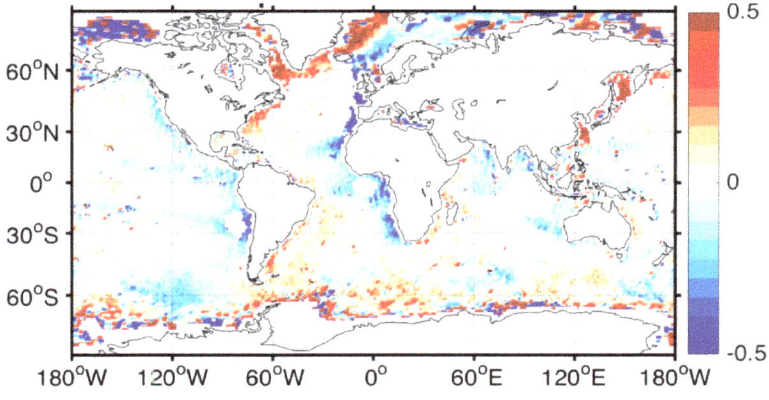

Figure 9. Forty-five-month average of ascending–descending data. Major bias regions are described in the text. The "white" regions in the North Atlantic, Western Pacific (China, Japan) or Indonesia are masked out due to suspected undetected radio frequency interference (RFI). See flagging/masking tables in [9] for details.

4. Discussion and Conclusions

In this paper, the quality of Aquarius SSS data has been evaluated with multiple approaches. Regarding the data accuracy on monthly $1^\circ \times 1^\circ$ scales, the results consistently demonstrate that the estimated errors are less than the mission requirement of 0.2 psu. The triple-point analysis resolved RMSE ~0.17 psu for point comparisons (no monthly averaging). On monthly time scales, triple-point analysis demonstrated a nominal RMSE ~0.128 psu, improving on the mission requirement by a substantial margin. Figure 3 shows that over most of the ocean, the RMSD between weekly SSS maps and collocated in situ data do not exceed 0.2 psu. Figure 6 demonstrates that the monthly, global average RMSD between gridded Aquarius and Argo data is smaller than 0.2 psu. All of these results present consistent evidence that the Aquarius Ocean Salinity Mission has met the measurement accuracy requirement.

However, the following notes of caution need to be provided to data users.

Localized persistent biases between ascending and descending passes exist in the areas close to the continental boundaries. They could be linked to radio frequency interference (RFI) that is not completely corrected by the RFI filter. The RFI will bias the brightness temperatures toward the positive, thus the salinity will be biased negative. These regions are primarily in the eastern North Atlantic adjacent to Europe where it is likely that the ascending pass is contaminated as the antenna faces the European subcontinent. Likewise, the western North Atlantic and Asia-Pacific regions are biased on the descending pass when the antenna views westward. The radiometer flag is added in the data files to exclude the area with unacceptable ascending and descending differences. This flag identifies areas where the ascending and descending difference is sufficiently large that the data from the out-of-bound pass (i.e., either ascending or descending) is discarded for purposes of calibration. Users should be very cautious with using ascending pass data in the eastern North Atlantic and descending pass data in the western North Atlantic and Asia-Pacific regions.

It is strongly recommended that data users use caution when analyzing or drawing conclusions about the annual salinity cycle in the Aquarius data, because the variability may be contaminated by spurious, non-oceanographic, error signals, especially in the Southern Hemisphere and higher latitudes, as shown in Figures 4 and 7. At high-latitude oceans, the L-band radiometer has poor sensitivity to SSS related to the low SST. This is a fundamental issue for L-band radiometry that needs technology innovation to solve [2,16]. Although most of the validation results shown in this paper are based on the Level-3 product, similar caution should be used when analyzing the along-track, Level-2 data, including RFI-affected regions, latitudinal biases, ascending and descending biases.

The rain filters (instantaneous rain rates >0.25 mm/h) are applied for the Aquarius calibration, so the strong precipitation events, which introduce large vertical stratification and large SSS biases, do not dominate the SSS variations on global average. If the users are interested in the Aquarius SSS under strong precipitation, the data without rain masks should be used. Otherwise, data with rain masks should be used for general studies. The users can tell if the data have been rain masked from the file titles.

Author Contributions: T.M. designed the basics of salinity retrieval algorithms for Aquarius, and implemented the changes that were developed for Aquarius Version 5. G.S.E.L. designed the AVDS and developed the triple-point analysis; H.-Y.K., T.L. and O.M. performed the Aquarius salinity validation; H.K. wrote the paper; P.H. provided input for the validation analysis and edition of the manuscripts.

Funding: This study was supported by NASA Ocean Salinity Science Team through grants NNX17AK16G, NNX17AK06G, and grant NNX14AJ02G. This paper is IPRC/SOEST contribution 10436/1338.

Acknowledgments: We thank David LeVine (NASA Goddard Space Flight Center), Liang Hong (Science Application Intl. Corp.), and Jorge Vazquez (Jet Propulsion Lab) for useful comments for the Aquarius validation document. We also thank the Aquarius cal/val team for the improvement of the Aquarius data.

Conflicts of Interest: The authors declare no conflicts of interest.

Appendix A

The satellite salinity measurement S_S and the in situ validation measurement S_V are defined by:

$$S_S = S \pm \varepsilon_S,$$

$$S_V = S \pm \varepsilon_V,$$

where S is the true surface salinity averaged over the Aquarius footprint area and microwave optical depth in seawater (~1 cm). ε_S and ε_V are the respective satellite and in situ measurement errors relative to S. The mean square of the difference ΔS between S_S and S_V is given by:

$$<\Delta S_{SV}{}^2> = <\varepsilon_S{}^2> + <\varepsilon_V{}^2>, \quad (1)$$

where $< >$ denotes the average over a given set of paired satellite and in situ measurements, and $<\varepsilon_S\ \varepsilon_V> = 0$. Likewise, define HYCOM salinity interpolated to the satellite footprint as $S_H = S \pm \varepsilon_H$, and mean square differences

HYCOM vs. in situ validation data

$$<\Delta S_{HV}{}^2> = <\varepsilon_H{}^2> + <\varepsilon_V{}^2> \quad (2)$$

satellite vs. HYCOM

$$<\Delta S_{SH}{}^2> = <\varepsilon_S{}^2> + <\varepsilon_H{}^2> \quad (3)$$

Equations (1)–(3) comprise three equations with three variables given by:

Satellite measurement

$$<\varepsilon_S{}^2> = \{<\Delta S_{SV}{}^2> + <\Delta S_{SH}{}^2> - <\Delta S_{HV}{}^2>\}/2 \quad (4)$$

HYCOM measurement

$$<\varepsilon_H{}^2> = \{<\Delta S_{SH}{}^2> + <\Delta S_{HV}{}^2> - <\Delta S_{SV}{}^2>\}/2 \quad (5)$$

In situ validation measurement

$$<\varepsilon_V{}^2> = \{<\Delta S_{SV}{}^2> + <\Delta S_{HV}{}^2> - <\Delta S_{SH}{}^2>\}/2 \quad (6)$$

References

1. Lagerloef, G.; Colomb, R.; Le Vine, D.; Wentz, F.; Yueh, S.; Ruf, C.; Lilly, J.; Gunn, J.; Chao, Y.; de Charon, A.; et al. The Aquarius/SAC-D mission—Designed to Meet the Salinity Remote Sensing Challenge. *Oceanography* **2008**, *21*, 68–81. [CrossRef]
2. Le Vine, D.M.; Dinnat, E.P.; Meissner, T.; Wentz, F.J.; Kao, H.-Y.; Lagerloef, G.; Lee, T. Status of Aquarius and salinity continuity. *Remote Sens.* **2018**. submitted.
3. Kao, H.-Y.; Lagerloef, G.S.E. Salinity fronts in the tropical Pacific Ocean. *J. Geophys. Res. Oceans* **2015**, *120*, 1096–1106. [CrossRef] [PubMed]
4. Lee, T.; Lagerloef, G.; Gierach, M.M.; Kao, H.-Y.; Yueh, S.; Dohan, K. Aquarius reveals salinity structure of tropical instability waves. *Geophys. Res. Lett.* **2012**, *39*, L12610. [CrossRef]
5. Grodsky, S.A.; Reul, N.; Lagerloef, G.; Reverdin, G.; Carton, J.A.; Chapron, B.; Quilfen, Y.; Kudryavtsev, V.N.; Kao, H.-Y. Haline hurricane wake in the Amazon/Orinoco plume: AQUARIUS/SACD and SMOS observations. *Geophys. Res. Lett.* **2012**, *39*, L20603. [CrossRef]
6. Melnichenko, O.V.; Amores, A.; Maximenko, N.A.; Hacker, P.; Potemra, J.T. Signature of mesoscale eddies in satellite sea surface salinity data. *J. Geophys. Res. Oceans* **2017**, *122*, 1416–1424. [CrossRef]
7. Kao, H.-Y.; Lagerloef, G.; Lee, T.; Melnichenko, O.; Hacker, P. Aquarius Salinity Validation Analysis; Data Version 5.0. 2017. Available online: Ftp://podaac-ftp.jpl.nasa.gov/allData/aquarius/docs/v5/AQ-014-PS-0016_AquariusSalinityDataValidationAnalysis_DatasetVersion5.0.pdf (accessed on 21 August 2018).
8. Meissner, T. Assessment of Uncertainties in Aquarius Salinity Retrievals. 2015. Available online: Ftp://podaac-ftp.jpl.nasa.gov/allData/aquarius/docs/v4/AQ-014-PS-0017_AquariusATBD_uncertainties_Addendum5_DatasetVersion4.0.pdf (accessed on 21 August 2018).
9. Le Vine, D.M.; Lagerloef, G.S.E.; Colomb, F.R.; Yueh, S.H.; Pellerano, F.A. Aquarius: An instrument to monitor sea surface salinity from Space. *IEEE Trans. Geosci. Remote Sens.* **2007**, *45*, 2040–2050. [CrossRef]
10. Piepmeier, J. Aquarius Radiometer Post-Launch Calibration for Product Version 2. 2013. Available online: Ftp://podaac-ftp.jpl.nasa.gov/allData/aquarius/docs/v2/AQ-014-PS-0015_AquariusInstrumentCalibratrionDescriptionDocument.pdf (accessed on 21 August 2018).
11. Lagerloef, G.; Kao, H.-Y.; Hacker, P.; Hackert, E.; Chao, Y.; Hilburn, K.; Meissner, T.; Yueh, S.; Liang, H.; Lee, T. 2013: Aquarius Salinity Validation Analysis; Data Version 2.0. Available online: https://aquarius.umaine.edu/docs/AQ-014-PS-0016_AquariusSalinityDataValidationAnalysis_DatasetVersion2.0.pdf (accessed on 21 August 2018).
12. Brasnett, B. The impact of satellite retrievals in a global sea-surface-temperature analysis. *Q. J. R. Meteorol. Soc.* **2008**, *134*, 1745–1760. [CrossRef]
13. Meissner, T.; Wentz, F.; Le Vine, D. Aquarius Salinity Retrieval Algorithm: Algorithm Theoretical Basis Document (ATBD) End of Mission Version. 2017. Available online: Ftp://podaac-ftp.jpl.nasa.gov/allData/aquarius/docs/v5/AQ-014-PS-0017_Aquarius_ATBD-EndOfMission.pdf (accessed on 21 August 2018).
14. Chassignet, E.P.; Hurlburt, H.E.; Metzger, E.J.; Smedstad, O.M.; Cummings, J.; Halliwell, G.R.; Bleck, R.; Baraille, R.; Wallcraft, A.J.; Lozano, C.; et al. GODAE: Global Ocean Prediction with the HYbrid Coordinate Ocean Model (HYCOM). *Oceanography* **2009**, *22*, 64–75. [CrossRef]
15. Lilly, J.; Lagerloef, G. Aquarius Level 3 Processing Algorithms Theoretical Basis Document. 2008. Available online: Ftp://podaac-ftp.jpl.nasa.gov/allData/aquarius/docs/v2/AquariusLevel3_GriddingSmoothingPaper_Lilly&Lagerloef2008.pdf (accessed on 21 August 2018).
16. Meissner, T.; Wentz, F.J.; Le Vine, D. The salinity retrieval algorithm for the NASA Aquarius Version 5 and SMAP Version 3 Releases. *Remote Sens.* **2018**, *10*, 1121. [CrossRef]
17. Boutin, J.; Chao, Y.; Asher, W.E.; Delcroix, T.; Drucker, R.; Drushka, K.; Kolodziejczyk, N.; Lee, T.; Reul, N.; Reverdin, G.; et al. Satellite and in situ salinity: Understanding near surface stratification and sub-footprint variability. *Bull. Am. Meteorol. Soc.* **2016**, *97*, 1391–1407. [CrossRef]
18. Vinogradova, N.T.; Ponte, R.M. Assessing temporal aliasing in satellite-based surface salinity measurements. *J. Atmos. Ocean. Technol.* **2012**, *29*, 1391–1400. [CrossRef]

19. Vinogradova, N.T.; Ponte, R.M. Small-scale variability in sea surface salinity and implications for satellite-derived measurements. *J. Atmos. Ocean. Technol.* **2013**, *30*, 2689–2694. [CrossRef]
20. Lee, T. Consistency of Aquarius sea surface salinity with Argo products on various spatial and temporal scales. *Geophys. Res. Lett.* **2016**, *43*, 3857–3864. [CrossRef]
21. Lee, T.G.; Lagerloef, H.-Y.; Kao, M.; Willis, J.; Gierach, M. The influence of salinity on tropical Atlantic instability waves. *J. Geophys. Res. Oceans* **2014**, *119*, 8375–8394. [CrossRef]
22. Grunseich, G.; Subrahmanyam, B.; Wang, B. The Madden-Julian oscillation detected in Aquarius salinity observations. *Geophys. Res. Lett.* **2013**, *40*, 5461–5466. [CrossRef]
23. Guan, B.; Lee, T.; Halkides, D.; Waliser, D.E. Aquarius surface salinity and the Madden-Julian Oscillation: The role of salinity in surface layer density and potential energy. *Geophys. Res. Lett.* **2014**, *41*, 2858–2869. [CrossRef]
24. Qu, T.D.; Yu, J.Y. ENSO indices from sea surface salinity observed by Aquarius and Argo. *J. Oceanogr.* **2014**, *70*, 367–375. [CrossRef]
25. Du, Y.; Zhang, Y. Satellite and Argo observed surface salinity variations in the Tropical Indian Ocean and their association with the Indian Ocean Dipole mode. *J. Clim.* **2015**, *28*, 695–713. [CrossRef]

remote sensing

MDPI

Article

Improving SMOS Sea Surface Salinity in the Western Mediterranean Sea through Multivariate and Multifractal Analysis

Estrella Olmedo [1,*], Isabelle Taupier-Letage [2], Antonio Turiel [1] and Aida Alvera-Azcárate [3]

1 Department of Physical Oceanography, Institute of Marine Sciences, CSIC, Barcelona Expert Center, Pg. Marítim 37-49, Barcelona E-08003, Spain; turiel@icm.csic.es

2 Aix Marseille Université, CNRS/INSU, Université de Toulon, IRD, Mediterranean Institute of Oceanography (MIO), F-83507 La Seyne, Marseille, France; isabelle.taupier-letage@univ-amu.fr

3 Département d'astrophys., géophysique et océanographie (AGO), GeoHydrodynamics and Environment Research (GHER), Université de Liège, Allée du 6 Août, 17 Sart Tilman, Liège 4000, Belgium; a.alvera@ulg.ac.be

* Correspondence: olmedo@icm.csic.es

Received: 23 January 2018; Accepted: 17 March 2018; Published: 20 March 2018

Abstract: A new methodology using a combination of debiased non-Bayesian retrieval, DINEOF (Data Interpolating Empirical Orthogonal Functions) and multifractal fusion has been used to obtain Soil Moisture and Ocean Salinity (SMOS) Sea Surface Salinity (SSS) fields over the North Atlantic Ocean and the Mediterranean Sea. The debiased non-Bayesian retrieval mitigates the systematic errors produced by the contamination of the land over the sea. In addition, this retrieval improves the coverage by means of multiyear statistical filtering criteria. This methodology allows obtaining SMOS SSS fields in the Mediterranean Sea. However, the resulting SSS suffers from a seasonal (and other time-dependent) bias. This time-dependent bias has been characterized by means of specific Empirical Orthogonal Functions (EOFs). Finally, high resolution Sea Surface Temperature (OSTIA SST) maps have been used for improving the spatial and temporal resolution of the SMOS SSS maps. The presented methodology practically reduces the error of the SMOS SSS in the Mediterranean Sea by half. As a result, the SSS dynamics described by the new SMOS maps in the Algerian Basin and the Balearic Front agrees with the one described by in situ SSS, and the mesoscale structures described by SMOS in the Alboran Sea and in the Gulf of Lion coincide with the ones described by the high resolution remotely-sensed SST images (AVHRR).

Keywords: sea surface salinity; remote sensing; mediterranean sea; smos; alboran sea; data processing; quality assessment

1. Introduction

The Mediterranean Sea is a hot spot for climate change [1]. The water balance in the basin is characterized by an excess of evaporation over precipitation and river runoff, which is compensated by the entrance of fresher water from the Atlantic. This Atlantic water (AW), which spreads through the Mediterranean Sea, determines the surface circulation [2]. In the Algerian Basin, AW forms an unstable current that generates fresh-core coastal eddies that propagate downstream [3]. The eddy activity in the region enhances the mixing of the recently entered fresher AW with the saltier resident ones, strongly affecting the spatial distribution of salinity and, therefore, playing a major role in the surface circulation of the Mediterranean Sea ([4,5] and references therein).

Additionally, during winter, in the Northwestern Mediterranean, deep water convection occurs under the influence of dry and cold northerly winds. The result is the Western Mediterranean Deep

Water, which contributes to the Mediterranean thermohaline circulation. The budget of salt involved in the Deep Water Formation (DWF) remains an open question (e.g., [6,7]), which is addressed by the HYMEX (HYdrological cycle in the Mediterranean EXperiment [8]) program [9].

Reliable estimates of budgets are indeed critically needed to improve the Mediterranean climate evolution simulations [10–12]. Finally, DWF also impacts the ecosystems: the associated winter vertical mixing is responsible for the nutrients enrichment of the surface layer, and therefore contributes to the following spring bloom ([13,14] and references therein).

In such a context, data from Soil Moisture and Ocean Salinity (SMOS) European Space Agency (ESA)'s mission [15–17] spanning more than eight years can help to gain a better understanding of the Sea Surface Salinity (SSS) dynamics in the Mediterranean Sea. Unfortunately, this critical area is strongly affected by Radio Frequency Interference (RFI) and systematic biases due to the coast contamination (also called Land-Sea Contamination, (LSC)). Both effects make it very difficult to retrieve SMOS SSS in these areas.

Recently, several studies have been presented showing that retrieving SMOS SSS in the Mediterranean is actually feasible. The first one [18] considers the domain North Atlantic Ocean and the Mediterranean Sea. They start from the official Level 2 Ocean Salinity (L2OS) product (version 550), and, by using DINEOF (Data Interpolating Empirical Orthogonal Functions), the authors reconstruct missing data using a truncated EOF basis. The methodology allows removing part of the biases (for example, the ones associated with the edge of the swath). However, the biases present near land masses (LSC), large scale biases and latitudinal biases are not corrected because they are persistent and retained in dominant EOFs.

In the second one [19], the probability of detecting and following eddies in the Algerian Basin from the SMOS SSS is computed. They also use the SMOS SSS to reconstruct currents in the basin. In this case, the SMOS SSS are retrieved from the Level 1 Brightness Temperature (TB) product (v620) following the methodology proposed in [20], which aims to mitigate the spatial systematic biases (such as the LSC). Furthermore, this methodology applies statistical filtering criteria, which allow for a better removal of those TB degraded by RFI. After that, they improve the spatial and temporal resolution of the products by using the multifractal vectorial fusion algorithm proposed in [21]. Despite the fact that the resulting SMOS SSS maps in the Mediterranean display coherent spatial structures (in comparison with other remotely sensed variables such as the sea surface temperature or height), the authors pointed out that some biases still remain in the Mediterranean when they compare with Argo floats.

The purpose of this work is to improve the quality of the SMOS SSS in the Mediterranean Sea by: (1) better retrieval and filtering of SMOS SSS [20]; (2) deletion of time-dependent residual biases by means of DINEOF [18], and (3) improvement of the spatial and temporal resolutions by use of multifractal fusion [21]. The structure of the work is the following: in Section 2, the different data sets used in our study are described. A brief description of the methodologies is presented in Section 3. The results are shown in Section 4. Finally, Section 5 includes the main findings and lists the prominent conclusions of our work.

2. Data Sets

2.1. Sea Surface Salinity SMOS Data

Three different SMOS data sets have been used in this study:

- Level 1 B product: the input data for the computation of the product in this work is the Level 1 Brightness Temperature product (L1B v620). This product is the output of the image reconstruction of the SMOS measurements and consists of the Fourier components of brightness temperatures in the antenna polarization reference frame. The latency of the products is 6–8 h. The L1B TB product is distributed by the ESA and it is freely available in [22].

- Objectively analyzed SMOS SSS maps provided by the Barcelona Expert Center: the global advanced debiased non-Bayesian L3 9-day SSS maps at 0.25 × 0.25° available at [23] have been used for assessing the improvements of our proposed methodology at Level 3.
- Level 4 SMOS SSS maps provided by the Barcelona Expert Center: The global L4 advanced SSS maps at 0.05 × 0.05° available at [23] have been used also for assessing the proposed methodology at Level 4, after applying the multifractal fusion with the OSTIA SST maps.

2.2. Sea Surface Salinity In Situ Data

Three different sources of in situ SSS data have been used in this study:

- Argo SSS: SSS data from Argo floats have been used in Section 3.2 for the characterization of the bias of the binned SMOS SSS products. After that, Argo data have been also compared with the resulting SMOS products in Section 4.1. The collocation of SMOS and Argo SSS has been performed as follows: we compare the uppermost SSS measurement provided by the Argo profile at the instant t_0 with the SMOS SSS field given by the 9-day map ($t_0 \pm 4$ days). The Argo floats present errors on the conductivity records when measurements are done shallower than 0.5 m due to air bubbles. Therefore, those records are not considered for the analysis. Additionally, Argo SSS deeper than 10 m have not been considered in our study.
- SSS time series from six moorings located in the Western Mediterranean have been considered as part of the independent validation process described in Section 4.2 (see their location in Figure 1):
    - Four moorings are operated by Puertos del Estado ([24]): the moorings number 61,198, 61,417, 61,281 and 61,280.
    - Two moorings are operated by SOCIB (Sistema de Observacion y prediccion Costero de las Islas Baleares): the moorings number 61,141 and 61,499. Data before 2016 have been collected from [25,26]. Data of 2016 have been collected directly from SOCIB website:
        * Data in the Ibiza chanel is provided at [27]
        * Data in the Palma Bahia is providad at [28]

  The collocation between mooring and SMOS SSS has been computed as follows: we have only considered data flagged as "good-quality". In addition, salinity values out of the range [36:39] PSU (Practical Salinity Units) have been removed from the comparison. A sliding averaging window of nine days has been applied to the mooring data in order to compare both data (satellite and mooring) in a more similar temporal scale. Therefore, the mooring acquisition at time t_0, which has been filtered with a centered 9-day window, has been compared with the SMOS SSS maps computed from the same 9-day period.
- Data from the TRANSMED system thermosalinometer, ([29]) on board the MV Marfret Niolon [30] have been used in this study as independent source of data for validation in Section 4.3. Sea Surface Temperature (TRANSMED SST) and Sea Surface Salinity (TRANSMED SSS) are recorded underway at ∼3 m deep, during weekly trips between Marseilles, France, and Algeria. The data has been post-processed following water samples results and yearly recalibration of the thermosalinometers. In this case, the collocation strategy between satellite and in situ data is the same as the one used for the Argo comparison: the TRANSMED SSS acquired at time t_0 has been compared with the SMOS SSS 9-day map with t_0 the mid-day of the 9-day period. Each TRANSMED SSS measurement has been compared with the SMOS SSS corresponding to the cell 0.25° × 0.25° (or 0.05° × 0.05° in the case of L4) corresponding to the TRANSMED SSS location.

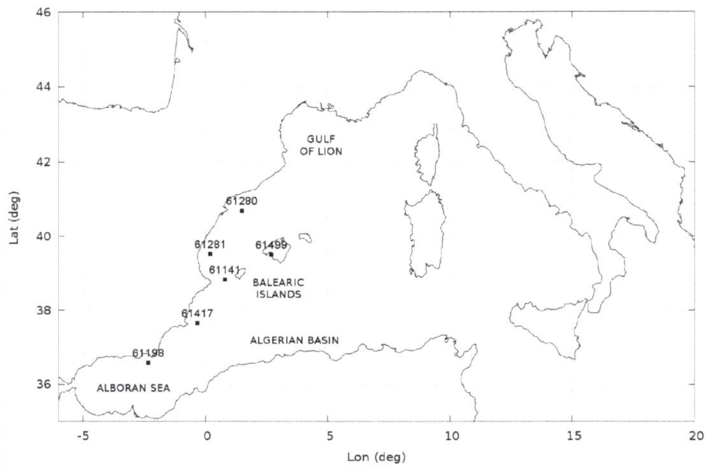

Figure 1. Location of the moorings that have been used for validation.

2.3. Sea Surface Temperature Data

Two different sources of Sea Surface Temperature have been used:

- The $0.05^\circ \times 0.05^\circ$ daily Operational Sea Surface Temperature and Sea Ice Analysis (OSTIA) product (see [31]) has been used for the generation of the SMOS SSS L4 products (hereafter OSTIA SST). The OSTIA system is part of the Group for High Resolution Sea Surface Temperature (GHRSST). Access to OSTIA data is available at [25]. The OSTIA output is a daily global coverage combined SST and sea-ice concentration product on a 1/20deg. grid, based on measurements from several satellite and in situ SST data sets. OSTIA uses SST data in the common format developed by GHRSST and makes use of the uncertainty estimates and auxiliary fields as part of the quality control and analysis procedure. Satellite derived sea ice products from the EUMETSAT Ocean and Sea Ice Satellite application Facility (OSI-SAF) provide sea-ice concentration and edge data to the analysis system. After quality control of the SST observations, a bias correction is performed using ATSR-2/AATSR data as a key component. To provide the final SST analysis, a multi-scale optimal interpolation (OI) is performed using the previous analysis as the basis for a first guess field.
- Images of Brightness Temperature (Advances Very High Resolution Radiometer (AVHRR) channel 4) from satellites National Atmospheric and Oceanic Administration (NOAA) / AVHRR sensors at 1×1 km have been used in Section 4.4 as reference data set in a qualitative assessment of the resulting spatial structures of the SMOS SSS products. Brightness temperature images (hereafter AVHRR images) are produced by the SATMOS (CNRS INSU/MeteoFrance) at the CMS Lannion from NOAA/AVHRR images, and processed at MIO (Available online: [32] (accessed on 19 March 2018)), (see [33]). Brightness Temperatures are relative temperatures, and the color scale is stretched to enhance the structures. These images do not include any compositing, and hence provide the best visualisation of the structures. As a result, the color scale is not relevant and it will not be included.

3. Methods

This section summarizes the steps realized to obtain fields of SSS over the North Atlantic Ocean and the Mediterranean Sea using a combination of a debiased non-Bayesian retrieval, DINEOF and multifractal fusion. The starting point of our study is the Level 1 Brightness Temperatures (L1B v620) product provided by the ESA.

3.1. Debiased Non-Bayesian Retrieval of SSS

The retrieval of SSS from SMOS TB is quite challenging in areas affected by the presence of systematic biases such as LSC or RFI. The standard retrieval scheme is based on a Bayesian approach, which is based in a Maximum Likelihood Estimator of the geophysical parameters obtained by assuming that the errors on TBs are independent and Gaussian-distributed, a hypothesis that poorly adapts to real conditions in the Mediterranean Sea and precludes obtaining quality retrievals in that region. A different approach is provided by the non-Bayesian approach [20,34], in which no hypothesis about the distribution or the correlation of the errors is made. Thus, each single TB measurement is used to derive a noisy estimate of SSS. In [20], these retrievals are later classified according to their geographical (latitude/longitude coordinates) and orbital characteristics (position in the antenna, overpass direction) to correct common biases. This classification allows to define a SMOS-based climatology for each acquisition condition which allows removing systematic biases and also defining proper filtering criteria (based on the statistical properties of each class of salinities), which allow discarding those measurements with a deviation from the reference value that cannot be explained by the radiometric variability. Therefore, SMOS-based anomalies are computed by subtracting their corresponding SMOS-based climatology to each retrieval. The debiased SSS field is obtained when an annual SSS reference (the annual World Ocean Atlas 2013 SSS at $0.25 \times 0.25°$ ([35]) is considered) is added to the SMOS-based anomalies. For each satellite overpass, at every point (i.e., each dwell line), the corrected measurements are thus averaged together to obtain a non-Bayesian L2 SSS. Contrary to the Bayesian approach, in which either the full dwell line is accounted for or discarded, the non-Bayesian approach allows using parts of a dwell line less affected by artifacts, once corrected for systematic biases. The non-Bayesian method allows therefore to retrieve SSS values in the Mediterranean sea.

We have used this methodology for the retrieval of debiased SMOS SSS. Daily 9-day maps of SMOS SSS have been generated by averaging those debiased SMOS SSS in a rectangular grid of $0.25 \times 0.25°$. After applying the debiased methodology, the SMOS SSS product still contains residual time-dependent biases. In [20], the mitigation of the time-dependent biases is done by means of the removal of the mean value of the SMOS-based anomaly of each 9-day map. This time-dependent bias correction provides quite accurate global maps (as shown in [20]). However, in some specific more challenging regions, such as the Mediterranean Sea, some residual time-dependent biases still remain (as observed in [19]). In this work, we propose a different approach. We have not applied this time correction to our 9-day maps, but we try to remove the residual biases by means of Data Interpolating Empirical Orthogonal Functions.

3.2. Mitigation of Time-Dependent Biases: Removal of EOFs from the DINEOF Basis

DINEOF ([36,37]) has been used to analyze the L3 SSS binned fields obtained using the debiased non-Bayesian approach detailed in Section 3.1. DINEOF is a technique to reconstruct missing data and reduce noise in geophysical data sets using an EOF basis. The most dominant EOFs, calculated iteratively to overcome the presence of missing data, are used to infer the missing information. As the amount of missing data in the 9-day L3 binned SSS data set used in this work is very low (5.3%), the main aim of the use of DINEOF with these data is to reduce noise and remove non-physical patterns. A filter has been applied to the covariance matrix, following [38]. This filter allows for ensuring a smooth transition between subsequent images, by providing information on the time step between them, which is otherwise not considered in the EOF decomposition. The filter length is of 1.98 days.

DINEOF has been successfully used with several variables and domains. Among others, Ref. [39] used sea surface temperature, chlorophyll-a and winds in a multivariate approach over the Gulf of Mexico, Ref. [40] analyzed total suspended matter in the North Sea, Ref. [41] reconstructed turbidity in the North Sea from a geostationary satellite, and, more recently, Ref. [18] worked with SSS from SMOS in the North Atlantic Ocean and the Mediterranean Sea.

Six years (2011–2016) of 9-day binned L3 SMOS SSS maps at 0.25° × 0.25° have been decomposed into 30 EOFs. Our spatial domain contains part of the North Atlantic Ocean and the full Mediterranean Sea. Let us call our spatial domain $\Omega = [26°, 50°]N \times [50°W, 38°E]$. Thus, the initial binned SMOS SSS ($SSS_{bin}(t, \varphi, \lambda)$) has been approximated by:

$$SSS_{bin}(t, \varphi, \lambda) \approx SSS_{mean}(t, \varphi, \lambda) + \sum_{i=1}^{30} s(i) * v(t, i) * u(\varphi, \lambda, i), \quad (1)$$

where φ, λ represent the latitude and longitude, respectively, t represents the time variable, $v(t, i)$ is the time-dependent part of the EOF decomposition, $u(\varphi, \lambda, i)$ is the spatial-dependent part of the EOF decomposition and $s(i)$ are the singularity values. For the same time period (2011–2016), we have collocated Argo float SSS data with the SMOS SSS maps (as described in Section 2.2). Then, at every day, t_0, we have a set of locations with Argo and SMOS SSS. Let us call $\Theta(t_0) \subset \Omega$ the set of locations where Argo and SMOS SSS have acquisitions at time t_0. We define the following time series:

$$\epsilon_0(t) = \sum_{\varphi,\lambda \in \Theta(t)} \left(SSS_{bin}(t, \varphi, \lambda) - SSS_{Argo}(t, \varphi, \lambda) \right). \quad (2)$$

If we assume that Argo covers the domain uniformly, this time series shows the time-dependent bias of the SMOS SSS. In particular, we can approximate it by:

$$\epsilon_0(t) \approx \sum_{\varphi,\lambda \in \Theta(t)} \left\{ \left(SSS_{mean}(t, \varphi, \lambda) + \sum_{i=1}^{30} s(i) * v(t, i) * u(\varphi, \lambda, i) \right) - SSS_{Argo}(t, \varphi, \lambda) \right\}. \quad (3)$$

We want to remove the EOFs that describe the majority of the time-dependent bias. Thus, we compute the correlation between $\epsilon(t)$ and each one of the 30 time series corresponding to:

$$\omega_i(t) = \sum_{\varphi,\lambda \in \Theta(t)} s(i) * v(t, i) * u(\varphi, \lambda, i), \quad (4)$$

such that each ω_i corresponds to the contribution of the i-th EOF to the SMOS SSS at the collocated locations with Argo SSS. Therefore, the $\omega_i(t)$ with the highest correlation with $\epsilon_0(t)$ will describe the highest part of the residual time-dependent bias of the initial SMOS SSS data. Let us call i_0 the EOF that provides the highest correlation with $\epsilon_0(t)$. We remove the EOF i_0 from the initial approximation and redefine:

$$\epsilon_1(t) = \epsilon_0(t) - \omega_{i_0}(t). \quad (5)$$

Then, we iterate the process. We compute the correlation of all the $\omega_i(t)$ (except for $\omega_{i_0}(t)$, which has been removed) with $\epsilon_1(t)$. The $w_i(t)$ with the highest correlation will describe the highest part of the residual time-dependent bias. We iterate until the correlation is not significant, or when the residual $\epsilon_n(t)$ is too small. In our case, as we are interested in the characterization and correction of the seasonal and other eventual time-dependent biases, we expect that the signal of this error will be concentrated in a few EOFs. The final selection of the number of EOFs to be removed has been done by considering the partial solutions obtained in each iteration (and shown in Table 1 and Figure 2). We explain it in more detail below.

In Table 1, the different iteration steps are shown. In the first step (first row), the mean value of the initial residual $\epsilon_0(t)$ is −0.07 PSU (fourth column) and the corresponding standard deviation (fifth column) is 0.23 PSU. The EOF that better correlates with ϵ_0 is the EOF number 2 (second column) with a correlation value of 0.69 (third column). The first plot of Figure 2 shows the time evolution of $\epsilon_0(t)$ (green squares) and $\omega_2(t)$ (violet triangles). In the second row of Table 1, the results of the second iteration are shown: the mean value of the residual $\epsilon_1(t) = \epsilon_0(t) - \omega_2(t)$ and the corresponding standard deviation are −0.04 PSU and 0.17 PSU (respectively) (the fourth and fifth columns of the second row of Table 1). The EOF with the largest correlation with $\epsilon_1(t)$ is the EOF number 1. In this case, the correlation between $\omega_1(t)$ and $\epsilon_1(t)$ is 0.46. The second plot of Figure 2 shows the time evolution of $\epsilon_1(t)$ (green squares) and $\omega_1(t)$ (violet triangles)). In the third iteration, the mean value and standard deviation of the residual $\epsilon_2(t) = \epsilon_1(t) - \omega_1(t)$ are -0.03 PSU and 0.15 PSU (respectively). The EOF with the largest correlation (0.27) is the EOF number 6. The third plot of Figure 2 shows the time evolution of $\epsilon_2(t)$ and the $\omega_6(t)$.

We have removed only these three EOFs. As observed in Table 1, from the third iteration onward (the table only includes the first 10 steps), the resulting residuals are almost the same, and they do not improve the correlation.

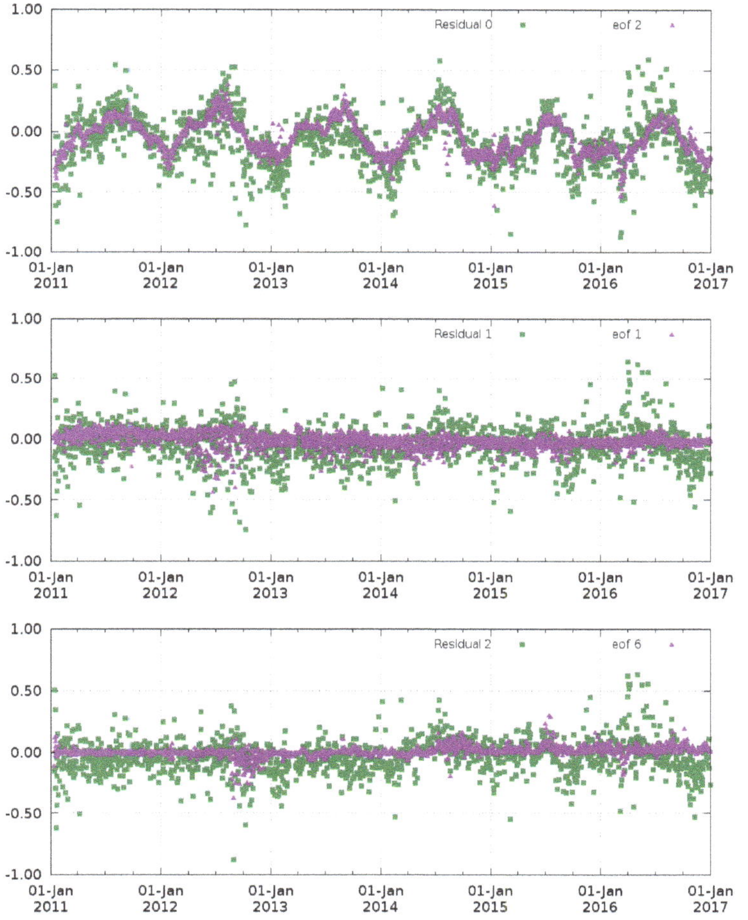

Figure 2. Time evolution of the mean difference between SMOS and Argo SSS ($\epsilon(t)$ green squares) and the contribution to the SSS of the EOF, which provides the largest correlation with the residual ($\omega(t)$ violet triangles). On the top the results corresponding to the first iteration step; in the middle, the ones corresponding to the second iteration; and, in the bottom, the results of the third and last iteration.

Table 1. Results of the first 10 iterations from the time-dependent bias deletion from EOF decomposition. The first column shows the residual remaining in each iteration. In the second column, the number of the EOF that provides the largest correlation with the residual in each step. The correlation between the different EOFs contributions to the SMOS SSS and the time-dependent bias is shown in the fifth column. In the third and fourth columns, the mean and standard deviation (respectevely) of the residual error before the correction.

Residual	EOF with Largest Correlation	Correlation	Mean	Std
ϵ_0	2	0.69	−0.07	0.23
$\epsilon_1 = \epsilon_0 - \omega_2$	1	0.46	−0.04	0.17
$\epsilon_2 = \epsilon_1 - \omega_1$	6	0.27	−0.03	0.15
$\epsilon_3 = \epsilon_2 - \omega_6$	3	0.28	−0.05	0.15
$\epsilon_4 = \epsilon_3 - \omega_3$	5	0.31	−0.04	0.14
$\epsilon_5 = \epsilon_4 - \omega_5$	7	0.21	−0.05	0.14
$\epsilon_6 = \epsilon_5 - \omega_7$	8	0.20	−0.05	0.14
$\epsilon_7 = \epsilon_6 - \omega_8$	4	0.19	−0.05	0.13
$\epsilon_8 = \epsilon_7 - \omega_4$	27	0.17	−0.05	0.13
$\epsilon_9 = \epsilon_8 - \omega_{27}$	9	0.16	−0.05	0.13

3.3. Reduction of White Noise by Means of Objective Analysis

We have applied a scheme of objective analysis to the resulting debiased and time-corrected L3 maps. The scheme is the one described in [20], which is also the same applied for the generation of World Ocean Atlas [35]. The algorithm consists of averaging the binned SSS fields in three iterative steps, such that in each step a different radius of influence is considered. In our case, the three different radii are 175, 125 and 75 km, respectively.

3.4. Improving the Spatial and Temporal Resolution by Mean of Multifractal Fusion

It has been shown that different ocean scalars possess a multifractal structure, which is due only to the action of the underlying ocean currents [42] and thus is shared by different scalars [43,44]. It has been shown that this degree of redundancy among different ocean scalars can be used to enhance the quality of one noisy ocean variable if a different, less noisy variable is known: this is the so-called multifractal fusion [45]. Multifractal fusion has been shown to be useful to increase the spatial- and temporal-resolution of ocean scalars (particularly, SMOS SSS) while, at the same time, reducing the level of noise ([21] in the case of SMOS, and [46] in the case of CO_2 satellite imagery).

The main idea of the algorithm is to assume that we have two different ocean variables measured by remote sensing means such that one of them, θ, is considerably less noisy than the other, s. The main goal of the fusion method is to use the more reliable information provided by θ to reduce the noise and improve the spatial structure of s. In our case, this is done by assuming that both θ and s have the same singularity exponents (SE).

Now, if both variables have the same SE, a relation must exist between the spatial variations of both variables because they behave the same way under changes of scale (they are ruled by the same SE). As discussed in [45], the correspondence between SE must be smooth as it cannot be an additional source of singular behavior. In particular, the relation between the gradient of s and that of θ can be expressed in terms of a smooth matrix Φ:

$$\nabla s(\vec{x}) = \Phi(\vec{x}) \nabla \theta(\vec{x}). \tag{6}$$

The estimation of the matrix $\Phi(\vec{x})$ from Equation (6) is not straightforward: this problem is ill-posed, as for each x the number of unknowns (the elements of the matrix $\Phi(\vec{x})$) is d^2 while the number of equations is d.

In [21], different strategies to estimate this matrix from real data are investigated, with the only constraint that the matrix $\Phi(\vec{x})$ must smoothly depend on $\vec{x}$. We use the scalar approach here that consists of assuming what is proportional to the identity matrix, i.e., the matrix can be simplified to a scalar a, as in [45]:

$$\nabla s(\vec{x}) = a(\vec{x})\theta(\vec{x}). \tag{7}$$

This expression can be integrated under the assumption that the gradient of a is negligible as compared to the gradients of s and θ, namely

$$s(\vec{x}) = a(\vec{x})\theta(\vec{x}) + b(\vec{x}) + \varepsilon(\vec{x}), \tag{8}$$

where the functions a, b have small gradients and $\varepsilon(\vec{x})$ is the additive error to be minimized.

4. Results

In this section, we have assessed four different SMOS SSS products:

- SMOS SSS L3 corrected products (L3 corrected): objectively analyzed 9-day maps at $0.25 \times 0.25°$ resulting from the steps described in Sections 3.1–3.3.
- SMOS SSS L3 products (L3): They are the global debiased non-Bayesian advanced products (see Section 2.1). They result from the methodologies described in Sections 3.1 and 3.3. The unique difference with respect to the corrected L3 product is the time-bias correction. Instead of applying DINEOF decomposition, these products are time-bias corrected by assuming that the mean value of the SMOS-based anomaly at each map is null (see [20] for more details).
- SMOS SSS L4 corrected products (L4 corrected): result from the multifractal fusion between the SMOS SSS L3 corrected maps and the OSTIA SST (step described in Section 3.4). They are provided at $0.05 \times 0.05°$ on a daily basis.
- SMOS SSS L4 products (L4): Multifractal fusion applied to the SMOS SSS L3 product (see Section 2.1).

4.1 Comparison with Argo Floats

Since we have used Argo floats for the generation of the time-dependent bias correction, this in situ data set is not an independent source of SSS data to be used in our assessment. However, we have included here a comparison with Argo not only for completeness, but also because our methodology uses the average in the full domain (North Atlantic and Mediterranean Sea) for removing the biases. We want to see here also the effect in the Mediterranean subdomain (which is the most challenging region in terms of processing). An independent validation is presented in the next sections. We also want to highlight that the number of Argo surface measurements available for validation in the Mediterranean Sea is low (less than 50) before 2014.

Thus, we will analyze the SMOS SSS errors with respect to Argo in two different regions:

- Full domain (DOM), which includes part of the North Atlantic and the Mediterranean Sea and is defined in $[26°, 50°]\text{N} \times [-50°, 38°]\text{E}$
- The Mediterranean Sea (MED) that is defined by the following rectangle $[30°, 44°]\text{N} \times [-6°, 30°]\text{E}$.

Figure 3 shows the time evolution of the mean difference between the L3 (green squares) and L3 corrected (violet triangles) products and Argo floats (plot in the first row and first column) in DOM and the corresponding standard deviation (in the plot of the second column in the first row of the same figure). As observed in the figure, the mean values of the differences of the L3 corrected product are closer to the 0 line than the ones of the L3 product. Additionally, the standard deviation of the difference corresponding to the L3 corrected product is lower than the one corresponding to the L3 product. The same plots for L4 products are shown in the lower row of Figure 3. The analysis of these plots leads to analogous conclusions. The corrected L4 reduces the errors with respect to the Argo measurements.

Figure 4 shows the same four plots as before but computed over the MED subdomain only. As in the case of the DOM, the corrected products are closer to the Argo measurements (both in bias and standard deviation). However, in this case, a seasonal bias is still present. Although the biases and the standard deviations in the Mediterranean basin have been noticeably reduced (see the numbers in Table 2), some residual (time-dependent) bias is still present.

Figures 5 and 6 show the spatial distribution of the mean and standard deviation of the differences between the different SMOS SSS products and Argo. As observed in Figure 5, the major reduction of the bias is obtained in the Mediterranean basin, as well as the major reduction of the standard deviation (Figure 6). In particular, the SSS in the non-corrected products in the Northwestern Mediterranean are strongly biased, while the corrected ones strongly reduce the bias in this region.

In these figures, no noticeable differences between corrected L3 and L4 products are observed. We want to also highlight that the accuracy in the Eastern Mediterranean is very poor. The Eastern Mediterranean region is strongly affected by RFI and the SMOS SSS retrieval there is even more challenging. Recent calibration and image reconstruction techniques [34,47,48] are providing promising results in these more degraded regions.

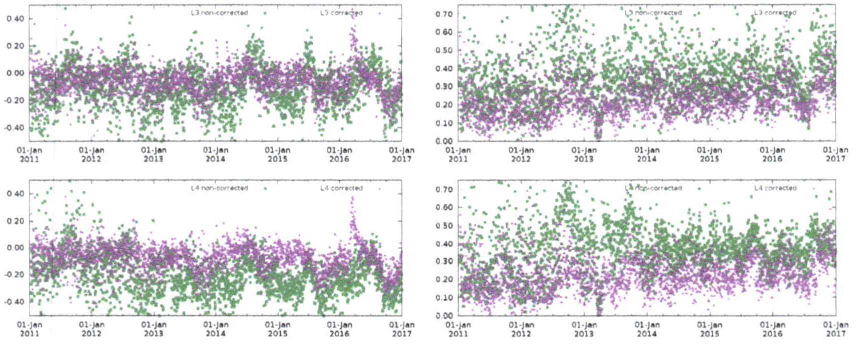

Figure 3. Time evolution of the mean differences (**left**) and standard deviation of the differences (**right**) between SMOS and Argo SSS computed over the full domain of the study. The top results correspond to the L3 products. The bottom results correspond to the L4 products. The non-corrected SMOS values are plotted with green squares and the corrected ones with violet triangles.

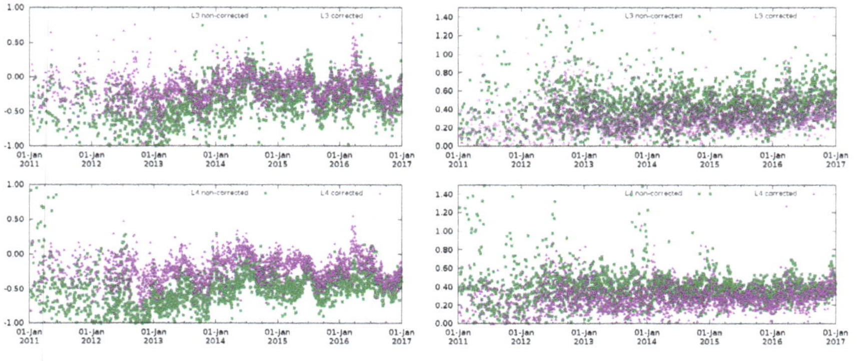

Figure 4. As in Figure 3 but for the Mediterranean Sea (MED subdomain).

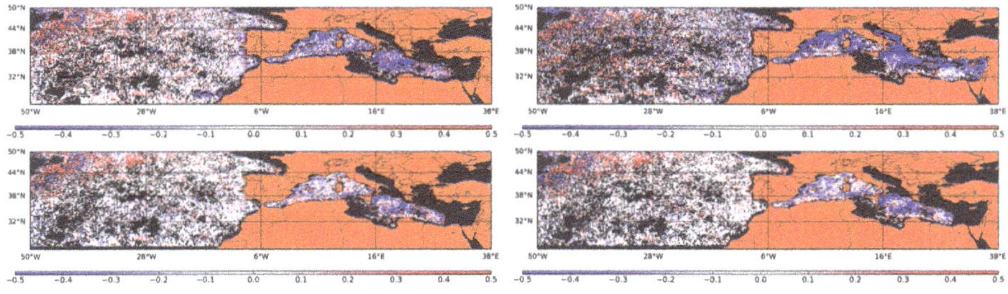

Figure 5. Spatial distribution of the differences between SMOS and Argo SSS. In the first row, the plots correspond to the non-corrected products (**left** L3 and **right** L4). In the second row, the plots correspond to the corrected products (**left** L3 and **right** L4).

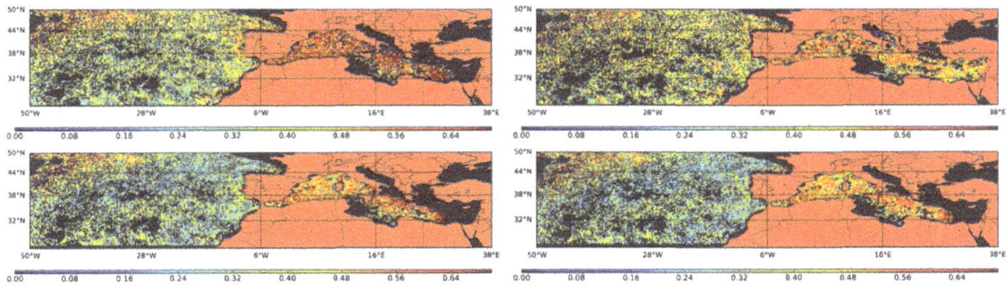

Figure 6. Spatial distribution of the standard deviation of the differences between SMOS and Argo SSS. In the first row, the plots correspond to the non-corrected products (**left** L3 and **right** L4). In the second row, the plots correspond to the corrected products (**left** L3 and **right** L4).

Table 2. Statistics of the differences between SMOS and Argo SSS. For each of the four SMOS SSS products, the mean, the standard deviation and the root mean squared are shown. The statistics are computed over two different domains: first row: DOM, which contains the Mediterranean Sea and part of the North Atlantic; and the second row: MED, which contains the Mediterranean Sea only.

	L3			L3 Corrected			L4			L4 Corrected		
	mean	std	rms	mean	std	rms	mean	std	rms	mean	std	rms
DOM	−0.12	0.35	0.41	−0.06	0.24	0.26	−0.20	0.38	0.46	−0.07	0.23	0.26
MED	−0.40	0.45	0.67	−0.16	0.34	0.43	−0.50	0.40	0.70	−0.19	0.29	0.39

4.2. Comparison with Moorings Data

In this section, we have compared the time series of the six years of SMOS SSS with the time series provided by moorings located in the coastal Western Mediterranean (see their location in Figure 1).

The spatial and temporal scales corresponding to the mooring and satellite measured SSS are very different. While the satellite observations correspond to SSS over an area of $0.25 \times 0.25°$ (or $0.05 \times 0.05°$ in the case of L4 products) and computed from an integrated time period of nine days, the mooring measurements represent SSS values of a pinpointed location at hourly time steps. These differences are even more noticeable in regions with high mesoscale activity. For example, moorings number 61,198, 61,141 61,281 and 6120 are located very close to the Mediterranean continental Spanish coast (see Figure 1). These locations are strongly affected by the general circulation (old AW coming from the North), sometimes by recent (fresher) AW coming from the South, and episodically by continental fresh water discharges. These produce complex mixing dynamics that form filaments and mesoscale and submesoscale structures with strong SSS gradients and fast dynamics. This is even more problematic in the vicinity of the Alboran Sea (mooring number 61,198), where the SSS variability amplitude can reach 2PSU within hours. Thus, in order to make both SSS data sets more comparable, we have applied a slide averaging window of nine days to the mooring data. This smoothing technique does not completely solve the temporal scale differences in the comparison between mooring and satellite SSS (as we will discuss below).

Figure 7 shows the time series corresponding to the moorings anchored in the Ibiza Channel and the Palma Bahia (buoys number: 61,141 and 61,499, respectively, see their location in Figure 1). Time series corresponding to the moorings are plotted in blue, L3 in yellow, L3 corrected in grey, L4 in green and L4 corrected in black. The in situ salinity range is relatively narrow (<1PSU) in the case of these two Balearic moorings. In their case, the improvement in the corrected SMOS SSS is not evident because the initial values of the L4 products (and L3 to a lesser extent) are already close. Indeed, although spurious oscillations are also corrected, the mean salinity value of the non-corrected SMOS products seem to be closer to the in situ than the mean salinity values of the corrected SMOS products.

Figure 8 shows the time series corresponding to the four moorings anchored on the Spanish continental coast (moorings number 61,198, 61,417 61,281 and 61,280, see their locations in Figure 1), where the in situ salinity range is higher (>1 PSU). In this case, although a slide averaging window of nine days has been applied to the mooring data, the time series of these moorings (blue lines) still present jumps with a fast dynamics that will not be captured by the satellite. When these jumps are persistent (for example January 2013 mooring number 6147), the satellite captures the jump. However, if the jump does not lead to a persistent change of the SSS (for example January 2012 mooring number 61,198), the satellite will not capture the dynamics. In spite of the temporal scale limitations, as observed in Figure 8, both corrected products (grey and black points) are closer to the in situ (blue points) than the non-corrected ones (yellow and green).

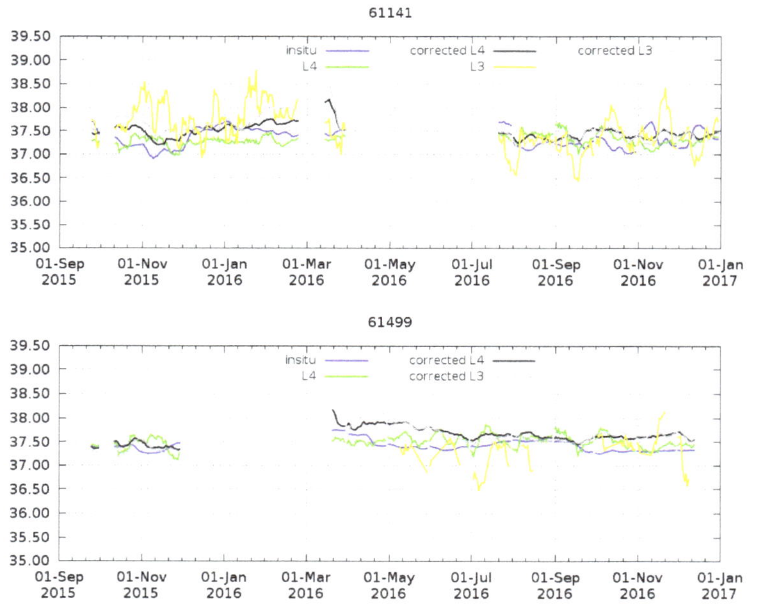

Figure 7. Moorings in the Ibiza Channel (**top**) and Palma bahia (**bottom**). Time evolution of the salinity: in blue the in situ SSS; L3 SMOS SSS in yellow; L4 SMOS SSS in green; corrected L3 SMOS SSS in grey; and corrected L4 SMOS SSS in black. The location of the buoys is shown in Figure 1.

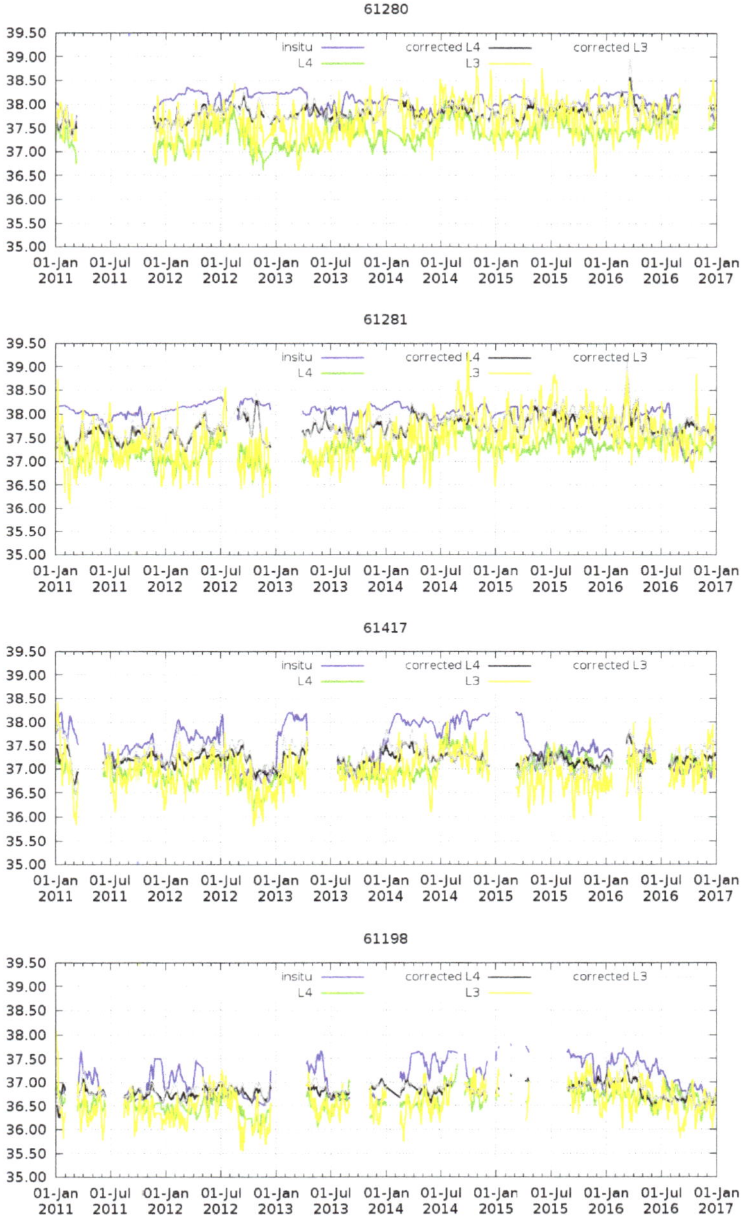

Figure 8. Moorings on the Spanish continental Mediterranean coast. Time evolution of the salinity: in blue the in situ SSS; L3 SMOS SSS in yellow; L4 SMOS SSS in green; corrected L3 SMOS SSS in grey; and corrected L4 SMOS SSS in black. The location of the buoys is shown in Figure 1.

Table 3 shows the mean and the standard deviation of the differences between SMOS SSS and in situ SSS. The comparison shows that, in general, we obtain a reduction of the bias that is more than two-fold for both L3 and L4. We also observe a reduction in the corrected SMOS SSS of the standard deviation. If we compare the results corresponding to the L3 corrected with the results corresponding to the L4 corrected products, the statistics are very similar. In general, a slight increase of the bias appears in the L4 product, which is compensated by a reduction of the standard deviation. In general, there is a residual negative bias still present in the corrected SMOS SSS products. The moorings anchored in the Baleares (61,141 and 61,499) have a different behaviour with respect to the rest of moorings. On one hand, the residual bias is positive. On the other hand, the bias in the Palma bahia (61,499) corresponding to the L4 corrected increases with respect to the bias of the L4 (as mentioned above).

This section shows that our corrections can improve the SMOS SSS retrieval over the whole range of salinity of the Western Mediterranean (36.5–38.5). However, the difference of residual offsets at the respective moorings locations has yet to be investigated.

Table 3. Differences between SMOS and in situ SSS. For each one of the four SMOS products, the mean and the standard deviation are shown. In the first column the mooring ID is shown.

Mooring ID	L3		L3 Corrected		L4		L4 Corrected	
	mean	std	mean	std	mean	std	mean	std
61,198	−0.50	0.40	−0.20	0.31	−0.46	0.35	−0.25	0.32
61,417	−0.52	0.49	−0.19	0.44	−0.46	0.46	−0.26	0.42
61,141	0.20	0.48	0.20	0.28	0.03	0.24	0.14	0.20
61,499	0.22	0.32	0.20	0.27	0.09	0.17	0.23	0.15
61,281	−0.51	0.56	−0.22	0.38	−0.73	0.36	−0.27	0.34
61,280	−0.38	0.43	−0.21	0.25	−0.66	0.34	−0.22	0.22

4.3. Comparison with TRANSMED SSS

In situ data from the TRANSMED program have been also used as an independent source of SSS data for validating the resulting SMOS products. These data cover from February 2012 to July 2014 and they are composed of SSS and SST acquired underway along repeated transects at $\sim$3 m deep along the Western Mediterranean (see Figure 9).

Table 4 shows the statistics corresponding to the four different SMOS SSS products and the in situ SSS. The time correction proposed in this work reduces the bias in both products L3 and L4 from −0.54 PSU to less than −0.1 PSU. On the other hand, a reduction of 0.12 (0.03) PSU in the standard deviation of the L3 (and L4 respectively) is also obtained.

In order to take a global look at the spatial biases of the SMOS SSS product with respect to TRANSMED SSS data, Figure 10 shows a latitudinal Hovmöller diagram of the different SSS data ordered in time. Thus, on the top of the figure, the Hovmöller corresponding to TRANSMED SSS is shown; in the second row, the Hovmöller corresponding to the L3 SSS product; in the third row, the corresponding to the L3 corrected; in the fourth, the L4; and in the last row, the L4 corrected SMOS SSS product is shown. The Hovmöller corresponding to the non-corrected SMOS SSS products are contaminated by a strong negative bias (as observed also in the statistics of Table 4). Additionally, the time correction proposed in this work has strongly mitigated some particular events that contaminated the initial SMOS product. For instance, in October and November 2012, the non-corrected plots show SSS values lower than 37 PSU practically at all the latitudes, when such low SSS values are limited to recent AW and thus confined to the southernmost part of the basin, as shown by the TRANSMED SSS. The Hovmöller plots corresponding to the corrected products show on the contrary a more similar behavior than the ones observed by the TRANSMED SSS. A similar effect is observed in October 2013 and April 2014, where spurious fresh events contaminate all the latitude range in the non-corrected Hovmöller diagrams but not in the corrected nor in the TRANSMED diagrams. In spite of the different spatial scales, which produce a more pixelated plot for SMOS, and in spite of a slight global fresh bias, which also shows the statistics of Table 4, the corrected SMOS SSS products allow for reducing the spatial noise and they also improve the depiction of the North-South SSS gradients. For instance, the fresher (blue) area moves further north around October 2013 in the TRANSMED SSS and both corrected SMOS products, whereas in March–April 2014, the salty water moves south.

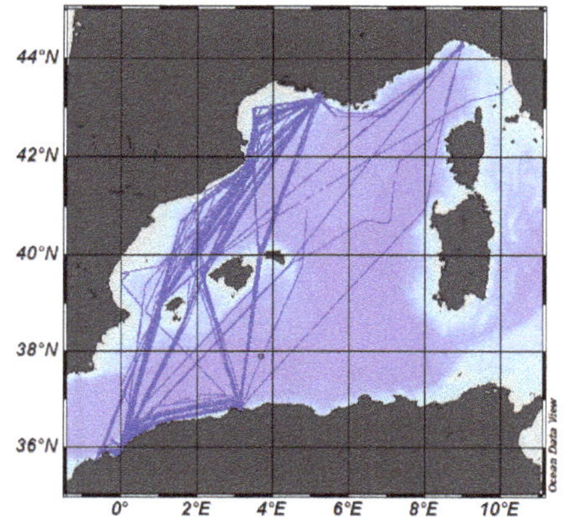

Figure 9. TRANSMED routes between 2012 and 2014.

Table 4. Differences between SMOS and TRANSMED SSS values. For each one of the four SMOS SSS products, the mean, the standard deviation and the total number of colocations between in situ and satellite used in the statistics are provided.

	L3	L3 Corrected	L4	L4 Corrected
Mean	−0.54	−0.03	−0.54	−0.09
Std	0.39	0.27	0.28	0.25
Nmeas	75,129	97,530	88150	91,113

In the northern part of the basin, the comparison between corrected SMOS SSS products and TRANSMED SSS is impaired mostly by the Rhone river plume. Its scales of variability in both space (along the shoreline or offshoreward) and time (hours under wind influence or heavy precipitating events) cannot be captured by SMOS. Further analysis on larger scales (northwestern part of the basin, season) would probably help with assessing the freshwater discharges observed by SMOS in the Gulf of Lion.

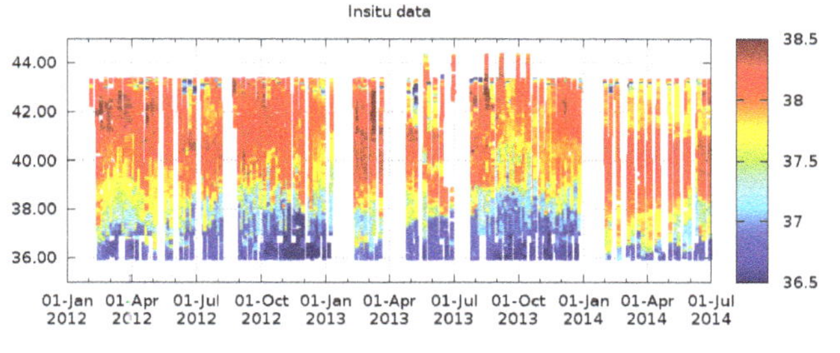

Figure 10. *Cont.*

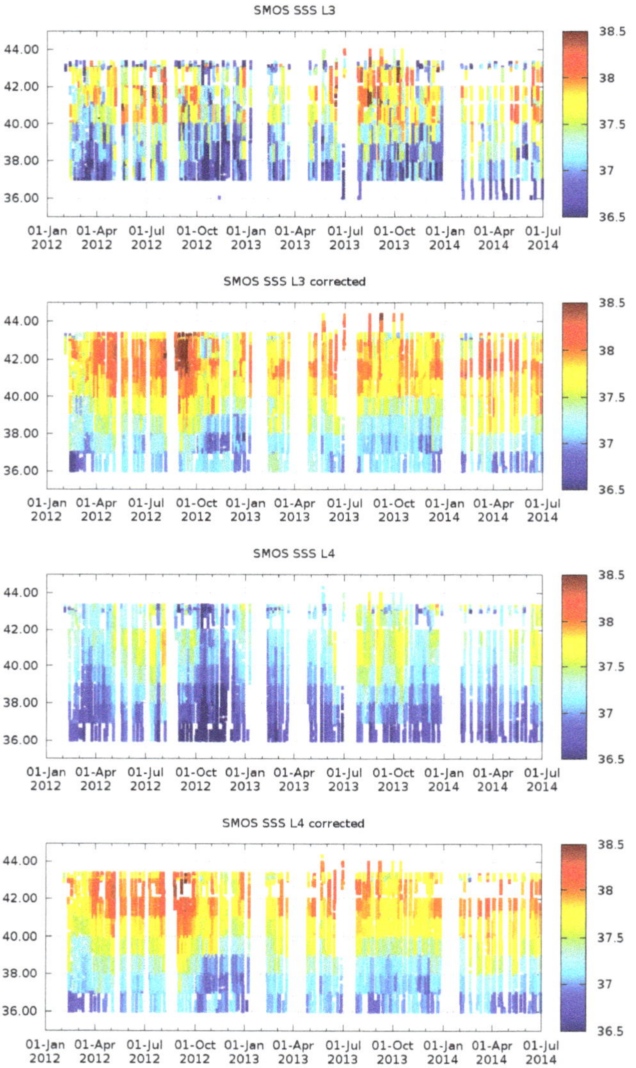

Figure 10. Latitudinal Hovmöller with the SSS values ordered in time: the first row corresponds to the TRANSMED SSS; the second row, the L3 SSS; the third row, L3 corrected; the fourth row, the L4; and the fifth shows the L4 corrected SMOS SSS values.

4.4. *Spatial Structures: Qualitative Comparison with AVHRR Images*

In the previous sections (Sections 4.1–4.3), we have shown that the corrected SMOS products have lower error than the non-corrected SMOS products (lower bias and std with respect to in situ measurements). However, the performances provided by the corrected L3 and L4 products in the comparison with in situ data are very similar. The main conclusions are that L3 typically provides a slightly lower bias, but L4 provides a slightly lower standard deviation. Therefore, the quality of both products as described by these metrics is very similar. In this section, we want to compare both SMOS

SSS corrected products with high resolution AVHRR images in order to assess the kind of spatial structures that each one describes (i.e., which is the added value of the L4 SMOS corrected product with respect to the L3 SMOS corrected one). Therefore, this section is just a qualitative comparison. We want to stress that, although the L4 are computed by means of multifractal fusion with OSTIA SST maps, in this section, we have not considered OSTIA maps, but the AVHRR images as mentioned in Section 2.3.

In Figures 11–13, AVHHR images of the Alboran Sea on 25 September of 2011, 1 March 2013 and 25 June 2014 are shown. In Figure 11, both Western (A) and Eastern (B) Alboran gyres are fully developed. Close to 1.5°W, a thermal front (C) clearly bounds the AW that enters into the Mediterranean. The corrected L3 SMOS SSS map also displays clearly the front close to the 1.5°W separating the Atlantic and the Mediterranean waters, and also recognizes the fresh water that enters from the Strait of Gibraltar. However, the spatial structure of the gyres is blurred. The corrected L4 SMOS SSS map shows clearly both gyres. In Figure 12, the Western gyre collapses. The spatial gradients shown by L4 SMOS SSS are not coherent this time, with the spatial structures shown by the AVHRR image. Indeed, during wintertime, AW is signified by higher SST (red for AW and blue for Mediterranean waters in AVHRR images). Thus, lower salinity values would have been expected in the Southern part of the SMOS SSS images, which is not the case. Until further investigations are carried out, it must be noted that gyres in the Alboran Sea are extremely variable in both space and time, and that comparison of SMOS data with instantaneous images has its limits.

In Figure 13, the Eastern gyre collapses. In this case, SMOS presents coherent values with what the AVHRR image is showing.

In Figure 14, the main feature of both the SMOS SSS L3 and L4 corrected maps is the frontal zone between the Atlantic and the Mediterranean waters north of the Baleares. Such a zonal gradient does not appear in the AVHRR image. The area of the Rhone river influence (A) is well marked on the three products, with colder SST and lower SSS west of $\sim$ 5°E in the gulf, and a tongue extending southward (the latter hardly visible though on the SMOS SSS L3 corrected map). A second colder plume extending southward off $\sim$ 6°E (B) in the AVHRR image also shows as a plume of lower SSS on the SMOS SSS L4 corrected map, but not on the SMOS SSS L3 corrected map.

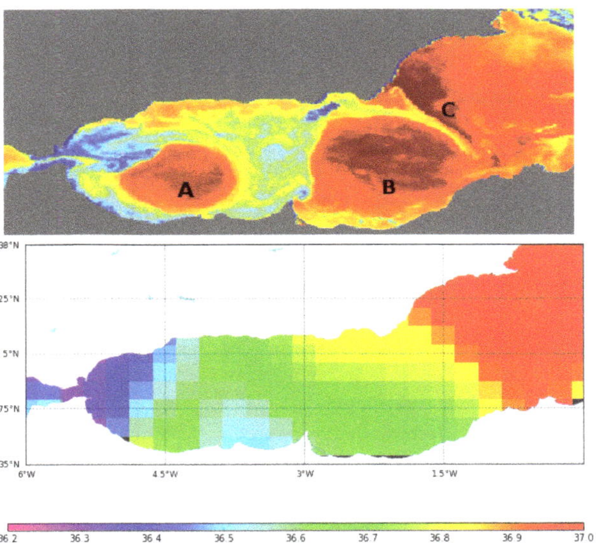

Figure 11. *Cont.*

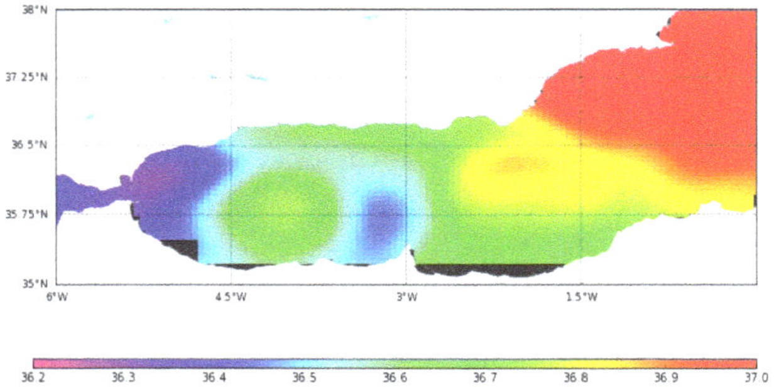

Figure 11. Images of the Alboran Sea on 25th September 2011 as observed by AVHRR satellite (**top**); the SMOS SSS L3 corrected product; and the SMOS SSS L4 corrected product.

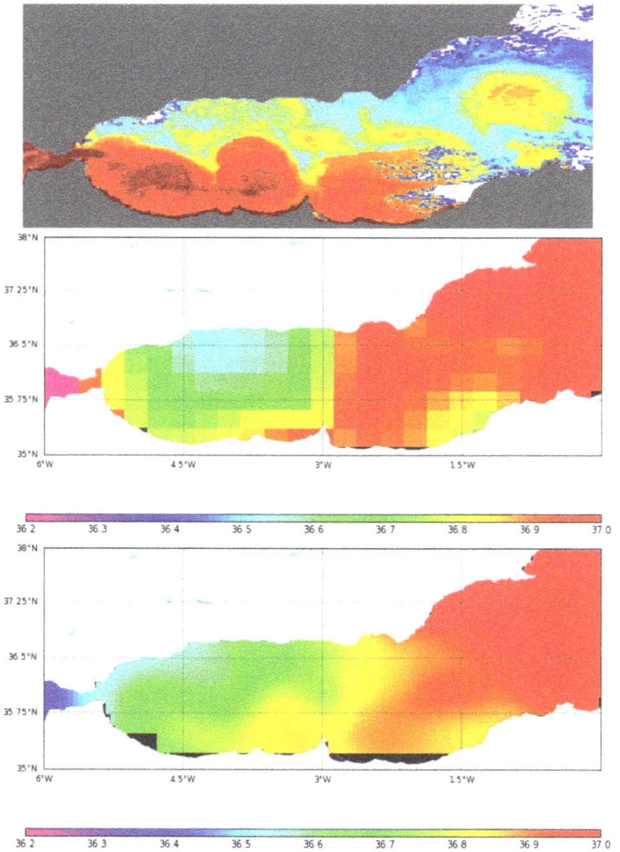

Figure 12. Images of the Alboran Sea on 1st March 2013 as observed by AVHRR satellite (**top**); the SMOS SSS L3 corrected product; and the SMOS SSS L4 corrected product.

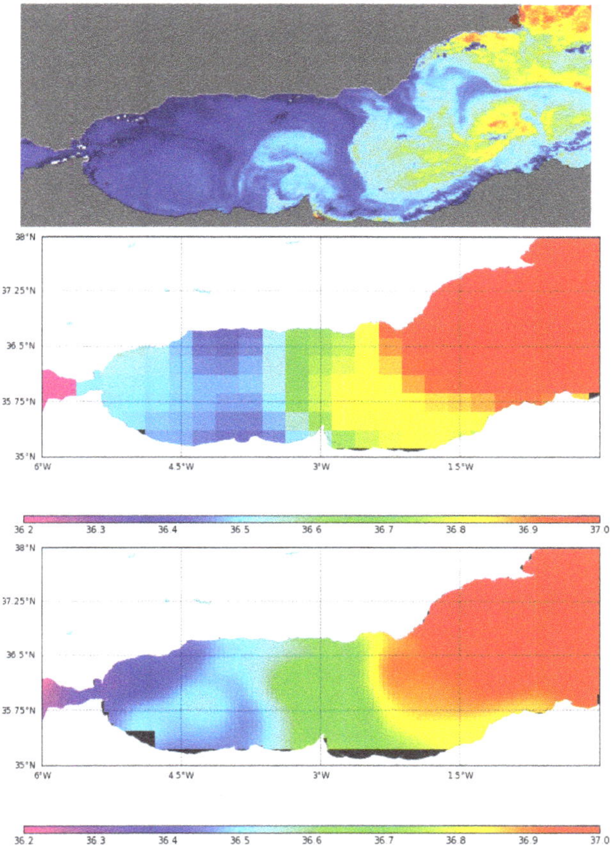

Figure 13. Images of the Alboran Sea on 25th June 2014 as observed by AVHRR satellite (**top**); the SMOS SSS L3 corrected product; and the SMOS SSS L4 corrected product.

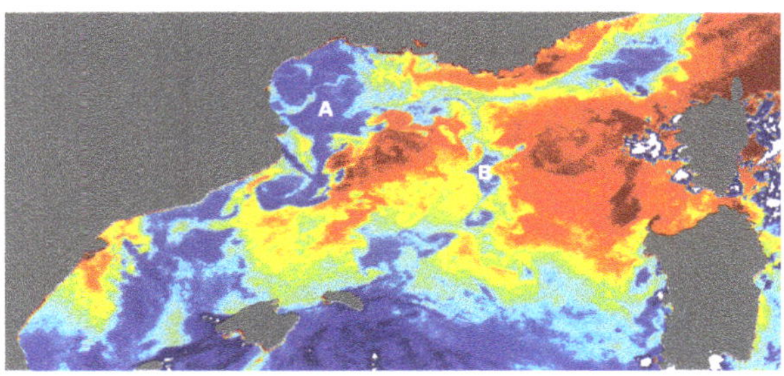

Figure 14. *Cont.*

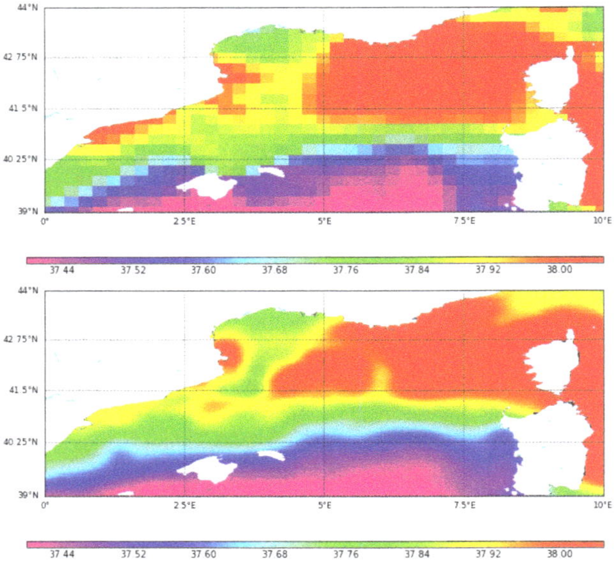

Figure 14. Images of the Gulf of Lion and Balearic Front on 28th September 2011 as observed by AVHRR satellite (**top**); the SMOS SSS L3 corrected product; and the SMOS SSS L4 corrected product.

5. Conclusions

A new methodology using a combination of the new retrieval debiased non-Bayesian algorithm, DINEOF and multifractal fusion has been used to obtain Sea Surface Salinity (SSS) fields over the North Atlantic Ocean and the Mediterranean Sea. The data used for the analysis is the 9-day 0.25 SMOS SSS L3 binned maps produced by the Barcelona Expert Centre (BEC), and it has been shown that the proposed methodology improves the accuracy with respect to the Objectively Analyzed maps and the L4 products that are currently being produced at the BEC. In particular, a reduction of the seasonal bias has been observed in the final data set as well as a reduction in the standard deviation of the error. The overall comparison of the new products with Argo floats provides an RMSE of 0.26 PSU in the full domain of study (North Atlantic and Mediterranean Sea) in comparison with an RSME of 0.41 PSU corresponding to the initial SMOS products. In the Mediterranean basin, the methodology also allows for reducing the error (0.39 PSU against 0.70 PSU in the case of the L4 products).

The comparison with the SSS TRANSMED in situ data shows that, in spite of the limitation of the differences in the spatial and temporal resolutions, the corrected SMOS SSS products not only provide a coherent description of the dynamics in the Algerian basin (as pointed out in [19]), but also a coherent description of the Balearic frontal zone. The short time and space scales of the freshwater discharges impair the detection of the latter by the SMOS SSS products, as seen in the gulf of Lion. An analysis at the seasonal scale might improve the comparison in this region. On the other hand, the L4 corrected products provide a description of the mesoscale salinity in regions so challenging as the Alboran Sea, where the water is surrounded by land a few kilometers away, and, therefore, the land sea contamination is especially strong.

The use of the DINEOF decomposition in combination to these new debiased non-Bayesian techniques has a high potential for improving the quality of the SMOS SSS maps not only in the Mediterranean, but also in other regions. On one hand, the debiased non-Bayesian approach mitigates the systematic biases (constant on time) like the land-sea contamination. In addition, its statistical filtering criteria improve the coverage with respect to official Level 2 Ocean Salinity products. On the other hand, the EOF decomposition allows for the characterization of the time-dependent biases. One of the major benefits of using EOFs decomposition in this context is that the technique not only allows to characterize the time-dependent biases with a regular (seasonal) bias (which is still an issue in the SMOS products, [49]), but the technique also allows the removal of some particular events, which could be caused by strong RFI. For example, in this study, we have shown that the proposed methodology has removed the spurious fresh event that contaminated the Mediterranean basin in October–November 2012. In addition, the use of multifractal fusion allows for improving the spatial resolutions and describing better the mesoscale dynamics.

Despite the encouraging results, there is still significant room for improvement. For instance, the EOFs to be removed should not be chosen one by one but as a single, optimized combination. This means assuming that the time-dependent bias is captured not by a single EOF, but by a combination of them. On the other hand, we have applied here a multifractal fusion scheme that is scalar, and more complex schemes can be investigated and tested in the Mediterranean.

All in all, this work has shown that these techniques are not only beneficial, but required to fully correct the SMOS SSS retrievals in some particular regions. Because of the capability of the combined debiased non-Bayesian retrieval-DINEOF-fusion approach to detect biases and reduce other sources of noise and error from the SMOS SSS data, the application of this technology to enclosed seas suffering from strong sources of contamination, such as the China Sea, the Baltic Sea and the Black Sea, will potentially provide promising results.

Acknowledgments: This work has been carried out within the project "SMOS sea surface salinity data in the Mediterranean Sea" ([50]), funded by the European Space Agency "Support to Science (STSE) PATHFINDERS" call. Part of this work was also supported by the Ministry of Economy and Competitiveness, Spain, through the National R+D Plan under L-Band Project ESP2017-89463-C3-1-R, PROMISES Project ESP2015-67549-C3 and previous grants. Argo data were collected and made freely available by the international Argo project and the national programs that contribute to it [51]. SSS time series from the moorings located in the Western Mediterranean were collected and made freely available by the Copernicus project and the programs that contribute to it (available online: [25] (accessed on 19 March 2018) and available online: [26] (accessed on 8 May 2012).). The TRANSMED system (available online: [29] (accessed on 19 March 2018)) was first fostered by the Commission Internationale pour l'Exploration de la Méditerranée, the Science Commission (CIESM) under the program TRANSMED/PartnerSHIPs (available online: [52] (accessed on 19 March 2018)), and achieved thanks to the support of the INSU MISTRALS /HyMeX program. The involvement of Gilles Rougier, the MIO's staff and from the DT INSU, and the longstanding support of the maritime companies Linea Messina (Genoa, Italy) and Marfret (Marseille, France) have been pivotal to the TRANSMED data collection. This work is a contribution to the ANR ASICS-MED project (grant ANR-12-BS06-0003). The MIO laboratory acknowledges the support received from the European FEDER Fund under project 1166-39417. OSTIA SST data were provided by GHRSST, UKMO and CMEMS.

Author Contributions: Estrella Olmedo, Aida Alvera-Azcárate and Antonio Turiel conceived and designed the algorithm for the SMOS SSS products generation. Estrella Olmedo and Aida Alvera-Azcárate implemented the new algorithm. Estrella Olmedo and Isabelle Taupier-Letage performed the validation of the new SMOS SSS product. Isabelle Taupier-Letage conducted the oceanographic interpretation of the results. The manuscript was prepared by Estrella Olmedo. All authors revised and approved the final manuscript.

Conflicts of Interest: The authors declare no conflict of interest.

References

1. Giorgi, F. Climate change hot-spots. *Geophys. Res. Lett.* **2006**, *33*, L08707, doi:10.1029/2006GL025734.
2. Millot, C.; Taupier-Letage, I. Circulation in the Mediterranean Sea. In *The Handbook of Environmental Chemistry, 5 Part K*; Saliot, A., Ed.; Springer: Berlin/Heidelberg, Germany, 2005; pp. 29–66, doi:10.1007/b107143. Available online: https://hal.archives-ouvertes.fr/hal-01191856 (accessed on 23 January 2018).
3. Taupier-Letage, I.; Puillat, I.; Millot, C.; Raimbault, P. Biological response to mesoscale eddies in the Algerian Basin. *J. Geophys. Res. Oceans* **2003**, *108*, 3245.
4. Isern-Fontanet, J.; Font, J.; Garcia-Ladona, E.; Emelianov, M.; Millot, C.; Taupier-Letage, I. Spatial structure of anticyclonic eddies in the Algerian basin (Mediterranean Sea) analyzed using the Okubo-Weiss parameter. *Deep Sea Res. II* **2004**, *51*, 3009–3028.

5. Schroeder, K.; Garcìa-Lafuente, J.; Josey, S.; Artale, V.; Buongiorno-Nardelli, B.; Carrillo, A.; Gacic, M.; Gasparin, G.; Herrmann, M.; Lionello, P.; et al. Circulation of the Mediterranean Sea and its variability. In *The Climate of the Mediterranean Region*; Lionello, P., Ed.; Elsevier: Oxford, UK, 2012; pp. 187–256.
6. Estournel, C.; Testor, P.; Damien, P.; D'ortenzio, F.; Marsaleix, P.; Conan, P.; Prieur, L. High resolution modelling of dense water formation in the north-western Mediterranean during winter 2012–2013: Processes and budget. *J. Geophys. Res. Oceans* **2016**, *121*, 5367–5392.
7. Waldman, R.; Somot, S.; Herrmann, M.; Testor, P.; Estournel, C.; Sevault, F.; Prieur, L.; Mortier, L.; Coppola, L.; Taillandier, V.; et al. Estimating dense water volume and its evolution for the year 2012–2013 in the Northwestern Mediterranean Sea: An observing system simulation experiment approach. *J. Geophys. Res. Oceans* **2016**, *121*, 6696–6716.
8. HYMEX (HYdrological Cycle in the Mediterranean EXperiment). Available online: https://www.hymex.org/ (accessed on 23 January 2018).
9. Drobinski, P.; Ducrocq, V.; Allen, J.; Alpert, P.; Anagnostou, E.; Béranger, K.; Borga, M.; Braud, I.; Chanzy, A.; Davolio, S.; et al. HyMeX, a 10-year multidisciplinary project on the Mediterranean water 1 cycle. *Bull. Am. Meteorol. Soc.* **2014**, *95*, 1063–1082.
10. Somot, S.; Sevault, F.; Déqué, M.; Crépon, M. 21st century climate change scenario for the Mediterranean using a coupled atmosphere-ocean regional climate model. *Glob. Planet. Change* **2008**, *63*, 112–126.
11. Adloff, F.; Somot, S.; Sevault, F.; Jordà, G.; Aznar, R.; Déqué, M.; Herrmann, M.; Marcos, M.; Dubois, C.; Padorno, E.; et al. Mediterranean Sea response to climate change in an ensemble of twenty first century scenarios. *Clim. Dyn.* **2015**, *45*, 2775–2802.
12. Somot, S.; Houpert, L.; Sevault, F.; Testor, P.; Bosse, A.; Taupier-Letage, I.; Bouin, M.; Waldman, R.; Cassou, C.; Sanchez-Gomez, E.; et al. Characterizing, modelling and understanding the climate variability of the deep water formation in the North-Western Mediterranean Sea. *Clim. Dyn.* **2016**, 1–32, doi:10.1007/s00382-016-3295-0.
13. Herrmann, M.; Diaz, F.; Estournel, C.; Marsaleix, P.; Ulses, C. Impact of atmospheric and oceanic interannual variability on the Northwestern Mediterranean Sea pelagic planktonic ecosystem and associated carbon cycle. *J. Geophys. Res. Oceans* **2013**, *118*, 5792–5813.
14. Herrmann, M.; Estournel, C.; Diaz, F.; Adloff, F. Impact of climate change on the Northwestern Mediterranean Sea pelagic planktonic ecosystem and associated carbon cycle. *J. Geophys. Res. Oceans* **2014**, *119*, 5815–5836.
15. Mecklenburg, S.; Drusch, M.; Kerr, Y.H.; Font, J.; Martin-Neira, M.; Delwart, S.; Buanadicha, G.; Reul, N.; Daganzo-Eurebio, E.; Oliva, R.; et al. ESA's Soil Moisture and Ocean Salinity Mission: Mission Performance and Operations. *IEEE Trans. Geosci. Remote Sens.* **2012**, *50*, 1354–1366.
16. Font, J.; Camps, A.; Borges, A.; Martín-Neira, M.; Boutin, J.; Reul, N.; Kerr, Y.; Hahne, A.; Mecklenburg, S. SMOS: The Challenging Sea Surface Salinity Measurement From Space. *Proc. IEEE* **2010**, *98*, 649–665.
17. Kerr, Y.; Waldteufel, P.; Wigneron, J.; Delwart, S.; Cabot, F.; Boutin, J.; Escorihuela, M.; Font, J.; Reul, N.; Gruhier, C.; et al. The SMOS mission: New tool for monitoring key elements of the global water cycle. *Proc. IEEE* **2010**, *98*, 666–687.
18. Alvera-Azcárate, A.; Barth, A.; Parard, G.; Beckers, J. Analysis of SMOS sea surface salinity data using DINEOF. *Remote Sens. Environ.* **2016**, *180*, 137–145.
19. Isern-Fontanet, J.; Olmedo, E.; Turiel, A.; Ballabrera-Poy, J.; García–Ladona, E. Retrieval of eddy dynamics from SMOS sea surface salinity measurements in the Algerian Basin (Mediterranean Sea). *Geophys. Res. Lett.* **2016**, *43*, 6427–6434.
20. Olmedo, E.; Martínez, J.; Turiel, A.; Ballabrera-Poy, J.; Portabella, M. Debiased non-Bayesian retrieval: A novel approach to SMOS Sea Surface Salinity. *Remote Sens. Environ.* **2017**, *193*, 103–126.
21. Olmedo, E.; Martínez, J.; Umbert, M.; Hoareau, N.; Portabella, M.; Ballabrera-Poy, J.; Turiel, A. Improving time and space resolution of SMOS salinity maps using multifractal fusion. *Remote Sens. Environ.* **2016**, *180*, 246–263.
22. ESA. Available online: https://earth.esa.int/web/guest/-/level-1b-full-polarization-6889 (accessed on 23 January 2018).
23. Global SMOS SSS Maps. Available online: http://bec.icm.csic.es/ocean-experimental-dataset-global/ (accessed on 23 January 2018).
24. Puertos del Estado. Available online: http://www.puertos.es (accessed on 23 January 2018).

25. Copernicus Marine Environment Monitoring Service. Available online: http://marine.copernicus.eu (accessed on 23 January 2018).
26. OceanSITES. Available online: http://www.oceansites.org (accessed on 23 January 2018).
27. SOCIB. Available online: http://www.socib.eu/?seccion=observingFacilities&facility=mooring&id=146 (accessed on 23 January 2018).
28. SOCIB. Available online: http://www.socib.eu/?seccion=observingFacilities&facility=mooring&id=143 (accessed on 23 January 2018).
29. TRANSMED. Available online: http://www.mio.univ-amu.fr/?TRANSMED (accessed on 23 January 2018).
30. Taupier-Letage, I.; Bachelier, C.; Rougier, G. Thermosalinometer TRANSMED, Marfret Niolon, definitive data set. *SEDOO OMP* **2014**, doi:10.6096/MISTRALS-HyMeX.1127.
31. Donlon, C.J.; Martin, M.; Stark, J.; Roberts-Jones, J.; Fiedler, E.; Wimmer, W. The operational Sea Surface Temperature and Sea Ice Analysis (OSTIA) system. *Remote Sens. Environ.* **2012**, *116*, 140–158.
32. MIO. Available online: www.ifremer.fr/osis_2014 (accessed on 19 March 2018).
33. Taupier-Letage, I. On the use of thermal infrared images for circulation studies: Applications to the eastern Mediterranean basin. In *Remote Sensing of the European Seas*; Barale, V., Gade, M., Eds.; Springer: Berlin/Heidelberg, Germany, 2008. Available online: https://hal.archives--ouvertes.fr/hal--01196705 (accessed on 23 January 2018).
34. González-Gambau, V.; Olmedo, E.; Turiel, A.; Martínez, J.; Ballabrera-Poy, J.; Portabella, M.; Piles, M. Enhancing SMOS brightness temperatures over the ocean using the nodal sampling image reconstruction technique. *Remote Sens. Environ.* **2016**, *180*, 202–220.
35. Zweng, M.; Reagan, J.; Antonov, J.; Locarnini, R.; Mishonov, A.; Boyer, T.; Garcia, H.; Baranova, O.; Johnson, D.; Seidov, D.; et al. *World Ocean Atlas 2013, Volume 2: Salinity*; Levitus, A., Ed.; Mishonov Technical; NOAA: Silver Spring, MD, USA, 2013; pp. 39–74.
36. Beckers, J.M.; Rixen, M. EOF calculations and data filling from incomplete oceanographic data sets. *J. Atmos. Ocean. Technol.* **2003**, *20*, 1839–1856.
37. Alvera-Azcárate, A.; Barth, A.; Rixen, M.; Beckers, J.M. Reconstruction of incomplete oceanographic data sets using Empirical Orthogonal Functions. Application to the Adriatic Sea surface temperature. *Ocean Model.* **2005**, *9*, 325–346.
38. Alvera-Azcárate, A.; Barth, A.; Sirjacobs, D.; Beckers, J.M. Enhancing temporal correlations in EOF expansions for the reconstruction of missing data using DINEOF. *Ocean Sci.* **2009**, *5*, 475–485.
39. Alvera-Azcárate, A.; Barth, A.; Beckers, J.M.; Weisberg, R.H. Multivariate Reconstruction of Missing Data in Sea Surface Temperature, Chlorophyll and Wind Satellite Fields. *J. Geophys. Res.* **2007**, *112*, C03008, doi:10.1029/2006JC003660.
40. Nechad, B.; Ruddick, K.; Park, Y. Calibration and validation of a generic multisensor algorithm for mapping of total suspended matter in turbid waters. *Remote Sens. Environ.* **2010**, *114*, 854–866.
41. Alvera-Azcárate, A.; Vanhellemont, Q.; Ruddick, K.; Barth, A.; Beckers, J.M. Analysis of high frequency geostationary ocean colour data using DINEOF. *Estuar. Coast. Shelf Sci.* **2015**, *159*, 28–36.
42. Turiel, A.; Solé, J.; Nieves, V.; Ballabrera-Poy, B.; García-Ladona, E. Tracking oceanic currents by singularity analysis of Microwave Sea Surface Temperature images. *Remote Sens. Environ.* **2008**, *112*, 2246–2260.
43. Isern-Fontanet, J.; García-Ladona, E.; Font, J. Microcanonical multifractal formalism: Application to the estimation of ocean surface velocities. *J. Geophys. Res.* **2007**, *112*, 2156–2202.
44. Nieves, V.; Llebot, C.; Turiel, A.; Solé, J.; García-Ladona, E.; Estrada, M.; Blasco, D. Common turbulent signature in sea surface temperature and chlorophyll maps. *Geophys. Res. Lett.* **2007**, L23602, doi:10.1029/2007GL030823.
45. Umbert, M.; Hoareau, N.; Turiel, A.; Ballabrera-Poy, J. New blending algorithm to synergize ocean variables: The case of SMOS sea surface salinity maps. *Remote Sens. Environ.* **2014**, *146*, 188–200.
46. Hernández-Carrasco, I.; Sudre, J.; Garcon, V.; Yahia, H.; Garbe, C.; Paulmier, A.; Dewitte, B.; Illig, S.; Dadou, I.; González-Dávila, M.; et al. Reconstruction of super-resolution ocean pco2 and airsea fluxes of co2 from satellite imagery in the southeastern atlantic. *Biogeosciences* **2015**, *12*, 5229–5245.
47. González-Gambau, V.; Turiel, A.; Olmedo, E.; Martínez, J.; Corbella, I.; Camps, A. Nodal Sampling: A New Image Reconstruction Algorithm for SMOS. *IEEE Trans. Geosci. Remote Sens.* **2016**, *54*, 2314–2328.

48. González-Gambau, V.; Olmedo, E.; Martínez, J.; Turiel, A.; Duran, I. Improvements on Calibration and Image Reconstruction of SMOS for Salinity Retrievals in Coastal Regions. *IEEE J. Sel. Top. Appl. Earth Obs. Remote Sens.* **2017**, *10*, 3064–3078.
49. Martín-Neira, M.; Oliva, R.; Corbella, I.; Torres, F.; Duffo, N.; Durán, I.; Kainulainen, J.; Closa, A.; Zurita, A.; Cabot, F.; et al. SMOS Instrument performance and calibration after 5 years in orbit. *Remote Sens. Environ.* **2016**, *180*, 19–39.
50. DINEOF Analysis of SMOS Sea Surface Salinity Data. Available online: http://www.gher.ulg.ac.be/WP/ (accessed on 23 January 2018).
51. Argo. Argo float data and metadata from Global Data Assembly Centre (Argo GDAC). *SEANOE* **2000**, doi:10.17882/42182.
52. CIESM. Available online: www.ciesm.org/marine/programs/partnerships.htm (accessed on 23 January 2018).

remote sensing

MDPI

Article

Seven Years of SMOS Sea Surface Salinity at High Latitudes: Variability in Arctic and Sub-Arctic Regions

Estrella Olmedo [1,*], Carolina Gabarró [1], Verónica González-Gambau [1], Justino Martínez [1], Joaquim Ballabrera-Poy [1], Antonio Turiel [1], Marcos Portabella [1] and Severine Fournier [2] and Tong Lee [2]

1 Department of Physical Oceanography, Institute of Marine Sciences, CSIC & Barcelona Expert Center, Pg. Marítim 37–49, E-08003 Barcelona, Spain; cgabarro@icm.csic.es (C.G.); vgonzalez@icm.csic.es (V.G.-G.); justino@icm.csic.es (J.M.); joaquim@icm.csic.es (J.B.-P.); turiel@icm.csic.es (A.T.); portabella@icm.csic.es (M.P.)

2 Jet Propulsion Laboratory, California Institute of Technology, 4800 Oak Grove Drive, Pasadena, CA 91109, USA; Severine.Fournier@jpl.nasa.gov (S.F.); tlee@jpl.nasa.gov (T.L.)

* Correspondence: olmedo@icm.csic.es

Received: 14 September 2018; Accepted: 3 November 2018; Published: 8 November 2018

Abstract: This paper aims to present and assess the quality of seven years (2011–2017) of 25 km nine-day Soil Moisture and Ocean Salinity (SMOS) Sea Surface Salinity (SSS) objectively analyzed maps in the Arctic and sub-Arctic oceans (50°N–90°N). The SMOS SSS maps presented in this work are an improved version of the preliminary three-year dataset generated and freely distributed by the Barcelona Expert Center. In this new version, a time-dependent bias correction has been applied to mitigate the seasonal bias that affected the previous SSS maps. An extensive database of in situ data (Argo floats and thermosalinograph measurements) has been used for assessing the accuracy of this product. The standard deviation of the difference between the new SMOS SSS maps and Argo SSS ranges from 0.25 and 0.35. The major features of the inter-annual SSS variations observed by the thermosalinographs are also captured by the SMOS SSS maps. However, the validation in some regions of the Arctic Ocean has not been feasible because of the lack of in situ data. In those regions, qualitative comparisons with SSS provided by models and the remotely sensed SSS provided by Aquarius and SMAP have been performed. Despite the differences between SMOS and SMAP, both datasets show consistent SSS variations with respect to the model and the river discharge in situ data, but present a larger dynamic range than that of the model. This result suggests that, in those regions, the use of the remotely sensed SSS may help to improve the models.

Keywords: sea surface salinity; remote sensing; Arctic ocean; SMOS; Arctic rivers; data processing; quality assessment

---

1. Introduction

In recent years, the Arctic Ocean has been under significant changes as shown by numerous in situ and remotely sensed measurements. The temperature of the upper layer of the Arctic Ocean has been increasing and more solar heat has been absorbed by the increasing ice-free areas [1–3].

Latest observational and modeling studies have documented changes in the upper Arctic Ocean hydrography [4]. In particular, an increase of liquid freshwater content over both the Canadian Basin and the central Arctic Ocean has been observed. This increase of freshwater has been linked to an intensification in the large-scale anticyclonic winds as well as sea level pressure changes [5]. An increased Bering Strait freshwater import to the Arctic Ocean, a decreased Davis Strait export, and the enhanced net sea ice melt could play an important role in the observed freshwater trend [6].

Rivers are important sources of freshwater and heat to the Arctic Ocean and changes in the river runoff or temperature could have a strong impact on the Arctic system. An increment of the global mean annual temperature will produce an increase in the discharge of Arctic rivers [7,8].

The 2015 update of the Arctic Report Card alerts that, in 2014, the combined discharge of the eight largest Arctic rivers was 10% greater than their average discharge during the 1980–1989 period [9]. However, the impact of this increase of freshwater runoff on the Arctic ocean dynamics remains unknown due to the lack of available salinity measurements in the Arctic.

Unfortunately, the number of surface salinity measurements is very scarce at high latitudes, especially in the Arctic Ocean. In such context, the three L-band missions—the Soil Moisture and Ocean Salinity (SMOS) mission [10–12]; the Aquarius mission [13,14]; and Soil Moisture Active Passive (SMAP) observatory [15]—can provide an unprecedented source of salinity information over the Arctic Ocean, which can help to improve the models.

The retrieval of sea surface salinity (SSS) from microwave radiometric measurements is based on the emissivity of the ocean surface, which depends on the dielectric constant of sea water that is a function of temperature and salinity, and on the sea surface roughness. The SMOS radiometer operating frequency (1.43 GHz, in the L-band) provides good sensitivity of the ocean-surface brightness temperature (T_B) to SSS in the tropics and subtropics [16]. In cold waters, however, the sensitivity of the T_B to salinity decreases rapidly [17]. As shown in [18], such sensitivity drops from 0.5 K/psu to 0.3 K/psu, when SST decreases from 15 °C to 5 °C. Moreover, some undesired effects in SMOS T_B and to lesser extent in Aquarius and SMAP T_B measurements, such as the land–sea and ice–sea contaminations, and Radio Frequency Interference (RFI) [19] make the Arctic region one of the most challenging regions for SMOS SSS retrieval.

Some previous works assessed the quality of SMOS SSS at high latitudes. For example, in Köhler et al. [20] he authors performed a comparison of previous versions of SMOS (salinity maps computed from the L2OS v550) and Aquarius products with in situ measurements and models for the north Atlantic region, but they did not perform any comparison inside the Arctic Basin. Despite the large biases (mainly produced by land–sea and ice–sea contaminations) that affected the SMOS L2OS v550, in Matsuoka et al. [21] this product was used to develop an algorithm for identifying surface water sources in the southern Beaufort Sea by using Aqua/MODIS ocean color along with SMOS SSS L2. Recently, the potential and challenges of monitoring the Arctic Ocean SSS by using SMAP data have been demonstrated in [22].

A recently developed SSS retrieval algorithm [23] has noticeably improved the coverage of the global SMOS SSS leading to retrievals in some critical areas where no-valid or few salinity retrievals were available before (for example in the Mediterranean Sea [24,25]). The Barcelona Expert Center (BEC) team used this methodology for the generation of three-year time series of SMOS SSS at high latitudes. In [26], a comparison of these SMOS SSS maps and three other SSS products provided by Aquarius with in situ data is performed. The authors concluded that SMOS SSS maps are consistent with ship and CORA5.0 data, although they also pointed out that the sea ice mask should be improved.

In this work, we generate seven years of SMOS SSS maps at high northern latitudes (beyond 50°N) by using the methodology described in [23]. Additionally, we improve the methodology in terms of the seasonal bias. The objectives of this work are the following: (i) to present seven-year time series of this new SMOS SSS product at high northern latitudes; (ii) to assess the quality of these new SMOS SSS maps at high latitudes by comparing them to different sources of in situ data; (iii) to compare the SMOS SSS with other available products in this region (model and other remotely sensed SSS products); and (iv) to show the potential of SMOS SSS to capture the SSS variability in the Arctic region.

The paper is structured as follows: In Section 2, we describe the different datasets that are used. In Section 3, the methodology used for the generation of the SMOS salinity maps is briefly presented. The assessment of the SMOS salinity maps is presented in Section 4. Variations of SSS shown by SMOS, Aquarius, SMAP and the model outputs from TOPAZ close to the mouth of the major Arctic rivers are shown in Section 5. A final discussion is provided in Section 6.

2. Datasets

2.1. SMOS Brightness Temperatures: Level 1B Product

The input data for the computation of the SMOS SSS maps are the Level 1 Brightness Temperature product (L1B v620). This product consists of the Fourier components of brightness temperatures in the antenna reference frame. The latency of the products is 6–8 h. The L1B T_B product is distributed by the European Space Agency (ESA) and is freely available at https://earth.esa.int/web/guest/-/how-to-obtain-data-7329.

2.2. Argo Salinity

We use Argo salinity [27] in Section 3 for the characterization of the SMOS SSS bias and for the generation of a time-dependent bias correction. After that, Argo data are also compared with the resulting SMOS products in Section 4.2.

We consider the uppermost salinity measurement provided by the Argo profiles (hereafter, Argo SSS) to be compared with the nine-day SMOS SSS maps. Thus, for every SMOS SSS nine-day map, the available Argo SSS during these nine days are compared with the corresponding fields of the SMOS SSS map. The cut-off depth for Argo profiles is taken at 10 m but no measurements shallower than 0.5 m are used due to the formation of bubbles and foam. In the case of SOLO and PROVOR Argo floats, only the data deeper than 5 m below the surface are used because their Conductivity, Temperature and Depth (CTD) probes stop pumping water at around 5 m below the surface. Profiles from BioArgo and those included in the greylist (i.e., floats which may have problems with one or more sensors) are discarded. In addition, we use World Ocean Atlas (WOA) 2013 as an indicator: Argo float profiles with anomalies larger than 10 °C in temperature or 5 PSU in salinity when compared to WOA are discarded. Only profiles having temperature close to surface between −2.5 and 40 °C and salinity between 2 and 41 PSU are used. In Figure 1 (top-left), the number of Argo SSS and their spatial distribution for the period of study is represented.

2.3. TARA Salinity

The Tara Polar Circle Expedition dataset (hereafter, TARA SSS) [28] is used to validate the SMOS nine-day maps in the Arctic. This campaign took place in the Arctic Ocean from June to October 2013, and a thermosalinograph (TSG) Seabird SB45 and a temperature sensor (SBE38) recorded sea surface temperature and salinity at 3 m depth during the whole cruise. Since TARA salinity data present a large range of spatial variability in the Arctic Ocean (≈26 to 35), they are a very valuable source for assessing the annual SSS reference used for the generation of the SMOS SSS product.

The collocation strategy between satellite and TARA SSS is the following: for a given time instant t_0 at which a value of TARA SSS was acquired, that value is compared with the nine-day SMOS SSS map centered around t_0. All the TARA SSS data that cross a single SMOS cell (25 × 25 km) are averaged and the resulting mean value is the one that is compared with the SMOS SSS. In Figure 1 (top-right), the TARA SSS data measured in the expedition are represented.

2.4. TSG Salinity Data

We use 86 transects provided by Copernicus (hereafter, TSG SSS) for assessing the SMOS SSS. These data are freely available on http://marine.copernicus.eu/services-portfolio/access-to-products/ and are labeled as INSITU_ARC_NRT_OBSERVATIONS_013_031. In Figure 1 (bottom-left), the locations of the measurements are shown. The collocation strategy between SMOS and TSG SSS is the same as that of SMOS and TARA. Only TSG SSS data flagged as "good quality" have been used. Measurements deeper than 10 m are discarded. In Section 4.3, this dataset is used as an independent reference for the SMOS SSS validation.

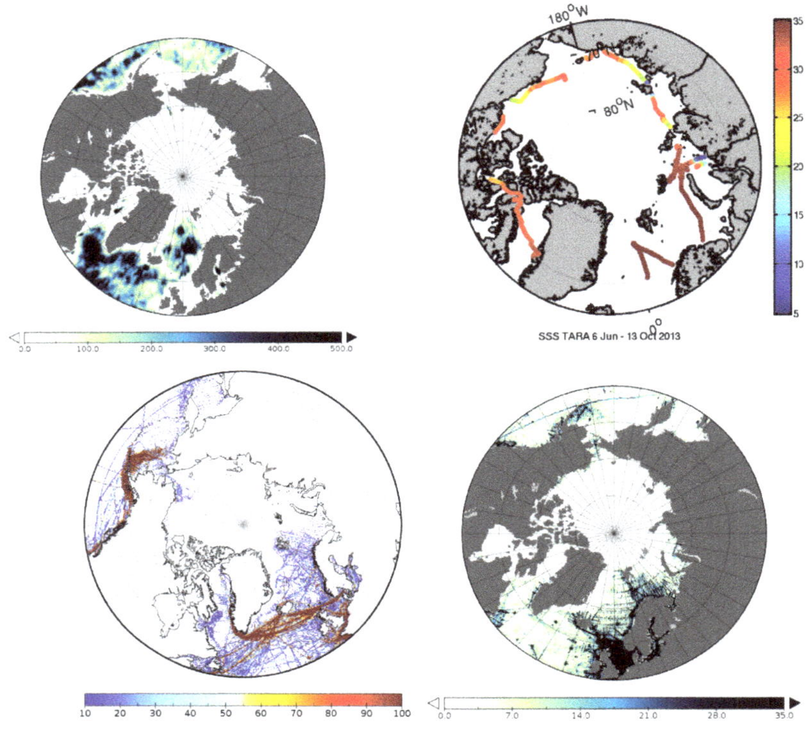

Figure 1. Number of measurements provided by Argo floats in the period of study 2011–2017 (**top-left**); salinity values (using the practical salinity scale) provided in the TARA campaign (**top-right**); number of SSS measurements provided by the TSG that have been used in this study for validation (**bottom-left**); and number of in situ measurements used in the computation of annual SSS climatology WOA (**bottom-right**).

2.5. TOPAZ Salinity

In Section 5, for those regions where no in situ measurements are available, we compare the variability shown by SMOS SSS with the one captured by TOPAZ (Towards an Operational Prediction system for the North Atlantic European coastal Zones) SSS. The TOPAZ Arctic Prediction system consists of a coupled ocean sea ice data assimilation system (Ensemble Kalman filter) for the North Atlantic and the Arctic Ocean using the Hybrid Coordinate Ocean Model (HYCOM [29]). Satellite and in situ observations (including Argo floats) are assimilated. The observations assimilated in the system are satellite-derived sea level anomaly, SST, sea-ice concentrations from AMSR-E, sea-ice drift products from CERSAT and Coriolis, and in situ temperature and salinity profiles [30]. Since no SMOS SSS can be derived over sea ice, only grid points with sea ice fraction lower than 30% have been used when comparing TOPAZ and SMOS SSS. It is important to mention that TOPAZ relaxes SSS to seasonal climatology [31]. This has implications for the dynamic range of SSS from TOPAZ in regions without sufficient in situ salinity measurements to constrain the model during the data assimilation. Data since October 2011 are available at the Copernicus web page http://www.copernicus.eu/.

2.6. Aquarius SSS L3 Maps

The Aquarius/SAC-D mission dataset consists of weekly gridded products of L-band (frequency 1.4 GHz) radiometer SSS [32]. This product contains the average SSS retrieved from all the three beams of Aquarius. Data are then gridded to the Equal-Area Scalable Earth version 2.0 grid [33], with a cell resolution of 36 km. Version 5 of Aquarius L3 Weekly Polar-Gridded Sea Surface Salinity utilizes Version 4 of the Level-2 Aquarius SSS as input data. The product is distributed by the National Snow and Ice Data Center (NSIDC, http://nsidc.org/data/aq3_sss).

2.7. SMAP SSS L3 Maps

We use the level-3 SMAP SSS version-4 dataset produced by the Jet Propulsion Laboratory [34] at 0.25° horizontal resolution and eight-day running average time window from 31 March 2015 to 31 December 2017. The horizontal resolution of SMAP SSS is 40 km. Data are then gridded to the Equal-Area Scalable Earth version 2.0 grid [33], with a cell resolution of 36 km. The data are available on the PO.DAAC website (https://podaac.jpl.nasa.gov/dataset/SMAP_JPL_L3_SSS_CAP_8DAY-RUNNINGMEAN_V4).

2.8. River Discharge In Situ Measurements

Discharge data, provided by river gauge measurements made under the Arctic Great Rivers project (http://www.arcticgreatrivers.org/) from 2011 to 2017, are used in Section 5 for completing the description of the inter and intra annual variability observed by SMOS SSS close to the major Arctic river mouths.

3. Methodology Used for the SMOS SSS Product Generation

Seven years (2011–2017) of the SMOS L1B T_B data product (v620), provided by the ESA, are processed to generate salinity maps at high latitudes (from 50°N to 90°N).

The galactic [35], sun glint [36] and surface roughness [37] contributions are corrected using auxiliary information provided by ECMWF [38], similar to what is done in the official ESA SMOS L2 SSS products. The dielectric constant model proposed by Meissner and Wentz (M&W) [39] is used instead of the model defined by Klein and Swift (K&S) [40], which is used in the official SMOS Ocean Salinity Level 2 product. The work presented in [41] shows that, when analyzing SSS from Aquarius, differences between M&W and K&S are small at low and mid latitudes, but they increase at high latitudes, i.e., in cold waters. The authors concluded that, for very cold waters (colder than 3 °C), retrieved salinities using M&W model are significantly closer to in situ floats measurements than those retrieved using K&S.

The T_B measurements are geo-referenced using a 25-km resolution Equal-Area Scalable Earth (EASE) North Pole grid [42]. To account for the SMOS residual spatial and temporal systematic errors, the SSS retrieval methodology presented in [23] is used. This methodology introduces important changes with respect to the standard processing [16,43] used in the ESA SMOS L2OS processor:

(a) Individual retrievals: The retrieval follows a non-Bayesian scheme, that is, for each SMOS T_B a single value of SSS is retrieved.

(b) Characterization of the systematic errors: All the SSS retrieved under the same acquisition conditions, i.e., the same geographical location, incidence and azimuth angles and satellite overpass direction (ascending/descending). throughout this seven-year period are accumulated in a SSS distribution. The systematic error associated to each acquisition condition is estimated by computing the central estimator of the corresponding SSS distributions. We use the mode of the distribution as the central estimator, i.e., as the SMOS climatological value for each specific acquisition condition. In this aspect, a relevant difference with respect to the official SMOS L2OS processor is that the T_B used for the ESA SMOS L2OS SSS retrieval are previously corrected by Ocean Target Transformation (OTT) [44]. The OTT is computed as the mean of the difference between the measured and modeled T_Bs (applying the Geophysical Forward Model) at a particularly stable region of the ocean. We do not apply an OTT since systematic errors are already accounted for, point by point, with the new methodology.

(c) Filtering criteria: The statistical properties of those SSS distributions are also used for filtering the non-accurate measurements. Two types of filters are applied to remove questionable values and outliers in the SSS retrievals: (i) all the SSS belonging to distributions having a large standard deviation (std larger than 10), defined by too few measurements (less than 100), or with a large skewness (larger than 1 in absolute value) or kurtosis (lower than 2) are all excluded (i.e., the distribution is marked as "bad" distribution, and all its salinities are discarded); and (ii) an additional outlier criterion is applied to the remaining retrieval values by further excluding any value that is farther than 10 (in absolute value) from the SMOS climatological value (see more details in [23]).

(d) Computation of SMOS anomalies: The SMOS-debiased SSS anomalies are computed by subtracting to each individually retrieved SSS value (corresponding to a specific acquisition condition) the corresponding SMOS climatological value (computed as explained in (b)), thus effectively removing local biases, especially those produced by the land–sea (or ice–sea) contamination and permanent RFIs.

(e) Computation of SMOS SSS: In [23], the SMOS SSS are generated by adding an annual SSS reference (annual WOA SSS, [45]) to the SMOS anomalies. This is an issue for the SMOS SSS values of the Arctic Ocean, since there are many zones in the Arctic with very few measurements of SSS (as shown in the bottom-right plot of Figure 1), and any reference could provide non-accurate SSS values there. For this reason, before generating the seven years of SMOS SSS, we analyze two test datasets by using two annual references: WOA; and the Polar science center Hydrographic annual Climatology (PHC) (version 3) [46] which is the usual reference for Arctic regions. We assess the quality of these two datasets by comparing the resulting SMOS SSS products—the SMOS SSS computed from WOA annual reference (SMOS woa) and SMOS SSS computed from PHC (SMOS phc)—with TARA SSS. In Section 4.1, a full discussion of this assessment is given. The conclusions of these comparisons are summarized in Figure 2. SMOS SSS has lower RMS with respect to TARA SSS than the corresponding annual references used for its generation (i.e., the blue and red lines are below the green and black lines, respectively, except in Buffin Bay where the SMOS-PHC has slightly larger RMS than the PHC product). The actual RMS values, together with the bias and standard deviation values, can be found in Table 1. However, many regions in the Arctic Ocean present large RMS values. These regions correspond to the areas where few or no in situ data were taken into account in the generation of the annual reference. In Section 4.1, we describe this analysis in more detail. Since both references provide similar results (WOA is slightly better), we use WOA for the generation of SMOS SSS product to be coherent with the global SMOS SSS product distributed by BEC.

(f) Objectively analyzed maps: Objectively analyzed nine-day SSS maps at 25-km resolution are generated daily. In [23], the same correlation radii used for the computation of WOA products were proposed for the generation of the SMOS SSS products: 321 km, 267 km, and 175 km (see [45]). These correlation radii do not seem to be the most appropriate for describing the dynamics in the Arctic region [47]. For this reason, we assess the impact on SSS quality of using different correlation radii, by means of the following experiment:

- We consider a finite set of candidates for the first correlation radius, R_1: 175 km, 200 km, 225 km, 250 km, 275 km, 300 km, and 325 km.
- For each one of the previous values, we consider the second R_2 and third R_3 radii of convergence such that:

$$R_2 = R_1 * \gamma \quad (1)$$

$$R_3 = R_2 * \gamma \quad (2)$$

with γ taking each one of the following values: $\gamma = 0.3, 0.4, 0.5, 0.6, 0.7, 0.8$.

- For each set of three convergence radii computed as before, we generate seven months of SMOS nine-day SSS maps: from the 1 April to the 9 November of 2013, one map every five days.
- We apply the time-bias correction proposed in [23], by removing the (spatial) mean anomaly between each SMOS SSS map and the annual WOA SSS.
- We compare the resulting SMOS SSS with Argo SSS. We use as a metric the global RMS of the difference between SMOS and Argo SSS, averaging the seven months for each one of the previous choices of R_1 and γ.

The configuration which provides the lowest error ($RMS \approx 0.25$) is the one with $R_1 = 325$ km and $\gamma = 0.8$, which is the closest one to the configuration proposed in [23]: 321 km, 267 km and 175 km. Notice that the global L3 SMOS SSS maps distributed by BEC (http://bec.icm.csic.es/ocean-experimental-dataset-global/) use a set of smaller correlation radii (175 km, 125 km and 75 km) for better describing the mesoscale. The results raised from this experiment suggest that, at high latitudes, the larger is the correlation radii, the larger is the smoothing effect, and therefore the lower is the noise. This is probably because individual SMOS SSS retrievals at high latitudes are noisier than in other regions of the globe. In other words, the generation of SMOS SSS maps with smaller correlation radii and the same level of noise as in the case of the global SSS maps requires SMOS SSS retrievals less noisy. In this sense, improvements at T_B level as the ones introduced in [48] and assessed at salinity level in [49,50] are providing promising results in terms of noise reduction in the SSS retrievals. The application of this technique will probably help to retrieve more accurate SSS in those regions and therefore to generate SMOS SSS maps with smaller correlation radii (more appropriate to capture the dynamics of this region).

(g) Mitigation of the seasonal bias: An additional time-dependent bias correction is needed to mitigate the effect of seasonal and other time-dependent biases which affect the SMOS T_B ([19]). In [23], the authors proposed subtracting the global mean of the SMOS SSS anomaly for each nine-day map. This assumption is appropriate for global SSS maps, as it implies that the total content of salt remains constant in time. However, the application of this hypothesis regionally, in particular at high latitudes, produces seasonal biases. In other words, there are net exchanges of salinity across region boundaries. In a recent study [25], a multivariate analysis is used to characterize and mitigate the time-dependent bias in the SMOS SSS maps in the Mediterranean Sea. In this work, we include a simpler time-dependent bias correction:

- We consider the Argo SSS available for the same nine-day period used in the generation of the nine-day SMOS SSS maps.
- We compute the median of the differences between the collocated SMOS SSS fields and the Argo SSS.
- We subtract this median from each nine-day SMOS SSS map.

Figure 3 shows the time-dependent correction resulting from this procedure, which has been applied to each map.

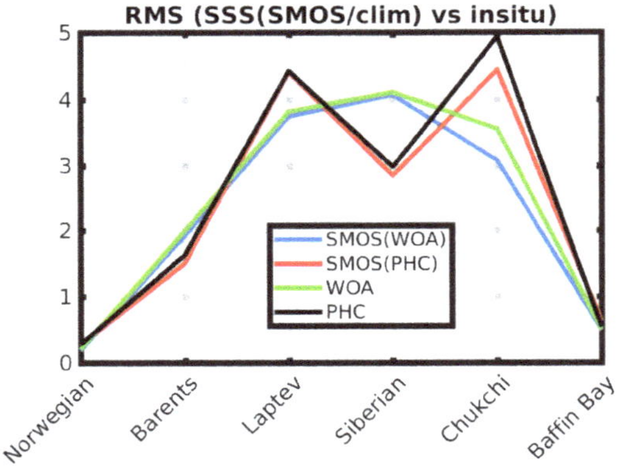

Figure 2. Assessment of the impact associated to the choice of the annual climatology in the SMOS SSS product generation. The blue line corresponds to the root mean square (RMS) of the differences between SMOS SSS derived with WOA annual mean and the TARA SSS; the red line corresponds to the RMS of the differences between SMOS SSS derived with PHC annual mean and TARA SSS; and the green (black) line represents the RMS differences between WOA (PHC) SSS and TARA SSS.

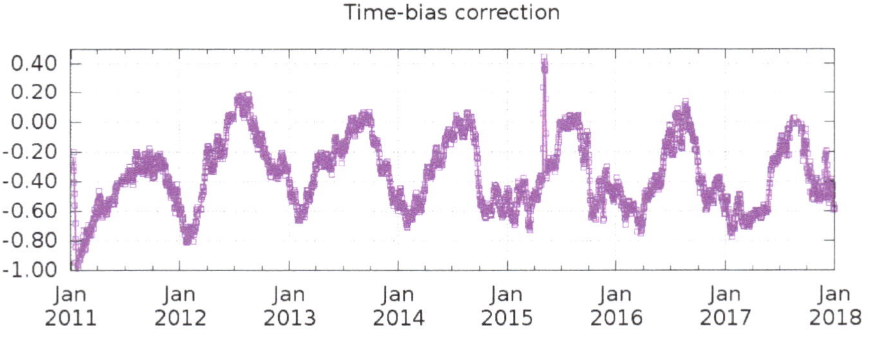

Figure 3. Time-bias correction applied to the nine-day SMOS SSS L3 maps. During 1–10 January 2011, SMOS provided degraded acquisitions due to a problem in the physical temperatures acquisition (reported in https://earth.esa.int/c/document_library/get_file?folderId=118493&name=DLFE-5407.pdf). The reason for the jump at the end of April 2015 is still under study, but is probably related to several degraded orbits that were reported in the data quality report (available on https://earth.esa.int/documents/10174/1785702/SMOS_Public_Monthly_Report_April_2015).

Table 1. Statistics in the comparison of SMOS SSS with TARA SSS.

Arctic Region	SMOS woa			WOA			SMOS phc			PHC		
	Mean	Std	RMS	Mean	Std	RMS	Mean	Std	RMS	Mean	Std	RMS
Norwegian sea	0.11	0.15	0.18	−0.06	0.19	0.20	−0.18	0.19	0.26	−0.13	0.24	0.28
Baffin Bay	0.16	0.46	0.49	0.04	0.49	0.50	0.29	0.55	0.63	0.17	0.51	0.54
Chukchi region	2.16	2.18	3.07	2.74	2.26	3.55	3.46	2.79	4.44	4.02	2.87	4.94
Barents sea	−1.35	1.37	1.93	−1.22	1.58	2.00	−0.86	1.24	1.51	−0.73	1.46	1.64
Laptev sea	1.37	3.48	3.74	1.78	3.37	3.81	1.46	4.16	4.41	1.87	4.02	4.43
Siberian region	−3.20	2.50	4.06	−2.88	2.93	4.11	−1.54	2.39	2.84	−1.21	2.72	2.98

4. Quality Assessment of the SMOS SSS Data

4.1. Comparison with TARA SSS: Impact Analysis of the SSS Annual Reference

As explained in Section 3, the computation of the final value of the SMOS SSS maps depends on the annual SSS reference. In this section, we compare with TARA SSS two SMOS SSS products which use different annual SSS references: the SMOS SSS using WOA annual reference (SMOS woa) and SMOS SSS using PHC (SMOS phc).

The validation against TARA SSS is done by separating the transects per different seas, for a better understanding of the regional quality of the SMOS SSS product. We divide the whole dataset into six regions: Norwegian Sea, Barents Sea, Laptev Sea, East Siberian Sea, Chukchi Sea and Baffin Bay.

Figure 4 (right) shows differences of TARA SSS and SMOS woa and SMOS phc SSS. To keep the figure readable, the WOA and PHC climatological values have been excluded from the right plots. Note that, although the differences of SMOS and in situ are lower than those of the corresponding climatologies and in situ, it is clear that, when the climatological values largely deviate from the in situ values (e.g., in the Chuckchi Sea, up to 8 PSU deviations are observed), the SMOS product also shows large deviations, therefore indicating that the annual climatology plays an important role in the quality of the SMOS-derived products. Let us comment in more detail the statistics in the different regions.

The comparison for the Norwegian Sea (June 2013) is shown in the first row of Figure 4. The RMS between SMOS and TARA in that transect is 0.18 (for SMOS woa) and 0.26 (for SMOS phc) (see Table 1). In the Norwegian Sea region, the annual WOA SSS was generated with a lot of in situ measurements (as shown in Figure 1 (bottom-right)). Thus, proper SSS values for the reference are expected.

The bottom-right plot of Figure 1 shows also some in situ measurements in the Greenland coast and in the Beaufort Sea. The Tara Polar circle Expedition crossed the Baffin Bay in October 2013. Plots in the second row of Figure 4 show the comparison between TARA and SMOS measurements with a mean RMS of 0.49 for SMOS woa and 0.63 for SMOS phc.

In September 2013, the expedition arrived to the Chukchi region. Large errors of SMOS measurements with respect to TARA are observed in the Canadian coast (see the plots in the third row of Figure 4) with a RMS of 3.07 and 4.44 for SMOS woa and SMOS phc respectively (Table 1). Although some in situ data are available for the computation of the annual reference (bottom-right plot of Figure 1), they may not be sufficient for generating an accurate annual reference in this region, which is affected by a large dynamic range of salinity due to the fresh water coming from the Mackenzie River.

The other three regions that we analyze (Barents, Laptev and Siberian Seas) correspond to regions where no or very few in situ data are available for the computation of the WOA (see bottom-right plot of Figure 1). In these three cases, the differences observed between SMOS and TARA SSS are larger than what is expected, with RMS greater than 1 psu in all cases. Neither WOA nor PHC seems to provide a proper annual reference for the computation of SMOS SSS. Let us comment each one of the three regions separately:

The comparison in the Barents Sea (July 2013) is shown in the forth row of Figure 4. The differences between TARA and SMOS measurements increase when the ship goes to the Kara Sea region and near the Yenesey River mouth, where SMOS always measures fresher waters than TARA. The Ob and Yenesey River plumes affect that region. This could explain part of the negative differences observed: while SMOS measures the salinity in the top cm of the ocean, TARA measures at 3 m depth. The stratification at these epochs of the year is usually strong, so, some negative differences are expected [51]. Another possible explanation for these differences is that these regions are typically affected by RFI. Although the methodology described in Section 3 aims to mitigate permanent effects of RFI, some residual RFI contamination may affect the SSS retrievals in this region. On the other hand, although differences of few psu are observed between satellite and in situ (1.93 in the case of SMOS woa and 1.51 in the case of SMOS phc), SMOS follows the full range of SSS observed by TARA in this region (also goes from 35 to 25 psu with both annual SSS references).

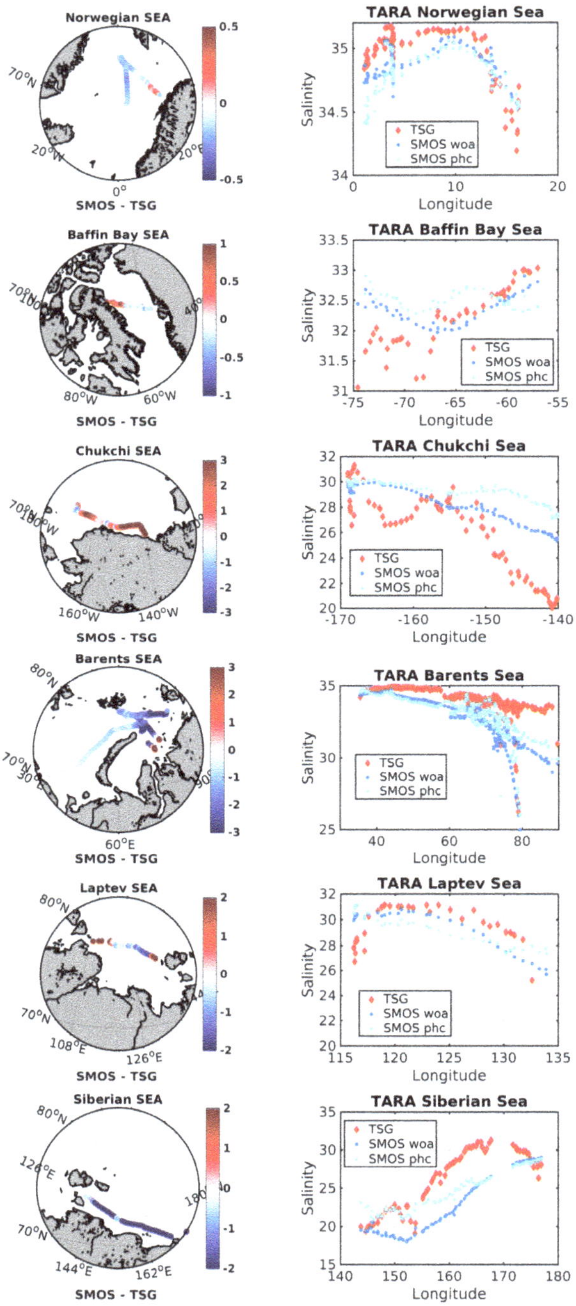

Figure 4. (**Left**) Difference between SMOS SSS and TARA SSS. (**right**) TARA, SMOS woa and SMOS phc SSS as function of the longitude. From top to bottom: Norwegian Sea in June 2013; Baffin Bay in October 2013; Chukchi Sea in September 2013; Barents Sea in July 2013; Laptev Sea in August 2013; and Siberian Sea in August 2013.

The expedition crossed Laptev Sea and Siberian region in August 2013. In both comparisons (see plots in the sixth and fifth rows, respectively, of Figure 4), as in the Barents Sea, SMOS measured fresher than TARA. In the Laptev Sea, although the mean RMS is about 3.74 for SMOS woa and 4.41 for SMOS phc, SMOS recovers the same SSS range as TARA (from 32 to 26 psu). In the Siberian region, the differences between satellite and TARA result in a RMS of 4.06 for SMOS woa and 2.84 for SMOS phc. Part of these negative differences can be explained because of the proximity to the coast and the river discharges (as occurs close to the Yenesey River mouth). Despite these large differences in the value of SSS, the dynamical range of SMOS SSS is similar to the one of TARA: TARA goes from 20 to 32 psu; SMOS woa from 18 to 30 psu; and SMOS phc from 20 to 30 psu. To better illustrate this, we include in Table 2 the correlation between the SMOS SSS products and TARA over the analyzed regions (similar values are obtained for the corresponding annual climatologies, not included). Correlation between SMOS phc and TARA in the Baffin Bay and Laptev Sea is very low (even negative). However, in the case of SMOS woa, despite the large biases between SMOS and TARA SSS, the correlation between both datasets is higher than 0.7 in most of the regions. This means that the major features of the spatial salinity gradients captured by the SMOS SSS in those regions are coherent with the ones captured by TARA SSS.

Figure 2 summarizes the main results discussed in this section. The conclusions are that there are some regions in the Arctic Ocean where no annual SSS reference is good enough and more efforts should be dedicated to the generation of a better reference. Since both analyzed references provide similar results in terms of RMS, but SMOS woa provides better correlation with TARA, we select WOA to develop the SMOS SSS products. This choice is also coherent with the annual reference selected for the global SMOS SSS products distributed at the BEC.

Table 2. Correlation of SMOS SSS with respect to TARA SSS.

	Norwegian Sea	Baffin Bay	Chukchi Region	Barents Sea	Laptev Sea	Siberian Region
SMOS woa	0.73	0.51	0.91	0.88	0.68	0.80
SMOS phc	0.56	−0.20	0.86	0.91	0.28	0.81

4.2. Comparison with Argo SSS

In this section, a statistical comparison between the nine-day SMOS SSS products and salinity provided by Argo floats (see Section 2.2) is presented. Since we use Argo floats for performing the time-dependent bias correction (see Section 3), this dataset is not an independent source of SSS data to be used for assessing the global mean of the product. However, this comparison is used to assess the residual spatial biases and the uncertainty (standard deviation of the differences SMOS-Argo SSS) of the SMOS SSS product.

In Figure 5, the mean of the differences between SMOS and Argo SSS during 2011–2017 are displayed (left), as well as the standard deviation of these differences (right). Large differences are observed in the Baffin Bay, the Labrador Sea and the eastern coast of Greenland (Greenland sea and Fram Strait). A possible factor contributing to these differences is the high-frequency and small-scale variability of SSS associated with the currents and the differences in temporal and spatial samplings between SMOS and the Argo SSS. SMOS maps are based on nine-day averages while Argo SSS represents instantaneous salinity values. Besides, SMOS SSS provides spatial average within a 40-km footprint, further smoothed by the Objective Analysis large correlation radii (see Section 3), while in situ data are instantaneous and point-wise measurements. The differences caused by these effects could be substantial if there are significant sub-footprint variability [51].

The northern coast of Alaska also presents large discrepancies between SMOS and Argo SSS. This region is strongly stratified, and the mixed layer is typically thinner than 3 m in this region. This implies also a limitation in the comparison of SMOS with Argo data since SMOS is measuring the first cm depth and Argo SSS are typically provided at some meters depth.

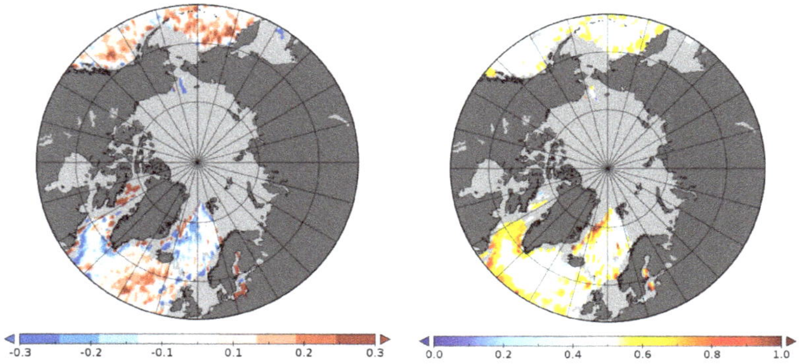

Figure 5. Spatial distributions of the differences between SMOS and Argo SSS: (**Left**) the mean of SMOS-Argo SSS; and (**Right**) the std of the difference.

Figure 6 shows temporal evolution of the the standard deviation of the differences SMOS-Argo SSS (top) and the number of collocations used in the statistics (bottom), such that, for a given time t_0 (x-axis), the point represents the std of the differences between SMOS and Argo SSS and the number of collocations, respectively, for all the collocations available with t_0 the first day of the nine-day period. We do not show the time evolution of the mean difference of SMOS and Argo SSS because, by definition (see Section 3), it is zero since the seasonal bias that was present in the previous version of this product has been mitigated. Large std values are observed in autumn 2012. The causes for such large std values (whether geophysical or instrumental) are currently being investigated. Our preliminary hypothesis indicates that they are due to a strong RFI episode. The std of the differences between SMOS and Argo SSS is between 0.24 and 0.35.

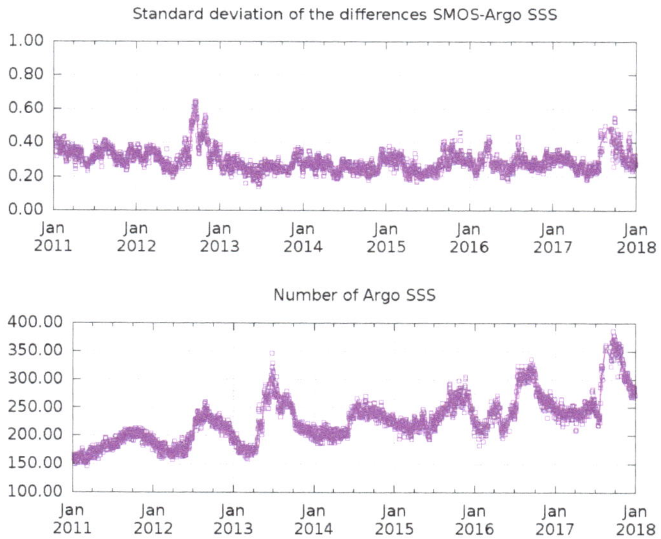

Figure 6. Time evolution of the standard deviation of the differences between SMOS and Argo SSS (**top**); and the number of measurements used in the comparison (**bottom**). The statistics are computed for data above 50°N. Every SMOS nine-day map is compared with the Argo SSS available for the same nine-day period.

4.3. Comparison with TSG Data from Copernicus

In this section, we compare SMOS SSS with SSS provided by 86 TSG transects distributed by Copernicus. The objective is to show that both SSS sources agree on the major features of the inter annual SSS dynamics, despite the different spatial and temporal resolutions of TSG and SMOS data. The statistics of the differences between SMOS and TSG SSS in several regions are provided in Table 3. In general, SMOS SSS has a positive bias with respect to TSG SSS. More detailed discussions of the results shown in Table 3 are provided below.

Table 3. Regional analysis on the differences between TSG and SMOS SSS (SMOS-TSG). Regional statistics of differences between SMOS and ARGO are also included.

Region	Latitude Range	Longitude Rage	SMOS-TSG			SMOS-ARGO		
			Meas	Mean	Std	Meas	Mean	Std
Denmark Strait	60N–65N	40W–25W	2322	0.22	0.28	25,153	0.02	0.21
North Atlantic	50N–60N	50W–20W	8028	0.19	0.45	102,575	0.01	0.35
Norwegian Sea	60N–70N	10W–5E	4587	0.02	0.67	33,841	−0.05	0.31
Northern Sea	55N–60N	0E–5E	53,366	0.16	0.99	0	-	-
Gulf of Alaska	55N–60N	175W–125W	604	0.26	0.72	14,841	0.09	0.27
Chuckchi Sea	70N–75N	170W–145W	315	0.01	1.42	1751	−0.99	0.37
Labrador Sea	55N–60N	55W–45W	737	0.14	0.71	33,711	−0.07	0.29
Baffin Bay	60N–65N	65W–55W	278	0.89	0.62	5919	−0.02	0.41

In the North Atlantic and Denmark Strait, the biases are 0.22 and 0.19, respectively, and the standard deviations are 0.28 and 0.45, respectively. These values of the standard deviation are in the expected range of error if we compare with the ones provided in the comparison with Argo (see Table 3). In the other regions, the standard deviations are larger than the standard deviations resulting from the comparison with Argo. Part of the increase in the standard deviation can be explained because the TSG SSS data reach more coastal regions than the Argo data do. Typically, these coastal regions are affected by complex circulation dynamics that could form filaments and mesoscale and submesoscale structures with strong SSS gradients and fast dynamics that cannot be resolved by SMOS, particularly after applying objective analysis. In those regions, we typically observe that the TSG captures strong SSS gradients with differences between consecutive coastal pixels greater than several PSU.

In Figures 7 and 8, TSG and collocated SMOS SSS are shown for 2013 and 2015, and 2012 and 2014, respectively, over two regions: the Norwegian and Northern Seas (Figure 7) and the Baffin Bay and the Labrador Sea (Figure 8). In both figures, near the coast, the TSG captures strong SSS gradients displaying consecutive salinity values of one lower than 33.5 and the next salinities saltier than 35 (in the Norwegian coast) and lower than 32 and saltier than 34 (in the Greenland coast). These strong SSS gradients are marked with red lines in Figures 7 and 8. Objectively analyzed SMOS SSS cannot fully capture these dynamics among other reasons because objectively analyzed maps are produced using correlation radii of 325–175 km. However, both sources of SSS are in good agreement regarding the major features of the spatial SSS gradients and the inter annual variability. For example, in 2013, TSG SSS captures a freshening in the southern part of the Norwegian Peninsula which spreads through the Northern Sea. SMOS SSS also shows fresher SSS in 2013 than in 2015 and it captures the spreading of the fresh water towards the Northern Sea (green circles in Figure 7). On the southern coast of Baffin Island, both sources of salinities agree on capturing fresher SSS in 2012 than in 2014 (green circles in Figure 8).

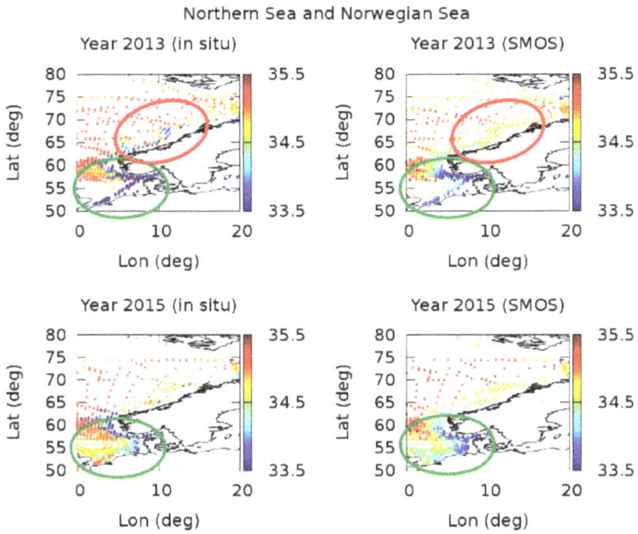

Figure 7. TSG (**left column**) and SMOS (**right column**) SSS in the Northern and Norwegian Seas for 2013 (**top**) and 2015 (**bottom**).

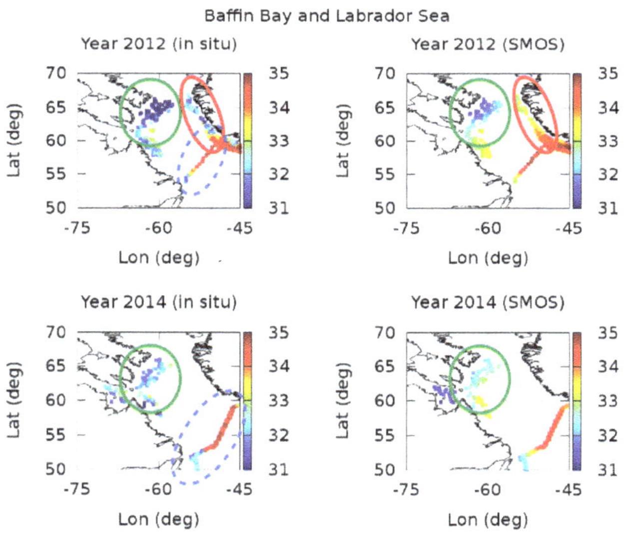

Figure 8. TSG (**left column**) and SMOS (**right column**) SSS in the Baffin Bay and Labrador Sea for 2012 (**top**) and 2014 (**bottom**).

To better analyze the inter annual variability, we study three routinely annual-performed transects: (i) one horizontal transect in the Gulf of Alaska around 58°N (first and second plots of Figure 9); (ii) one vertical transect in the Baffin Bay with longitude fixed around 79°W and latitudes in a range of 57°N and 67°N (third and forth plots of Figure 9); and (iii) one diagonal transect in the Labrador Sea (encircle with a dashed blue line in Figure 8) with a latitude range from 50°N to 60°N; and longitude range from 55°W to 50°W (fifth and sixth plots of Figure 9).

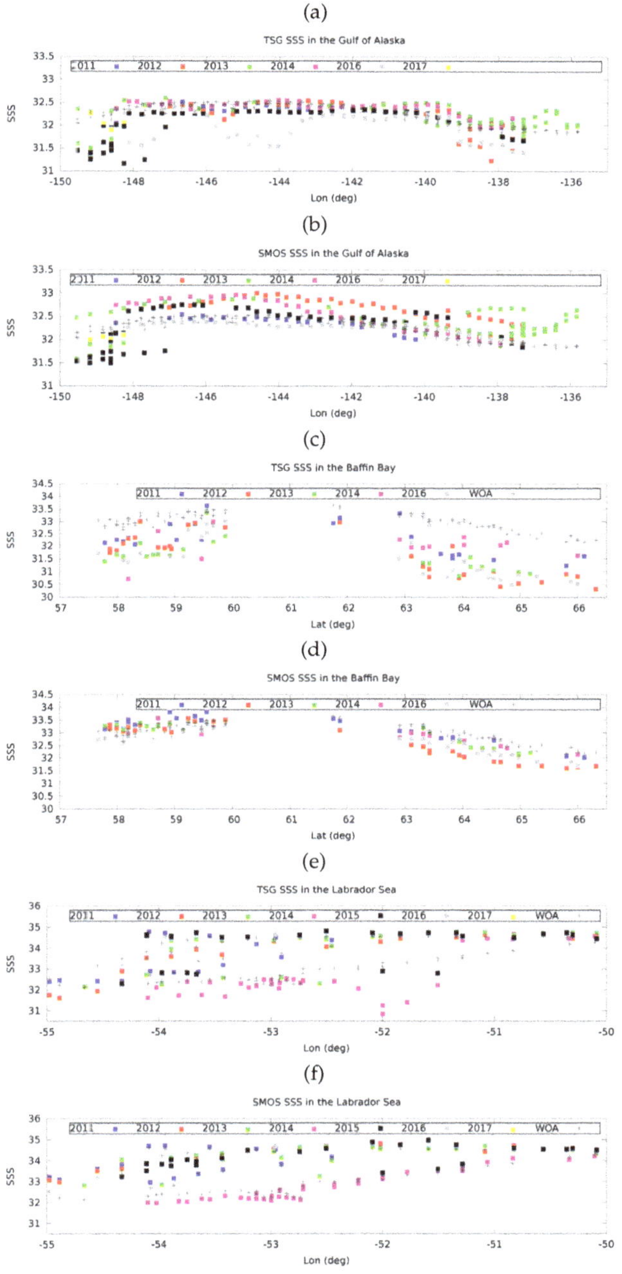

Figure 9. Routinely performed transects during 2011–2017 in: (i) the Gulf of Alaska as observed by TSG ((**a**) plot) and SMOS ((**b**) plot) in a zonal transect around 58°N and 150°W–135°W; (ii) the Baffin Bay as observed by TSG ((**c**) plot) and SMOS ((**d**) plot) in a meridional transect around 79°W and 57°N–67°N; and the Labrador Sea as observed by TSG ((**e**) plot) and SMOS ((f) plot) in a diagonal transect in the range of latitudes from 50°N to 60°N and in the longitude range from 55°W to 50°W.

In the central part of the Gulf of Alaska ((a) and (b) plots of Figure 9, longitudes between 146°W and 143°W), TSG captures a fresh anomaly in 2016. The freshest year measured by SMOS in this range of longitudes is also 2016, although SMOS does not capture the sudden freshening. The inter-annual variability captured in the eastern part of the gulf (longitudes between 137°W and 135°W) for 2013 and 2015 (green and black points, respectively) is also coherent between both sources: both SMOS and TSG show 2013 being saltier than 2015.

Close to the coast of Baffin Island ((c) and (d) plots of Figure 9, latitudes between 63°N and 65°N), the two freshest years as observed by TSG are 2012 and 2016 (red and grey points). These are also the two freshest years captured by SMOS. On the coast of Canada (latitudes between 57°N and 60°N), TSG captures a large fresh anomaly in 2014. In this case, SMOS does not capture the freshening. The (c) plot of Figure 9 shows that, in the range of 58.25°N–58.75°N, consecutive pink points jump from 30.5 to 32.5. This strong SSS gradient occurs over a very short distance, which is smaller than the correlation radii used in the objectively analyzed SMOS SSS fields. Therefore, this SMOS SSS product cannot capture this gradient. In the Labrador Sea ((e) and (f) plots of Figure 9), both SSS sources agree on the fact that the 2014 (pink pixels) is the freshest one.

Figure 10 shows the scatter plots of SMOS SSS and TSG SSS (red) and WOA and TSG SSS (blue), for the three previous cases of study. To better illustrate the added value of SMOS with respect to the selected annual reference (WOA), several statistical parameters (number of points, linear regression coefficient, RMS of the regression residuals and Pearson correlation) as well as the corresponding linear regression lines are also shown. In all three regions, the SMOS SSS data are in better agreement with TSG than the WOA SSS, as shown by the different statistical parameters. This is notably the case for both the Gulf of Alaska and the Baffin Bay regions, where the SMOS SSS correlations are, respectively, 0.73 and 0.79, while those of the WOA are 0.42 and 0.68. Moreover, the SMOS SSS regression lines (red lines) are closer to the diagonal than those from the WOA data set (blue lines). In the Labrador Sea the added value of SMOS SSS with respect to WOA is less pronounced.

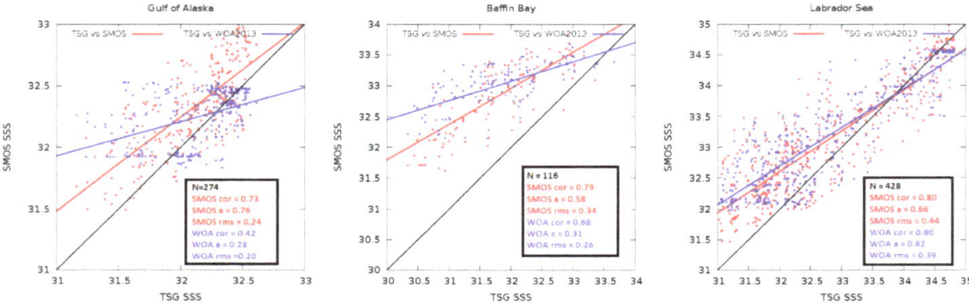

Figure 10. Correlation between SMOS SSS and TSG SSS in the three transects presented in Figure 9: Gulf of Alaska (**left**); Baffin Bay (in the **middle**); and Labrador Sea (**right**). Red line is the linear regression between SMOS SSS and TSG SSS (red points) and blue line is the linear regression between WOA SSS and TSG SSS (blue points). The number of measurements (N), the correlation coefficient (cor), the coefficient of linear regression (a) and the RMS of the residuals are shown in the legend for SMOS SSS (red) and WOA SSS (blue).

5. Sea Surface Salinity Variability Observed by SMOS at the Mouth of the Main Arctic Rivers

The largest intra-annual variability observed by SMOS is located near the mouth of the main Arctic rivers. In Section 4.3, we show that the inter annual variations of SSS described by SMOS agree with the ones described by TSG SSS in sub-Arctic regions. However, inside the Arctic basin, very few in situ measurements are available. In this section, we show that SMOS SSS variability is consistent with the SSS dynamics of the region, in the Arctic basin. In particular, we analyze the SSS variability close to the mouth of the Mackenzie and Ob Rivers. We compare the SMOS SSS maps with the output of the TOPAZ model and with the remotely sensed SSS provided by Aquarius and SMAP. We also look at the discharge data provided by Arctic Great Rivers project to correlate the freshening observed by SMOS to the river discharge events.

In Figure 11, monthly SMOS SSS maps (July, August and September) close to the mouth of the Mackenzie and Ob Rivers are shown (left and right plots, respectively). In Figure 12, TOPAZ SSS maps are displayed for the same months and regions. In Figure 13, the same regions and months are also used for representing Aquarius SSS maps (for years 2012 and 2014) and SMAP SSS maps (for years 2016 and 2017 in the case of the maps close to the Mackenzie River and 2015 and 2017 in the case of the maps close to the Ob River). Ice mask thresholds are different: the SMAP products use the limit of 3% of Sea Ice Concentration (SIC) computed with the Bootstrap algorithm [52], above this threshold that pixel is not considered water and is filtered out; Aquarius uses a threshold of 15% of the same algorithm; SMOS considers water pixels those with a SIC lower than 15% by using the EUMETSAT Ocean and Sea Ice Satellite application Facility (OSI-SAF) product.

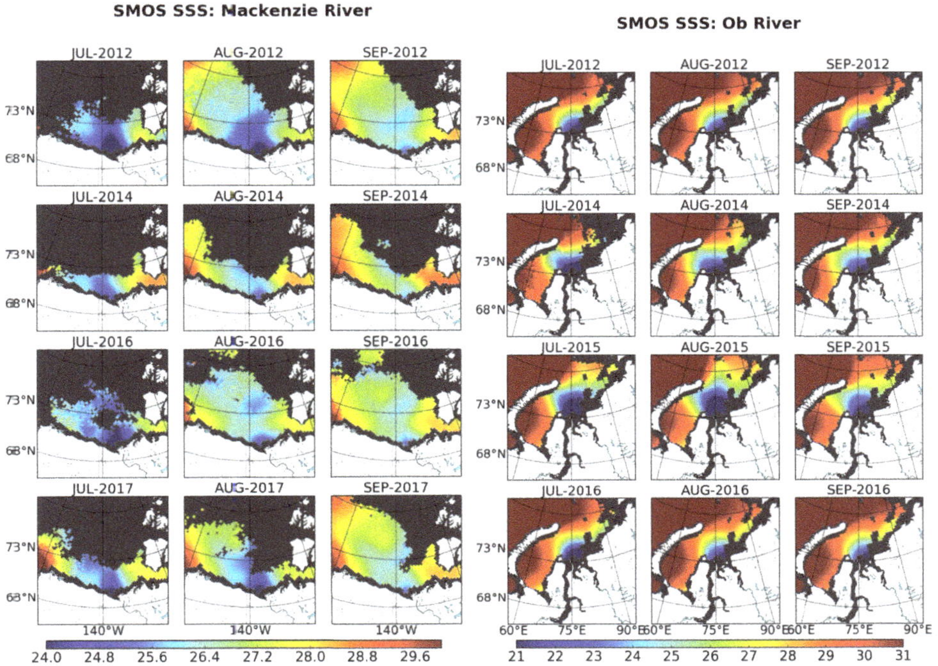

Figure 11. Monthly SMOS SSS maps close to the mouth of Mackenzie River: (**left**); and Ob River (**right**).

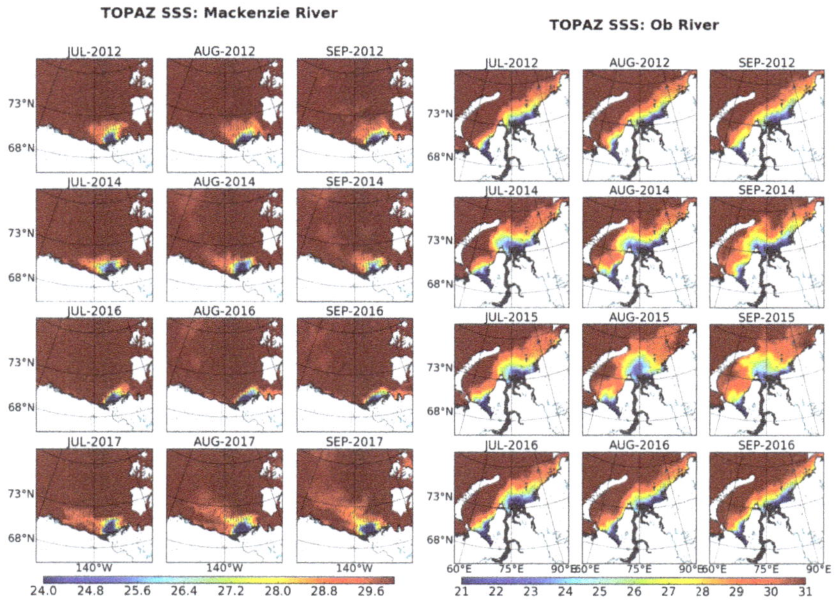

Figure 12. Monthly TOPAZ SSS maps close to the mouth of Mackenzie River: (**left**); and Ob River (**right**).

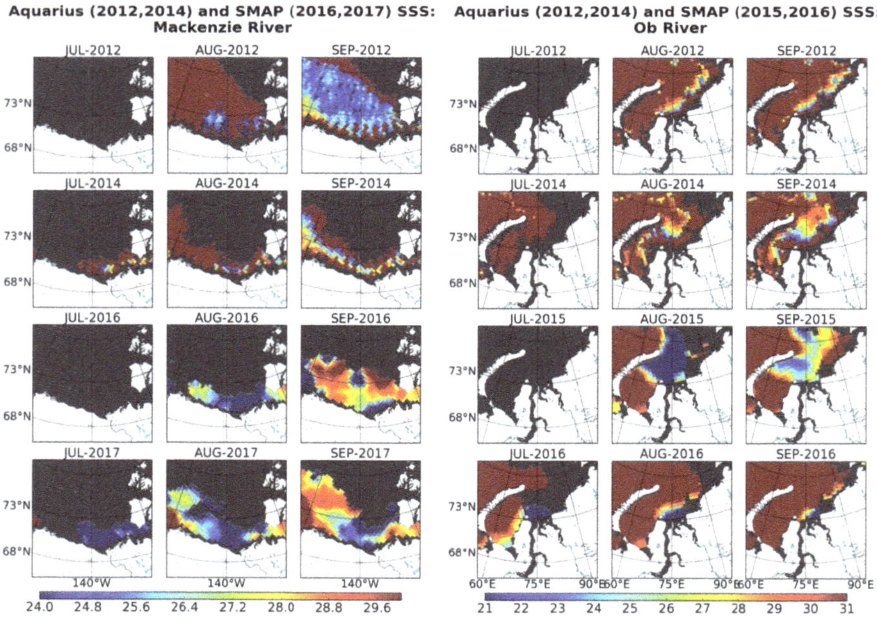

Figure 13. Monthly Aquarius and SMAP SSS maps close to the mouth of: Mackenzie River (**left**); and Ob River (**right**).

Figure 14 shows the daily river discharge of the Mackenzie and Ob Rivers. The Mackenzie River presents the maximum discharge by the end of May, except in 2012 when two maximums were observed (Figure 14 left). At that time, there is still a high percentage of sea ice in the region. Sea ice can be considered melted (less than 30% of ice concentration) by mid-July, except for 2013, when it melted slightly later (not shown). Due to the strong density stratification of the Arctic Ocean, the newly supplied fresh water tends to stay at the surface in the absence of enough strong wind-driven stirring. The persistence of the river plume on surface is well observed in the temporal evolution (July–August) of SMOS SSS maps in the Mackenzie River (left plots of Figure 11). The output SSS from the TOPAZ model in the mouth of the Mackenzie River (left plots of Figure 12) shows a smaller plume than the one displayed by SMOS. It does not change noticeably during the different years, while SMOS shows important inter annual differences. Moreover, variability for different months is not observed with TOPAZ outputs, while it is observed very clearly with SMOS. Aquarius SSS (2012 and 2014 of the left plots of Figure 13) suffers from strong positive biases with respect to the SSS captured by SMOS and TOPAZ at the ice edge. Although the ice mask in SMAP is more restrictive than for SMOS, both satellites observe coherent plumes structures. However, SMAP SSS gradients are larger than the ones captured by SMOS.

Daily river discharge data illustrates that the maximum discharge of the Ob River occurs by the end of May (in Figure 14 right), and that the greatest discharge happened in 2015. Since the region is almost melted around the beginning of July, the maximum Ob River discharge occurs before the sea water is free of ice.

Despite of the differences in the grid resolution: 12.5 km for TOPAZ and 25 km for SMOS, the plume of the Ob River described by TOPAZ (right plots of Figure 12) seems more consistent with the one described by SMOS (right plots of Figure 11) than in the case of the Mackenzie plume (in terms of SSS variability and the spatial coverage of the plume). Aquarius maps (right plots of Figure 13) display saltier SSS than SMOS, SMAP and TOPAZ close to the Ob River mouth, which does not seem geophysically reasonable. The inter annual variations shown by SMAP and SMOS (right plots of Figures 11 and 13) are also coherent. For example, 2015 is the year that both satellites display the major extension of fresh water in July and August, even though SMAP shows a larger plume than SMOS. This is also coherent with the in situ river discharge data shown in Figure 14 (right). TOPAZ maps also show a large extension of fresh water in 2015.

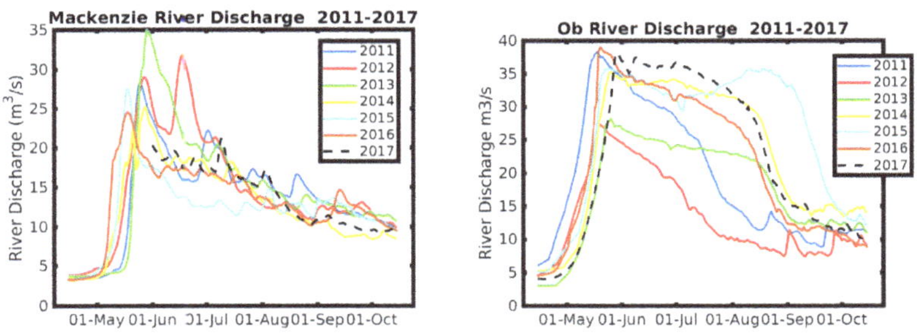

Figure 14. Daily river discharge from 2011 to 2017 from Arctic Great Rivers project: Mackenzie River (**left**); and Ob River (**right**).

In Figure 15, we compare monthly SMOS SSS anomalies with respect to monthly river discharge anomalies. These anomalies are computed as follows. First, the daily river discharge data are monthly averaged. Since these regions are frozen in winter months, and we are interested in comparing these data with SSS, we consider as reference the average of the monthly discharge of June, July, August and September for the seven years of study (2011–2017). The monthly discharge anomalies (green points in Figure 15) are generated by subtracting this reference to the monthly discharge data. Consistently, we consider as SMOS SSS reference the average of the monthly SMOS SSS maps of June, July, August and September for 2011–2017. The SMOS SSS anomalies shown in Figure 15 are the differences between the monthly SMOS SSS maps and the mentioned reference. The purple points represent the spatial average of these SMOS SSS anomalies in two regions close to the mouth of the Mackenzie River (latitudes between 69°N and 73°N and longitudes between 150°W and 130°W) and the Ob River (latitudes between 73°N and 77°N and longitudes between 60°E and 90°E). The relationship between the SSS anomaly and the discharge anomaly is not straightforward. The geophysical phenomena that play a role on the modification of the SSS in those regions are diverse and complex, for example the melting of the ice, the mixing due to wind, water advection, etc. A complete understanding of the interactions of these phenomena and the salinity are out of the scope of this work. At this stage, a more qualitative analysis is carried out, i.e., to check whether decreasing trends of SMOS SSS anomalies are linked to positive trends of river discharge anomalies and vice versa. As already mentioned, in both cases (Mackenzie and Ob Rivers), the maximum discharge happens in May when the region is usually frozen. That is why the green lines in Figure 15 usually show a decreasing trend. Consequently, increasing trends of SMOS SSS anomalies (purple lines) are expected. Figure 15 shows that, typically, SMOS SSS anomalies increase from July. However, from June to July, both the SMOS SSS and river discharge anomalies mainly show a decreasing trend. This can be explained by several mechanisms: on the one hand, the discharge measurements are not performed in the mouth of the rivers, therefore a delay between the discharge variations and the SSS response is expected, while, on the other hand, as already discussed, these regions are typically strongly stratified and, when the wind is not strong enough, the newly supplied fresh water tends to stay at the surface. Additionally, in July, the ice melting still occurs and refreshes the water in those regions. For the rest of months, an anti-symmetric (anti-correlated) behavior of the SSS with respect to the river discharge is observed (as expected).

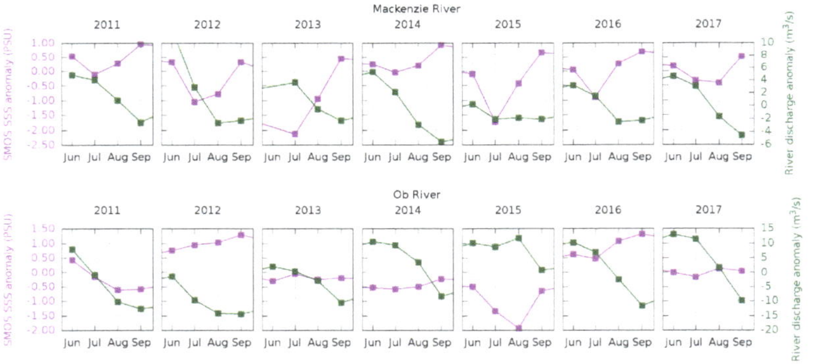

Figure 15. Monthly SMOS SSS anomaly (purple) versus monthly river discharge anomaly (green) from 2011 to 2017 from Arctic Great Rivers project: Mackenzie River (**top**); and Ob River (**bottom**).

6. Conclusions

This paper demonstrates the capability of SMOS-derived SSS to follow the vast salinity spatial and temporal variability in Arctic and sub-Arctic regions. This new product is produced using the retrieval algorithm proposed by Olmedo et al. [23] and includes some improvements with respect to the initial data set distributed by the Barcelona Expert Center. In particular, an enhanced time-dependent bias correction is applied. We also analyze the impact of considering different annual SSS references for the generation of the product. The conclusion of this analysis is that there are regions in the Arctic Ocean where the analyzed references are of poor quality, and, therefore, the corresponding SMOS SSS products suffer from spatial biases (which are constant in time). However, the dynamical range of the SSS described by these annual references (and also by the SMOS product) is geophysically consistent. Additionally, we analyze which is the most appropriate correlation radii for the generation of the objectively analyzed SSS fields. The conclusion of this analysis is that correlation radii greater than what is expected by the dynamics of the region provide the lowest error with respect to Argo floats, probably because the SMOS SSS retrievals have a larger error than in other regions (of warmer waters).

Validation of SMOS salinity maps at high latitudes against Argo floats shows that the product has a std in the range of 0.24–0.35. We have also used a dataset of 86 TSG provided by Copernicus. Besides the statistics of the differences between SMOS and TSG, we analyze the inter annual variations described by each one of the two SSS sources. SMOS and TSG SSS are in agreement in the major features of the SSS dynamics, although, as expected, the spatial scales that are resolved by SMOS are blurred with respect to the ones resolved by the TSG. It is important to notice that most of the measurements used in these comparisons are located outside the Arctic Ocean. Therefore, this comparison cannot be used to project a realistic quality assessment of SMOS SSS maps inside the Arctic Circle.

To complete the overview of the quality of this SMOS SSS product in the Arctic regions, we compare the SMOS SSS with TOPAZ, Aquarius and SMAP SSS close to two of the main rivers of the Arctic Ocean: the Mackenzie and Ob Rivers. We observe that the output from TOPAZ underestimates the plume of the Mackenzie River. Aquarius SSS maps are probably affected by ice–sea (and maybe also by some residual land–sea) contamination, showing saltier waters than the other satellite datasets do. Despite the differences in the ice mask, the inter- and intra-annual variations described by SMAP and SMOS are quite consistent. Although a more extensive work on the comparison between the remote sensed salinity products is still required, these results suggest that both SMOS and SMAP have great potential to routinely monitor the extension of the surface freshwater fluxes in the Arctic Ocean.

Even though the original resolution of the SMOS SSS is $\approx$40 km, the selected correlation radii are likely too large to monitor some mesoscale features of these Arctic and sub-Arctic regions. The reduction of the correlation radii will be driven by the reduction of the noise of the salinity retrievals. Future improvements of this SMOS SSS product are aimed at reducing the error of the salinity retrievals at high latitudes. Enhanced image reconstruction techniques as the one introduced in [48] have been assessed in [49,50] reaching promising results. The decrease of the error in the salinity retrieval would allow generating SMOS SSS maps with smaller correlation radii. Then, a richer mesoscale dynamics is expected to be captured by the future SMOS SSS products.

Finally, we would like to underline the need of implementing more in situ measurements in Arctic regions. On the one hand, they are absolutely required for a proper assessment of the satellite products in these regions. On the other hand, they are also required for the computation of a better annual reference in the Arctic Ocean. There are still many regions in the Arctic Ocean where very few measurements have been used for the computation of the salinity climatology. This in particular means that the knowledge of the salinity of those regions is really limited.

Author Contributions: Estrella Olmedo and Carolina Gabarró conceived and designed the idea of the manuscript. Estrella Olmedo conceived and implemented the algorithm for the SMOS SSS products generation. Estrella Olmedo and Carolina Gabarró performed the validation of the new SMOS SSS product. The manuscript was prepared by Estrella Olmedo and Carolina Gabarró. All authors contributed to the scientific discussions, revised and approved the final manuscript.

Funding: This research was funded by the Spanish R+D plan under projects L-BAND (ESP2017-89463-C3-1-R) and PROMISES (ESP2015-67549-C3-2) and from European Space Agency by means of the contracts SMOS ESL L2OS, CCI+ SSS and Arctic+ SSS.

Acknowledgments: This work was supported by the Spanish R+D plan under projects L-BAND (ESP2017-89463-C3-1-R) and PROMISES (ESP2015-67549-C3-2) and from European Space Agency by means of the contracts SMOS ESL L2OS, CCI+ SSS and Arctic+ SSS. We would like to thanks James McClelland, for offering us the daily river discharge data. We acknowledge Copernicus and Met Office Hadley Centre for providing TOPAZ and ESA for the L1B SMOS dataset. The authors would also like to thank LOCEAN for providing the Arctic TARA campaign dataset. Argo data were collected and made freely available by the International Argo Program and the national programs that contribute to it (http://www.argo.ucsd.edu,http://argo.jcommops.org). The Argo Program is part of the Global Ocean Observing System. The authors are grateful to the anonymous reviewers for their helpful suggestions.

Conflicts of Interest: The authors declare no conflict of interest.

References

1. Serreze, M.C.; Holland, M.M.; Stroeve, J. Perspectives on theArctic's shrinking sea-ice cover. *Science* **2007**, *315*, 1533–1536. [CrossRef] [PubMed]
2. Haas, C.; Pfaffling, A.; Hendricks, S.; Rabenstein, L.; Etienne, J.L.; Rigor, I. Reduced ice thickness in Arctic transpolar drift favors rapid ice retreat. *Geophys. Res. Lett.* **2008**, *35*. [CrossRef]
3. Comiso, J.C. Large Decadal Decline of the Arctic Multiyear Ice Cover. *J. Clim.* **2012**, *25*, 1176–1193. [CrossRef]
4. Haine, T.; Curry, B.; Gerdes, R.; Hansen, E.; Karcher, M.; Lee, C.; Rudels, B.; Spreen, G.; de Steur, L.; Stewart, K.D.; et al. Arctic freshwater export: Status, mechanisms, and prospects. *Glob. Planet. Chang.* **2015**, *125*, 13–35. [CrossRef]
5. Zhang, J.; Steele, M.; Runciman, K.; Dewey, S.; Morison, J.; Lee, C.; Rainville, L.; Cole, S.; Krishfield, R.; Timmermans, M.L.; et al. The Beaufort Gyre intensification and stabilization: A model-observation synthesis. *J. Geophys. Res. Oceans* **2016**. [CrossRef]
6. Rabe, B.; Karcher, M.; Kauker, F.; Schauer, U.; Toole, J.M.; Krishfield, R.A.; Pisarev, S.; Kikuchi, T.; Su, J. Arctic Ocean basin liquid freshwater storage trend 1992–2012. *Geophys. Res. Lett.* **2014**, *41*, 961–968. [CrossRef]
7. Peterson, B.; Holmes, R.; McClelland, J.; Vörösmarty, C.; Lammers, R.; Shiklomanov, A.; Shiklomanov, I.; Rahmstorf, S. Increasing river discharge to the Arctic Ocean. *Science* **2002**, *298*, 2171–2173. [CrossRef] [PubMed]
8. Mulligan, R.P.; Perrie, W.; Solomon, S. Dynamics of the Mackenzie River plume on the inner Beaufort shelf during an open water period in summer. *Estuar. Coast. Shelf Sci.* **2010**, *89*, 214–220. [CrossRef]
9. Jeffries, M.; Richter-Menge, J.; Overland, J.E. Arctic Report Card 2015. Technical Report, NOAA Reports. Available online: https://www.arctic.noaa.gov/Report-Card (accessed on 8 November 2018).
10. Font, J.; Camps, A.; Ballabrera-Poy, J. *'Microwave Aperture Synthesis radiometry: Setting the Path for Sea Surface Salinity Measurements from Space' in Remote Sensing of European Seas*; Springer: Berlin, Germany, 2008; ISBN 978-1-4020-6771-6.
11. Mecklenburg, S.; Wright, N.; Bouzina, C.; Delwart, S. Getting down to business—SMOS operations and products. *ESA Bull.* **2009**, *137*, 25–30.
12. Kerr, Y.; Waldteufel, P.; Wigneron, J.; Delwart, S.; Cabot, F.; Boutin, J.; Escorihuela, M.; Font, J.; Reul, N.; Gruhier, C.; et al. The SMOS mission: New tool for monitoring key elements of the global water cycle. *Proc. IEEE* **2010**, *98*, 666–687. [CrossRef]
13. Le Vine, D.; Lagerloef, G.; Colomb, F.; Yeh, S.; Pellerano, F. Aquarius: an instrument to monitor sea surface salinity from space. *IEEE Trans. Geosci. Remote Sens.* **2007**, *45*, 2040–2050. [CrossRef]
14. Lagerloef, G.; Colomb, F.; LeVine, D.; Wentz, F.; Yueh, S.; Ruf, C.; Lilly, J.; Gunn, J.; Chao, Y.; de Charon, A.; et al. The Aquarius/Sac-D mission: designed to meet the salinity remote-sensing challenge. *Oceanography* **2008**, *21*, 68–81. [CrossRef]
15. Entekhabi, D.; Njoku, E.; O'Neill, P.; Kellogg, K.; Crow, W.; Edelstein, W.; Entin, J.; Goodman, S.; Jackson, T.; Johnson, J.; et al. The Soil Moisture Active Passive (SMAP) mission. *Proc. IEEE* **2010**, *98*, 704–716. [CrossRef]
16. Zine, S.; Boutin, J.; Font, J.; Reul, N.; Waldteufel, P.; Gabarro, C.; Tenerelli, J.; Petitcolin, F.; Vergely, J.; Talone, M.; et al. Overview of the SMOS Sea Surface Salinity Prototype Processor. *IEEE Trans. Geosci. Remote Sens.* **2008**, *46*, 621–645. [CrossRef]
17. Swift, C.; McIntosh, R. Considerations for Microwave Remote Sensing of Ocean-Surface salinity. *IEEE Trans. Geosci. Electron.* **1983**, *GE-21*, 480–491. [CrossRef]
18. Yueh, S.; West, R.; Wilson, W.; Li, F.; Nghiem, S.; Rahmat-Samii, Y. Error Sources and Feasibility for Microwave Remote Sensing of Ocean Surface Salinity. *IEEE Trans. Geosci. Remote Sens.* **2001**, *39*, 1049–1059. [CrossRef]
19. Martín-Neira, M.; Oliva, R.; Corbella, I.; Torres, F.; Duffo, N.; Duran, I.; Kainulainen, J.; Closa, A.; Zurita, A.; Cabot, F.; et al. SMOS Instrument performance and calibration after six years in orbit. *Remote Sens. Environ.* **2016**, *180*, 19–39. [CrossRef]
20. Köhler, J.; Sena Martins, M.; Serra, N.; Stammer, D. Quality assessment of spaceborne sea surface salinity observations over the northern North Atlantic. *J. Geophys. Res. Oceans* **2015**, *120*, 94–112. [CrossRef]
21. Matsuoka, A.; Babin, M.; Devred, E. A new algorithm for discriminating water sources from space: A case study for the southern Beaufort Sea using MODIS ocean color and SMOS salinity data. *Remote Sens. Environ.* **2016**, *184*, 124–138. [CrossRef]

22. Tang, W.; Yueh, S.; Yang, D.; Fore, A.; Hayashi, A.; Lee, T.; Fournier, S.; Holt, B. The Potential and Challenges of Using Soil Moisture Active Passive (SMAP) Sea Surface Salinity to Monitor Arctic Ocean Freshwater Changes. *Remote Sens.* **2018**, *10*. [CrossRef]
23. Olmedo, E.; Martinez, J.; Turiel, A.; Ballabrera-Poy, J.; Portabella, M. Debiased non-Bayesian retrieval: A novel approach to SMOS Sea Surface Salinity. *Remote Sens. Environ.* **2017**, *193*, 103–126. [CrossRef]
24. Isern-Fontanet, J.; Olmedo, E.; Turiel, A.; Ballabrera-Poy, J.; García–Ladona, E. Retrieval of eddy dynamics from SMOS sea surface salinity measurements in the Algerian Basin (Mediterranean Sea). *Geophys. Res. Lett.* **2016**, *43*, 6427–6434. [CrossRef]
25. Olmedo, E.; Taupier-Letage, I.; Turiel, A.; Alvera-Azcárate, A. Improving SMOS Sea Surface Salinity in the Western Mediterranean Sea through Multivariate and Multifractal Analysis. *Remote Sens.* **2018**, *10*. [CrossRef]
26. Garcia-Eidell, C.; Comiso, J.; Dinnat, E.; Brucker, L. Satellite observed salinity distributions at high latitudes in the Northern Hemisphere: A comparison of four products. *J. Geophys. Res. Ocean* **2017**, *122*, 7717–7736. [CrossRef]
27. Argo. Argo float data and metadata from Global Data Assembly Centre (Argo GDAC). *SEANOE* **2000**. [CrossRef]
28. Reverdin, G.; Le Goff, H.; Tara Oceans Consortium, Coordinators; Tara Oceans Expedition, Participants. Properties of seawater from a Sea-Bird TSG temperature and conductivity sensor mounted on the continuous surface water sampling system during campaign TARA_20090913Z of the Tara Oceans expedition 2009–2013. *PANGAEA* **2014**. . [CrossRef]
29. Chassignet, E.P.; Hurlburt, H.E.; Metzger, E.J.; Smedstad, O.M.; Cummings, J.; Halliwell, G.R.; Bleck, R.; Baraille, R.; Wallcraft, A.; Lozano, C.; et al. Global Ocean Prediction with the HYbrid Coordinate Ocean Model (HYCOM). *Oceanography* **2009**, *22*, 64–75. [CrossRef]
30. Sakov, P.; Counillon, F.; Bertino, L.; Lisæter, K.A.; Oke, P.R.; Korablev, A. TOPAZ4: An ocean-sea ice data assimilation system for the North Atlantic and Arctic. *Ocean Sci.* **2012**, *8*, 633–656. [CrossRef]
31. Korosov, A.; Counillon, F.; Johannessen, A. Monitoring the spreading of the Amazon freshwater plume by MODIS, SMOS, Aquarius and TOPAZ. *J. Geophys. Res. Oceans* **2015**, *120*, 268–283. [CrossRef]
32. Brucker, L.; Dinnat, E.; Koenig, L. *Aquarius L3 Weekly Polar-Gridded Sea Surface Salinity, Version 6*; NASA National Snow and Ice Data Center Distributed Active Archive Center: Boulder, CO, USA, 2015. [CrossRef]
33. Brodzik, M.J.; Billingsley, B.; Haran, T.; Raup, B.; Savoie, M.H. EASE-Grid 2.0: Incremental but Significant Improvements for Earth-Gridded Datasets. *ISPRS Int. J. Geo-Inf.* **2012**, *1*, 32–45. [CrossRef]
34. Fore, A.G.; Yueh, S.H.; Tang, W.; Stiles, B.W.; Hayashi, A.K. Combined Active/Passive Retrievals of Ocean Vector Wind and Sea Surface Salinity With SMAP. *IEEE Trans. Geosci. Remote Sens.* **2016**, *54*, 7396–7404. [CrossRef]
35. Tenerelli, J.; Reul, N.; Mouche, A.; Chapron, B. Earth-Viewing L-Band Radiometer Sensing of Sea Surface Scattered Celestial Sky Radiation—Part I: General Characteristics. *IEEE Trans. Geosci. Remote Sens.* **2008**, *46*, 659–674. [CrossRef]
36. Reul, N.; Tenerelli, J.; Guimbard, S.; Collard, F.; Kerbaol, V.; Skou, N.; Cardellach, E.; Tauriainen, S.; Bouzinac, C.; Wursteisen, P.; et al. Analysis of L-band radiometric measurements conducted over the North Sea during the CoSMOS-OS airborne campaign. In Proceedings of the IEEE International Geoscience and Remote Sensing Symposium (IGARSS2007), Barcelona, Spain, 23–27 July 2007.
37. Guimbard, S.; Gourrion, J.; Portabella, M.; Turiel, A.; Gabarro, C.; Font, J. SMOS Semi-Empirical Ocean Forward Model Adjustment. *IEEE Trans. Geosci. Remote Sens.* **2012**, *50*, 1676–1687. [CrossRef]
38. Sabater, J.; De Rosnay. P. *Tech Note—Parts 1/2/3: Operational Pre-processing Chain, Collocation Software Development and Offline Monitoring Suite*; Technical Report; ECMWF: Reading, Uk, 2010.
39. Meissner, T.; Wentz, F. The complex dielectric constant of pure and sea water from microwave satellite observations. *IEEE Trans. Geosci. Remote Sens.* **2004**, *42*, 1836–1849. [CrossRef]
40. Klein, L.; Swift, C. An Improved Model for the Dielectric Constant of Sea Water at Microwave Frequencies. *IEEE Trans. Antennas Propag.* **1977**, *AP-25*, 104–111. [CrossRef]
41. Dinnat, E.P.; Boutin, J.; Yin, X.; Vine, D.M.L. Inter-comparison of SMOS and Aquarius Sea Surface Salinity: Effects of the dielectric constant and vicarious calibration. In Proceedings of the 2014 13th Specialist Meeting on Microwave Radiometry and Remote Sensing of the Environment (MicroRad), Pasadena, CA, USA, 24–27 March 2014; pp. 55–60. [CrossRef]

42. Brodzik, M.J.; Knowles, K.W. EASE-Grid: A Versatile Set of Equal-Area Projections and Grids. In *Discrete Global Grids*; National Center for Geographic Information & Analysis: Santa Barbara, CA, USA, 2002.
43. Gabarro, C.; Portabella, M.; Talone, M.; Font, J. Toward an Optimal SMOS Ocean Salinity Inversion Algorithm. *IEEE Geosci. Remote Sens. Lett.* **2009**, *6*, 509–513. [CrossRef]
44. Tenerelli, J.; Reul, N. *Analysis of L1PP Calibration Approach Impacts in SMOS Tbs and 3-Days SSS Retrievals over the Pacific Using an Alternative Ocean Target Transformation Applied to L1OP Data*; Technical Report; IFREMER/CLS: Paris, France, 2010.
45. Zweng, M.; Reagan, J.; Antonov, J.; Locarnini, R.; Mishonov, A.; Boyer, T.; Garcia, H.; Baranova, O.; Johnson, D.; Seidov, D.; et al. *World Ocean Atlas 2013, Volume 2: Salinity*; A. Mishonov Technical, Levitus, Eds.; NOAA Atlas NESDIS 74; NOAA: Washington, DC, USA, 2013; p. 39.
46. Steele, M.; Morley, R.; Ermold, W. PHC: A global ocean hydrography with a high-quality Arctic Ocean. *J. Clim.* **2001**, *9*, 2079–2087. [CrossRef]
47. Nurser, A.; Bacon, S. The Rossby radius in the Arctic Ocean. *Ocean Sci.* **2014**, *10*, 967–975. [CrossRef]
48. González-Gambau, V.; Turiel, A.; Olmedo, E.; Martínez, J.; Corbella, I.; Camps, A. Nodal Sampling: A New Image Reconstruction Algorithm for SMOS. *IEEE Trans. Geosci. Remote Sens.* **2016**, *54*, 2314–2328. [CrossRef]
49. González-Gambau, V.; Olmedo, E.; Turiel, A.; Martínez, J.; Ballabrera-Poy, J.; Portabella, M.; Piles, M. Enhancing SMOS brightness temperatures over the ocean using the nodal sampling image reconstruction technique. *Remote Sens. Environ.* **2016**, *180*, 202–220. [CrossRef]
50. González-Gambau, V.; Olmedo, E.; Martínez, J.; Turiel, A.; Duran, I. Improvements on Calibration and Image Reconstruction of SMOS for Salinity Retrievals in Coastal Regions. *IEEE J. Sel. Top. Appl. Earth Obs. Remote Sens.* **2017**, *10*, 3064–3078. [CrossRef]
51. Boutin, J.; Chao, Y.; Asher, W.E.; Delcroix, T.; Drucker, R.; Drushka, K.; Kolodziejczyk, N.; Lee, T.; Reul, N.; Reverdin, G.; et al. Satellite and In Situ Salinity: Understanding Near-Surface Stratification and Subfootprint Variability. *Bull. Am. Meteorol. Soc.* **2016**, *97*, 1391–1407. [CrossRef]
52. Comiso, J.C.; Meier, W.N.; Gersten, R. Variability and trends in the Arctic Sea ice cover: Results from different techniques. *J. Geophys. Res. Oceans* **2017**, *122*, 6883–6900. [CrossRef]

remote sensing

MDPI

Correction

Correction: Olmedo, E., *et al.* Seven Years of SMOS Sea Surface Salinity at High Latitudes: Variability in Arctic and Sub-Arctic Regions. *Remote Sensing* 2018, 10, 1772

Estrella Olmedo [1,*], Carolina Gabarró [1], Verónica González-Gambau [1], Justino Martínez [1], Joaquim Ballabrera-Poy [1], Antonio Turiel [1], Marcos Portabella [1], Severine Fournier [2] and Tong Lee [2]

1 Department of Physical Oceanography, Institute of Marine Sciences, CSIC & Barcelona Expert Center, Pg. Marítim 37–49, E-08003 Barcelona, Spain; cgabarro@icm.csic.es (C.G.); vgonzalez@icm.csic.es (V.G.-G.); justino@icm.csic.es (J.M.); joaquim@icm.csic.es (J.B.-P.); turiel@icm.csic.es (A.T.); portabella@icm.csic.es (M.P.)

2 Jet Propulsion Laboratory, California Institute of Technology, 4800 Oak Grove Drive, Pasadena, CA 91109, USA; Severine.Fournier@jpl.nasa.gov (S.F.); tlee@jpl.nasa.gov (T.L.)

* Correspondence: olmedo@icm.csic.es

Received: 10 April 2019; Accepted: 15 April 2019; Published: 18 April 2019

The authors wish to make the following corrections to this paper [1]:

The authors have detected an error in the program which computes the statistics with ARGO. The definition of some domains were wrong. The following table contains the correct statistics and substitute (Table 3).

Original Table 3.

Table 3. Regional analysis of the differences between TSG and SMOS SSS (SMOS-TSG). Regional statistics of differences between SMOS and ARGO are also included.

Region	Latitude Range	Longitude Range	SMOS-TSG			SMOS-ARGO		
			Meas	Mean	Std	Meas	Mean	Std
Denmark Strait	60N–65N	40W–25W	2322	0.22	0.28	34,177	0.03	0.21
North Atlantic	50N–60N	50W–20W	8028	0.19	0.45	99,177	0.01	0.35
Norwegian Sea	60N–70N	10W–5E	4587	0.02	0.67	32,906	−0.04	0.30
Northern Sea	55N–60N	0E–5E	53,366	0.16	0.99	11,855	−0.05	0.29
Gulf of Alaska	55N–60N	175W–125W	604	0.26	0.72	14,252	0.09	0.27
Chuckchi Sea	70N–75N	170W–145W	315	0.01	1.42	1751	−0.99	0.37
Labrador Sea	55N–60N	55W–45W	737	0.14	0.71	32,853	−0.08	0.29
Baffin Bay	60N–65N	65W–55W	278	0.89	0.62	5791	−0.02	0.42

Updated Table 3.

Table 3. Regional analysis of the differences between TSG and SMOS SSS (SMOS-TSG). Regional statistics of differences between SMOS and ARGO are also included.

Region	Latitude Range	Longitude Range	SMOS-TSG			SMOS-ARGO		
			Meas	Mean	Std	Meas	Mean	Std
Denmark Strait	60N–65N	40W–25W	2322	0.22	0.28	25,153	0.02	0.21
North Atlantic	50N–60N	50W–20W	8028	0.19	0.45	102,575	0.01	0.35
Norwegian Sea	60N–70N	10W–5E	4587	0.02	0.67	33,841	−0.05	0.31
Northern Sea	55N–60N	0E–5E	53,366	0.16	0.99	0	-	-
Gulf of Alaska	55N–60N	175W–125W	604	0.26	0.72	14,841	0.09	0.27
Chuckchi Sea	70N–75N	170W–145W	315	0.01	1.42	1751	−0.99	0.37
Labrador Sea	55N–60N	55W–45W	737	0.14	0.71	33,711	−0.07	0.29
Baffin Bay	60N–65N	65W–55W	278	0.89	0.62	5919	−0.02	0.41

The corrected table does not change the scientific results of the paper. The authors would like to apologize for any inconvenience caused to the readers by these changes.

Reference

1. Olmedo, E.; Gabarró, C.; González-Gambau, V.; Martínez, J.; Ballabrera-Poy, J.; Turiel, A.; Portabella, M.; Fournier, S.; Lee, T. Seven Years of SMOS Sea Surface Salinity at High Latitudes: Variability in Arctic and Sub-Arctic Regions. *Remote Sens.* **2018**, *10*, 1772. [CrossRef]

remote sensing

MDPI

Article

Intercomparison of In-Situ and Remote Sensing Salinity Products in the Gulf of Mexico, a River-Influenced System

Jorge Vazquez-Cuervo [1,*], Severine Fournier [2], Brian Dzwonkowski [1] and John Reager [2]

1 University of South Alabama, Dauphin Island Sea Lab, Dauphin Island, AL 36528, USA; briandz@disl.org
2 Jet Propulsion Laboratory, California Institute of Technology, Pasadena, CA 91109, USA; severine.fournier@jpl.nasa.gov (S.F.); john.reager@jpl.nasa.gov (J.R.)
* Correspondence: jorge.vazquez@jpl.nasa.gov; Tel.: +1-818-354-6980

Received: 7 August 2018; Accepted: 1 October 2018; Published: 4 October 2018

Abstract: The recent emergence of satellite-based sea surface salinity (SSS) measurements provides new opportunities for oceanographic research on freshwater influence in coastal environments. Several products currently exist from multiple observing platforms and processing centers, making product selection for different uses challenging. Here we evaluate four popular SSS datasets in the Gulf of Mexico (GoM) to characterize the error in each product versus in-situ observations: Two products from NASA's Soil Moisture Active Passive (SMAP) mission, processed by Remote Sensing Systems (REMSS) (40 km and 70 km); one SMAP 60 km product from the Jet Propulsion Laboratory (JPL); and one 60 km product from ESA's Soil Moisture Ocean Salinity (SMOS) mission. Overall, the four products are remarkably consistent on seasonal time scales, reproducing dominant salinity features. Towards the coast, 3 of the 4 products (JPL SMAP, REMSS 40 km SMAP, and SMOS) show increasing salty biases (reaching 0.7–1 pss) and Root Mean Square Error (RMSD) (reaching 1.5–2.5 pss), and a decreasing signal to noise ratio from 3 to 1. REMSS 40 km generally shows a lower RMSD than other products (~0.5 vs. ~1.1 pss) in the nearshore region. However, at some buoy locations, SMOS shows the lowest RMSD values, but has a higher bias overall (>0.2 vs. <0.1 pss). The REMSS 70km product is not consistent in terms of data availability in the nearshore region and performs poorly within 100 km of the coast, relative to other products. Additional analysis of the temporal structure of the errors over a range of scales (8/9-day to seasonal) shows significantly decreasing RMSD with increasing timescales across products.

Keywords: sea surface salinity; SMAP; SMOS; Gulf of Mexico; validation

1. Introduction

Salinity dynamics are a critical component of marine systems that shape the physical environment and consequently influence ecosystem function by indirectly mediating biogeochemical processes. The structure and evolution of salinity patterns often have a significant role in setting stratification levels as well as influencing circulation patterns by affecting density gradients. This is particularly true in regions that are strongly influenced by river discharge. For example, freshwater associated with river plumes can have an influence on air–sea interaction through the formation of barrier layers, a layer near the surface with salinity stratification, but uniform temperature [1,2]. Barrier layers separate the density mixed layer above from the thermocline below, thus limiting the vertical mixing of heat between the mixed layer and the thermocline [3,4]. Rivers also supply a large amount of nutrients to the ocean, which can have an impact on biogeochemistry, ecology, and the carbon cycle [5,6]. As a result, improving the understanding of the impacts of river discharge on continental shelves and adjacent open ocean represent an active area of research in the oceanographic community.

Due to the importance of the upper ocean salinity structure, there have been notable efforts to improve the ability to map salinity patterns in the global ocean. In particular, satellite missions have provided unprecedented mapping capabilities of sea surface salinity (SSS) with spatiotemporal resolutions far beyond traditional in-situ measurement programs. Since 2010, SSS observations have been available from the European Space Agency's (ESA's) Soil Moisture and Ocean Salinity (SMOS) mission. During the period 2011–2015, the NASA's Aquarius/SAC-D satellite mission was providing SSS measurements as well. Finally, since 2015, NASA's Soil Moisture Active Passive (SMAP) mission also provides continuous SSS observations. From these satellite missions, several data products using different retrieval algorithms are currently available, with a range of spatial and temporal resolutions.

These products have been traditionally validated with in-situ data primarily in open ocean environments, for example, the Argo drifter program. Previous studies on the validation of SSS have focused on global validation using in-situ data [7–9]. Overall, Tang et al. [7] found that, on a monthly time scale, the Root Mean Square Error (RMSD) of SMAP SSS, when compared with Argo data, was approximately 0.2 pss. This is consistent with the global study of Lee, [8] which showed an accuracy of 0.2 pss for Aquarius SSS. Reference [9] also shows that the precision of the monthly SMOS SSS measurements is on the order of 0.2 pss globally. However, these statistics are based on global comparisons and there are only a few regional studies that evaluate the SSS observations with in-situ data in a coastal, river-influenced environment.

There have been some limited efforts to evaluate the capability of satellite SSS measurements in a few basin-scale studies. Reference [9] presents comparisons done in the Mediterranean with thermosalinograph (TSG) data and shows an average regional RMSD of 0.52 pss for SMOS SSS. Reference [10] evaluates SMAP SSS measurements in the Bay of Bengal and finds an average regional RMSD of 0.54 pss and a correlation coefficient of 0.81 when comparing SMAP to in-situ values. They also found the RMSD was dependent on distance from the coast, with RMSD increasing to 2 pss near the coast. Reference [11] also compares an early version of SMAP SSS measurements with in-situ data in the Gulf of Mexico (GoM) and finds an average regional RMSD of 1.36 pss over only a few months in 2015. Some other regional studies assessed SSS by comparison with other remote sensing datasets, such as altimetry, as well as in-situ data [12]. Reference [13] directly compares measurements of SMOS derived SSS with sea surface temperature from the National Oceanographic and Atmospheric Administration's (NOAA's) Advanced Very High-Resolution Radiometer in the Agulhas Retroflection Region. The comparison was done to examine the responses of salinity and sea surface temperature at the interannual time scales. Despite being less common, the regional evaluation of these products is critically needed, as modern research efforts push the limits of these products from global scale to regional and coastal science applications. However, to our knowledge, no regional study evaluates and intercompares different SSS products and their error characteristics for a river-influenced region.

With the current availability of remote sensing datasets that measure SSS, it is imperative to conduct regional validations that allow users to make decisions about the dataset which is best for their application. Here, we focus on assessing contemporary SSS products in a representative river-influenced system, namely the GoM. The GoM presents an opportunity to evaluate these products in a semi-enclosed basin with large temporal and spatial variability due to river discharge and advection from regional circulation. Typical of many riverine influence systems, the strong salinity gradients and highly variable conditions present in the GoM represent a markedly different environment than open ocean conditions that have slower temporal scales and larger spatial scales of variability. We highlight the performance of different satellite derived SSS data products from the SMAP and SMOS missions versus in-situ data over the period 2015–2017 in the GoM. In particular, this study assesses how well all the products reproduce the known seasonal and interannual cycles associated with river runoff and if there is a relationship between the quality of the SSS retrievals and distance from shore. This type of validation exercise, and its relationship to applications, are part of the crucial documentation needed by the user community for both research and applications to ensure the use of the most desirable dataset for their needs. Furthermore, understanding the performance of SSS products in the GoM is an important undertaking for regional interests, but more generally, it serves as an assessment of SSS product performance in a representative marine system strongly influenced by river discharge. The goal of this work is to provide statistics that can be used to conclude what products perform well for specific applications in the study region of the GoM.

Study Region

The GoM is a region that is influenced by large sources of both fresh and salty waters, resulting in large seasonal and interannual variability as well as strong horizontal and vertical salinity gradients within the system. The predominant source of freshwater in the northern GoM is the Mississippi River outflow, being the 6th large river system in the world [14]. In addition, there are other large freshwater sources across the northern GoM, including the Atchafalaya River, the Mobile Bay river system, the Apalachicola River, as well as numerous smaller rivers from Texas to Florida that can individually and collectively impact the freshwater inputs into the coastal ocean region [15–17]. For example, the effects of the Mississippi River discharge on the transport and fate of oil from the Deepwater Horizon Oil Spill have been a major topic of interest in the Gulf of Mexico [18]. On the opposite end of the salinity spectrum, the Loop Current brings high salinity waters (>35 pss) into the GoM at the southern boundary of the system via the channel between the Yucatan Peninsula and Cuba [19]. While the Loop Current flow path is generally in the eastern portion of the GoM, the system does shed westward propagating eddies, typically every 6–11 months [20,21]. As a result, the high salinity inputs from the Loop Current can impact large areas of the GoM. The interaction between coastal sources of freshwater and the Loop Current (and its associated eddies) leads to highly variable salinity fields that change over time at temporal scales of days to weeks to months [10,11,22,23]. Furthermore, this region of the ocean is typically underrepresented by the Argo drifter program, which has limited the SSS community's ability to assess the quality of data products that are readily available. Overall, the different physical phenomena that affect salinity in the GoM make it an ideal basin for the validation of SSS products in a semi-enclosed basin.

2. Materials and Methods

2.1. Data

Our study focuses on SSS from a SMAP mission which is relatively new (beginning in 2015), and the longer term-SMOS satellite mission (2010–present) in our analysis to provide a comparison between different sensor products. In total, we examine 4 different satellite SSS products during the period 2015–2017 and details on each of these products are provided below. In addition, in-situ measurements of SSS from buoys and the World Ocean Database are also used as ground truth data in assessing the quality of the satellite SSS products. While there are inherent differences in the satellite SSS and the in-situ measurements, for example, depth of measurement (satellite SSS represent salinity in the top few cm of the ocean while in-situ salinity measurements are typically deeper than 1 m), spatial foot print, and so on, the combination of these two datasets provides a reasonable means of evaluating SSS product performance [12].

2.1.1. REMSS 40 km SMAP and REMSS 70 km SMAP

We used the Remote Sensing Systems SMAP Data Version 2.0 [24] from April 2015 to December 2017, distributed by Remote Sensing Systems (REMSS). Two versions of this data were used, a 40 km resolution product, and a 70 km resolution product, named REMSS SMAP 40 km and REMSS 70 km SMAP respectively in the following. Both products come as 8-day running means combining both ascending and descending orbits gridded on a 0.25 $\times$ 0.25 grid. More details on this SMAP product can be found in [24]. Briefly, the difference between the 40 km and 70 km is based on the smoothing radius. The smoothing is applied to the level 2 data and these values are then gridded onto the same 0.25° grid as the 40 km data. Global comparisons with HYCOM indicate that biases and RMSD are reduced in the 40 km dataset to 0 and 0.26 pss [24]. However, differences may exist regionally. Thus, both of these datasets were implemented in the validation study.

2.1.2. JPL SMAP

A third SMAP SSS product used in this study is the Level 3 Version 4.0, 8-day running mean SMAP SSS product distributed by the Jet Propulsion Laboratory, named JPL SMAP in the following. This product is a combination of ascending and descending orbits and also comes gridded on a 0.25 × 0.25 grid with an effective resolution of 60 km. More information on the data and their use may be found in the user's guide [25], but the primary differences between this product and the REMSS 40 km relates to the application of different approaches to land contamination and smoothing of the overall product. The JPL SMAP product uses a land climatology to flag values for possible land contamination. As the GoM is an enclosed basin, this could lead to potential differences between the datasets. Quality flagging, gridding schemes based on the fore and aft views are also different between the two datasets. In the dual-view geometry, there is a forward scan from the satellite ground track position as well as an aft scan. Both scans are used to derive salinity measurements.

2.1.3. SMOS

We used the L3_DEBIAS_LOCEAN_v3 SMOS Sea Surface Salinity, produced by LOCEAN/IPSL (UMR CNRS/UPMC/IRD/MNHN) laboratory and ACRI-st company that participate in the Ocean Salinity Expertise Center (CECOS) of (French Ground Segment for the SMOS Level 3 and Level 4 data) Centre Aval de Traitement des Donnees SMOS (CATDS). This product is distributed by the Ocean Salinity Expertise Center (CECOS) of the CNES-IFREMER Centre Aval de Traitement des Donnees SMOS (CATDS), at IFREMER, Plouzane (France) [26]. The 0.25-degree maps are provided every 4 days from January 2010 to December 2017 and are derived from a combination of ascending and descending orbits. Debiased SSS are temporally averaged using a slipping Gaussian kernel with a full width at half maximum of 9 days. A median filtering over nearest neighbors is applied. Based on the smoothing and filtering, the effective resolution of the SMOS is 60 km, similar to the JPL SMAP product. This similarity in smoothing is important for interpreting the statistical comparisons with the buoys and the World Ocean Database (WOD).

2.1.4. In-Situ Data

The availability of coastal in-situ data is critical for these types of comparisons. In regional basins, the temporal and spatial coverage of the Argo drifter program is often not sufficient to resolve shelf and shelf-open ocean processes. As a result, this study focuses on two sets of in-situ data, measurements from continuous surface buoys, as well as observations of opportunity from the WOD, which address, to some extent, the limitation of typical SSS validation efforts. The buoy data are collected from 7 buoys in the eastern and western shelves of the GoM. Measurements were made every 0.5–1.0 h, from which daily averages were produced for comparisons with the satellite SSS data. The data are part of the NOAA National Data Buoy Center (NDBC) and may be retrieved from (https://www.ndbc.noaa.gov/) and the Texas Automated Buoy System (TABS) (http://tabs.gerg.tamu.edu/). The west Florida shelf buoys acquire data at 1 m deep and the Texas buoys at 1.8 m [27]. In the following we refer to these buoy data as buoy SSS.

Data from the WOD were retrieved directly from NOAA's National Oceanographic Data Center (NODC) (now part of the National Centers for Environmental Information (NCEI) https://www.nodc.noaa.gov/OC5/WOD/pr_wod.html). Data from the WOD include conductivity/temperature/depth (CTDs), gliders, profilers, and drifting buoys and provide a unique opportunity for validation in a coastal/regional basin. Here, we use the salinity in-situ measurements from the 1–9 m depth range in order to have more in-situ near-surface observations. In the following, we refer to these WOD in-situ near-surface measurements as WOD SSS. The WOD2013 contains the full set of quality controls. Updates from April 2013 to the present implement the initial quality controls.

2.2. Methodology

To get a broad sense of the systematic performance between products, means and their associated standard deviations are derived from the REMSS 40 km SMAP, REMSS 70 km SMAP, JPL SMAP, and SMOS products. Record-length and seasonal means are calculated from the start of the SMAP mission in April 2015 through the end of 2017. Winter is defined as the monthly averages for January, February, and March; spring as the monthly averages for April, May, and June; summer as the monthly averages for July, August, and September; fall as the monthly averages for October, November, and December. Standard deviations are calculated based on the same periods of time.

To validate the accuracy of the satellite SSS products, comparisons were made directly with the seven buoys and data from the WOD. Figure 1 shows a map of the density of the WOD SSS from April 2015 through the end of 2017. Most available data are located in the northern part of the GoM and the density is higher along the shelf. Overlaid on the map are the locations of the 7 buoys, 4 on the Texas shelf and 3 along the coast of Florida.

We collocated the 7 buoys with satellite SSS using the nearest neighbor location. Comparisons between the SSS products and the buoys were made at three-time scales: Daily/8-day, 31-day, and 91-day time scales. SMAP SSS data have a temporal resolution (repeat orbit) of 8-day. SMOS has a temporal resolution of 9-day. It has an exact repeat of 149 days with a subcycle of 18 days and a revisit time of 3 to 5 days. Nine days was chosen because it is half of the SMOS subcycle and close to one week.

These were compared with the daily buoy SSS data, as the complete coverage of satellite SSS is completed in 8–9 days. For the 31- and 91-day time scales, SSS and buoys SSS data were smoothed using a running mean with a window of +/−15 and +/−45 days, respectively. Statistics were then calculated for the daily/8-day running means, as well as 31-day and 91-day smoothed results. The rationale for different averaging schemes was to examine whether smoothing over the monthly and seasonal time scales significantly change product quality.

Concerning the WOD SSS, the subsetting capability was used to extract all the data that overlapped from the start of the SMAP mission data availability (April 2015) through the end of 2017. All the data were extracted in the region from 18°N to 32°N and 100°W to 80°W. The co-location between satellite and WOD in-situ SSS was made using a nearest neighbor approach as well. Note that near-surface salinity stratification in the upper few meters can contribute to differences between satellite (for the top cm) and in-situ SSS at these depths [12].

Below are the definitions we used for the root mean square difference (RMSD) and signal to noise ratio. RMSD was based on the co-located satellite derived products and in-situ data. RMSD is then calculated as:

$$RMSD = \sqrt{\frac{\sum(satellite\ SSS - in\ situ\ SSS)^2}{n\ total}} \quad (1)$$

where *n total* is the total number of points for the summation. The signal to noise ratio (*SN*) can then be defined as:

$$SN = \frac{STDs}{RMSD} \quad (2)$$

where *STDs* is the standard deviation of the satellite product defined as:

$$STDs = \sqrt{\frac{\sum(satellite - mean\ sat)^2}{n\ total}} \quad (3)$$

where *mean sat* is the mean of the SSS derived from the satellite product over the n total points. It is important that we used the STD of the satellite product to define the signal. As defined by the variability measured by the satellite, the signal will not include missing subpixel scale variability. It will include, of course, noise in the data, due to instrument error, land contamination, and other factors in determining SSS from the satellite. Thus, the signal to noise estimate is a best estimate, based on using the satellite measurements and the WOD.

Figure 1 shows the density of the WOD salinity data. A considerable amount of the northern GoM is covered for the period of time between April 2015 and December 2017. Overlaid on the map are also 7 independent buoys that are used in comparisons with the 4 SSS products. Following the naming convention of the buoys, the 7 buoys are labeled C10, C12, C13, F, X, V, and K. Each buoy consists of a time series of surface salinity, which spanned the 2015–2017 time frame. However, the starting time, as well as data gaps, vary for each buoy. As a result, the 3 buoys (2 western and 1 eastern gulf locations) with the most complete time records are highlighted in the section below. However, the overall statistics for all the buoys are shown in Table 1. Results will be elaborated on in the Discussion section. The issue of the quality of retrievals based on proximity to land will also be addressed in the Discussion section. The issue of stratification and biases in the SMOS data has been addressed by Reference [12]. Results here build on those presented in Reference [28] in the Gulf of Mexico.

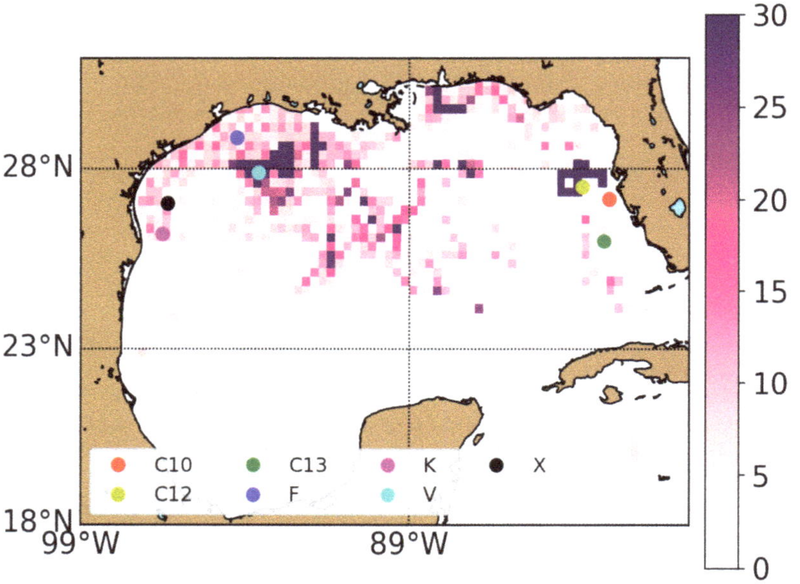

Figure 1. Density plot of the World Ocean Database (WOD) salinity in-situ data per bin of 0.25 degree from April 2015 to December 2017. The colored circles show the locations of the buoys (C10: Red, C12: Yellow, C13: Green, F: Blue, K: Magenta, V: Cyan, X: Black).

3. Results

3.1. Climatologies and Seasonal Variability

Figures 2 and 3 show the seasonal means and standard deviations of the REMSS 40 km and 70 km SMAP, JPL SMAP, and SMOS products, respectively, for the four datasets in winter, spring, summer, and fall (see Materials and Methods section). All four datasets clearly show maximum freshening in the northern GoM in the summer time frame. The seasonal cycle is consistent with previous results reported by Fournier et al. [11]. Freshening starts in the spring, reaching its maximum in summer, with a large tongue of freshwater extending offshore in the eastern GoM associated with the Mississippi River plume. However, differences in the datasets do exist primarily in the coastal regions. Freshening is clearly visible, starting in the spring time frame for the JPL SMAP, REMSS SMAP 40 km and 70 km, and SMOS. However, freshening in spring is difficult to identify in the REMSS

70 km SMAP product. This can be explained by the smoothing that occurs in the REMSS 70 km SMAP product and the proximity to land. White in the images indicates missing data in the product due to land contamination. The area associated with land contamination in the REMSS 70 km SMAP product extends further from the coast, thus masking the majority of spring time freshening. The same issue is seen in the fall means, where the fresher conditions along the northern GoM coast are masked out. Overall, all four datasets show remarkably good comparisons.

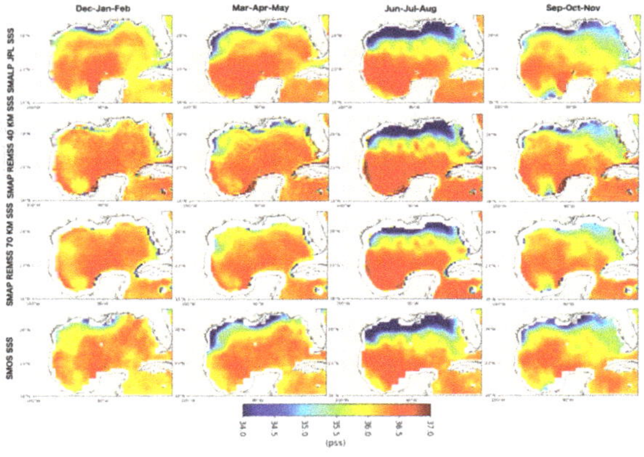

Figure 2. 2015–2017 seasonal maps of sea surface salinity (SSS) from the Jet Propulsion Laboratory's Soil Moisture Active Passive (JPL SMAP) (first row), Remote Sensing Systems Soil Moisture Active Passive (REMSS SMAP) 40 km (second row), REMSS SMAP 70 km (third row), and Soil Moisture Ocean Salinity (SMOS) (last row). The columns correspond to seasons defined in the Materials and Methods section.

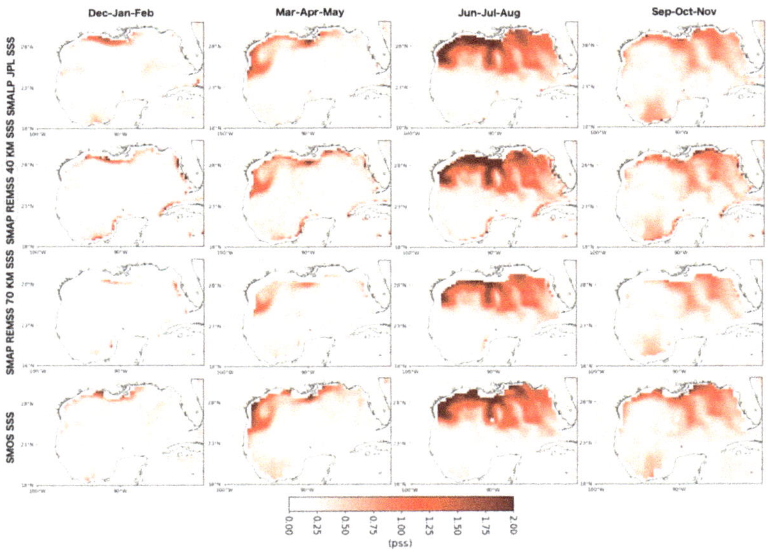

Figure 3. 2015–2017 seasonal maps of the standard deviation of the SSS from JPL SMAP (first row), REMSS SMAP 40 km (second row), REMSS SMAP 70 km (third row), and SMOS (last row). The columns correspond to seasons defined in the Materials and Methods section.

Figure 3 shows standard deviation (STD) for the seasonal means. The dominant features occur during spring through fall with a maximum variability extending offshore in summer. All four datasets show three large areas of variability, corresponding to several river plumes associated with multiple river basins, such as the Mississippi River Basin, the Mobile Bay river system basin, and the Texas river basin [28]. The largest differences are again seen closest to the coast. However, all four datasets are consistent in showing climatological variability. For example, there is a high variability in the winter in the western GoM, as well as a high variability in both summer and fall that covers much of the central and western GoM.

The remarkable consistency of the mean and STD seasonal means between the four datasets is a confirmation that, on the seasonal time scale, they are all reproducing the dominant features associated with salinity changes in the GoM, especially those associated with river discharge. Both the mean and STD establish that these datasets are consistent in reproducing climatological temporal and spatial variability, and so these datasets were further examined for possible biases and RMSD when compared with in-situ SSS from independent buoys and the WOD.

3.2. Comparisons with Buoys

In Figures 4–6, we present comparisons between satellite SSS and buoy SSS at selected buoys for three different time scales. Figure 4a–c shows the time series at buoy C12 for the 4 SSS products JPL SMAP, REMSS 40 km SMAP, REMSS 70 km SMAP, and SMOS. Figure 4 shows (a) the daily/8-day time series, (b) averaged monthly, and (c) averaged seasonally. Referencing Figure 1, buoy C12 is located approximately 100 km offshore from the west Florida coast. All the datasets show a freshening of approximately 2 pss in the summer to fall of 2016. Figure 4b,c shows similar results for smoothing at 31-day running means and 91-day running means. Results on the statistics are summarized in Table 1.

Figure 5a–c shows similar results for Buoy V, which is located approximately 400 km offshore of the Texas Coast. As with buoy C12, all four SSS datasets show the large freshening event in summer to fall of 2015 and 2016, reproducing the seasonal cycle due to river discharge [11]. Figure 5a shows the daily time series, Figure 5b the monthly means, and Figure 5c the seasonal means. The freshening seen at Buoy V of approximately 4 pss is consistent with previous reports by [11] on the extreme freshening event of 2015 associated with the flood along the Texas Coast. Similar results are shown in Figure 6a–c for Buoy F, which is located approximately 100 km from the Texas Coast. All datasets show a freshening in the summer of 2016 of approximately 5 pss.

Statistics for all the buoys are presented in Table 1. For buoy C10, it is the REMSS 40 km product that shows the best results, with fresh biases of 0.1 pss and RMSD of 0.35 at the seasonal time scale, while the JPL SMAP product has a 0.2 pss bias and a RMSD decreasing from 0.78 to 0.5 pss at seasonal time scales. Neither the REMSS 70 km product or the SMOS product have co-locations retrieved for the C10 buoy.

At buoy C12, overall SMOS shows the best results, with biases close to zero and RMSD reduced to 0.1 pss at the seasonal time scale from 0.43 pss at 9-day, similar to the JPL SMAP product. However, the JPL SMAP product shows larger fresh biases of 0.2 pss. The REMSS 40 km SMAP product has fresh biases of 0.1 pss, with RMSD reduced to 0.25 pss at the seasonal time scale from 0.5 pss. At buoy C12, the REMSS 70 km SMAP product shows slightly reduced fresh biases (approximately 0.09 pss), but the REMSS 70 km SMAP product has larger RMSD at the 8-day, monthly, and seasonal time scales (0.6, 0.5 and 0.37 pss respectively).

At buoy C13, overall SMOS has the best results, with biases close to zero, with RMSD reduced to 0.1 pss at the seasonal time scale from 0.3 pss at 9-day. The two REMSS products perform similarly, with biases close to zero and RMSD reduced to 0.25 at the seasonal time scale from 0.5 pss at 8-day. The JPL SMAP product shows fresh biases of 0.1 pss, with RMSD reduced to 0.2 pss at the seasonal time scale from 0.4 pss at 8-day.

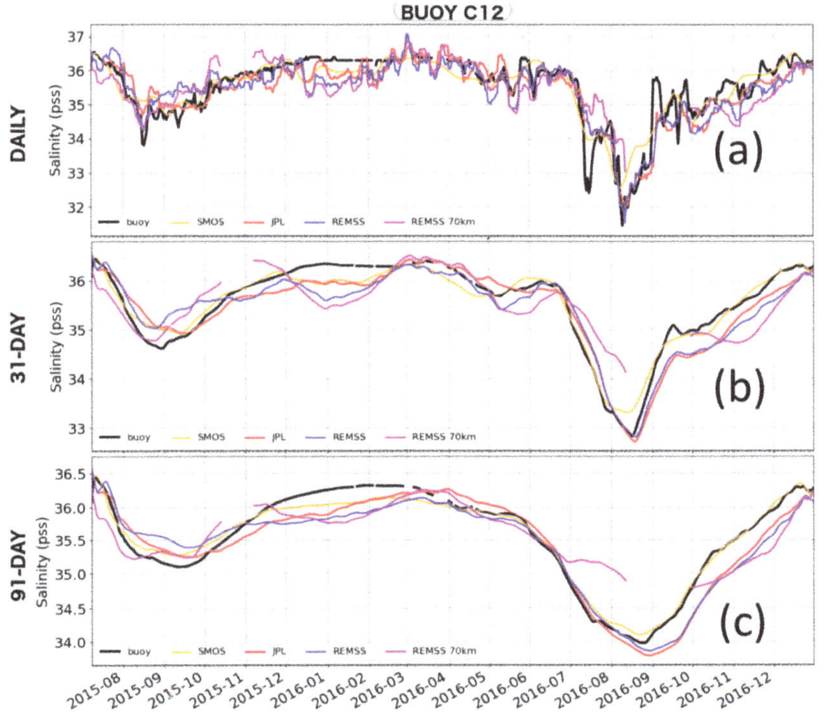

Figure 4. Time series of buoy C12 salinity (black), SMOS (orange), JPL SMAP (red), REMSS SMAP 40 km (blue), and REMSS SMAP 70 km (magenta) sea surface salinity. Panel (**a**) shows the daily buoy time series, the 8-day SMAP time series and the 9-day SMOS time series. Panels (**b**,**c**) respectively represent the time series smoothed using a 31-day and 91-day running mean window.

In the western GoM, results are shown for Buoys K, F, V, and X with buoys K, F, and X being close to shore. At Buoy F. SMOS and SMAP JPL show larger salty biases of approximately 0.6 and 0.5, respectively, and similar large RMSD from 1.1 and 1.2 pss at 8 and 9-day, respectively, which reduce to 0.39 and 0.43 pss at the seasonal time scale. The REMSS 40 km SMAP product shows biases close to zero pss for the 8-day and monthly time scales, increasing to 0.1 pss at seasonal time scale. RMSD reduces from 0.9 pss at the 8-day time scale to 0.6 at the monthly time scale, before increasing to 0.89 pss at the seasonal time scale. No values are retrieved for the REMSS 70 km SMAP product, with the buoy being too close to land. Overall, the REMSS 40 km SMAP has the lowest bias whatever the time scale and the lowest RMSD at 8-day, while JPL SMAP and SMOS have lower RMSD at monthly and seasonal time scales.

At buoy K, SMOS performs better, with a fresh bias of 0.1 pss and RMSD decreasing from 1 pss at 9-day to 0.26 pss at seasonal scale. Both REMSS 40 km SMAP and JPL SMAP have biases around 0.2 pss at 8-day and monthly time scales and RMSD decreasing from 1.1 to 0.5 pss. However, while the JPL SMAP RMSD decreases to 0.4 pss at seasonal time scales, the REMSS 40 km SMAP increases to 0.6 pss. Similarly, the REMSS 70 km SMAP increases to 0.45 pss at seasonal scales from a low RMSD of 0.25 pss at monthly time scales.

At buoy V, SMOS has a large fresh bias of 0.66 pss, while JPL SMAP has an 0.4 pss fresh bias and REMSS 40 and 70 km have a fresh 0.2 pss bias. SMOS has the lowest RMSD values at every time scale, from 0.6 pss at 9-day to 0.2 pss at seasonal time scales. The three SMAP products have larger RMSD around 0.8–0.9 pss at 8-day, 0.5–0.6 at monthly and 0.4 pss at seasonal time scales.

At buoy X, the furthest offshore, SMOS shows the best results in terms of RMSD, with values of 0.24 pss at 9-day decreasing to 0.14 and 0.18 pss at monthly seasonal time scales. However, like buoys V and F, SMOS has a large fresh bias of 0.5–0.6 pss. REMSS 40 km SMAP, REMSS 70 km SMAP, and JPL SMAP products show similar results, with RMSD decreasing from 0.3–0.4 pss at 8-day to 0.25 pss at seasonal time scales. However, the REMSS SMAP product shows a bias close to zero, whereas the JPL SMAP and REMSS SMAP 70 km products show biases close to 0.2 pss.

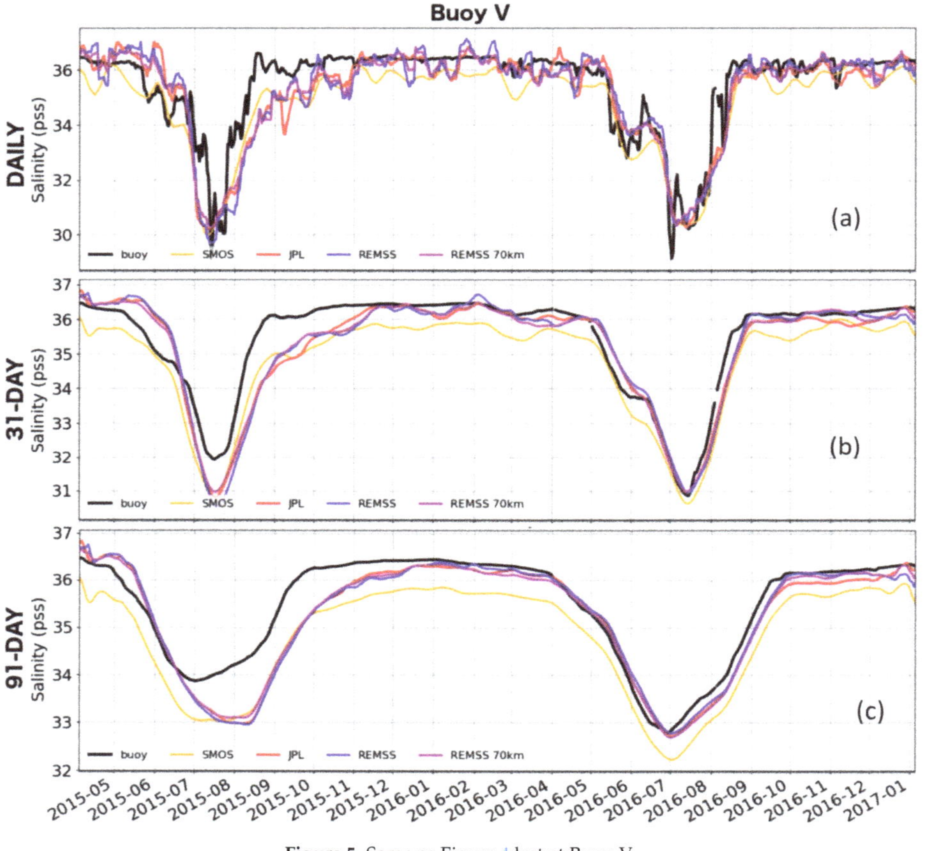

Figure 5. Same as Figure 4 but at Buoy V.

3.3. Comparisons with the WOD

In addition to the time series, the WOD allows us to perform comparisons between satellite and in-situ SSS at many point locations, covering a large span of the GoM over different seasons. The point locations do not allow for a spatial variability study of the GoM. However, they do allow for direct comparisons with the SSS products. Figure 1 shows that the density of the WOD measurements is not sufficient to resolve the subpixel scale variability (<25 km). This type of analysis has been done by Boutin et al. [12]. Thus, interpretations of the statistics must be done considering the natural variability of the GoM. The WOD extended dataset does allow us to study the biases and RMSD of the four SSS datasets, depending on the distance from the coast or time of year or salinity range. Figure 7 shows the spatial distribution of the bias (first column), root mean square difference (RMSD) (second column), and scatter plots for the four SSS datasets co-located with the WOD (third and fourth columns). Overall correlations for the three products, JPL SMAP, REMSS 40 km SMAP, and SMOS had similar correlations of approximately 0.80. The REMSS 70 km SMAP product had lower correlations of 0.70. Because one cannot attribute the differences to errors due to unresolved spatial variability, the statistics have been labeled as root mean square differences (RMSD). For the WOD data comparison, there were two options available in terms of the intercomparison of multiple data sets. We chose to explore both options, namely a validation based on the temporally and spatially co-located in-situ data for each individual satellite product, as well as a validation based only on co-locations common to all of the satellite products. Thus, two scatter plots are presented: One for the analysis that takes into account the co-locations for each separate dataset (third column) and the other for the analysis based on only the common co-locations (fourth column).

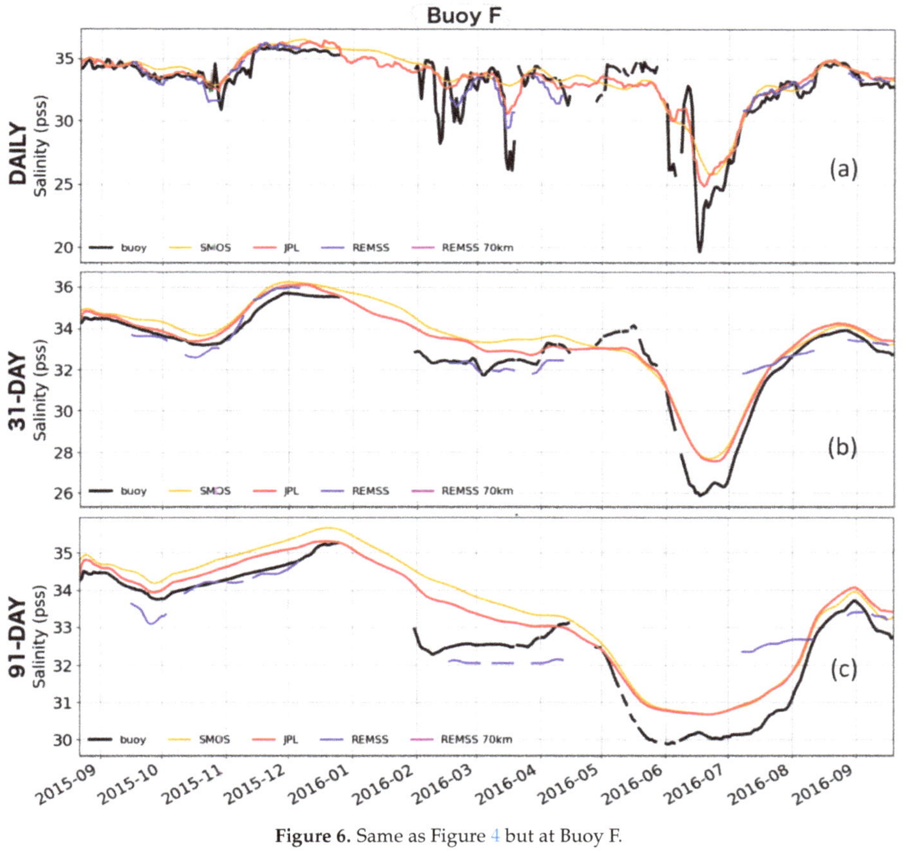

Figure 6. Same as Figure 4 but at Buoy F.

Overall, when considering only the co-locations common to all datasets, RMSD values are very similar across the datasets, with a value of 0.7 for REMSS 40 km SMAP, JPL SMAP, and SMOS and 0.99 for REMSS 70 km SMAP. The bias is larger for SMOS (0.3 pss), whereas it is lower for REMSS 70 km SMAP (0.27 pss), JPL SMAP (0.25 pss), and REMSS 40 km SMAP (0.15 pss), biases being fresher in all cases. When considering all the co-locations possible for each individual dataset, the RMSD increases to 0.9 for JPL SMAP and SMOS, and to 0.8 for REMSS 40 km SMAP. The fresh biases also decrease to 0.25 for SMOS and JPL SMAP and 0.11 for REMSS 40 km SMAP. An interesting feature is the difference in the statistics based on using all the co-locations and only those common to all the products. This indicates that the major differences in the statistics (confirmed later) are occurring close to land where the dominant differences in the number of co-locations are found. Additionally, the lower correlations for the SMAP 70 km product are most likely reflective of the smoother product not resolving mesoscale and submesoscale variability associated with the Gulf of Mexico. Overall, the JPL SMAP and SMOS products had the highest correlation. This will also be discussed later in the context of the signal to noise ratio.

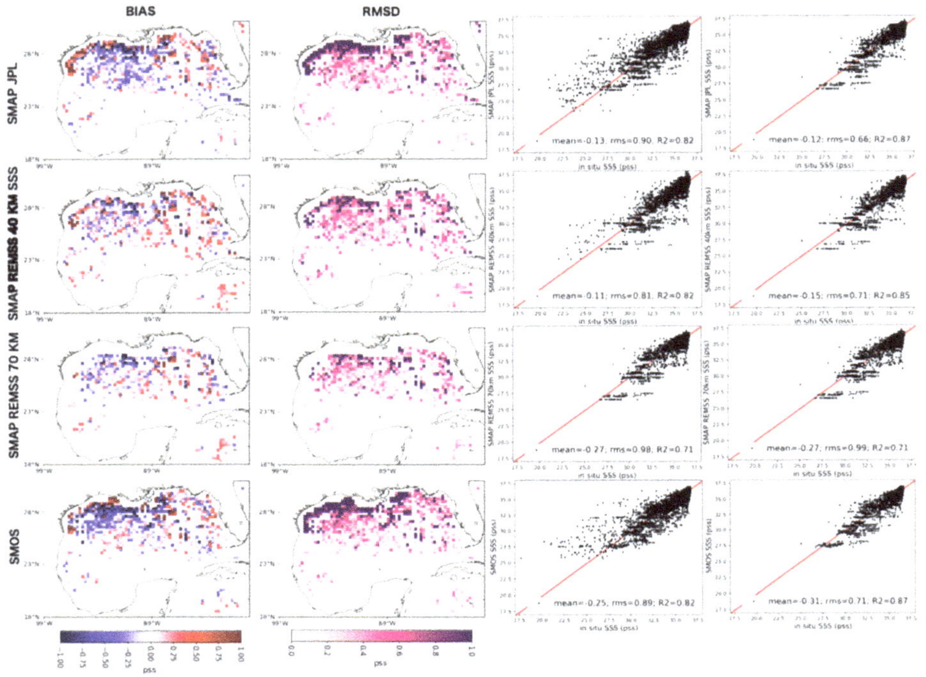

Figure 7. Maps of the bias (first column) and root mean square difference (RMSD) (second column) between satellite SSS and WOD in-situ salinity in every 0.25-degree bin from April 2015 to December 2017. The third and fourth columns show the scatter plots between satellite and in-situ SSS co-localized. The third column considers all the co-locations available for each data set, whereas the fourth column only considers the co-location common to all the dataset. Each row corresponds to a satellite product (JPL SMAP, REMSS SMAP 40 km, REMSS SMAP 70 km, and SMOS).

The top panels of Figure 7 show the bias, RMSD, and scatter plots for the JPL SMAP product. Overall, the largest biases and differences are seen close to shore. Nearshore biases reach 1 pss, with the JPL SMAP product being saltier than the WOD. RMSD reaches 1 pss in the nearshore area. The second and third panels from the top show the statistics for the REMSS 40 km SMAP and REMSS 70 km SMAP products, respectively. The same general trend exists with saltier biases and larger RMSD near the coast. In the spatial distribution plots, one notices that the 70 km product lacks values nearer to the coast than those of the REMSS 40 km SMAP or JPL SMAP products. A reason for the possible differences in the biases between the 40 km and 70 km product is most likely a combination of both the mesoscale variability in the GoM and the REMSS 40 km SMAP product resolving features closer to land. The bottom panel shows the statistics for the SMOS product. SMOS data show large negative biases (1 pss) along the coast, consistent with the large RMSD in the other SMAP datasets. However, SMOS data show slightly more values closer to the coast. RMSD are also greater than 1 pss close to the coast.

The selection of only the co-located values common to all of the datasets generally impacts the RMSD, which decreases to approximately 0.7 pss from the individual comparisons, mostly due to the elimination of stations within 70 km of the coast. The results (which will be discussed in the next section) appear to indicate that the largest discrepancies across satellite data products appear close to land or at low salinity during summer (when the region is mostly influenced by riverine waters). To fully understand the impact of the distance from land on the statistics, the next step was to derive the statistics from the WOD based on distance to land, salinity values, and period of the year. In the GoM, as in other regional basins, assessing the quality of the SSS near to land is critical for quantifying the effects of river discharge on seasonal to interannual time scales.

Summarizing, three major contributors to the differences between the SSS products and in-situ data may be evaluated based on the comparisons of the satellite products with the buoys and the WOD SSS: (1) How different do the four SSS products behave close to land? (2) Are there seasonal or discharge event effects on product performance? (3) What is the impact of faster time-scale processes on apparent error in salinity products? These questions address critical information needed for understanding SSS applications for coastal processes and dynamics. Here we highlight key differences between the four products, with the goal of helping future science users to better understand the existing SSS products. To address these questions, one needs to examine in more detail the four SSS products and their relationship to the WOD and buoy data. The Discussion section will address the three above issues.

4. Discussion

4.1. Impact of Distance from Land on Differences

We quantify the dependence of bias and RMSD of the four SSS products with respect to distance from shore (Figure 8). The top left panel (Figure 8a) shows the number of co-locations for each of the products. As expected, the REMSS 70 km SMAP product has the fewest number of co-locations. Additionally, the 70 km product does not retrieve values closer to land than 70 km, consistent with the spatial distribution maps. SMOS and SMAP JPL have the maximum number of co-locations within the first 100 km, while REMSS 40 km SMAP has fewer co-locations, suggesting that there are fewer product data near the coast. The maximum number of co-locations occurs at approximately 100 km from shore, decreasing in both the shoreward and offshore directions. At distances greater than 100 km from shore, the JPL SMAP, REMSS 40 km SMAP, and SMOS converge on the number of co-locations. This is consistent with the decreasing influence of land contamination on the three different products. Additionally, it shows that a large percentage of the WOD is concentrated close to land, consistent with Figure 7.

Figure 8b, the top right panel, shows the bias as a function of distance from shore for the four SSS products. All the products, for distances less than 100 km, show increasing biases. For JPL SMAP, REMSS 40 km SMAP, and SMOS, saltier biases occur close to land, indicating the satellite-based products are saltier than the WOD, which is consistent with what is seen on the scatter plots in Figure 7. SMOS has biases that increase to 1 pss near the coast (based on location of in-situ data). The JPL SMAP and REMSS 40 km products overall have smaller biases, approximately 0.6 and 0.7 pss. The REMSS 70 km SMAP product, however, shows large fresh biases, reaching 2.5 pss within 70 km from shore. Negative biases would be consistent with possible land contamination. However, an explanation for this is beyond the scope of this article, but should be explored in future work. For distances greater than 100 km from the coast, all four products converge to fresh biases of less than 0.5 pss. The change in the statistics at 100 km is very similar to results found by Fournier et al. [10].

Figure 8c shows the signal to noise ratio as a function of distance from the coast. For all four products, the signal to noise ratio reaches a maximum of approximately 3–3.5 at 140 km from shore, decreasing to 1–2 closer to shore for SMOS, JPL SMAP, and REMSS 40 km SMAP. The REMSS 70 km SMAP product shows the smallest signal to noise ratio of less than 1 within the first 100 km. At less than 70 km, the JPL SMAP and SMOS products have the largest signal to noise ratios. SMOS has the largest signal to noise of approximately 2.0 at distances less than 100km while the JPL SMAP has a signal to noise ratio of approximately 1.5.

Figure 8d shows the RMSD as a function of distance from shore. Close to shore, less than 100 km, the REMSS 40 km SMAP has maximum values of 1.4 pss, whereas SMOS and JPL SMAP have significantly larger RMSD values of 0.9–2.5 pss. From 50 km and further, the RMSD values of SMAP JPL, REMSS 40 km SMAP, and SMOS converge. REMSS 70 km SMAP, however, has larger RMSD values up to 100 km from the coast, reaching 2.7 pss at around 70 km. At distances greater than 100 km, all four products converge to values between 0.5 pss and 1.0 pss. Thus, comparing the RMSD with the signal to noise ratio indicates that the larger SMOS and JPL SMAP signal to noise ratios are likely due to larger STD values of the individual products and ability of the SMOS and JPL SMAP to resolve features closer to land. This is also consistent with the JPL SMAP and SMOS having the overall highest correlations.

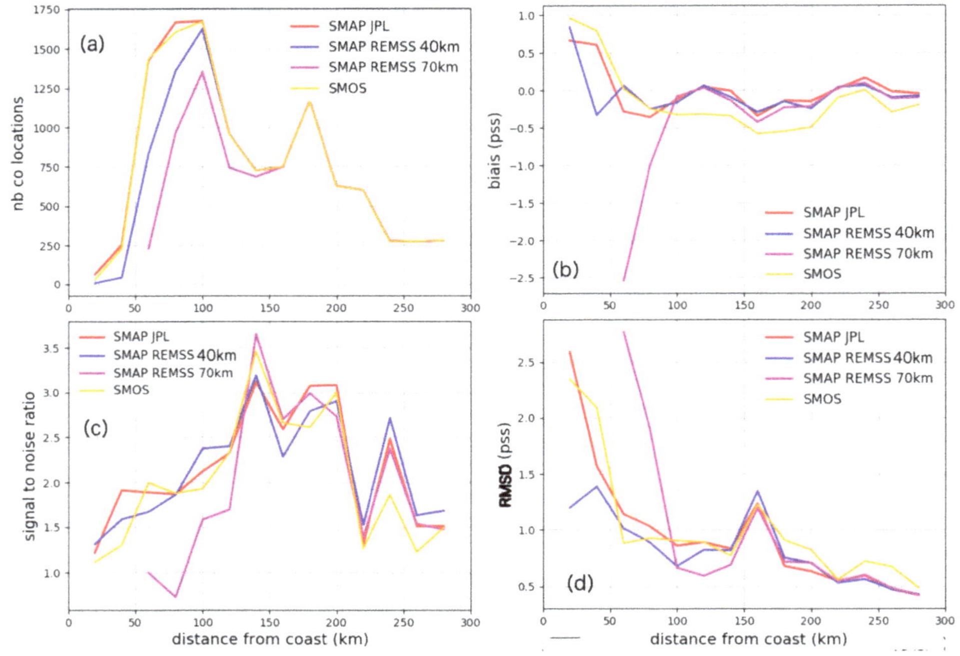

Figure 8. Comparisons between satellite SSS and WOD in-situ salinity as a function of the distance from the coast from April 2015 to December 2017. (**a**) Number of co-locations, (**b**) bias, (**c**) signal to noise ratio, and (**d**) RMSD between satellite SSS (SMOS, JPL SMAP, REMSS SMAP 40 km, and REMSS SMAP 70 km) and in-situ salinity per bin of 20 km from the coast.

Considering the comparisons of satellite SSS to the in-situ SSS from the buoys (statistics presented in Table 1), the buoys that are closest to shore (buoys C10, K, F, and X) show the largest biases and RMSD. REMSS 40 km SMAP has the lowest RMSD values (0.6 pss) at buoys C10, the closest to shore, while neither SMOS nor REMSS 70 km SMAP retrieve data. At buoys K and X, SMOS has the lowest RMSD values of 1 and 0.24 pss, respectively. However, at buoy X, SMOS has the largest bias, which is consistent with Figure 8b. On the Florida shelf, buoy C12 is further from the coast and showed improved statistics over buoy C10 for every product. Similarly, on the Texas shelf, buoy V, further offshore than buoy F, shows improvements in biases and RMSD values for all the products.

Overall, all three JPL SMAP, REMSS 40 km SMAP, and SMOS products behave comparably as a function of the distance from the coast, with larger biases and RMSD closer to the coast and lower signal to noise ratios. SMOS has, however, the largest bias closer to the coast, while REMSS 40 km SMAP has a lower RMSD value. At some locations, SMOS has lower RMSD values, even close to the coast, as shown by comparisons with buoys. All four products have a maximum in the signal to noise ratio at approximately 150 km from shore.

4.2. Impact of Riverine Freshwater on Differences

A remaining question is how the errors might be related to riverine freshwater discharge into the GoM. Figure 9 shows the statistics based on the same datasets used as Figure 8, but as a function of salinity (first column) and time of year (second column). Figure 9 confirms what is shown in Figure 8, which is that REMSS 40 and 70 km do not capture the lowest salinity near the coast. While REMSS 40 km SMAP does not capture any salinity lower than 22 pss (20 pss for JPL SMAP and SMOS), REMSS 70 km SMAP only captures salinity higher than 26 pss. Because a majority of the WOD database in the area are CTDs, point measurements, an analysis of the spatial variability is not feasible and beyond the scope of the work. Future work, incorporating new measurements, would help to quantify the differences that are due to the spatial variability of the GoM.

From Figure 9a, all four products show significant saltier biases of greater than 2 pss for salinity values less than 25 pss and can reach as high as 6–7 pss. For salinity values greater than 26 pss, biases are reduced to near 0 pss. Figure 9d shows the bias as a function of the time of year, May to September being the period of time after the maximum Mississippi River discharge. Overall, the JPL SMAP and REMSS 40 km show no significant differences based on time of year, with biases of less than 0.2 pss. SMOS tends to exhibit a larger bias in winter from January to April (−0.6 in January and February against −0.1 to 0.1 for the other three products). The REMSS 70 km SMAP product has a significant large fresh bias of 1 pss during the summer time frame (when the region is impacted by riverine fresh water). One possible explanation is that the smoother resolution is not fully resolving the spatial scales of the discharge. A more detailed explanation needs to be examined in future work.

Figure 9b shows the same results for the RMSD. For salinity values less than 25 pss, RMSD values increase rapidly to greater than 3 pss (reaching 6–7 pss), with RMSD being slightly lower for the REMSS 40 km SMAP product than for the others. Figure 9e shows that the REMSS 70 km SMAP product has a maximum RMSD of greater than 1.6 pss in June, while the other three products reach maxima of 1.2 pss in July. Thus, all four products show maxima in RMSD during seasonal maxima in freshwater discharge. Large biases in the REMSS 70 km SMAP in June are also consistent with a maximum in the RMSD. Here, the RMSD values are also slightly lower for REMSS 40 km SMAP all year long except in June, during the peak RMSD value.

Figure 9c,f shows similar results, but for the signal to noise ratio. There is a slight trend of increasing signal to noise ratio with increasing salinity for the three products, JPL SMAP, REMSS 40 km SMAP, and SMOS. The REMSS 70 km SMAP shows a sharp minimum of 0.6 signal to noise ratio at 25 pss and has, overall, a lower signal to noise ratio all year long. All four products show a shared maximum of 2.4 in the signal to noise ratio in July.

Overall, all the products indicate that RMSD, biases, and signal to noise ratio depend on the salinity values and time of year. Large freshwater biases, large RMSD, and lower signal to noise ratio occur in the summer time frame, when the area is affected by riverine freshwater. While JPL SMAP, REMSS 40 km SMAP, and SMOS have comparable results, the REMSS 70 km SMAP product has larger biases and RMSD associated with the summertime freshwater river discharge.

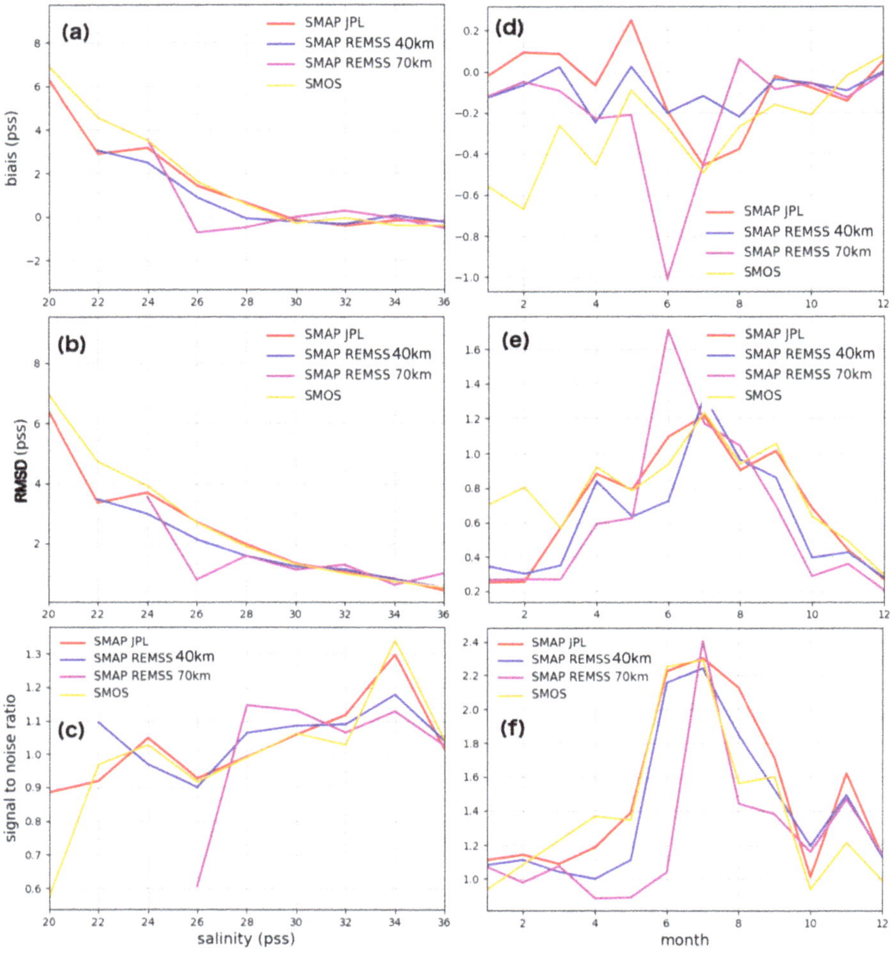

Figure 9. Comparisons between satellite SSS and WOD in-situ salinity as a function of salinity (left panel) and the month of the year (right panel) from April 2015 to December 2017. (**a**,**d**) bias, (**b**,**e**) RMSD, and (**c**,**f**) signal to noise ratio between satellite SSS (SMOS, JPL SMAP, REMSS SMAP 40 km and REMSS SMAP 70 km) and in-situ salinity. (**a**–**c**) The data are plotted per bin of 2 pss.

4.3. Impact of Temporal Sampling on Differences

Because co-located satellite and in-situ observations may not truly be co-located at a precise spatial point and may also have different sampling frequencies, it is possible that the undersampling of high-frequency variability in either data source may lead to a type of "aliasing" noise, which decreases the skill of a specific validation. While we are limited in the format of the available data, we can look more closely at the possible temporal effects on our results. These effects could indicate results or trends in validation that may be influenced by forces commonly referred to as "luck". As such, we examine temporally smoothed results to check for consistency and to examine the gross behavior of the products and their ability to capture major dominant signals. Overall, interpreting the smoothing impact on the statistics is critical for understanding the possible impacts of spatial variability on the statistics. This is especially true in the GoM, where mesoscale and sub-mesoscale variability can dominate during the formation of the loop current and associated eddy detachments.

Table 1 summarizes the comparison statistics of the JPL SMAP, REMSS 40 km SMAP, REMSS 70 km SMAP, and SMOS with the 7 buoys at three different time scales. Overall, with temporal smoothing from 8/9-day to monthly to seasonal time scales, there is a significant improvement in the RMSD for all the products. On average, at 8/9-day across products, the RMSD is about 0.7 pss, whereas it decreases to 0 4 at monthly and 0.3 at seasonal time scales.

From 8/9-day to monthly time scales, the improvement in RMSD extends across products, with a decrease of 0.1 reaching 1 pss, with an average of 0.28 pss. The largest improvements are seen at buoys F and K near shore on the Texas shelf, with a decrease of 1 pss for REMSS 70 km, 0.7 pss for SMOS, 0.6 pss for REMSS 40 km SMAP, and 0.55 pss for JPL SMAP at buoy K. JPL SMAP RMSD improves, with a decrease of 0.65 pss at buoy F, and SMOS RMSD with a decrease of 0.5 pss. From monthly to seasonal time scales, the improvement in RMSD is lower, with a decrease of 0.05 to 0.3 pss and an average of 0.1 pss. However, for the REMSS 40 and 70 km, there is an increase in the RMSD of 0.3 pss at buoy F and 0.1 pss at buoy K for REMSS 40 km SMAP and an increase of 0.2 pss at buoy K for REMSS 70 km. Consistent improvement also occurred with respect to correlations in going from the 8-day to seasonal time scale, indicative of the reduction in noise. The overall trend shows the JPL SMAP and SMOS products had the highest correlations. The SMOS product showed a correlation of 0.98 at Buoy V (furthest offshore) for the seasonal time scale. REMSS 70 km SMAP had negative correlations (consistent with the large negative biases near land) at Buoy X.

Table 1. Statistics of the comparisons between each satellite product (SMOS, JPL SMAP, REMSS SMAP, and REMSS SMAP 70 km) and buoy salinity data. The locations of the buoys are indicated in Figure 6.

Buoy	Satellite Dataset	Bias (pss)	RMSD (pss)	Correlation
Buoy C10	**JPL 8-day**	0.24	0.78	0.62
	JPL 31-day	0.24	0.66	0.67
	JPL 91-day	0.26	0.50	0.69
	REMSS 40 km 8-day	−0.23	0.61	0.13
	REMSS 40 km 31-day	−0.24	0.42	0.30
	REMSS 40 km 91-day	−0.10	0.35	0.50
	REMSS 70 km 8-day	N/A	N/A	N/A
	REMSS 70 km 31-day	N/A	N/A	N/A
	REMSS 70 km 91-day	N/A	N/A	N/A
	SMOS 9-day	N/A	N/A	N/A
	SMOS 31-day	N/A	N/A	N/A
	SMOS 91-day	N/A	N/A	N/A
Buoy C12	**JPL 8-day**	−0.21	0.44	0.89
	JPL 31-day	−0.21	0.26	0.94
	JPL 91-day	−0.20	0.19	0.95
	REMSS 40 km 8-day	−0.12	0.49	0.86
	REMSS 40 km 31-day	−0.12	0.33	0.92
	REMSS 40 km 91-day	−0.11	0.25	0.93
	REMSS 70 km 8-day	−0.10	0.62	0.69
	REMSS 70 km 31-day	−0.09	0.51	0.74
	REMSS 70 km 91-day	−0.08	0.37	0.80
	SMOS 9-day	−0.01	0.43	0.90
	SMOS 31-day	−0.02	0.19	0.98
	SMOS 91-day	−0.01	0.12	0.99
Buoy C13	**JPL 8-day**	−0.13	0.40	0.81
	JPL 31-day	−0.13	0.25	0.91
	JPL 91-day	−0.12	0.20	0.92
	REMSS 40 km 8-day	−0.09	0.47	0.72
	REMSS 40 km 31-day	−0.09	0.31	0.75
	REMSS 40 km 91-day	−0.07	0.23	0.88
	REMSS 70 km 8-day	−0.09	0.54	0.56
	REMSS 70 km 31-day	−0.09	0.40	0.68
	REMSS 70 km 91-day	−0.08	0.25	0.83
	SMOS 9-day	−0.03	0.33	0.85
	SMOS 31-day	−0.03	0.19	0.94
	SMOS 91-day	−0.02	0.14	0.95

Table 1. *Cont.*

Buoy	Satellite Dataset	Bias (pss)	RMSD (pss)	Correlation
Buoy F	JPL 8-day	0.43	1.25	0.91
	JPL 31-day	0.45	0.58	0.97
	JPL 91-day	0.51	0.43	0.98
	REMSS 40 km 8-day	0.04	0.90	0.85
	REMSS 40 km 31-day	0.02	0.65	0.87
	REMSS 40 km 91-day	0.13	0.89	0.75
	REMSS 70 km 8-day	N/A	N/A	N/A
	REMSS 70 km 31-day	N/A	N/A	N/A
	REMSS 70 km 91-day	N/A	N/A	N/A
	SMOS 9-day	0.57	1.12	0.87
	SMOS 31-day	0.60	0.61	0.96
	SMOS 91-day	0.66	0.39	0.96
Buoy K	JPL 8-day	0.25	1.13	0.80
	JPL 31-day	0.28	0.54	0.93
	JPL 91-day	0.29	0.45	0.90
	REMSS 40 km 8-day	0.23	1.10	0.83
	REMSS 40 km 31-day	0.21	0.52	0.94
	REMSS 40 km 91-day	0.14	0.60	0.89
	REMSS 70 km 8-day	−0.06	1.23	0.68
	REMSS 70 km 31-day	−0.25	0.25	0.94
	REMSS 70 km 91-day	−0.25	0.45	0.70
	SMOS 9-day	−0.16	1.04	0.83
	SMOS 31-day	−0.15	0.33	0.97
	SMOS 91-day	−0.13	0.26	0.97
0.Buoy V	JPL 8-day	−0.43	0.78	0.88
	JPL 31-day	−0.43	0.54	0.93
	JPL 91-day	−0.43	0.39	0.93
	REMSS 40 km 8-day	−0.22	0.87	0.87
	REMSS 40 km 31-day	−0.22	0.61	0.92
	REMSS 40 km 91-day	−0.23	0.45	0.93
	REMSS 70 km 8-day	−0.25	0.77	0.89
	REMSS 70 km 31-day	−0.25	0.53	0.94
	REMSS 70 km 91-day	−0.25	0.38	0.94
	SMOS 9-day	−0.66	0.59	0.93
	SMOS 31-day	−0.66	0.29	0.98
	SMOS 91-day	−0.66	0.20	0.98
Buoy X	JPL 8-day	−0.18	0.37	0.30
	JPL 31-day	−0.19	0.27	0.50
	JPL 91-day	−0.18	0.24	0.56
	REMSS 40 km 8-day	0.07	0.33	0.38
	REMSS 40 km 31-day	0.06	0.21	0.55
	REMSS 40 km 91-day	0.05	0.25	0.42
	REMSS 70 km 8-day	−0.17	0.39	−0.01
	REMSS 70 km 31-day	−0.19	0.25	−0.30
	REMSS 70 km 91-day	−0.16	0.22	−0.57
	SMOS 9-day	−0.50	0.24	0.69
	SMOS 31-day	−0.55	0.14	0.88
	SMOS 91-day	−0.64	0.18	0.77

5. Conclusions

This work examines the performance of several satellite SSS products in a representative marine system that is significantly impacted by river discharge. The comparisons of four satellite derived SSS products (REMSS 40 km SMAP, REMSS 70 km SMAP, JPL SMAP, and SMOS) in the GoM demonstrate that these products generally perform well compared to in-situ observations in a basin that has large temporal and spatial variability in the surface salinity structure. In interpreting the results, it is important to consider not only how the differences between the satellite products and in-situ data affect the correlations and signal to noise ratio, but also how the inherent spatial variability of the

GoM might not be resolved by the individual products, depending on smoothing. The remarkable consistency of the seasonal mean and STD between the four datasets confirms that, on the seasonal time scales, they are all reproducing the dominant features associated with salinity changes in the GoM, especially those associated with river discharge.

The satellite datasets were compared with buoys salinity data on the Florida and Texas shelves and with a large amount of diverse in-situ measurements from the WOD. We computed biases, RMSD, and signal to noise ratio to address three issues: (1) The impact of distance from shore, (2) the impact of freshwater river discharge, and (3) the impact of temporal variability.

Overall, JPL SMAP, REMSS 40 km SMAP, and SMOS products are comparable, independent of distance from the coast, when compared with WOD in-situ data. Within the first 100 km from the coast, the three products showed increasing biases and RMSD and a decrease in the signal to noise ratio towards the shore. Saltier biases close to shore are likely due to some residual land contamination and possible overestimation of the land correction. Additional error arises due to the fact that, closer to the coast, larger horizontal and vertical salinity gradients can be observed, altering the comparison between a 40 to 70 km satellite footprint with a point-location from an in-situ measurement. These three products exhibit a similar low signal to noise ratio close to shore that peaks at 150 km from the coast. They also show a larger RMSD within the first 100 km from the coast, before decreasing to 0.5–1.5 pss further than 100 km.

Even if the products behave similarly overall, some differences can be observed between the products near the coast. The REMSS 40 km SMAP product shows the lowest bias overall (−0.15 pss against −0.25 for JPL SMAP and −0.31 for SMOS). It also shows the lowest RMSD (1.2 to 1.4 pss) within the first 60 km, while the SMOS and JPL SMAP product show a RMSD of 1.5 to 2.5 pss. However, REMSS 40 km SMAP retrieves fewer data closer to the coast than JPL SMAP and SMOS do and may not support an equivalent comparison. This might be due to more stringent land corrections applied.

The SMOS product exhibits the largest RMSD values near the coast and at low salinities. It also exhibits the largest bias overall (−0.31 pss), especially at low salinity and during winter independent of the distance from the coast. However, our results on the comparisons with buoys suggest that at some locations, even near the coast, SMOS has the lowest RMSD values of all of the products. However, at buoys on the Texas shelf, SMOS exhibits the largest bias, reaching 0.7 pss against a bias of 0 to 0.4 for the other products. These results are important to consider in light of the fact that the SMOS product provides the longest satellite SSS record currently available. The JPL SMAP product gets similar results to the SMOS product when compared with the WOD, especially near the coast. However, SMAP JPL doesn't exhibit large bias values (−0.25 overall versus −0.31 for SMOS).

Because of the required spatial smoothing, the REMSS 70 km SMAP product does not offer any data within 70 km of the coast, reducing its usefulness in coastal applications. Moreover, within the first 100 km of the coast, this product exhibits a much larger bias, reaching −2.5 pss. It also exhibits a lower signal to noise ratio within the first 140 km of the coast and a higher RMSD value. Results are consistent with assumed error characteristics based on the smoothness of the REMSS 70 km SMAP product that reduces accuracy close to land.

Temporal variability demonstrates a clear effect on error characteristics across the products. At seven buoys located in the eastern and western GoM, smoothing over the 8/9-day to seasonal time scales decreased the RMSD significantly, with an average RMSD at 8/9-day across the product of 0.7 pss, 0.4 pss at monthly and 0.3 pss at seasonal time scales. While temporal sampling is clearly a factor for each of the products tested, the validation results described above hold and are consistent in terms of which products perform best even after temporal smoothing is applied, with performance generally improved at longer timescales.

Because of the number of SSS products now available, the goal of this study was to present a validation in a coastal region based on independent in-situ data. The results are presented here in a format that we hope will give the users an overall assessment of what datasets are best for different applications in the GoM. Generally, close to land (less than 100 km), the REMSS 40 km SMAP product performs well and exhibits the smallest RMSD values. However, it has fewer co-locations, suggesting that there are fewer product data near the coast, likely due to some filtering of lower quality, nearshore observations. Additionally, at some specific locations, when compared with buoys, SMOS shows lower RMSD values whatever the distance from the coast. JPL SMAP and SMOS have similar overall RMSD values near the coast, with JPL SMAP having a much lower bias value.

The broadening of the footprint of in-situ ocean observing systems, as well as the increased archival efforts required by many funding agencies, is increasing the availability of data, and we hope this work can serve as a template for other such regional comparisons in a coastal, semi-enclosed basin. Future work should focus on applying similar methodologies to other regions, while taking advantage of coastal datasets.

Author Contributions: All the authors played a critical part in the preparation of the manuscript. Each author brought their expertise, including scientific expertise in the Gulf of Mexico (B.D. and S.F.). S.F. did all co-location of the buoys and the WOD. All the authors contributed extensively to the writing of the manuscript.

Funding: This research was supported by the NASA Science Utilization of the Soil Moisture Active-Passive Mission (NNH15ZDA001N-SUSMAP) program.

Acknowledgments: The research described in this paper was carried out at the Jet Propulsion Laboratory, California Institute of Technology, under a contract with NASA. This research was supported by the NASA Science Utilization of the Soil Moisture Active-Passive Mission (NNH15ZDA001N-SUSMAP) program. Data for this paper are available at the following data centers, which we gratefully thank: The CNES-IFREMER Centre Aval de Traitement des Données SMOS (CATDS) for SMOS SSS data (http://catds.ifremer.fr); the NASA Physical Oceanography Distributed Active Archive Center (PO.DAAC) for REMSS and JPL SMAP SSS data (https://podaac.jpl.nasa.gov/dataset/SMAP_RSS_L3_SSS_SMI_8DAY-RUNNINGMEAN_V2, https://podaac.jpl.nasa.gov/dataset/SMAP_RSS_L3_SSS_SMI_8DAY-RUNNINGMEAN_V2_70km, https://podaac.jpl.nasa.gov/dataset/SMAP_JPL_L3_SSS_CAP_8DAY-RUNNINGMEAN_V4); the NOAA NODC World Ocean Database (WOD) select for in-situ salinity data (https://www.nodc.noaa.gov/cgi-bin/OC5/SELECT/dbsearch.pl); the NOAA National Data Buoy Center (https://www.ndbc.noaa.gov/) and the Texas Automated Buoy System (TABS) (http://tabs.gerg.tamu.edu/) for the buoys salinity data.

Conflicts of Interest: The authors declare no conflict of interest.

References

1. Lindstrom, E.; Lukas, R.; Fine, R.; Firing, E.; Godfrey, S.; Meyers, G.; Tsuchiya, M. The western equatorial Pacific Ocean circulation study. *Nature* **1987**, *330*, 533–538. [CrossRef]
2. Pailler, K.; Bourles, B.; Gouriou, Y. The barrier layer in the western tropical Atlantic Ocean. *Geophys. Res. Lett.* **1999**, *26*, 20692072. [CrossRef]
3. Carton, J.A. Effect of seasonal surface freshwater ux on sea sur-face temperature in the tropical Atlantic Ocean. *J. Geophys. Res. Oceans* **1991**, *96*, 12593–12598. [CrossRef]
4. Grodsky, S.A.; Reverdin, G.; Carton, J.A.; Coles, V.J. Year-to-year salinity changes in the Amazon plume: Contrasting 2011 and 2012 Aquarius/SACD and SMOS satellite data. *Remote Sens. Environ.* **2014**, *140*, 14–22. [CrossRef]
5. Muller-Karger, F.E.; Walsh, J.J.; Evans, R.H.; Meyers, M.B. On the seasonal phytoplankton concentration and sea surface temperature cycles of the gulf of mexico as determined by satellites. *J. Geophys. Res. Oceans* **1991**, *96*, 12645–12665. [CrossRef]
6. DelCastillo, C.E.; Coble, P.G.; Conmy, R.N.; Mueller-Karger, F.E.; Van-derbloomen, L.; Vargo, G.A. Multispectral in-situ measurements of organic matter and chlorophyll fluorescence in seawater: Documenting the intrusion of the mississippi river plume in the west Florida shelf. *Limnol. Oceanogr.* **2000**, *46*, 1836–1843. [CrossRef]
7. Tang, W.; Fore, A.; Yueh, S.; Lee, T.; Hayashi, A.; Sanchez-Franks, A.; Martinez, J.L.; King, B.; Baranowski, D. Validating SMAP SSS with in-situ measurements. *Remote Sens. Environ.* **2017**, *200*, 326–340. [CrossRef]
8. Lee, T. Aquarius sea surface salinity with Argo products on various spatial and temporal scales. *Geophys. Res. Lett.* **2016**, *43*, 3857–3964. [CrossRef]
9. Boutin, J.; Martin, N.; Kolodziejczyk, N.; Reverdin, G. Interannual anomalies of SMOS sea surface salinity. *Remote Sens. Environ.* **2016**, *180*, 128–136. [CrossRef]

10. Fournier, S.; Vialard, J.; Lengaigne, M.; Lee, T.; Gierach, M.M.; Chaitanya, A.V.S. Modulation of the Ganges-Brahmaputra River Plume by the Indian Ocean Dipole and Eddies Inferred From Satellite Observation. *J. Geophys. Res. Oceans* **2017**, *122*, 9591–9604. [CrossRef]
11. Fournier, S.; Lee, T.; Gierach, M.M. Seasonal and interannual variations of sea surface salinity associated with the Mississippi River plume observed by SMOS and Aquarius. *Remote Sens. Environ.* **2016**, *180*, 431–439. [CrossRef]
12. Boutin, J.; Chao, Y.; Asher, W.E.; Delcroix, T.; Drucker, R.; Drushka, K.; Kolodziejczyk, N.; Lee, T.; Reul, N.; Reverdin, G.; et al. Satellite and in situ salinity: Understanding near-surface stratification and subfootprint variability. *Bull. Am. Meteorol. Soc.* **2016**, *97*, 1391–1407. [CrossRef]
13. Paris, M.L.; Subrahmanyam, B. Role of El Nino Southern Oscillation (ENSO) Events on Temperature and Salinity Variability in the Agulhas Leakage Region. *Remote Sens.* **2018**, *10*, 127. [CrossRef]
14. Suzuki, T.; Yamazaki, D.; Tsujino, H.; Komuro, Y.; Nakano, H.; Urakawa, S. A dataset of continental river discharge based on JRA-55 for use in a global ocean circulation model. *J. Oceanogr.* **2018**, *74*, 421–429. [CrossRef]
15. Fournier, S.; Reager, J.T.; Lee, T.; Vazquez-Cuervo, J.; David, C.H.; Gierach, M.M. SMAP observes flooding from land to sea: The Texas event of 2015. *Geophys. Res. Lett.* **2016**, *43*, 19. [CrossRef]
16. Morey, S.L.; Dukhovskoy, D.S.; Bourassa, M.A. Connectivity of the Apalachicola River flow variability and the physical and bio-optical oceanic properties of the northern west Florida Shelf. *Cont. Shelf Res.* **2009**, *29*, 1264–1275. [CrossRef]
17. Dzwonkowski, B.; Park, K.; Collini, R. The coupled estuarine-shelf response of a river-dominated system during the transition from low to high discharge. *J. Geophys. Res. Oceans* **2015**, *120*, 6145–6163. [CrossRef]
18. Kourafalou, V.H.; Androulidakis, Y.S. Influence of Mississippi River induced circulation on the Deepwater Horizon oil spill transport. *J. Geophys. Res. Oceans* **2013**, *118*, 3823–3842. [CrossRef]
19. Boyer, T.; Levitus, S.; Garcia, H.; Locarnini, R.A.; Stephens, C.; Antonov, J. Objective analyses of annual, seasonal, and monthly temperature and salinity for the World Ocean on a 0.25 grid. *Int. J. Climatol.* **2005**, *25*, 931–945. [CrossRef]
20. Sturges, W.; Evans, J.C. On the variability of the Loop Current in the GoM. *J. Mar. Res.* **1983**, *41*, 639–653. [CrossRef]
21. Sturges, W.; Leben, R. Frequency of ring separations from the Loop Current in the GoM: A revised estimate. *J. Phys. Oceanogr.* **2000**, *30*, 1814–1819. [CrossRef]
22. Morey, S.L.; Martin, P.J.; O'Brien, J.J.; Wallcraft, A.A.; Zavala-Hidalgo, J. Export pathways for river discharged fresh water in the northern GoM of Mexico. *J. Geophys. Res. Oceans* **2003**, 108. [CrossRef]
23. Gierach, M.M.; Vazquez-Cuervo, J.; Lee, T.; Tsontos, V.M. Aquarius and SMOS detect effects of an extreme Mississippi River flooding event in the GoM. *Geophys. Res. Lett.* **2013**, *40*, 5188–5193. [CrossRef]
24. Meissner, T.; Wentz, F.J. *Remote Sensing Systems SMAP Ocean Surface Salinities [Level 2C, Level 3 Running 8-day, Level 3 Monthly], Version 2.0 Validated Release*; Remote Sensing Systems: Santa Rosa, CA, USA, 2016.
25. Fore, A.; Yueh, S.; Tang, W.; Hayashi, A. *SMAP Salinity and Wind Speed Data User's Guid*; V3; Jet Propulsion Laboratory/California Institute of Technology: Pasadena, CA, USA, 2016.
26. Boutin, J.; Vergely, J.L.; Marchand, S.; D'Amico, F.; Hasson, A.; Kolodziejczyk, N.; Reul, N.; Reverdin, G.; Vialard, J. New SMOS Sea Surface Salinity with reduced systematic errors and improved variability. *Remote Sens. Environ.* **2018**, *214*, 115–134. [CrossRef]
27. Bender, L.C., III; Guinasso, J.; Walpert, N.L.; Lee, J.N., III; Martin, L.L.; Hetland, R.D.; Baum, S.K.; Howard, M.K. Development, operation, and results from the Texas Automated Buoy System. *Gulf Mex. Sci.* **2007**, *25*, 4. [CrossRef]
28. Fournier, S.; Reager, J.T.; Dzwonkowski, B.; Vazquez-Cuervo, J. Mapping freshwater origin and fate signatures as land/ocean 'regions of influence' in the Gulf of Mexico. *J. Geophys. Res. Oceans.* **2018**. submitted.

remote sensing

MDPI

Article

Remote Sensing of Sea Surface Salinity: Comparison of Satellite and In Situ Observations and Impact of Retrieval Parameters

Emmanuel P. Dinnat [1,2,*], David M. Le Vine [1], Jacqueline Boutin [3], Thomas Meissner [4] and Gary Lagerloef [5]

1 NASA Goddard Space Flight Center, Greenbelt, MD 20771, USA; david.m.levine@nasa.gov
2 Center of Excellence in Earth Systems Modeling & Observations (CEESMO), Chapman University, Orange, CA 92866, USA
3 Sorbonne Université, CNRS, IRD, MNHN, Laboratoire d'Océanographie et du Climat: Expérimentations et Approches Numériques (LOCEAN), 75005 Paris, France; jb@locean-ipsl.upmc.fr
4 Remote Sensing Systems, Santa Rosa, CA 95401, USA; meissner@remss.com
5 Earth & Space Research, Seattle, WA 98121, USA; Lager@esr.org
* Correspondence: emmanuel.dinnat@nasa.gov; Tel.: +1-301-614-6871

Received: 14 February 2019; Accepted: 24 March 2019; Published: 28 March 2019

Abstract: Since 2009, three low frequency microwave sensors have been launched into space with the capability of global monitoring of sea surface salinity (SSS). The European Space Agency's (ESA's) Microwave Imaging Radiometer using Aperture Synthesis (MIRAS), onboard the Soil Moisture and Ocean Salinity mission (SMOS), and National Aeronautics and Space Administration's (NASA's) Aquarius and Soil Moisture Active Passive mission (SMAP) use L-band radiometry to measure SSS. There are notable differences in the instrumental approaches, as well as in the retrieval algorithms. We compare the salinity retrieved from these three spaceborne sensors to in situ observations from the Argo network of drifting floats, and we analyze some possible causes for the differences. We present comparisons of the long-term global spatial distribution, the temporal variability for a set of regions of interest and statistical distributions. We analyze some of the possible causes for the differences between the various satellite SSS products by reprocessing the retrievals from Aquarius brightness temperatures changing the model for the sea water dielectric constant and the ancillary product for the sea surface temperature. We quantify the impact of these changes on the differences in SSS between Aquarius and SMOS. We also identify the impact of the corrections for atmospheric effects recently modified in the Aquarius SSS retrievals. All three satellites exhibit SSS errors with a strong dependence on sea surface temperature, but this dependence varies significantly with the sensor. We show that these differences are first and foremost due to the dielectric constant model, then to atmospheric corrections and to a lesser extent to the ancillary product of the sea surface temperature.

Keywords: sea surface salinity; microwave radiometry; remote sensing; calibration; retrieval algorithm; validation; Aquarius; SMOS; SMAP; sea surface temperature

---

1. Introduction

Sea surface salinity (SSS) is a key parameter for physical oceanography and the study of the hydrological cycle. Together with sea surface temperature (SST), it determines surface water density which impacts the vertical flow through the thermohaline component of the ocean circulation. SSS is also a tracer of the water cycle, because it is impacted by precipitations and evaporation, rivers outflow, and ice formation and melt. Until the early 2000's, SSS had been observed sparsely, largely along commercial shipping lanes or during spatially or temporally limited oceanographic campaigns, resulting in spatial and temporal coverage inadequate for many applications. The sampling largely improved during the second half of the 2000's due to the rapid expansion of the deployment of the Argo network of free drifting profiling floats. But the spatial and temporal coverage of in situ measurements cannot rival that of satellite observations which constitute a critical component of the ocean observation system. In this context, two satellite missions were launched with the objective of monitoring SSS at a global scale with weekly to monthly temporal resolution. In November 2009, the European Space Agency (ESA) launched the Soil Moisture and Ocean Salinity (SMOS) mission [1,2] dedicated to monitoring SSS and soil moisture. In June 2011, the National Aeronautics and Space Administration (NASA) launched the Aquarius instrument [3–5] with a sole focus on SSS remote sensing. SMOS and Aquarius use L-band (1.4 GHz) radiometers as their primary instrument to retrieve SSS. Aquarius also uses an L-band scatterometer to help correct for the impact of sea surface roughness on radiometric measurements. The Aquarius mission ended on June 7, 2015 due to spacecraft failure. SMOS is still ongoing. In January 2015, NASA launched the third spaceborne L-band radiometer with the Soil Moisture Active Passive (SMAP) mission. While the science objective of SMAP is to measure soil moisture and monitor landscape freeze/thaw, its radiometer is also capable of retrieving SSS. The timeline of SSS products available from the three missions is reported in Figure 1; SMOS and Aquarius overlap for almost four years (the whole Aquarius mission lifetime); SMAP overlaps Aquarius during the last two months of Aquarius life; SMOS and SMAP have been overlapping since April 2015.

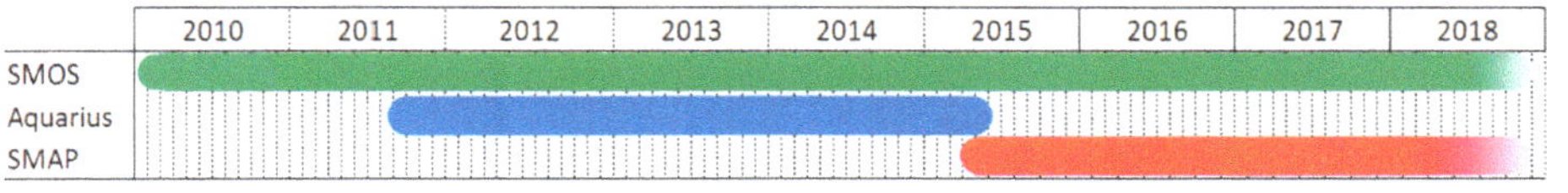

Figure 1. Timeline of satellite sea surface salinity (SSS) products availability. The Soil Moisture and Ocean Salinity (SMOS) SSS is available starting Jan 2010 (onward). The Aquarius product is available from Aug 25, 2011 until June 7, 2015 when the spacecraft bus failed. The Soil Moisture Active Passive (SMAP) product is available starting April 2015 (onward).

The main objectives of this paper are to (1) compare the satellite SSS from the three aforementioned sensors and (2) assess the potential impact of calibration approaches and retrieval algorithms on the differences between these satellite SSS products. We assess the impact of the model for the dielectric constant of sea water, the choice of ancillary product for sea surface temperature and the model for correcting for atmospheric emission and absorption. This is done by reprocessing the Aquarius retrievals using the same parameters as those used in SMOS retrievals. In doing so Aquarius measurements are taken as a reference, as they are less subject to contamination by land-sea transition, sun emission, radio-frequency interferences and as Aquarius processing is much lighter than SMOS processing. This will give us insight into the causes for the differences between SMOS and Aquarius SSS. We expect that our conclusions would be the same if we had reprocessed the SMOS data using the Aquarius retrieval parameters, but we have not done so given the complexity of the SMOS processing. The SMAP data are not reprocessed. As shown in the paper, the recently released version 3 of the SMAP product provided some significant improvements but was released too late to include a reprocessing. We use a comparison of satellite SSS retrievals with in situ measurements to assess the accuracy of the various algorithms.

Significant differences have been reported in SSS retrieved by SMOS and Aquarius at regional scales [6–10] and at global scales [11,12]. Differences between Aquarius and SMOS have also been reported at the brightness temperatures (TB) level over selected targets over ocean, land and ice [13], suggesting calibration inconsistencies between instruments or errors in the corrections applied to the antenna temperatures measurements in order to retrieve top of atmosphere brightness temperatures. There are important differences in calibration and retrieval approaches for both instruments that could explain the differences in TB and SSS. For example, pre-launch studies have assessed uncertainties due to forward and retrieval models for L-band remote sensing for Aquarius [14] and for SMOS [15–18].

Amongst the error source, the uncertainty on the sea water dielectric constant model was found to result in uncertainties in TB between 0.2 K and 1.5 K (depending on the model and incidence angle) mostly dependent on SST, with a smaller dependence on SSS. Because SMOS and Aquarius use different dielectric constant models, it is expected that some of the differences in SSS will be the result of the dielectric constant model differences. We quantify this impact by reprocessing Aquarius observation using the dielectric constant model used for SMOS. Another important input for estimating SSS from TB measurements is the SST. We assess the impact of the difference in ancillary SST used by both SMOS and Aquarius missions also by reprocessing Aquarius retrievals using the SST used for SMOS. Finally, the latest Aquarius product (version 5) uses an updated atmospheric model [19]. We will show the impact of this change on the SSS errors and the comparisons with SMOS SSS. Another difference between SMOS and Aquarius is the treatment of surface roughness. Both missions use significantly different approaches to correct for roughness, SMOS using multi-incidence angle passive measurements and Aquarius using collocated scatterometer observations. This issue is out of the scope of our study. Most of the assessments reported in this paper concern SMOS and Aquarius, but we also present some comparisons with the recently developed SMAP SSS product. However, the overlap period of SMAP with other missions is shorter (~3 years) and the algorithm development is less mature and largely inherited from Aquarius. It is also worth noting that SMAP is not a mission specifically designed for SSS remote sensing and its capabilities have been hindered by the loss of its radar early in the mission (July 2015). Still, we will show that SMAP is already a very capable instrument for SSS remote sensing and further evolution of the retrieval algorithm will likely mitigate current issues.

The paper is structured as follows. We first describe the data used in the study and how there are processed in Section 2 (Materials and Methods). The processing includes preparing the various datasets for inter-comparisons and the reprocessing of Aquarius radiometric data into new SSS products using a modified retrieval algorithm. In Section 3 (Results), we start by discussing the impact of the differences in the dielectric constant model (Section 3.1) and ancillary SST (Section 3.2) on L-band TB and retrieved SSS. Then we present comparisons of SSS from the three satellites products and in situ measurements. The comparisons follow two approaches: An analysis of long term average spatial patterns and statistical distributions (Section 3.3.1); and temporal analysis of SSS variations at selected regions of interest (Section 3.3.2). We discuss the dependence of SSS bias on SST in Section 3.3.3. Then, we analyze the performances of a new Aquarius product that we obtain from reprocessing radiometric observation using the dielectric constant model and ancillary SST used by SMOS (Section 3.3.4). We assess their impact on the retrievals in terms of the differences between SMOS and Aquarius, as well as the difference with in situ Argo SSS for validation. We conclude by discussing the known source of errors for SMOS, Aquarius and SMAP and the path forward to mitigate them.

2. Materials and Methods

All the satellite and the in-situ data are averaged monthly on a Cartesian grid in latitude and longitude at 1° resolution between 180°W and 180°E and 90°S and 90°N. In each grid cell, we compute the mean, median, and standard deviation of salinity, as well as the number of observations. Satellite SSS from ascending and descending orbits are combined. For Aquarius, SSS are averaged together regardless of which beam they are retrieved from. For SMAP, SSS from fore and aft looks are combined. The data sources, their resolution and data versions are described below.

2.1. Satellite Products

2.1.1. Aquarius

We use the Level 2 of Aquarius data, which are reported along track for 1.44 s temporal integration footprints. The data are grouped as full orbits (ascending and descending) with a duration of ~98 min at a rate of ~15 orbit files per day. We grid the Level 2 data into our own Level 3 which is a latitude/longitude gridded product using a drop-in-the-bucket method (i.e., all data falling in a grid cell are averaged together with equal weighting). We do not use existing Aquarius Level 3 products because we need the ability to modify the retrieval algorithm which is applied to the individual 1.44 s footprints. We provide a comparison between our gridded product and the official Level 3 Aquarius product in the supplemental materials. The differences are small and do not impact our conclusions. The Level 2 product contains data related to observations, calibration and retrieval processing, as well as quality control information (see below). It contains individual brightness temperatures for Aquarius' three beams which we use to reprocess our own SSS using alternative dielectric constant model and ancillary SST products. We use product versions 3.0, 4.0 and 5.0 produced by the Aquarius Data Processing System (ADPS) at the NASA Goddard Space Flight Center (GSFC) and distributed by the NASA Jet Propulsion Laboratory (JPL) Physical Oceanography Distributed Active Archive Center (PO.DAAC, ftp://podaac-ftp.jpl.nasa.gov/allData/aquarius/L2/). The three product versions are used to assess the impact of various changes in the retrieval algorithm. Version 3.0 is the last version that did not include an empirical correction for SST-dependent biases in SSS. Version 5.0 is the latest version available, and the last product delivered for the end of the mission [20].

We apply the following quality control (QC) filters to exclude data which would degrade SSS retrieval before performing the averaging into the gridded product (the number at the end refers to an entry in Table 1):

- Instrument is not in science mode (1);
- Observation time is during a reported mission event (such events include Moon interferences, spacecraft maneuvers) (1);
- Land fraction is larger than 0.01 (2) or ice fraction larger 0.001 (3) (both parameters are between 0 and 1 and represent the gain weighted fractions of land/ice in the antenna field of view);
- Antenna temperature, top of ionosphere temperature (TOI), or surface brightness temperature in V-pol or H-pol is unphysical (less than 0 K or larger than 300 K) (1);
- Expected antenna temperature computed with the forward radiative transfer model is unphysical (less than 0 K or larger than 300 K) (1);
- Retrieved SSS is less than 0 (4);
- Footprint center is in a region known for frequent radio frequency interference (RFI) contamination (5):
    - $30°N \leq$ latitude $\leq 60°N$ and $330° \leq$ longitude $\leq 360°$ for ascending orbits
    - $25°N \leq$ latitude $\leq 50°N$ and $290° \leq$ longitude $\leq 310°$ for descending orbits
    - $15°N \leq$ latitude $\leq 50°N$ and $120° \leq$ longitude $\leq 160°$ for ascending orbits
- RFI correction applied to antenna temperature in V-pol or H-pol is larger than 1 K (6);
- Large brightness temperature of the celestial sky along the direction of the reflected beam at the surface (above 5.18 K) (7).

The impact of the QC filtering on the amount of L2 samples is reported in Table 1. The percentage reduction applies to the amount of ocean data. Are considered ocean data the samples with less than 50% land fraction in the instrument's field of view (samples with even large amount of sea ice are considered ocean for the purpose of these statistics). We start with 248,856,189 L2 samples for all Aquarius observations from the three beams. They are reduced to 171,113,817 samples for ocean data only (68.76% of data remaining). The table reports the additional reduction in samples for the various flags. The various percentages are not cumulative because there is significant overlap between some filters. The reduction figures are for the number of L2 samples which are 1.44 s samples along satellite track. They are not equivalent to a reduction in the percentage of the ocean surface covered due to an uneven sampling of beam tracks on Earth surface. We also point the reader to a detailed study on the impact of the various flags and their thresholds (ftp://podaac-ftp.jpl.nasa.gov/allData/aquarius/docs/v5/AQ-014-PS-0017_Performance_Degradation_and_QC_Flagging_of_Aquarius_L2_Salinity_Retrievals_V4.pdf).

This study was conducted on a pre-version 3 of the product and covers only part of the Aquarius mission. The numbers reported in Table 1 are for V5 over the whole Aquarius mission.

Table 1. Reduction in the amount of Aquarius L2 samples over ocean due to filtering based on quality control flags.

QC Flag	Data Reduction
(1) Non science mode, event, anomalous TA, TOI TB or surface TB	1.92%
(2) Land contamination	13.50%
(3) Sea ice contamination	24.62%
(4) SSS less than 0	24.51%
(5) Regions of severe radio frequency interference (RFI)	2.67%
(6) Large RFI correction applied	2.72%
(7) Celestial Sky Contamination	5.94%
(8) All flags	39.69%

The Aquarius retrieval algorithm includes corrections for contamination by land, RFI and reflected galactic radiation. The filters above are intended to mitigate the uncertainty of these corrections. For example, the regions of severe RFI have been identified as having large and persistent errors in retrieved SSS that could not be mitigated with sufficient accuracy by the RFI filter on Aquarius (note: SMAP benefits from much improved RFI filtering capabilities). Another Aquarius SSS product, the Combined Active Passive (CAP) [21,22], is produced by JPL. It uses an empirical adjustment to the dielectric constant in order to minimize differences between measured and simulated TB using input SSS from the HYCOM numerical model [23]. Because one of our objectives is to assess the impact of the dielectric constant model on the retrievals and the differences between SMOS and Aquarius products, we do not consider this product in our analysis.

2.1.2. SMOS

We use the Level 3 SMOS SSS product distributed by the Centre Aval de Traitement des Données SMOS (CATDS; http://www.catds.fr/) based on the version 5 of the reprocessing identified as CPDC (Centre de Production des Données du CATDS) RE05 MIR_CSF3A. The original maps are monthly averages oversampled on a ~25 km Equal Area Scalable Earth Grid available for the period 01/2010–03/2017 (operational processing started 04/2017 and is still ongoing). We resample the data spatially using a drop-in-the-bucket method (i.e., all data located inside a grid cell are averaged with equal weight irrespective of their distance to the grid cell center) at $1^\circ \times 1^\circ$ resolution in latitude and longitude. SSS are from combined ascending and descending passes. There are several other SMOS SSS products distributed either by the CATDS or the Barcelona Expert Center (BEC, http://bec.icm.csic.es/data/available-products/). These products likely provide higher quality SSS retrievals because they use advanced techniques for error reduction. For example, the CATDS distributed product processed by Ifremer (CEC—Ifremer Dataset V02) applies an empirical adjustment to the SSS in order to calibrate to a monthly climatology using a $5^\circ \times 5^\circ$ spatial filter in latitude and longitude [24]. Other CATDS products (CEC—LOCEAN Debias v0 through v3) [25,26] consider systematic corrections to CATDS RE05 retrieved SSS derived from the self-consistency of SMOS SSS low frequency variations at various locations across swaths in order to reduce latitudinal and coastal biases. BEC employs non Bayesian SSS retrieval and systematic corrections on SSS retrieved from individual TBs [27]. We do not use these products in the present study because we seek to identify the impact of differences in retrieval algorithm parameters largely based on physics-based models (e.g., sea water dielectric constant, atmospheric attenuation) or physical properties (e.g., ancillary sea surface temperature). Calibration or error mitigation techniques mentioned above are designed to reduce spatial (e.g., latitudinal) and temporal (e.g., seasonal) biases and would likely remove some errors caused by differences in retrieval parameters and their dependence on SST and SSS that we seek to assess.

2.1.3. SMAP

We use the SMAP SSS product produced by Remote Sensing System and distributed by the NASA PO.DAAC (ftp://podaac-ftp.jpl.nasa.gov/allData/smap/L3/) in versions 2.0 and the recently released (Nov 2018) version 3.0. We use the level 3 products at a 70 km spatial resolution that are distributed resampled onto a 0.25° fixed Earth grid using a Backus-Gilbert type optimum interpolation (OI) in order to reduce random noise. As for Aquarius, there is a SMAP CAP product that we do not use here due to its empirical corrections.

2.2. *In Situ Products*

We use in-situ observations from the Argo network of free drifting profiling floats distributed by the Ifremer at ftp://ftp.ifremer.fr/ifremer/argo/dac/. As of 2016, there were ~4000 active floats measuring vertical profiles of pressure, temperature and salinity from 2000 m deep to a few meters deep every 10 days (for most floats). We select Argo measurements having their shallowest observations at depth of 10 m or less, a quality control value of 1 (good) or 2 (probably good) for pressure, salinity and date [28]. Because Argo measurements go through delayed quality control, they sometimes have adjusted values present in the data file. When an adjusted value exists, and its quality control flag is good or probably good, we replace the original salinity, pressure or temperature with the adjusted one. A map of the long-term average SSS from Argo and their variability (i.e., standard deviation over time) is reported in Figure 2. The map is derived from 1,175,056 samples over the period of January 2011–June 2018.

2.3. *Calibration and Retrieval Algorithm for Satellite SSS*

In the following paragraphs we summarize the approaches used for the calibration and SSS retrievals of SMOS, Aquarius and SMAP. While the different approaches share a lot in common, differences in model and instrument configuration exist. For detailed discussions about instrument calibration for the three instruments, we refer the reader to References [29–32]. The SSS retrieval algorithm is discussed in References [21,33,34] among others. In Figure 3, we report a simplified schematic of the Aquarius processing to support the following discussion. This discussion should also be helpful to understand our reprocessing of the Aquarius data with alternative models and ancillary data discussed in Section 2.4.

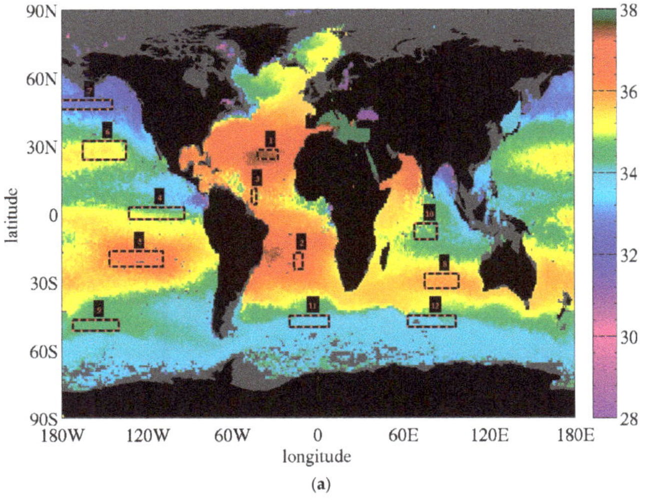

(a)

Figure 2. *Cont.*

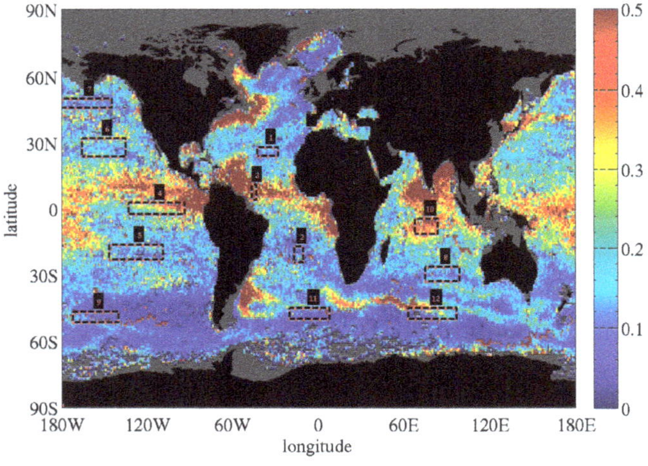

(b)

Figure 2. Map of long-term SSS patterns and variability derived from in situ Argo measurements. (**a**) Map of the average Argo SSS in 1° × 1° grid cells in latitude and longitude over the period Jan 2011–June 2018. (**b**) Standard deviation over time of monthly 1° × 1° average Argo SSS maps. The numbered red and black boxes report the limits for the regions of interest used to assess temporal variability of SSS (see text).

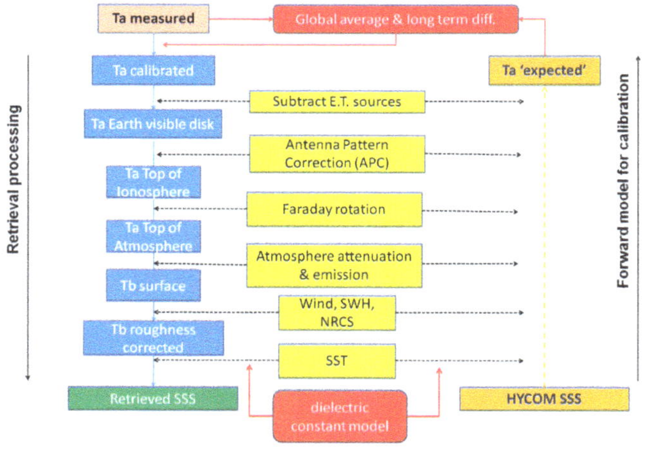

Figure 3. Schematic of Aquarius empirical calibration and SSS retrieval algorithm. The first step (top left) is to calibrate the antenna temperatures (TA) measured by the sensor using the forward radiative transfer model that will also be used in the retrieval to compute expected TA (right hand side of the diagram). This step assumes a reference SSS (e.g., from a numerical model) and uses the same ancillary data (e.g., sea surface temperature, wind, atmospheric parameters, . . .) that are used in the retrieval step. Global averages over seven-day periods for TA are used in the calibration to mitigate the impact of uncertainty in the reference SSS and other errors. The retrieval steps are illustrated on the left side, starting with TA calibrated at the top, going down the chain to remove unwanted contributions and ultimately retrieve TB at the ocean surface corrected for surface roughness. The last step is retrieving the SSS from the roughness-corrected TB. Other SSS product algorithm will differ slightly. More details are given in Section 2.3.

2.3.1. Calibration

All three instruments use hardware calibration and empirical calibration using environmental targets (e.g., celestial sky and reference target over Earth). For example, Aquarius' hardware calibration uses a combination of Dicke reference load and noise diode (for gain calibration), and alternate observations of the reference load and antenna with or without the added noise diode contribution to convert antenna counts to temperatures [35]. It was expected pre-launch that biases of a few Kelvin were likely to remain after hardware observation due to the uncertainty on the noise diode temperature and antenna pattern. To remove the residual bias, an empirical calibration is performed after the hardware calibration using the global ocean as a reference target, and the celestial sky to validate the adjusted calibration [30,35]. The ocean calibration uses comparisons between measured and simulated antenna temperatures (TA) averaged over seven days to have global coverage and avoid biasing the calibration to a particular region. The forward model uses the same components (e.g., dielectric constant, atmospheric model) and the same ancillary data as for the retrieval, with the addition of SSS reference fields produced by the HYCOM numerical model [23] (until version 4) and a combination of a gridded Argo product and HYCOM where Argo data is missing (version 5). The gridded Argo product is produced by the Scripps Institution of Oceanography and is distributed at www.argo.ucsd.edu/Gridded_fields.html. It was assumed that calibrating on the global average of SSS would mitigate uncertainty in the reference SSS. The impact of the change in reference SSS from V4 to V5 is relatively minor and does not impact our results. Many of the largest differences between the Argo and HYCOM products are in regions excluded during the calibration process (e.g., too close to land or in the high southern latitudes). The impact of the remaining differences on the global bias and its temporal variation is small compared to the differences discussed here. The accuracy of the calibration is further assessed by comparing retrieved SSS to individual in situ Argo SSS [36]. SMOS also uses an empirical approach for its SSS products (this does not apply to its soil moisture product) to remove significant TB bias residuals in its field-of-view after the hardware and cold sky calibration have been applied [37]. SMOS products below Level 2 rely only on hardware calibration and post launch calibration over cold sky [31]. Early in the mission, comparisons between SMOS TB and simulations showed very large biases (several Kelvins) with persistent patterns in the instrument's field of view [37]. Such biases would severely compromise the retrieval of SSS. They are removed based on the comparisons between measurements and simulations. This is similar in principle to what is done for Aquarius, with a few differences: 1/ SMOS calibrates top of atmosphere brightness temperatures (i.e., apparent temperatures) instead of antenna temperatures (i.e., integrated by the antenna and including all off-Earth contributions); 2/ SMOS uses a region of the Pacific Ocean as its reference target, instead of the global ocean for Aquarius; 3/ the forward model and ancillary data differ between the two sensors (e.g., SMOS reference salinity is from the World Ocean Atlas, WOA). SMAP uses an approach similar to Aquarius in that it uses comparisons of observations and simulations over open ocean and the celestial sky to adjust the calibration [32]. An important point here is that the empirical calibration uses the same radiative transfer model that is used for the retrieval. If models differ between sensors, like for SMOS and Aquarius, it is expected that the TB calibration will differ. When one changes components of the retrieval algorithm (e.g., changing the model for the dielectric constant or the ancillary SST data, see Section 2.4), the calibration needs to be changed accordingly to maintain consistency with the retrieval.

2.3.2. Salinity Retrieval Algorithm

The SSS retrieval algorithm uses the same radiative transfer model as the calibration to remove undesired contributions to the signal (left side of Figure 3 illustrates the process in the case of Aquarius). The Aquarius retrieval process starts with calibrated TA and goes through the following steps: Remove off-Earth contributions (celestial sky, Sun, Moon), correct for the effects of the antenna pattern, correct for Faraday rotation, remove atmospheric effects. At that point, one is left with TB at the ocean surface. The next step is to correct for the surface roughness. Retrievals for SMAP follow a similar approach by first removing unwanted contributions to the signal and then inverting SSS and other parameters using a radiative transfer model. For SMOS, the retrieval uses simulated TB containing all the components projected in the antenna frame. The retrieval is performed with measured and simulated TB in the antenna frame. All three missions retrieve SSS through an iterative algorithm that minimizes an estimator, or cost function, having the following general expression,

$$\chi^2 = \sum_{\theta,p} \frac{(TB_{obs} - TB_{mod})^2}{\sigma_T^2} + \sum \frac{(P_{rtr} - P_{anc})^2}{\sigma_P^2}, \quad (1)$$

with TB_{obs} the observed TB, TB_{mod} the modelled TB using the radiative transfer model, and the normalization parameter σ_T^2 which is the estimated variance on TB. The sum of squared differences in TB combines multi-incidence angles θ (for SMOS) and multiple polarizations p (all). For the initial iteration, the difference between observations and model uses ancillary estimation for geophysical parameters, such as SST, SSS, or wind, if needed. Further iterations adjust one or several geophysical parameters to achieve minimization of χ^2. Additional constraints can be introduced to the retrieval to prevent the process from diverging too far from first guess values and to allow retrieval of additional parameters. The second term of the right member in the equation introduces the difference between retrieved ancillary parameters (P_{rtr}) and their prior value derived from ancillary data (P_{anc}). SMOS uses such constraints and retrieves SST and wind speed at the same time as SSS. The Aquarius CAP algorithm uses constraints on wind speed and radar cross section from the Aquarius scatterometer and retrieves SSS and wind speed at the same time [21]. The Aquarius ADPS algorithm used in the product that we assess in this study retrieves SSS from a roughness-corrected TB and does not introduce the wind in the minimization estimator for SSS [38]. Empirical wind speed is derived in a prior step that includes scatterometer observations. The TB used in the minimization varies, with SMOS using the top of the atmosphere TB, Aquarius ADPS and SMAP RSS the surface roughness corrected TB, and the CAP algorithm the surface TB that includes the wind effects.

The flat sea surface TB is computed as,

$$TB = SST \times (1 - R), \quad (2)$$

with the sea water emissivity assumed to be $(1 - R)$ and the Fresnel reflectivity R at vertical and horizontal polarization given by,

$$R_v = \left| \frac{\varepsilon_r \cos\theta - \sqrt{\varepsilon_r - \sin^2\theta}}{\varepsilon_r \cos\theta + \sqrt{\varepsilon_r - \sin^2\theta}} \right|^2, \quad (3)$$

$$R_h = \left| \frac{\cos\theta - \sqrt{\varepsilon_r - \sin^2\theta}}{\cos\theta + \sqrt{\varepsilon_r - \sin^2\theta}} \right|^2, \quad (4)$$

where θ is the incidence angle and ε_r is the sea water dielectric constant. The TB dependence on SSS is from ε_r and its dependence on SST is from both ε_r and the transformation from emissivity to TB. Theoretical models predict a small dependence of the roughness contribution to TB on SSS and SST [16], but it is of second order compared to the flat surface component.

2.4. Reprocessing of Aquarius SSS with Modified Model and Ancillary Data

To assess the impact of differences in processing of the SMOS and Aquarius data on the SSS products, we reprocess the Aquarius retrievals changing some of the retrieval parameters. We start with the Aquarius version 3 product because it is the last version to not include empirical tuning designed to mitigate SST-dependent biases in SSS. In the reprocessing, we use:

- The Klein and Swift (KS) model [39] for the sea water dielectric constant, instead of the Meissner and Wentz (MW) model [40,41];
- The Operational Sea Surface Temperature and Sea Ice Analysis (OSTIA) SST produced by the U.K. Met Office [42]; instead of the National Oceanic and Atmospheric Administration (NOAA) Optimally Interpolated SST Version 2 [43] (NOAA OI V2) for the ancillary SST;
- The atmospheric attenuation model by Liebe et al. [44] instead of the model by Wentz and Meissner [45].

The KS model and the OSTIA ancillary SST are used in the SMOS processing. The MW model is used in the Aquarius product and the NOAA OI V2 ancillary SST was used for Aquarius until V4. As of V5, Aquarius uses the SST product by the Canadian Meteorological Center [46]. The ancillary SST is used to compute ε_r and TB. The dielectric constant ε_r is used to compute the Fresnel reflectivity for a flat surface. Because its impact on the roughness component of TB is small, and Aquarius and SMAP retrievals use an empirical correction for roughness, this effect is not accounted for here. We reprocess the flat surface emissivity only.

We also change the atmospheric attenuation model used in Aquarius until V4, replacing it with the one recently introduced in V5. The latter model is a return to the original model by Liebe et al. [44]. As discussed in Reference [38], a modified atmospheric model was being used in the Aquarius SSS algorithm from the beginning of the mission until recently, namely for products V1 through V4. The modification was introduced for microwave radiometers operating at higher frequencies as it was shown to improve performances. The temperature dependence of the oxygen component in the absorption model is given by the expression $(300/T_{air})^{\alpha}$, with T_{air} the atmospheric temperature in Kelvin, and α the exponent that was changed from 0.8 to 1.5 in Reference [45]. Recent assessments suggested that this change was not warranted at L-band, so the exponent was changed back to 0.8 in V5, which is also consistent with the model used for the SMOS algorithm. The changes are the largest (−0.2 K) at the high Northern and Southern latitudes.

When changing the model for the SSS retrieval, it is also necessary to recompute the Aquarius TB calibration over oceans. Therefore, we have recomputed the full time series (~46 months) of Aquarius surface TB (roughness corrected) at each footprint (1.44 s temporal resolution) for each change (i.e., using the KS model, the OSTIA SST, and both the KS model and OSTIA SST). We applied a correction to TB calibration derived from the 7-day sliding average of the global difference TB[new]–TB[old] with the new TB using the modified model and the old TB the one found in the Aquarius product (Table 2). In the case where only the dielectric constant model is changed to the KS model, the calibration correction is −0.113K, −0.120K, −0.130K in V-pol and −0.095K, −0.089K, −0.083K in H-pol for the inner, middle and outer beams, respectively. It is almost constant in time with a small seasonal signal about +/−0.015K. Recalibration due to the change of ancillary SST to OSTIA results in a much smaller adjustment of about −0.01 K or less (Table 2) with a small interannual variation (~ 0.005 K) and no clear seasonal signal. This small value is due to the small difference in ancillary SST products' global average and the small dependence of TB on SST at the most common SST (15–20 °C). Larger differences exist at regional scales (Section 3.2). The impact of changing both the dielectric constant model and the ancillary SST is also reported in Table 2.

Table 2. Recalibration offset for Aquarius brightness temperatures in Kelvin applied when changing the dielectric constant model to Klein and Swift (KS) and/or the ancillary SST product to Operational Sea Surface Temperature and Sea Ice Analysis (OSTIA) in the retrieval algorithm.

	KS		**OSTIA SST**		**KS and OSTIA**	
	V-pol	**H-pol**	**V-pol**	**H-pol**	**V-pol**	**H-pol**
Beam 1	−0.113	−0.095	−0.009	−0.007	−0.121	−0.102
Beam 2	−0.120	−0.089	−0.010	−0.007	−0.128	−0.094
Beam 3	−0.130	−0.083	−0.010	−0.006	−0.138	−0.088

The recalibration offsets reported above are indicative of the intrinsic uncertainties on the absolute calibration of L-band TB. Other model parameters are also likely to add to the calibration uncertainty, such as atmospheric attenuation, impact of surface roughness or antenna spillover. Therefore, each sensor's TB calibration is dependent on the forward model assumed to a certain extent. Because the same forward model is also used in the retrieval algorithm, the calibration of the SSS product should be minimally impacted by the model constant biases. But it is somewhat dependent on the reference SSS field used, both the data source (e.g., HYCOM or WOA) and the reference period and space (e.g., monthly climatology, global or regional SSS).

3. Results

We first discuss how the differences in dielectric constant impact modeled TB and retrieved SSS for a large range of SSS and SST (Section 3.1). In Section 3.2 we present the differences in ancillary SST products and their variability for four years covering the Aquarius era and discuss the potential impact on retrieved SSS. Finally, we present the comparison between SSS products of SMOS, Aquarius and SMAP, and report the impact of changing the dielectric constant model and ancillary SST on these comparisons (Section 3.3).

3.1. Brightness Temperature and SSS Differences Due to Differences in Dielectric Constant Models

Differences in TB due to differences in the dielectric constant model are reported in Figure 4a as a function of SSS and SST. We report the vertical polarization only as it is the most sensitive to SSS. The results in horizontal polarization exhibit a similar pattern but with differences in TB about 25% less. On the right side of the figure we report the corresponding differences in retrieved SSS. To compute the error in SSS, the KS model is used for the forward model that computes TB given assumed SSS_0 and SST_0 (reported along the x- and y-axis) and the MW model is used to retrieve SSS_1 from TB and the same SST_0. The difference between SSS_1 and SSS_0 is reported on the figure. The SSS and SST ranges reported ((28–40) psu and [−2, +31] °C) cover most oceanic waters, especially when considering open oceans. Fresher SSS can be found close to large river mouths and higher SSS in concentration basins, such as the Red Sea. The green dashed curve reports the limits of the likely SSS/SST combinations. The limits are computed from all Argo SSS and SST measurements between 2000 and 2018 totaling 1,840,082 observations. The number of observations per combination of SST/SSS is accumulated in 0.1 °C by 0.1 psu grid cell and a combination of SST/SSS is considered likely if at least 5 Argo observations occurred during the whole Argo history. As an example of the unlikely combination, SSS much larger than 35 psu are seldom observed in water colder than 12 °C. High salinities usually result from excess evaporation that requires high insolation and limited precipitations and vertical mixing (i.e., low wind) occurring mostly at low to mid latitudes and in small basins like the Mediterranean Sea. Very fresh waters with salinity lower than 32 psu are found in warmer (>24 °C) or colder (<7 °C) waters and usually result from large river runoff (e.g., cold waters of the Arctic basins, Amazon, Congo and Ganges river plumes in warm waters) or ice melt.

The differences in TB are between −0.25 K and +0.40 K with the KS model predicting TB smaller than MW at most SST (>4 °C) and SSS. The horizontal stratification of the contours in Figure 4 shows that the difference is mostly a function of SST (reported on the y-axis) and varies much less with SSS (reported on the x-axis). For the most common oceanic conditions (33.75 psu < SSS < 35.75 psu), TB from the KS model is ~0.1 K cooler than TB from MW in warm waters, and 0.2 K cooler in more temperate waters (around 10–15 °C). The differences are the largest (in absolute values) in the cold and fresh waters, at or below 0 °C for SSS fresher than 35 psu, with TB from the KS model being larger than MW by ~0.4 K. The change of sign in TB differences occur around 3.5 °C where both KS and MW models predict the same TB. The average global difference in TB (as reported previously in Table 2) is about 0.12 K and can be inferred from Figure 4a considering that most global SSS are around 35 psu and an effective SST of ~18 °C.

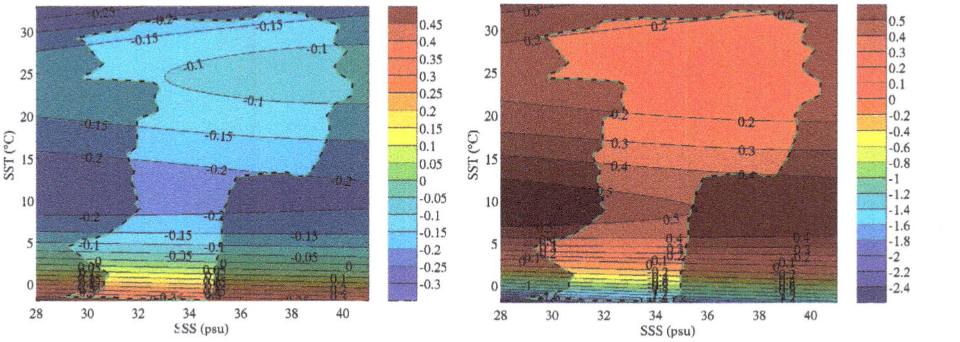

(a) (b)

Figure 4. Differences in TB and SSS due to differences between the dielectric constant model by Klein and Swift [39] and Meissner and Wentz [41]. (**a**) Differences in brightness temperature at L-band for a flat surface and an incidence angle of 38° (similar to Aquarius middle beam). (**b**) Differences in retrieved SSS when assuming KS the truth (i.e., used to compute TB from assumed SST_0 and SSS_0) and using MW for retrieving SSS from TB and the same SST_0. The shaded area reports unlikely combinations of SSS and SST (i.e., less than 5 Argo records over the last period 2000–2018).

The differences in retrieved SSS resulting from the TB differences discussed above are reported in Figure 4b. The pattern in SSS difference is very similar to the pattern in TB difference, with mostly a dependence on SST and a much lower dependence on SSS. For the warmer waters, the SSS difference is ~0.2 psu, or about 30% larger than the value of TB difference in Kelvin. When the temperature decreases, the sensitivity of TB to SSS decreases and the ratio of SSS difference to the TB difference increases. In temperate waters around 15 °C, the SSS difference is 0.4 psu, twice the difference in TB of 0.2 K. In the cold waters around 0 °C, the error in SSS increases very significantly (1 psu or more) due to two confounding factors: Larger differences in TB and a reduction of the TB sensitivity to SSS to a third of what it is in warm waters (~0.25 K/psu). It is important to note that these are not the differences that will be observed between two SSS products using different dielectric constant models because of the re-calibration process discussed in Section 2.4. This process effectively shifts the zero difference in SSS occurring around 3.5 °C in Figure 4b (consistent with the 0 difference in TB on the left side figure) to the SST and SSS most representative of the region used for the calibration. In the case of Aquarius, which uses global data to calibrate, this will be around the global average SST and SSS.

3.2. Differences in Ancillary Sea Surface Temperature Products

Comparisons between OSTIA [42] and NOAA OI V2 SST [43] products are reported in Figure 5. We use daily products and average them over seven days on global maps at 0.25° resolution in latitude and longitude. SST in grid cells where the sea ice fraction from the OSTIA product is 0.15 or larger are excluded from the comparisons. (We do not use the sea ice fraction from NOAA due to known inconsistencies with L-band observations [47]). Using maps between Sept 2011 and August 2015 we compute times series for the followings statistical parameters for weekly global SST: 1) median and mean of the SST differences; 2) percentiles of the absolute value of the SST differences from 70% to 99%. We compute x as the ith percentile of X if the probability of X to be equal to or less than x is i%:

$$P(X \leq x) = i, \quad (5)$$

with

$$X = |SST_O - SST_N|, \quad (6)$$

where X is the absolute value of the difference between OSTIA (O) and NOAA OI V2 (N) SST on a global weekly map. Computing the percentiles using the absolute value of the SST differences prevents cancellation of areas of positive and negative differences. The 70% percentile is close to the standard deviation of the difference for normal distribution and is more robust to large outliers for non-normal distributions. The higher percentiles provide information of less common but potentially much larger differences in SST products.

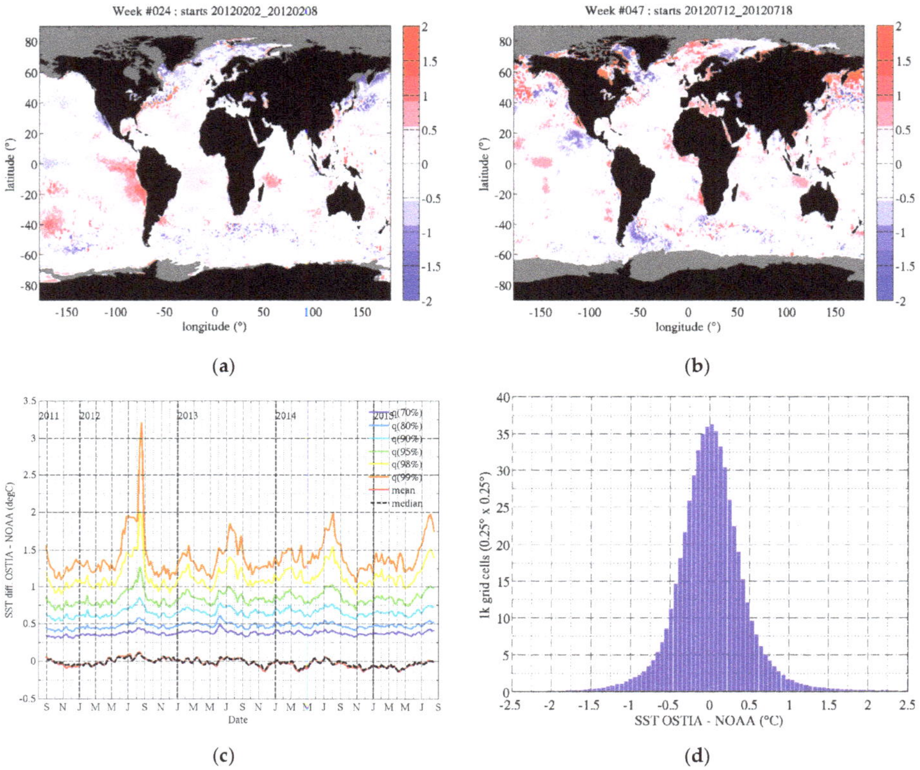

Figure 5. Differences in weekly average SST between OSTIA and National Oceanic and Atmospheric Administration (NOAA) OI V2 products. Difference maps are averaged weekly from daily products at 0.25° resolution in latitude/longitude during the period September 2011–August 2015. The top row (**a**,**b**) shows two examples of SST differences maps. The dates are chosen to show examples of (**a**) smaller (Feb 2012) and (**b**) larger (July 2012) differences in SST. Grey areas report sea ice fraction (from OSTIA) of 0.15 or above. The temporal evolution of SST differences is reported in panel (**c**) through several statistical indicators. The curves report the time series of (red) mean and (dashed-black) median difference in global weekly maps and the percentiles of the absolute value of the SST differences between 70% and 99% (see text in Section 3.2). Panel (**d**) reports the histogram of SST difference for the map reported in (**b**) (y-axis unit is 1000 grid cells).

The top row in Figure 5 reports difference maps between the OSTIA and NOAA products for two different weeks chosen to be representative of somewhat smaller and larger differences in SST products (as reported in Figure 5c). The histogram in Figure 5d is computed from the map in Figure 5b. The time series of statistical indexes are reported in Figure 5c. The mean and median global differences are small, usually between ±0.1 °C, and stable in time. Most of the differences (70%) are below 0.5 °C, varying between 0.3 °C and 0.45 °C. There are always a few percent (2–5% depending on the season) of the ocean with differences larger than 1 °C. Differences larger than 3 °C are seldom seen but still represent ~1% of the globe in the summer of 2012. There are seasonal variations visible in all the percentile curves, with the larger variations at the higher percentiles exceeding 0.5 °C. The largest differences occur during the northern hemisphere (NH) summer between July and September. It should be noted that not removing ice fractions above 0.15 would increase the differences by up to 0.15 °C (mean) and 0.5 °C (99th percentile) due to the very large differences in products in the Arctic ocean under the ice cover, but these data are not relevant for our application. As reported in the top row of Figure 5, SST differences exhibit large spatial patterns with positive and negative differences that tend to cancel each other when averaged globally. In NH Winter (Figure 5a), OSTIA SST tend to be colder at higher latitudes and warmer in the mid and lower latitudes (with notable exceptions). In NH summer, the patterns tend to be more mixed, and positive and negative large differences can be observed next to each other in the mid and high latitudes. The histogram in Figure 5d shows that most of the differences are below 0.5 °C and that differences larger than 1 °C are in the wings of the distribution that extend to substantial SST differences with rare occurrences.

The impact of SST differences on the SSS retrievals will depend on (1) the sensitivity of TB to SST and (2) the sensitivity of SSS to TB. Because both sensitivities change with SST, the impact of SST differences will show regional and seasonal dependencies. We show an example of the impact on retrieved SSS for one week in Figure 6. The first plot (left, a) is the SST difference interpolated at each Aquarius footprint location combining all beams for the week reported in Figure 5b (ascending passes only). The figure in the middle (b) shows the resulting difference in modeled surface TB in V-pol. For most mid-latitudes, the impact on TB is relatively small, less than 0.1 K, because TB sensitivity to SST is small around 15–20 °C. The impact on TB is larger in the cold waters at high latitudes and at low latitudes where SST is warmer than 20 °C, due to increased TB sensitivity to SST. TB sensitivity to SST also change signs: In the cold waters a positive SST difference leads to increased TB while in the warm waters of the lower latitudes a positive SST difference lead to decreased TB. The impact of SST differences on retrieved SSS is shown in the figure on the right (c). It can be very significant (larger than 1 psu) at high latitudes due to the combined effect of increased TB sensitivity to SST and reduced TB sensitivity to SSS in cold waters. The latter amplifies the impact of errors in TB. Both these effects result in differences of up to 1 psu at high latitudes when the same differences in SST result in much smaller differences in SSS of −0.2 psu at lower latitude.

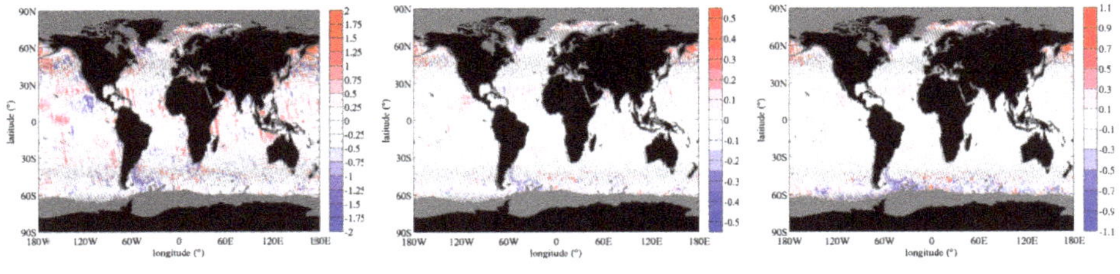

(a) (b) (c)

Figure 6. Impact of SST differences on TB and retrieved SSS. (**a**) Map of SST differences (°C) between OSTIA and NOAA OI interpolated at the Aquarius ground tracks (same week as in (**b**) in Figure 5, ascending orbits). (**b**) Differences in TB (K) for a flat surface due to the differences in SST in the left figure. (**c**) Differences in retrieved SSS (psu) due to differences in TB reported in the middle figure.

While the differences in SST ancillary product result in significant regional changes in TB, it does not impact significantly Aquarius calibration which uses observations averaged globally over seven days. Figure 7 reports changes in TB due to the differences in SST products for the duration of the Aquarius mission. We interpolate both SST products at each Aquarius 1.44 s footprint and compute the corresponding TB for a flat surface. The grey dots report the TB difference for each footprint, and the red curve reports the 7-day global average difference. The latter is very small (≪0.1 K, see average values in Table 2, middle columns) and temporally stable. While not covering the whole globe, daily averages (not shown) are close to weekly average with an additional small noise of less than 5 mK. The per footprint TB difference shows significantly larger differences, of a few tenths of a Kelvin and up to ~1 K occasionally (outer beam, V-pol). The large variability is due to differences in SST products that vary with time and location and due to the large changes in the sensitivity of TB to changes in SST over the range of observed SST. Figure 7 is for the middle beam at V-pol. All channels show similar result regarding the global average; the scatter in the per-footprint data is larger in V-pol and increases (decreases) with increasing incidence angle at V-pol (H-pol).

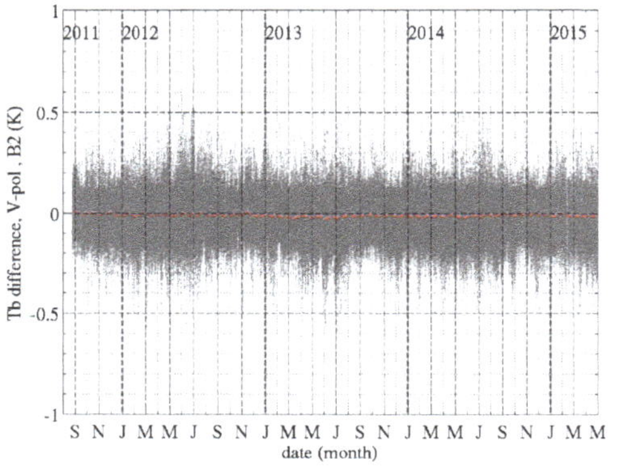

Figure 7. Differences in Aquarius TB due to differences between OSTIA and NOAA OI versus time, for the middle beam in vertical polarization. The TB difference is (grey) for each Aquarius 1.44 s footprint and (red) averaged globally over seven days.

3.3. Comparison of SSS Products

We follow a two parts approach to compare the SSS from the various satellite sensors and the in situ SSS. First in Section 3.3.1 we discuss the global spatial patterns and statistical distribution averaged over long periods of time. Second, in Section 3.3.2 we address SSS temporal variability by looking at time series over a few regions of interest. In Section 3.3.3 we report on the observed relationship between SSS bias and SST for various versions of the Aquarius and SMAP products. In Section 3.3.4 we analyze the impact of the model parameters on the comparisons between SMOS, Aquarius and Argo SSS products.

3.3.1. Spatial Patterns and Statistical Distribution

Maps of global SSS average over long time periods are reported in Figure 8. We use the Aquarius era (Sept 2011–May 2015) for Argo, SMOS and Aquarius on the left, and the period April 2015–June 2018 for Argo, SMOS and SMAP on the right. SMAP SSS is not available until April 2015 so it has a short overlap with the Aquarius era, and a direct comparison with Aquarius is not possible. Some of the differences observed between SMAP and Aquarius will be due to the different time period they cover, but the major patterns are expected to be similar.

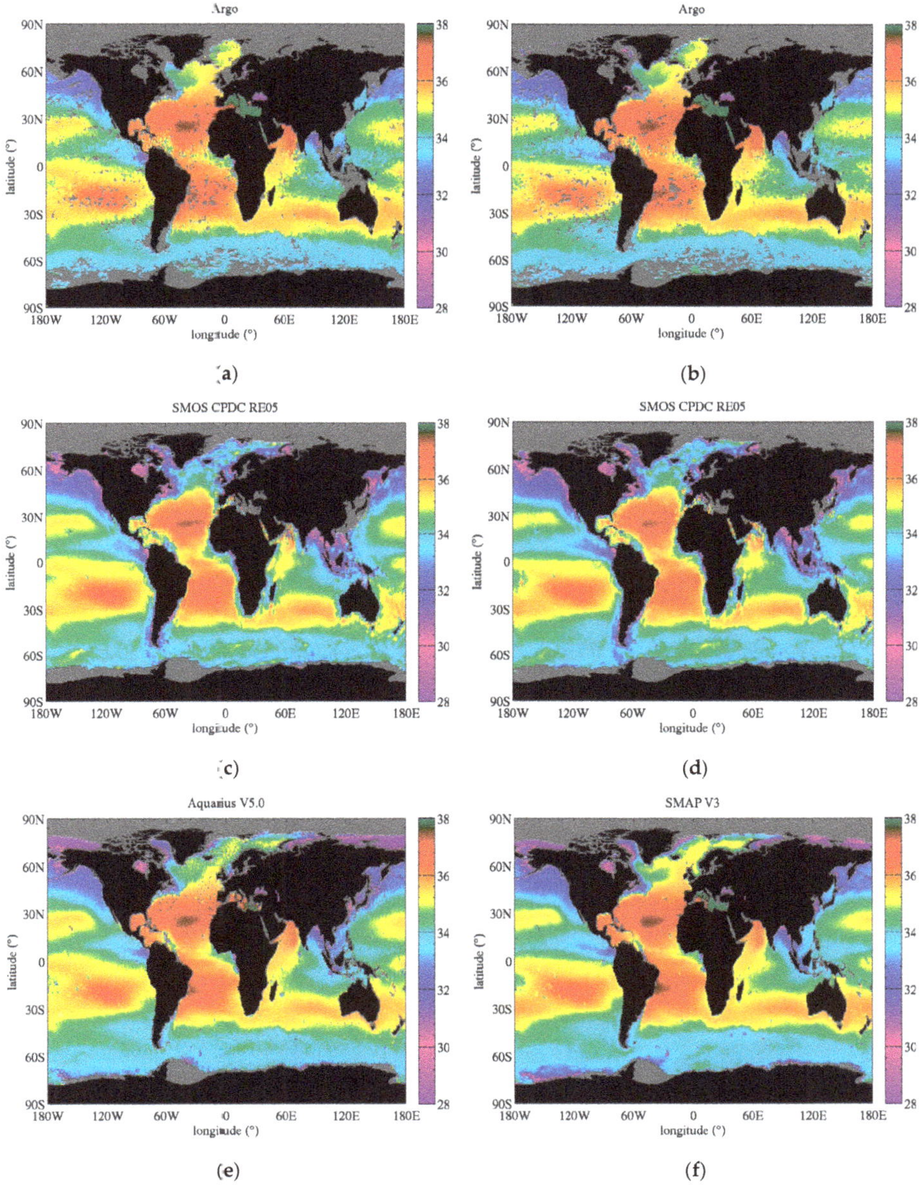

Figure 8. Global map of SSS from satellite sensors and in situ measurements averaged over several years. On the left: (**a**) Argo SSS, (**c**) SMOS SSS, and (**e**) Aquarius SSS are averaged over the Aquarius era (i.e., Sept 2011–May 2015). On the right: (**b**) Argo SSS, (**d**) SMOS SSS, and (**f**) SMAP SSS are averaged over the period April 2015–June 2018. Grey color reports missing data.

All products show similar large-scale structures and very similar overall dynamic range with SSS mostly between 32.5 psu and 37 psu (see also SSS histograms, statistics in Table 3 and discussion below). There are only a few regions with SSS less than 30 psu (e.g., close to large river mouths and in the Arctic Ocean) or larger than 38 psu (e.g., Mediterranean Sea, Red Sea). For all products, the Atlantic

Ocean is in general saltier than the Pacific Ocean, and fresher regions include high precipitation and upwelling regions, such as the equatorial Pacific Ocean, the Bay of Bengal, and the Indonesian Archipelago. The ocean gyres, marked with high SSS and low temporal variability ($\leq$0.13 psu, see below) at their center are clearly visible in the Atlantic North and South, the Pacific North and South and the Indian Ocean.

Table 3. SSS distribution statistics for satellite products and in situ observation from the Argo network (all values in psu). The SMOS, Aquarius and the top Argo row are computed for the Aquarius era (Sept 2011–May 2015), the second Argo row and SMAP results are computed during the SMAP era (04/2015–06/2018).

	Median	2.50%	16%	84%	97.50%
Argo	34.90	32.75	33.92	35.74	36.85
SMOS CPDC (filtered land)	34.82	33.23	33.99	35.76	36.90
Aquarius V5	34.85	32.71	33.97	35.69	36.85
Argo (SMAP period)	34.83	32.65	33.90	35.84	37.00
SMAP V3	34.78	32.61	33.85	35.85	37.00
SMAP V2	34.78	32.41	33.78	35.86	37.23

Some differences between the SSS products already appear clearly in the maps in Figure 8, but more subtle differences can be identified from the map of difference between satellite and Argo SSS reported in Figure 9. SMOS SSS exhibits much fresher SSS in coastal regions, a feature extending hundreds of kilometers away from land and ice boundaries. It is particularly visible in Figure 8 around Australia and the southern parts of Africa and South America. This coastal freshening is spurious and is due to the larger brighter temperature of land (and ice) compared to the ocean, entering the instrument's main beam or side lobes. In the case of SMOS, this contamination also occurs through the complex process of image reconstruction from its spatial interferometer measurements. In addition to land and ice contamination, RFI can be a significant contributor to coastal spurious freshening in the Northern hemisphere (e.g., near Alaska, Greenland, the Arabian Sea and Sea of China). Lower SSS also occur to a smaller extend around small islands, further away from large land masses, such as around Hawaii south of the North Pacific Gyre and the Reunion and Mauritius islands east of Madagascar. Too fresh salinities along the coasts has been an ongoing problem in previous versions of SMOS products and it has been partially mitigated in the latest versions by improving the flagging of land contamination. This problem is still the focus of research and further improvements are likely to come in future versions thanks to improved image reconstruction (see Figure 5 in Reference [31]). In addition, as discussed previously, other SMOS products use empirical land contamination corrections on TB or on SSS which extend to more than 1000 km from continents [25,26]; these corrections are not considered here in order to avoid side effects. Improving the correction is the focus of research for the next versions of the SSS products. Aquarius and SMAP use a correction on TB to mitigate the impact of land contamination that is derived from radiative transfer simulations of the sidelobe contribution. It is efficient at correcting the land contamination to the first order, but fresh biases persist at some locations (too little correction), and salty biases are created possibly due to over-correction (e.g., around Hawaii, Indonesian archipelago and South and East of Australia). Improving the correction is also the focus of research for the next versions of the Aquarius and SMAP products. To this date, there is no correction for the impact of sea ice contamination. A model is used to compute the fraction of ice in the observations, and data suspected to be impacted are flagged and removed from the product [47]. Correcting for the ice impact requires accounting for sea ice TB at L-band, which is a complex issue as ice TB could depend on poorly known factors, such as sea ice thickness and ice age, or the presence of snow at the surface [47–49].

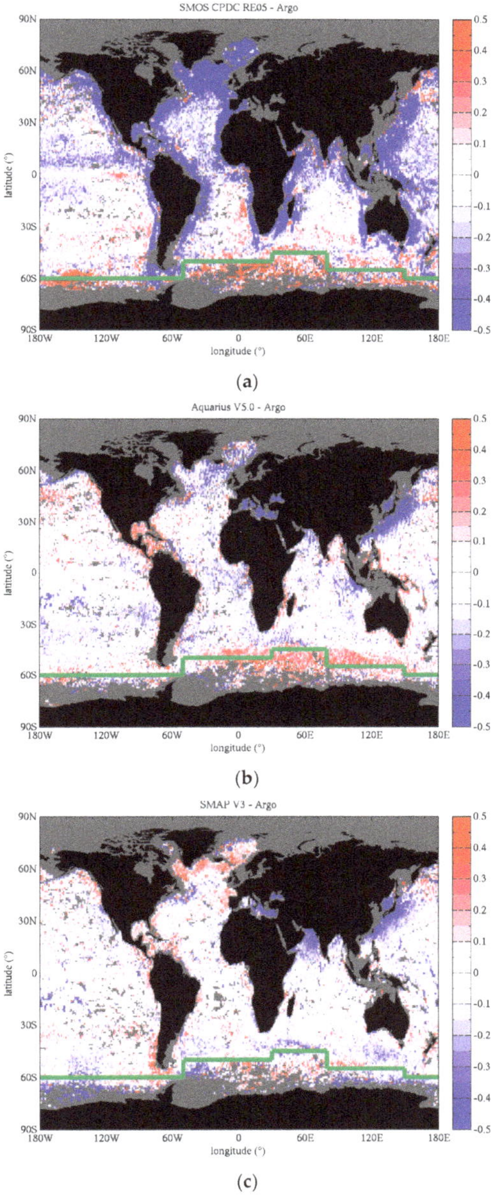

Figure 9. Map of SSS difference between satellite retrievals and in situ Argo observations for (**a**) SMOS, (**b**) Aquarius V5, and (**c**) SMAP V3. The maps are computed from long-term averages of monthly SSS maps over the Aquarius era (i.e., Sept 2011–May 2015) for SMOS and Aquarius, and the period April 2015–June 2018 for SMAP. The green dashed line between 60°S and 45°S marks the average location of the Antarctic convergence that parts cold and dense surface Antarctic waters and sub-Antarctic regions North of it.

The Atlantic Ocean North of 45°N (e.g., Labrador Sea and West of Northern Europe) and the Norwegian Sea are significantly fresher in the SMOS product than in the other products (Argo, Aquarius and SMAP). In the other high Northern latitudes waters not covered by Argo, SMOS tends to be fresher than SMAP and Aquarius in the Barents Sea, in the Baffin Bay. In the high Southern latitudes, below the Antarctic convergence (45–60°S), SMOS SSS are significantly saltier than the other products. These large SSS differences are associated with very cold waters. The possible impacts of the ancillary SST products and dielectric constant model on these differences are assessed in Section 3.3.4. At lower latitude, the SMOS product exhibits a less salty peak in the southern Atlantic gyre off the coast of Brazil. Argo, Aquarius and SMAP show peaks of 37.8 psu, while SMOS peak is 0.5 psu fresher. The map of the difference with Argo (Figure 9a) shows that the fresher peak in SMOS is likely due to land contamination that extends as far as the southern Atlantic gyre. In the warmer waters at low latitudes, the three satellites tend to exhibit small fresh biases, except in the Eastern equatorial Pacific were salty biases appear with differing intensity and latitudinal width depending on the product. SMOS exhibit the largest salty biases over a narrow band around the equator. SMAP and Aquarius salty biases are smaller and more extended north and south of the equator. The salty-biased equatorial region and the fresh-biased inter-tropical regions around it (at 10–20° latitude) are characterized by strong and highly variable precipitations changing seasonally. Large precipitations can create salinity vertical gradients [50] that could partly explain the discrepancies between satellite and in situ observations (e.g., fresh biases), because measurements by satellite are sensitive to the first few centimeters of water and Argo measurement usually occur a few meters under the surface. Previous studies have found effects between −0.1 and −0.4 psu/(mm/hr) varying with location and wind speed [50]. It is worth noting that the impact of evaporation has been found to have a very small impact on salinity stratification and is unlikely to impact significantly the large scales and long term comparisons presented here [50]. Differences in spatial and temporal sampling between satellites and in situ measurements are also likely causes for differences near the equator.

SMAP and Aquarius largely share the same retrieval algorithm. Consequently, their SSS products tend to exhibit similar features, especially compared to SMOS. Both SMAP and Aquarius SSS are saltier in the Eastern North Atlantic, west of Europe, compared to the fresher waters extending from the Labrador Sea in the west North Atlantic, in agreement with Argo in situ data. They also exhibit saltier waters in the Southern Atlantic gyre and in the Arabian Sea, compared to SMOS. There are also some notable differences between SMAP and Aquarius, for example in the equatorial Pacific, where SMAP fresh water is more extended westward than Aquarius. Similarly, the fresh water patterns around the Indonesian archipelago differ noticeably. These regions have substantial seasonal and inter-annual variability, and because the data are reported for different periods and length of time for SMAP and Aquarius, the differences are due to natural variability [51]. This is supported by the Argo maps for both periods, which exhibit similar differences as Aquarius and SMAP. The maps of SSS difference between SMAP and Argo computed for the SMAP period in Figure 9c exhibits similar or smaller errors in these regions (e.g., southwest of Sumatra and Java islands) than the difference map with Aquarius (Figure 9b).

One common feature of the three satellite products is the larger SSS biases in the Southern Ocean, close to or south of the Antarctic Convergence (AC) whose approximate location is reported in Figure 9 by the green line. While these biases have been reduced with recent revisions of the various SSS products, they have been a persistent issue that has not yet been completely mitigated and whose cause is still being investigated. The AC is the region separating the cold Antarctic waters from the warmer sub-Antarctic waters south of the Atlantic, Pacific and Indian oceans, where southern waters sink below waters in the North. The AC and other fronts a few degrees north of it (Subtropical Front and Subantarctic Front [52]) exhibit sharp SSS or SST spatial gradients. We computed SSS and SST meridional gradients from the Argo monthly maps used in this study and reported their long-term average in Figure 10. SSS and SST gradients are particularly large at the Antarctic convergence for longitudes between 12°E and 80°E. Because Argo sampling can be spatially sparse, we have also assessed the consistency of these gradients with cruise measurements from the RV Polarstern around the longitude 12°E and have found gradients of ~0.75 psu per degree of latitude and 3.5 °C per degree of latitude in SSS and SST respectively. We postulate that part of the difference between satellite and in situ SSS around the AC could be due to the difference in the spatial scale of the measurements. The potential impact of the very cold temperatures south of the AC will be investigated in the next sections.

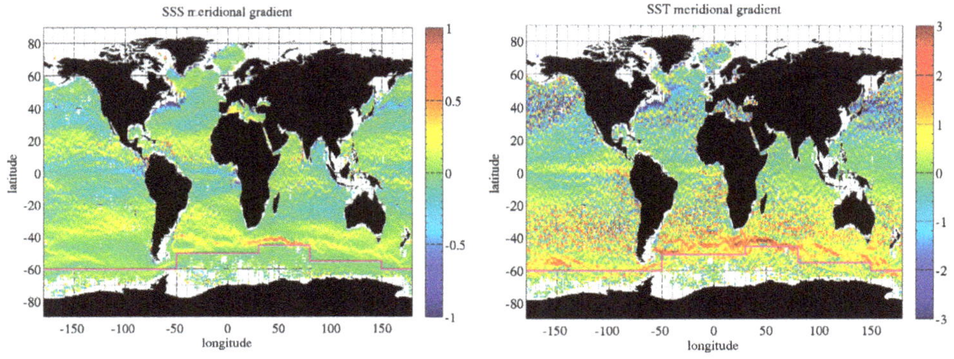

(a) (b)

Figure 10. Maps of meridional gradients of (**a**) SSS in psu per degree of latitude and (**b**) SST in °C per degree of latitude computed from monthly Argo maps. The magenta line reports the approximate location of the Antarctic Convergence.

Histograms for the monthly SSS maps of SMOS, Aquarius, SMAP and Argo are reported in Figure 11. Associated statistical measures (median, percentiles) are reported in Table 3 and are computed over the periods reported in Table 4. For SMOS we filter out data closer than 1000 km from the coast to avoid land contamination (the histogram is renormalized to have the same number of samples as Argo so that the area below the Argo and SMOS curves is the same). Because the histogram for SMAP SSS was computed over a different period (04/2015–06/2018) compared to the other histograms (Aquarius era: 09/2011–05/2015), we compare it to an Argo histogram computed for the SMAP period. There is good agreement between the various products in general. While SSS can span a large range, most SSS are confined to a relatively narrow 2 psu window around ~35 psu. Small values of 25 psu or less can be found in the Arctic and very close to large river mouths (not reported on the histogram). Large values of 39–42 psu are found in small concentration basins (e.g., Red Sea). However, most SSS (68%) are between 33.8 psu and 35.8 psu (Table 3, columns 4 and 5). The median of the global SSS is close between Argo and Aquarius, which is expected because Aquarius brightness temperature calibration relies on globally gridded optimally interpolated Argo fields (or previously on a numerical model for SSS, which assimilates the Argo measurements). SMOS median SSS is just a little bit fresher, a difference that could be due to the removal of coastal data. Reintroducing these data would further increase the fresh bias by ~0.11 psu however, which is expected as these data are clearly freshly biased by large amounts. Another possible cause for the difference is that SMOS is calibrated regionally, using only a portion of the South Pacific Ocean [53]. Aquarius and SMAP use global data. SMAP median SSS is also close to Argo's for the same period. The other statistical measures are also in good agreement between the latest SMAP and Aquarius products and Argo, with the differences being 0.05 psu or less. The agreement is also good for SMOS except for the 2.5 percentile where SMOS is higher by ~0.5 psu, likely owing to the absence of the plateau around 33 psu (arrow #3 in Figure 11).

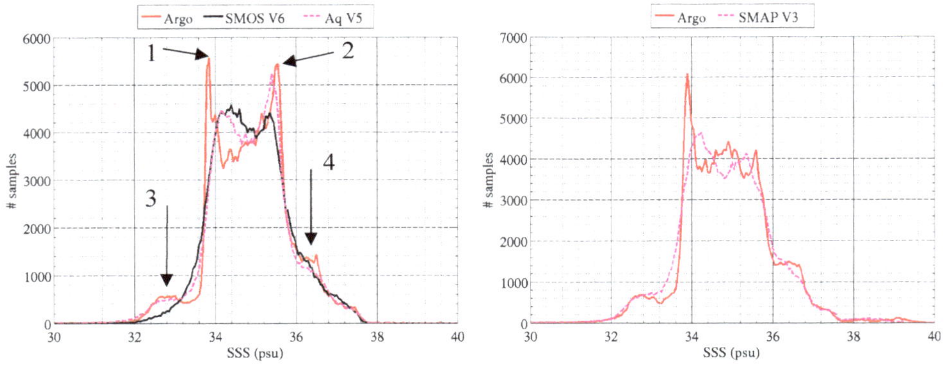

(a) (b)

Figure 11. Histogram of SSS from (red) Argo in situ measurements and satellite SSS measurements by (**a**) SMOS and Aquarius, and (**b**) SMAP. They are computed from global monthly maps at 1° × 1° resolution in latitude and longitude over the period (**a**) Sept 2011–May 2015 and (**b**) April 2015–June 2018. The x-axis sampling is 0.05 psu; the x-axis is cropped between 30 and 40 psu.

Table 4. Start and end date for the statistics of the differences between satellite products and Argo in situ data.

	SMOS CPDC	Aquarius V3	Aquarius V5	SMAP V2	SMAP V3
Start Date	Jan-2011	Sep-2011	Sep-2011	Apr-2015	Apr-2015
End Date	Mar-2017	Apr-2015	May-2015	Feb-2018	Jun-2018

There are some important differences in the shape of the histograms for the various SSS products. The Argo histogram for the Aquarius period (Figure 11a) exhibits 2 peaks (arrows #1 and #2 in the plot) at 33.85 psu and 35.55 psu and two plateaus (arrows #3 and #4 in the plot) at 32.5–33.5 psu, and 36.5 psu that are not matched by all satellite products. The right-hand peak at 35.5 psu (arrow #2) is reproduced only by the Aquarius product. It is missing in both SMOS and SMAP products. A clue to the cause for this peak is given by the absence of the peak in the Argo histograms for the SMAP period (Figure 11b). The grid cells contributing to the peak are mostly located around the South African coasts, eastward into the Southern Indian Ocean, along the Southern and Eastern Australian coasts and eastward into the Southern Pacific Ocean. The large fresh biases, due to land contamination in most of these regions, explains the absence of this peak from the SMOS histogram (either because the data are excluded or at a different location in the histogram if they are included). The absence of the peak for the SMAP histogram is consistent with the Argo histogram over the SMAP era indicating a change in sampled locations coupled with possible changes in geophysical conditions. The peak in Argo SSS around 33.85 psu (arrow #1) occurs near and south of the AC (green line in Figure 9 between 45°S and 60°S) and in high precipitation regions in the equatorial Pacific and around the Indonesian archipelago. Near and south of the AC, SMOS and Aquarius are significantly saltier than Argo (>0.5 psu) which explains the absence of the peak at 33.85 psu in their histograms and the increased density around 34.3 psu compared to Argo. The plateau around 32.5 psu (arrow #3 for Argo, SMAP and Aquarius) is missing from the SMOS histogram, due to the biases in the North Pacific Ocean above 40°N (e.g., Gulf of Alaska, East of Japan, West of Canada) and in the Bay of Bengal. The plateau around 36.5 psu (arrow #4) is missing in the SMOS histogram, due to biases in the Arabian sea and along the East Northern American coast (north of 20°N).

3.3.2. Temporal Variability

We selected 12 regions of interest (ROI) in which we average the monthly SSS from Argo and the satellite products. The boundaries of the ROI are reported in Table 5. Because the coverage of an ROI by Argo will change in time, we compute the satellite average using only grid cells in the ROI that have a corresponding Argo sample. The ROI are reported in Figure 2 over a map of Argo SSS (top) long term average and (bottom) standard deviation over the period 01/2011–06/2018. They cover various SSS and SST average value, as well as various temporal variability (Table 6). The five oceanic gyres in the Atlantic (#1, 2), Pacific (#5, 6) and South Indian (#8) oceans have SSS which is relatively high (35.3–37.5 psu) and stable (STD 0.07–0.11 psu). Highly variable regions, such as the equatorial Pacific upwelling region (#4), the Amazon plume (#3) and the Tropical Indian Ocean (#10) show seasonal variations of about 0.5–1.0 psu driven by changing fresh water influx from river outflow and large precipitations, and advection by coastal currents and the South Equatorial Current in the Indian Ocean. Low SSS (32.7–33.9 psu) are persistently observed in cold waters at high latitudes (#7, 11, 12), and occur seasonally in big river mouths and plumes (#3). The time series of Argo, SMOS, Aquarius and SMAP SSS in the ROI are reported in Figures 12 and 13. The SMAP and Aquarius SSS are those distributed at the PO.DAAC (Section 2.1).The temporal average and standard deviation of the difference between satellite and Argo SSS are reported for each region in Tables 7 and 8, respectively. Because SMAP V3 was recently released, we also report results for V2 to emphasize the changes in the latest version of the product. This is discussed at the end of the section. The first paragraph focusses on the latest SMAP, Aquarius and SMOS products.

Table 5. Longitude and latitude (in °) limits for the regions of interest in Figure 2.

Region name	Long. West	Long. East	Lat. North	Lat. South
North Atlantic Gyre	−42.5	−28	24	28
South Atlantic Gyre	−16.8	−10.5	−24.7	−17.3
Amazon Plume	−46.5	−43	4	11
Equatorial Pacific	−132.8	−93.4	−2.6	2.6
South Pacific Gyre	−146.3	−108.6	−23.4	−16.6
North Pacific Gyre	−165.1	−134.8	23.7	31.8
North Pacific High latitudes	−180.2	−144.1	46	50
South Indian Ocean Gyre	75.2	99.2	−33.1	−26.7
South Pacific High latitudes	−172	−140	−51.5	−46.8
Tropical Indian Ocean	68	84	−11.5	−4.5
South Atlantic High latitudes	−20	8	−50	−45

Table 6. Argo SSS and SST median and STD in regions of interest.

ROI #	Region Name	Median SSS (psu)	STD SSS (psu)	Median SST (°C)	STD SST (°C)
1	North Atlantic Gyre	37.42	0.09	23.96	1.77
2	South Atlantic Gyre	36.70	0.13	23.23	1.67
3	Amazon Plume	35.82	0.63	27.87	0.94
4	Equatorial Pacific	34.78	0.19	25.07	1.77
5	South Pacific Gyre	36.37	0.08	25.90	1.29
6	North Pacific Gyre	35.26	0.11	22.53	1.73
7	North Pacific High latitudes	32.64	0.10	7.41	2.82
8	South Indian Ocean Gyre	35.85	0.07	20.89	1.77
9	South Pacific High latitudes	34.48	0.05	9.97	1.30
10	Tropical Indian Ocean	34.61	0.39	28.51	0.85
11	South Atlantic High latitudes	33.83	0.05	5.39	1.19
12	South Indian Ocean High latitudes	33.91	0.13	7.19	1.35

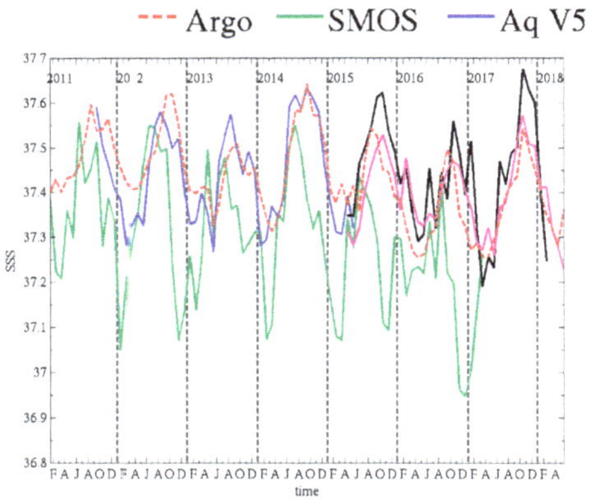

(**a**) #1: North Atlantic Gyre

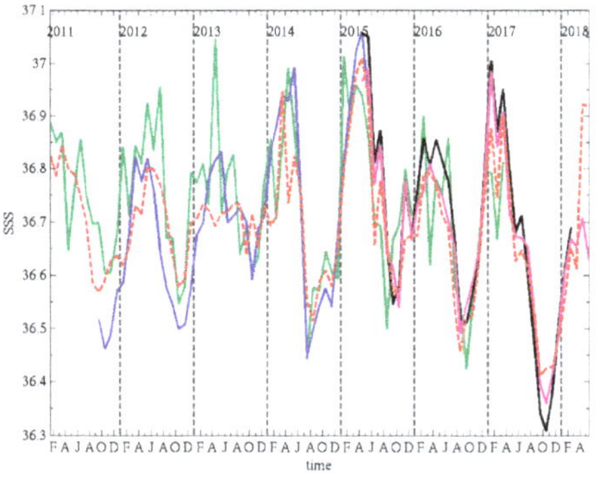

(**b**) #2: South Atlantic Gyre

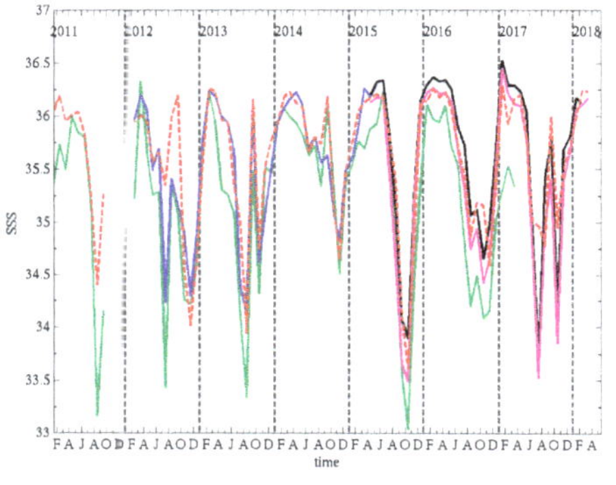

(**c**) #3: Amazon Plume

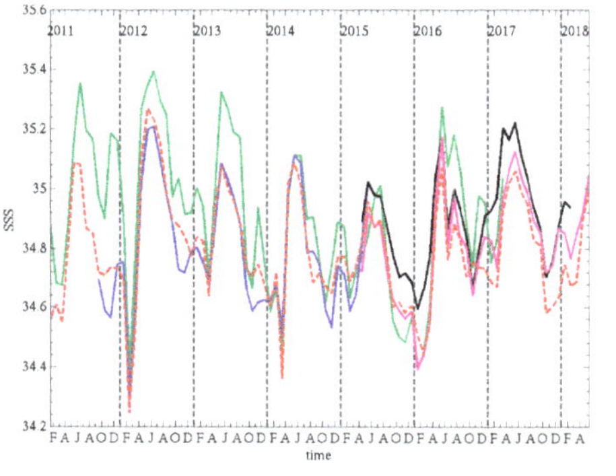

(**d**) #4: Equatorial Pacific

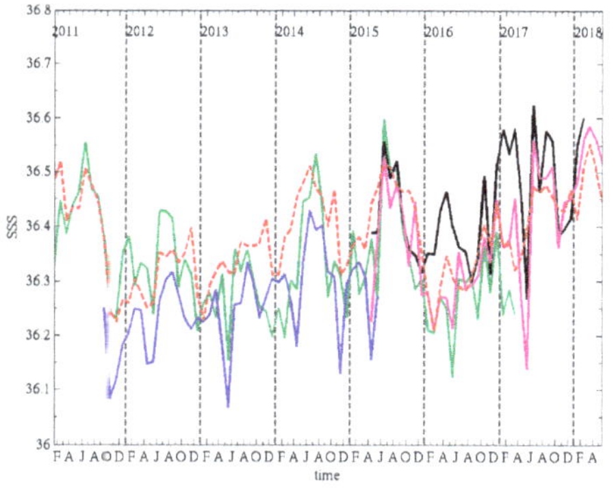

(**e**) #5: South Pacific Gyre

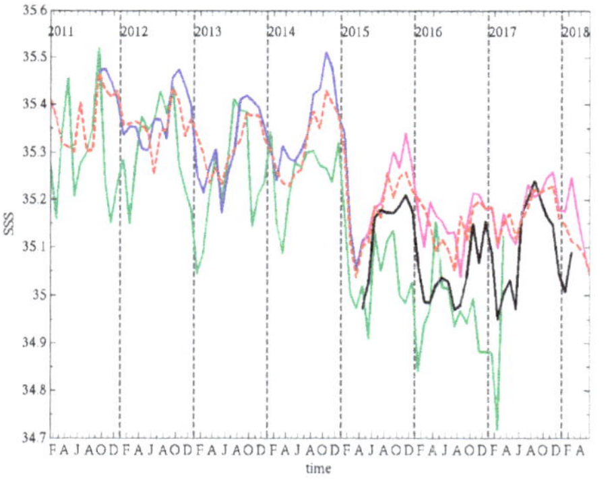

(**f**) #6: North Pacific Gyre

Figure 12. Time series of SSS from (red) Argo, (green) SMOS, (blue) Aquarius and (black) SMAP for regions of interest 1–6 reported in Figure 2.

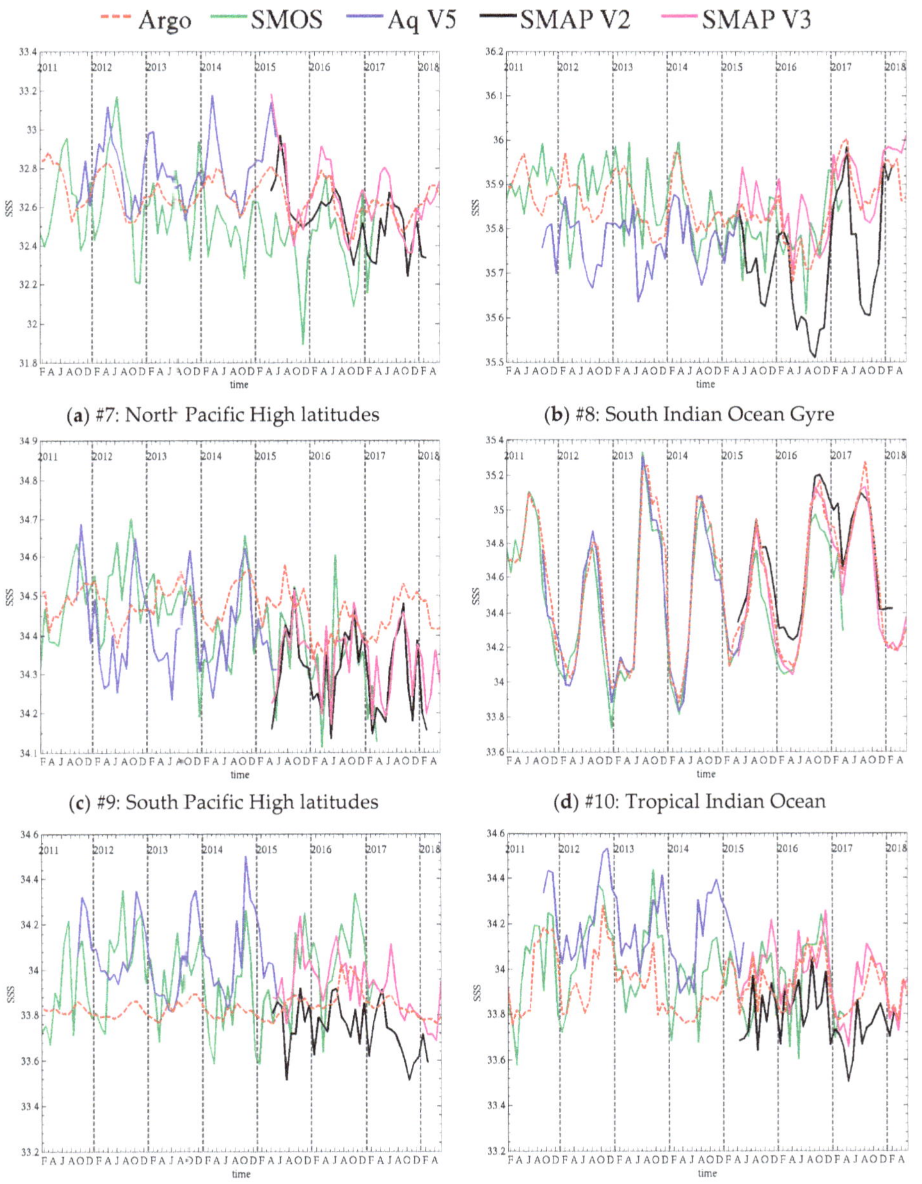

(**a**) #7: North Pacific High latitudes
(**b**) #8: South Indian Ocean Gyre
(**c**) #9: South Pacific High latitudes
(**d**) #10: Tropical Indian Ocean
(**e**) #11: South Atlantic High latitudes
(**f**) #12: South Indian Ocean High latitudes

Figure 13. Time series of SSS from (red) Argo, (green) SMOS, (blue) Aquarius and (black) SMAP for regions of interest 7–12 reported in Figure 2.

Table 7. Average (median) difference (psu) between satellite SSS and Argo SSS.

ROI#	Region Name	SMOS CPDC	Aquarius V5	SMAP V2	SMAP V3
1	North Atlantic Gyre	−0.14	−0.03	0.06	0.01
2	South Atlantic Gyre	0.04	−0.02	0.05	0.02
3	Amazon Plume	−0.31	−0.01	0.13	−0.06
4	Equatorial Pacific	0.10	−0.03	0.12	0.04
5	South Pacific Gyre	−0.04	−0.08	0.07	0.00
6	North Pacific Gyre	−0.10	0.02	−0.08	0.02
7	North Pacific High latitudes	−0.16	0.11	−0.10	0.02
8	South Indian Ocean Gyre	0.01	−0.09	−0.12	0.03
9	South Pacific High latitudes	−0.04	−0.09	−0.14	−0.12
10	Tropical Indian Ocean	−0.07	−0.02	0.11	−0.03
11	South Atlantic High latitudes	0.12	0.22	−0.14	0.07
12	South Indian Ocean High latitudes	0.07	0.22	−0.14	0.03

Table 8. Standard deviation (psu) of the difference between satellite and Argo SSS.

ROI#	Region Name	SMOS CPDC	Aquarius V5	SMAP V2	SMAP V3
1	North Atlantic Gyre	0.134	0.062	0.082	0.065
2	South Atlantic Gyre	0.097	0.086	0.062	0.078
3	Amazon Plume	0.400	0.317	0.357	0.367
4	Equatorial Pacific	0.129	0.069	0.078	0.075
5	South Pacific Gyre	0.074	0.069	0.095	0.077
6	North Pacific Gyre	0.118	0.045	0.060	0.047
7	North Pacific High latitudes	0.212	0.111	0.142	0.119
8	South Indian Ocean Gyre	0.079	0.046	0.081	0.053
9	South Pacific High latitudes	0.116	0.104	0.098	0.092
10	Tropical Indian Ocean	0.117	0.093	0.106	0.073
11	South Atlantic High latitudes	0.171	0.155	0.120	0.116
12	South Indian Ocean High latitudes	0.170	0.134	0.107	0.100

In both Atlantic gyres (#1, #2, Figure 12a,b), SSS seasonal variations are of the order of 0.2–0.3 psu. SMAP V3 and Aquarius show the best performances with overall good agreement in timing and amplitude of the cycles. In the North Atlantic gyre (#1, Figure 12a) SMOS peaks too early in the year compared to Argo and is biased fresh, with too strong freshening events in winter. In the South Atlantic gyre (#2, Figure 12b), SMOS timing is better but the signal is noisier. It does not exhibit excessive freshening like in the North, but there are a few salty peaks (e.g., April 2013) that are not consistent with Argo. The noise is likely due, in part, to our averaging only grid cells where Argo samples exist. Averaging SMOS SSS over the whole ROI would likely smooth out the curves. All the satellites products manage to capture sharp variations, such as the 0.5 psu drop in May–Aug 2014 and 0.4 psu increase in Dec 2014–Apr 2015. Smaller features, such as the peak in Oct 2014, are also reproduced by both SMOS and Aquarius. The Pacific gyres (#5, #6, Figure 12e,f) have smaller seasonal variations, of the order of 0.1 psu, but show larger interannual variations. In the Southern Pacific gyre (#5, Figure 12e), SSS drops twice by 0.25 psu, once in the second half of 2011 and once in 2015. After both drops, SSS follows an increasing trend over a few years. Both the drops and trends are well captured by the satellite products. SMOS in particular shows good agreement with Argo in 2011 and limited bias over the whole time series. Aquarius is biased fresh by almost 0.1 psu. SMAP V3 has a good match in phase and amplitude, albeit with a little noise in the signal. In the Northern Pacific gyre (#6, Figure 12f), the seasonal cycles are also small (~0.1 psu) and the largest signal is a 0.35 psu drop in between November 2014 and April 2015. SMOS and Aquarius reproduce accurately the timing of the drop, but the amplitude differs slightly. Both satellites show larger seasonal cycles than Argo, and SMOS is biased fresh (0.1 psu) and tends to peak ahead of Argo during the period 2011–2013. SMAP V3 has the best match to Argo with very small bias (0.02 psu) and good agreement in the seasonal cycles. In the South Indian gyre (#8, Figure 13b), Argo shows similar seasonal cycles almost every year (excluding 2015 and 2016) with peaks early in the year, and peak to peak variations

~0.12 psu. SMOS exhibits a small bias (0.01 psu) but large discrepancies in seasonal variations, except in 2014 and during the 0.2 psu increase from 2016 to 2017. Aquarius has much better seasonal variations, peaking early in the year when Argo does, but is freshly biased by 0.09 psu. SMAP V3 shows the best overall agreement with a small bias and good seasonal variations, although it shows amplified peaks in 2015 and 2016. In the highly variable regions that are the Amazon Plume (#3, Figure 12c), the Equatorial Pacific (#4, Figure 12d) and the Tropical Indian Ocean (#10, Figure 13d), where seasonal variations commonly reach and exceed 0.5 psu, satellite products agree well with Argo overall. In the Amazon Plume (#3, Figure 12c), which shows the largest variations up to 2.5 psu, the timing between satellite and in situ is good with a few exceptions. SMOS tends to be fresh compared to Argo, it has lower SSS peaks and much larger freshening events (up to 1 psu fresher) in the periods July-October. SMAP agrees well with Argo except for the 2017 freshening where it shows a much larger freshening. Overall, Aquarius bias is the smallest and it appears to track seasonal changes the best (e.g., peaks in May and Sept 2013). In a region where SSS is so variable in space and time, it is likely that mismatches between satellite and in situ observation times, locations and scales are causing some of the discrepancies being observed. In the Equatorial Pacific (#4, Figure 12d), seasonal changes are about 0.5 psu, with a larger change of 1 psu in early 2012. All the variations are well captured by SMAP and Aquarius. SMOS tends to overestimate the peak SSS, but the cycles timing is good. The Tropical Indian Ocean (#10, Figure 13d) has a seasonal variation of 0.5 psu–1 psu. All satellites show a very good match to Argo both in timing and amplitude. However, SMOS performances appear to decline starting in 2015 when it starts to underestimate the peaks by 0.1–0.2 psu. All high latitudes ROI (#7, #9, #11, #12, Figure 13a,c,e,f) show decreased performances of the satellite products, with increased bias or standard deviation of the differences with Argo. In the North Pacific high latitudes (#7, Figure 13a), SMAP and Aquarius reproduce well the timing of the seasonal cycles, but the peaks are too high by a few tenths of a psu. SMOS hardly shows seasonal cycles, except maybe in 2011, 2012 and 2016, with large errors in timing or amplitude. In the high latitudes of the South Pacific (#9, Figure 13c) and South Atlantic (#11, Figure 13e), both timing and amplitude of the satellite cycles show significant discrepancies with Argo and SSS is biased by −0.1–+0.2 psu depending on the product. In the high latitudes of the Indian Ocean (#12, Figure 13f), the agreement in the timing and amplitude of the seasonal cycles of all satellites is much better than in the other Southern high latitudes, especially for SMAP V3, but some biases and too salty peaks still occur.

The latest version of the SMAP RSS product (V3) shows improvements over V2 on multiple aspects. Both versions generally have a good agreement with Argo regarding the phase of seasonal variations, including small seasonal cycles in the gyres (#1, #6, Figure 12a,f) and the larger variations in the meanders of the Amazon plume (#3, Figure 12c). However, V2 tends to overestimate the amplitude of these variations at several locations (Gyres #1, #2 and equatorial Pacific #4, Figure 12a,d,f), with peaks too salty by ~0.1 psu. V3 shows substantial improvement in the agreement of the peaks with Argo. In the North Pacific Gyre (#6, Figure 12f), V2 is freshly biased (0.08 psu) and exhibits too large drops in SSS during the first half of the year. V3 improves the bias by 0.06 psu (now slightly too salty) and substantially improves the dynamic range of the seasonal variation by mitigating the drops. In the Amazon Plume (#3, Figure 12c), both versions are generally in agreement, but V2 over-estimates the high salinities in the first half of the year, where V3 matches Argo better. However, the drops in V3 in July and October 2017 are still too large with SSS too low by more than 1 psu compared to Argo. As discussed previously, it is likely that differences in space and time sampling between satellite and in situ observations contribute to the differences in this highly variable region. The Tropical Indian Ocean (#10, Figure 13d) is another variable region and V3 improves on the seasonal variations with SSS decreasing faster and lower after reaching peak value compared to V2 which has peaks that are too wide and underestimates the freshening early in the year by ~0.15 psu. There are a few locations where the seasonal cycles differ between satellite and in situ. In the South Pacific gyre (#5, Figure 12e), V2 exhibits large peaks (Apr 2016, March 2017) not present in Argo. V3 improves SMAP SSS here, but there are still some large variations (e.g., 0.25 psu drop in May 2017) not agreeing with Argo. In the

North Pacific high latitude region (#7, Figure 13a), V2 seasonal cycles peak too late compared to V3 and Argo. V3 cycles are in better phase but peaks are too salty. The South Indian gyre (#8, Figure 13b) also shows very substantial improvements in the seasonal cycles of V3 compared to V2 that shows largely overestimated drops in SSS during the southern hemisphere spring, and too fresh peaks early in the year. Finally, all the southern high latitude locations exhibit a significant discrepancy between SMAP SSS (both versions) and Argo, with satellite SSS more variable than in situ. V3 improves on V2 in the high latitudes of the Atlantic (#11, Figure 13e) and Indian (#12, Figure 13f) oceans in terms of bias and slightly in terms of variations. Over all the ROI, V3 reduces the average difference between SMAP SSS and Argo by 0.02–0.11 psu compared to V2 (Table 7), with the mean reduction being 0.07 psu. Its impact on the standard deviation of the difference between SMAP and Argo is relatively small, a reduction of ~0.01 psu on average with larger reductions of 0.023 and 0.033 psu in the North Pacific High latitudes (#7) and Tropical Indian Ocean (#10). This small value is to be compared to the small natural variation of SSS itself, which shows standard deviations of less than 0.2 psu at most locations. In addition, SMAP V3 either improves peaks and troughs, that involve few data points, or corrects biases, limiting its impact on a metric like the standard deviation.

3.3.3. SST-Dependent Bias

A recurring feature of various revisions of SSS products has been the dependence of the SSS biases on SST. This signature is not obvious in the products reported in Figure 9 because of semi-empirical corrections that have been applied to mitigate it in recent years. We report here a summary of various versions of Aquarius and SMAP product that illustrate the issue before examining possible causes in the next section. The SMAP and Aquarius SSS are those distributed at the PO.DAAC (Section 2.1).

The last version of the Aquarius product to not include an empirical correction for the SST-dependent bias was version 3. A map of the SSS difference with Argo for this product (Aquarius SSS-Argo SSS) is reported in Figure 14 (top left, a) and the SSS bias as a function of SST for V3 through V5 is reported in Figure 14 (bottom left, c). SSS errors in Aquarius V3 clearly exhibit a very strong dependence on SST, with SSS too fresh by up to 0.2 psu in warm (>22 °C) waters and too salty by ~0.25 psu in waters between 5 °C and 15 °C. Because of the strong latitudinal dependence of SST, the SSS difference map shows a dipole pattern with too fresh waters at low latitudes and too salty waters at high latitudes, with the separation around 20°N and 30°S. Subsequent versions 4 and 5 have included algorithm enhancements designed to empirically mitigate the SST-dependent bias. The latest version 5 includes a dependence on SST in the surface roughness correction and a modified atmospheric model that substantially decrease the SST-dependent bias. Similar plots for SMAP V2 and V3 products are reported in Figure 14 (top right (b) and bottom right (d)). SMAP V2 also showed substantial SST-dependent bias. However, the dipole pattern is inverted compared to Aquarius, with lower latitude biased salty and higher latitudes biased fresh. The latest version of the SMAP product (V3) is based on the Aquarius V5 algorithm with specific adjustments for SMAP that significantly mitigate the SST-dependent bias compared to V2.

A comparison of the latest SMOS, Aquarius and SMAP products is reported in Figure 15. The left (a) panel reports the SSS bias between satellite and in situ computed as the median of the SSS difference over SST bins 1 °C wide. The right (b) panel reports the variable error computed as the 68% quantile (Q68) of the absolute value of the SSS difference (which matches the standard deviation in case of a normal distribution). Biases in satellite products (Figure 15a) compared to Argo are small over a large range of temperatures. SMOS shows the best overall performance between 4 °C and 24 °C, with a small bias (<0.1 psu) fairly constant in SST. However, the bias dramatically increases for very cold waters (below 4 °C) to exceed +0.8 psu at −1 °C. This bias is much larger than with any other products at any temperature, even including products with strong SST-dependent biases reported in Section 3.3.3. SMOS bias also increases to −0.2 psu in warm waters (30 °C). The increase is not unusual among satellite products which tend to be fresher than in situ observations, and this could reflect in part a difference between shallow (~2 cm) satellite measurement and deeper (5–10 m) Argo measurements.

But the amplitude of this bias is larger by 0.1–0.15 psu compared to SMAP and Aquarius and may include additional sources of errors. The latest SMAP and Aquarius products also have small biases over a large range of SST and match the Argo data much better for waters warmer than 24 °C and colder than 4 °C. This is in part due to the empirical adjustments to mitigate the SST-dependent biases. However, not all biases are removed and investigation of the sources of the bias should prove useful in improving the algorithm corrections.

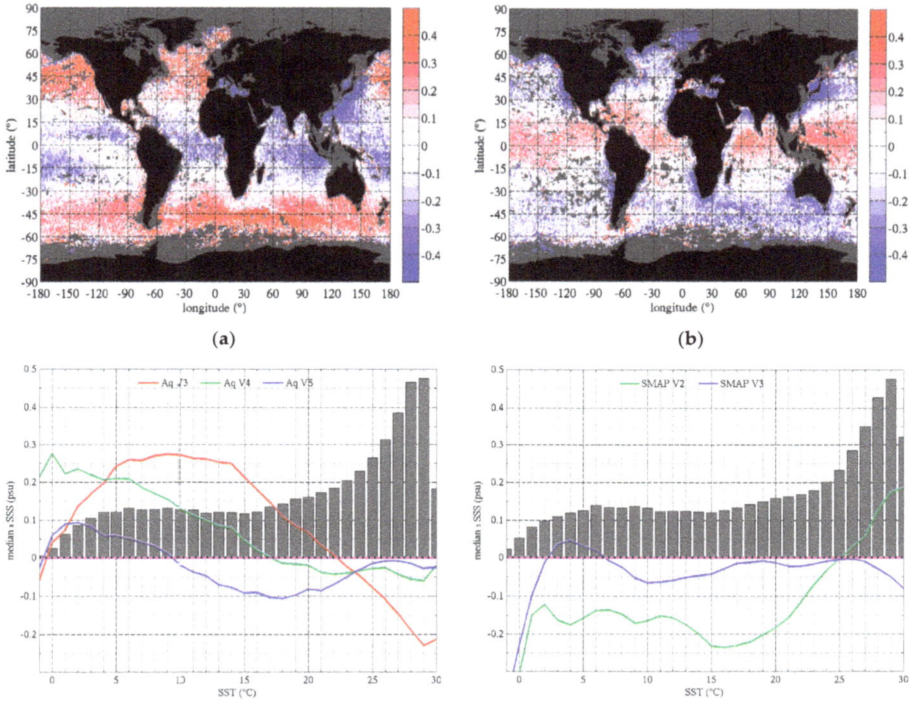

Figure 14. SST dependence of SSS bias for various versions of SMAP and Aquarius products. Top row (**a**,**b**) reports maps of SSS difference between satellite and in situ for (**a**) Aquarius V3 and (**b**) SMAP V2 averaged over several years from monthly products. The bottom row (**c**,**d**) reports the SSS difference between satellite and in situ SSS as a function of SST for (**c**) Aquarius V3 (red), V4 (green) and V5 (blue) and (**d**) SMAP V2 (green) and V3 (blue). The vertical bars in the background report the statistical distribution of the samples as a function of SST (amplitude is normalized to the amplitude of SSS curves).

The variable error (Figure 15b) is very similar between the various retrievals. The Q68 is ~0.2 psu for temperatures warmer than 15 °C. There is an increasing trend in the error with decreasing temperature between 22 °C (0.15–0.18 psu) and −2 °C (0.4–0.7 psu). An increase in retrieval random error with decreasing SST is expected because of the reduced radiometric sensitivity to SSS in cold waters. This is illustrated by the inverse of the radiometric sensitivity to SSS reported as a function of SST in Figure 15b (dashed magenta curve labelled dS/dTB, scaled to match the other curves at 18 °C). The decrease in radiometric sensitivity makes the retrievals more sensitive to all source of errors, be it radiometric noise or uncertainty on ancillary parameters, such as SST. In addition, the increase in sensitivity to SST in cold waters amplifies the error due to SST uncertainties. SMAP and Aquarius show the best performances in waters colder than 15 °C, with SMOS error being 0.22 psu larger around 2 °C. Above 10 °C, all satellite products show performance closer to each other, with SMAP having errors smaller by less than 0.03 psu compared to Aquarius, and SMOS being at most 0.07 psu higher. All products also show a small upward trend in warm waters that could be partially due to the impact of precipitation-driven stratification [22,50]. The radiometric sensitivity to SST also increases again in warm waters and there could be an associated impact of uncertainty in SST. The large reduction of samples at the cold end of the temperature range could also contribute to the increase in variable error.

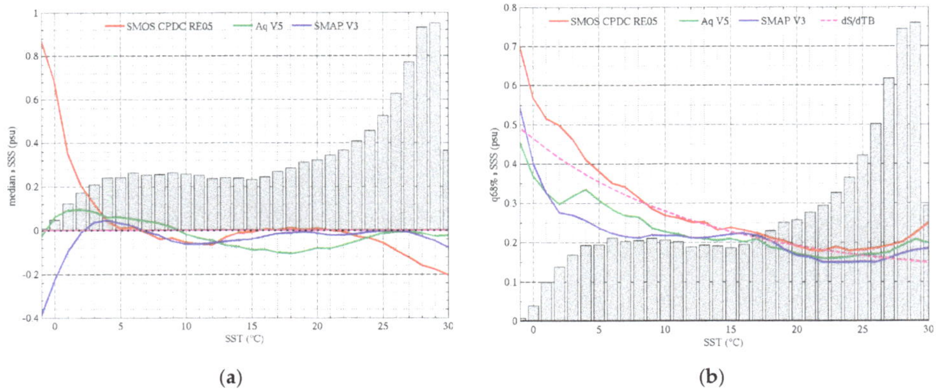

Figure 15. SSS (a) bias and (b) variable difference between satellite retrievals and in situ Argo measurements as a function of SST for SMOS CPDC RE05, and the latest Aquarius (V5) and SMAP (V3) versions. SMOS includes filtering of land contamination by removing data closer than 1000 km from the coast. The vertical bars in the background report the statistical distribution of the samples as a function of SST. The magenta dashed curve in the right (b) panel reports the relative change in the inverse of the radiometric sensitivity to SSS as a function of SST, scaled to match SSS error curves at 18 °C.

While the empirical algorithm adjustments manage to largely remove the SST-dependent biases, they tend to lack a physical explanation and do not completely resolve the issue. For that reason, we investigate possible causes for the SST-dependent bias in the next section. A better understanding should provide clues for improving further the algorithm.

3.3.4. Impact of the Dielectric Constant Model and Ancillary Temperature on SSS Differences

In the previous section, we showed that differences between the satellite SSS products and in situ SSS exhibit strong regional and seasonal signals that correlate significantly with SST. In this section we quantify the potential impact of some of the models and ancillary data used in the SSS retrieval algorithm. We assess the impact of the differences in the dielectric constant model, ancillary SST products and of the atmospheric model, recently modified for Aquarius, by reprocessing the Aquarius data using alternative parameterizations. We start the reprocessing from the Aquarius top of atmosphere TB from Version 3 because it is the last version to not include an empirical correction for the SST-dependent bias. We successively introduce changes to V3 of the retrieval algorithm, one at a time, to identify their impact. We first change the dielectric constant model to the one by Klein and Swift [39], then change to the ancillary OSTIA SST (both used in the SMOS processing). Finally, we change the atmospheric attenuation model to the one used in V5 which is a return to the original model by Liebe et al. [45].

Each time a change to the retrieval algorithm is applied, the TB is recalibrated as discussed in Section 2.4 In the following, we report the impacts of algorithm changes on the SST-dependent bias (Figure 16). SMOS differences with Argo are computed for distances to the coast of 1000 km or more to mitigate the impact of land contamination on the SSS differences. This does not change substantially our results and essentially shifts the SMOS curves in Figure 16 upwards by 0.03 psu (warm waters) to 0.15 psu (cold waters) compared to the results including also coastal data.

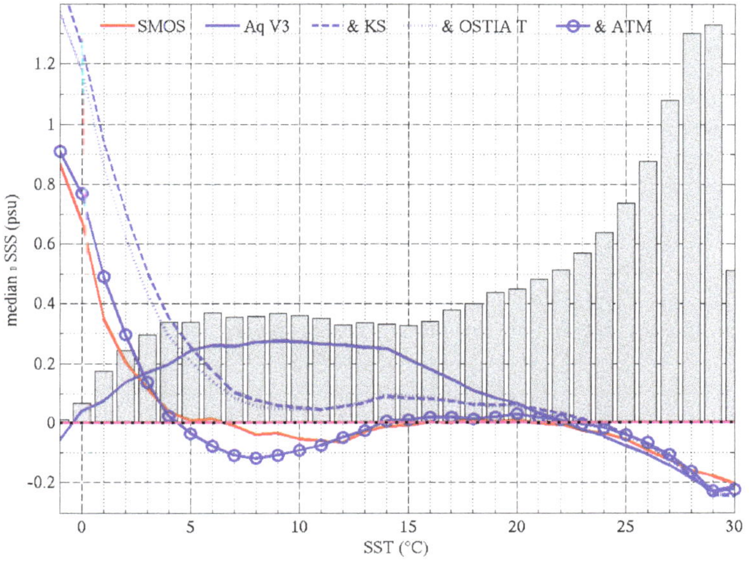

Figure 16. Comparison of SST-dependent bias between SMOS and Aquarius and impact of retrieval parameters. The plot reports the difference between satellite and in situ SSS binned as a function of SST and computed from monthly SSS maps over the period Sept 2011–May 2015. Satellite SSS are (Red) SMOS, and (Blue) Aquarius V3. The Aquarius V3 curves are for (plain) the nominal product, (dashed) V3 reprocessed using the KS dielectric constant model, (dotted) using KS model and OSTIA ancillary SST, and (plain and circles) using the KS model, the OSTIA SST and the adjusted atmospheric model from Aquarius V5.

Changing the dielectric constant model for the Aquarius SSS retrievals has a very large impact (changes from solid blue to dashed blue curve in Figure 16). Using the KS model reduces biases for SST between 5 °C and 22 °C compared to the MW model (used in Aquarius V3) and significantly reduces the dependence on SST. There is little change above 22 °C as both models agree and result in fresher SSS compared to Argo measurements. One striking feature of the new curve is the very large increase in bias below 5 °C. In the very cold waters, below 0 °C, the bias now exceeds 1.4 psu. Overall, the KS model brings the Aquarius SST-dependent bias in much better agreement with SMOS bias. But now the Aquarius bias in cold waters is too large, even compared to the already significant SMOS bias of 0.8 psu. Next, we introduce the OSTIA SST in the Aquarius retrieval (blue dotted curve in Figure 16). As illustrated by the example reported in Section 3.2, the impact of the OSTIA SST on the SST-dependent bias is relatively small because of the averaging over large regions and long periods (4 years of data in Figure 16). The impact of differences in ancillary SST can be large locally (>1 psu), but spatial and temporal averaging over large scales, such as latitudinal bands and several seasons, will mitigate the differences. The impact on the average SSS difference is small (±0.01 psu) above 10 °C; it increases slightly in very cold water to result in a bias reduced by ~0.05 psu at 5 °C and ~0.1 psu at 0 °C. The small reduction in bias brings Aquarius results slightly closer to SMOS. An assessment of several SST products for the Aquarius processing identified the CMC product as the one producing the smaller SSS errors compared to in situ data; this product was adopted for the Aquarius final release and the latest SMAP products [46]. Finally, we change to the Liebe atmospheric model in the Aquarius retrieval (blue curve with circles). The new atmospheric model decreases the retrieved SSS by 0.45 psu on average at the cold end (0 °C) and increases it by 0.07 psu at the warm end (30 °C). The bias is very significantly reduced in cold waters. Introducing these three changes (KS model for dielectric constant, OSTIA SST, and Liebe model for atmospheric attenuation) brings the Aquarius retrieval very close to SMOS at almost every temperature. This illustrates the importance of the ancillary data on the performance of the algorithm and suggests that models used in the algorithm (e.g., for the dielectric constant and atmosphere) contribute to the SST-dependent bias. But both sensors still produce too salty retrievals compared to Argo measurements in cold waters, which is likely due to inaccuracies in the KS dielectric constant model. Comparatively, the MW model performs much better in cold waters, but results in more variability of the SSS bias with SST for waters warmer than 5 °C. A new model for the dielectric constant based on extensive laboratory measurements at 1.4 GHz is being developed at George Washington University [54] and may help resolve this issue.

4. Discussion

The latest versions of the SMOS, Aquarius and SMAP SSS products show improved performances with only small biases remaining in open oceans and good accuracy on seasonal and interannual variability. The improvements come in part from empirical adjustments in the retrieval algorithms. The SST-dependent bias correction used by Aquarius and SMAP significantly improves SSS products, but issues remain in cold waters. High southern latitudes still suffer from significant biases, and seasonal variability is still not accurately captured by the satellites. SMOS SSS also exhibit large biases and increased noise in cold waters. While other factors may contribute to the performance issues in cold water, such as sea ice whose presence and quantity can be uncertain [47] and rougher sea surfaces, we find that the model for the dielectric constant of sea water is still questionable and has a significant impact on the retrievals. The KS model used by SMOS results in smaller biases at many temperatures but lead to very large and temperature dependent biases in cold waters. Our results with the KS model are consistent with assessments of microwave radiometers at higher frequencies which found that corrections to TB at the cold end (0 °C) needed to be larger than at the warm end (30 °C) by 0.5 K and 3.7 K at 6 GHz [55] and 37 GHz [56] respectively. It is therefore critical to improve the current dielectric constant models. Recent laboratory measurements performed at L-band at the George Washington University (GWU) [57] have led to a new model that shows promising results with the Aquarius data [54]. These measurements are currently being enhanced by adding more samples at more SST and SSS. When the analytical model using GWU measurements is updated to include the latest measurements, we plan to assess its performances with Aquarius and SMAP data. For SMOS, a product called CEC-LOCEAN DEBIAS v3 includes SST-dependent corrections to the KS model derived from assessments reported here. The capability of the updated GWU model to replace the empirical correction for SMOS should also be tested.

5. Conclusions

We compare satellite SSS products from SMOS, Aquarius and SMAP with in situ measurements from the Argo network of drifting profiling floats. The latest versions of the products show a significant reduction in bias and variable errors in satellite SSS and convergence in the performances of the three sensors. The satellites mostly reproduce the same large-scale patterns as Argo and have small biases in the open ocean overall. Coastal regions benefit from improved corrections compared to previous versions but exhibit larger errors than open oceans. Salty bias at some coastal locations is likely the result of over correction of land contamination. Improving land correction is the focus of current research for future versions of SSS products. The dynamic range of satellite SSS is similar to Argo, with most values between 33.85 psu and 35.85 psu. SMOS shows more prevalent SSS in the range 33.2–33.6 psu than the other products, and all sensors fail to reproduce some of the Argo SSS histogram features due to various regional biases. Time series at various locations around the globe show the good performance of the satellites in reproducing seasonal and interannual changes reported by the in situ data, but disagreements in phase and amplitude of the signals are not uncommon, especially in the high latitudes. Algorithm improvements are still needed for these locations and an example of possible gains is illustrated with SMAP. We have assessed the recently released version 3 of the SMAP product and show that it reduces significantly previous large-scale biases that were correlated with SST and that it improves substantially the seasonal and inter-annual variability, which are now in better agreement with in situ observations. Among the main discrepancies remaining between SSS products is the SST-dependent bias. It varies significantly between products, and results in latitudinal patterns in the satellite SSS error maps when compared to in situ observations. The SST-dependent bias is mitigated through empirical corrections in the retrieval algorithms, but the actual causes for it are not clearly established. We have assessed three possible causes: The sea water dielectric constant model, the ancillary product for SST and the atmospheric model. We find that the impact of the ancillary SST product can be large at small space and time scales. In the cold water of the high latitudes the large SST discrepancies between product are amplified by a lower sensitivity of TB to SSS which results in differences of 1 psu or more in retrieved SSS. However, the impact of the SST ancillary product is relatively small when errors are averaged at large scales or over time. As such, its impact is small on the SST-dependent bias and is mostly noticeable in very cold water (~0.1 psu at 0 °C). We find that the SST-dependent bias is mostly impacted by the dielectric constant model and the atmospheric attenuation model. When substituting the models used for SMOS into the processing of Aquarius observations, we find a much-improved match in the dependence of SSS bias to SST between both sensors. Using the KS dielectric constant model instead of the MW model reduces the mean SSS error by up to 0.2 psu and reduces its variability in the range 5–22 °C. In waters warmer than 20 °C, both dielectric constant models perform similarly to each other. In waters colder than 5 °C, the KS model exhibits very problematic performances. Considering the large errors it produces (>1 psu), and the fast change in the bias with SST (1 psu over a 5 °C range), it appears that the accuracy of the KS model is questionable below 5 °C. This result is consistent with the assessment of microwave radiometers at higher frequencies which found an increase in the KS model bias in cold waters [55,56]. It is doubtful that all the SST-dependent error can be attributed to the dielectric constant model, as other factors can commingle their error and its dependence on SST (e.g., roughness model). In fact, we find that the impact of the atmospheric attenuation and emission model in the latest Aquarius product has a strong correlation with SST. The impact of the new model is larger in cold waters and improves significantly the match with SMOS and generally reduces Aquarius biases. Future research on sea water dielectric constant and atmospheric attenuation at low microwave frequencies will be essential in order to move away from empirical adjustments that are currently necessary to ensure good and consistent performance across all ocean temperatures.

Author Contributions: Conceptualization, E.P.D.; Data curation, E.P.D., J.B. and T.M.; Formal analysis, E.P.D., D.M.L.V., J.B., T.M. and G.L.; Funding acquisition, E.P.D. and D.M.L.V.; Investigation, E.P.D., D.M.L.V., J.B. and T.M.; Methodology, E.P.D.; Software, E.P.D.; Supervision, D.M.L.V. and G.L.; Validation, E.P.D.; Visualization, E.P.D.; Writing—original draft, E.P.D.; Writing—review & editing, E.P.D., D.M.L.V., J.B., T.M. and G.L.

Funding: This research was funded by the National Aeronautics and Space Administration, grant numbers NNX14AR31G and 80NSSC18K1443, and by the CNES/CATDS (Centre Aval de Traitement des Données SMOS) and ESA Climate Change Initiative projects.

Acknowledgments: We would like to thank the Aquarius Cal/Val and Science teams for the useful discussions concerning this study.

Conflicts of Interest: The authors declare no conflict of interest. The funders had no role in the design of the study; in the collection, analyses, or interpretation of data; in the writing of the manuscript, or in the decision to publish the results.

References

1. Font, J.; Camps, A.; Borges, A.; Martin-Neira, M.; Boutin, J.; Reul, N.; Kerr, Y.H.; Hahne, A.; Mecklenburg, S. SMOS: The challenging sea surface salinity measurement from space. *Proc. IEEE* **2010**, *98*, 649–665. [CrossRef]
2. Kerr, Y.H.; Waldteufel, P.; Wigneron, J.-P.; Delwart, S.; Cabot, F.; Boutin, J.; Escorihuela, M.-J.; Font, J.; Reul, N.; Gruhier, C.; et al. The SMOS Mission: New Tool for Monitoring Key Elements of the Global Water Cycle. *Proc. IEEE* **2010**, *98*, 666–687. [CrossRef]
3. Lagerloef, G.; Colomb, F.R.; Le Vine, D.; Wentz, F.; Yueh, S.; Ruf, C.; Lilly, J.; Gunn, J.; Chao, Y.; deCharon, A.; et al. The Aquarius/SAC-D Mission: Designed to Meet the Salinity Remote-Sensing Challenge. *Oceanography* **2008**, *21*, 68–81. [CrossRef]
4. Le Vine, D.M.; Lagerloef, G.S.E.; Torrusio, S.E. Aquarius and remote sensing of sea surface salinity from space. *Proc. IEEE* **2010**, *98*, 688–703. [CrossRef]
5. Le Vine, D.M.; Dinnat, E.P.; Meissner, T.; Yueh, S.H.; Wentz, F.J.; Torrusio, S.E.; Lagerloef, G. Status of Aquarius/SAC-D and Aquarius Salinity Retrievals. *IEEE J. Sel. Top. Appl. Earth Obs. Remote Sens.* **2015**, *8*, 5401–5415. [CrossRef]
6. Grodsky, S.A.; Reul, N.; Lagerloef, G.; Reverdin, G.; Carton, J.A.; Chapron, B.; Quilfen, Y.; Kudryavtsev, V.N.; Kao, H.Y. Haline hurricane wake in the Amazon/Orinoco plume: AQUARIUS/SACD and SMOS observations. *Geophys. Res. Lett.* **2012**, *39*, 4–11. [CrossRef]
7. Grodsky, S.A.; Reverdin, G.; Carton, J.A.; Coles, V.J. Year-to-year salinity changes in the Amazon plume: Contrasting 2011 and 2012 Aquarius/SACD and SMOS satellite data. *Remote Sens. Environ.* **2014**, *140*, 14–22. [CrossRef]
8. Gierach, M.M.; Vazquez-Cuervo, J.; Lee, T.; Tsontos, V.M. Aquarius and SMOS detect effects of an extreme Mississippi River flooding event in the Gulf of Mexico. *Geophys. Res. Lett.* **2013**, *40*, 5188–5193. [CrossRef]
9. Hernandez, O.; Boutin, J.; Kolodziejczyk, N.; Reverdin, G.; Martin, N.; Gaillard, F.; Reul, N.; Vergely, J.L. SMOS salinity in the subtropical North Atlantic salinity maximum: 1. Comparison with Aquarius and in situ salinity. *J. Geophys. Res. C Ocean.* **2014**, *119*, 8878–8896. [CrossRef]
10. Garcia-Eidell, C.; Comiso, J.C.; Dinnat, E.; Brucker, L. Satellite observed salinity distributions at high latitudes in the Northern Hemisphere: A comparison of four products. *J. Geophys. Res. Ocean.* **2017**, *122*, 7717–7736. [CrossRef]
11. Ebuchi, N.; Abe, H. Evaluation of sea surface salinity observed by Aquarius and SMOS. In Proceedings of the International Geoscience and Remote Sensing Symposium (IGARSS), Melbourne, Australia, 21–26 July 2013; pp. 656–659.
12. Dinnat, E.; Boutin, J.; Le Vine, D.; Yin, X. Inter-Comparison of Aquarius and SMOS Calibration and Impact on Sea Surface Salinity Products. In Proceedings of the American Geophysical Union, Fall meeting, San Francisco, CA, USA, 15–19 December 2014.
13. Pablos, M.; Piles, M.; Gonzalez-Gambau, V.; Vall-Llossera, M.; Camps, A.; Martinez, J. SMOS and Aquarius radiometers: Inter-comparison over selected targets. *IEEE J. Sel. Top. Appl. Earth Obs. Remote Sens.* **2014**, *7*, 3833–3844. [CrossRef]
14. Yueh, S.H.; West, R.; Wilson, W.J.; Li, F.K.; Njoku, E.G.; Rahmat-Samii, Y. Error sources and feasibility for microwave remote sensing of ocean surface salinity. *IEEE Trans. Geosci. Remote Sens.* **2001**, *39*, 1049–1060. [CrossRef]
15. Dinnat, E.P.; Boutin, J.; Caudal, G.; Etcheto, J.; Waldteufel, P. Influence of sea surface emissivity model parameters at L-band for the estimation of salinity. *Int. J. Remote Sens.* **2002**, *23*, 5117–5122. [CrossRef]
16. Dinnat, E.P.; Boutin, J.; Caudal, G.; Etcheto, J. Issues concerning the sea emissivity modeling at L band for retrieving surface salinity. *Radio Sci.* **2003**, *38*. [CrossRef]
17. Boutin, J.; Waldteufel, P.; Martin, N.; Caudal, G.; Dinnat, E. Surface salinity retrieved from SMOS measurements over the global ocean: Imprecisions due to sea surface roughness and temperature uncertainties. *J. Atmos. Ocean. Technol.* **2004**, *21*, 1432–1447. [CrossRef]
18. Waldteufel, P.; Floury, N.; Dinnat, E.P.; Caudal, G. Ionospheric effects for L-band 2-D interferometric radiometry. *IEEE Trans. Geosci. Remote Sens.* **2004**, *42*, 105–118. [CrossRef]
19. Meissner, T.; Wentz, F.J.; Le Vine, D.M. *Aquarius Salinity Retrieval Algorithm Algorithm Theoretical Basis Document (ATBD) End of Mission Version. RSS Technical Report 120117*; Remote Sensing Systems: Santa Rosa, CA, USA, 2017; p. 104.

20. Le Vine, D.M.; Dinnat, E.P.; Meissner, T.; Wentz, F.J.; Kao, H.Y.; Lagerloef, G.; Lee, T. Status of Aquarius and salinity continuity. *Remote Sens.* **2018**, *10*, 1585. [CrossRef]
21. Yueh, S.; Tang, W.; Fore, A.; Hayashi, A.; Song, Y.; Lagerloef, G. Aquarius geophysical model function and combined active passive algorithm for ocean surface salinity and wind retrieval. *J. Geophys. Res. Ocean.* **2014**, *119*, 5360–5379. [CrossRef]
22. Tang, W.; Yueh, S.; Fore, A.; Neumann, G.; Hayashi, A.; Lagerloef, G. The rain effect on Aquarius' L-band sea surface brightness temperature and radar backscatter. *Remote Sens. Environ.* **2013**, *137*, 147–157. [CrossRef]
23. Chassignet, E.; Hurlburt, H.; Metzger, E.J.; Smedstad, O.; Cummings, J.; Halliwell, G.; Bleck, R.; Baraille, R.; Wallcraft, A.; Lozano, C.; et al. US GODAE: Global ocean prediction with the HYbrid Coordinate Ocean Model (HYCOM). *Oceanography* **2009**, *22*, 64–75. [CrossRef]
24. Boutin, J. CATDS Ocean Salinity Level 3 products–Differences & Pros/Cons. Available online: http://www.catds.fr/content/download/68781/file/OS_L3_products_Differences_and_ProsCons.pdf (accessed on 26 March 2019).
25. Kolodziejczyk, N.; Boutin, J.; Vergely, J.L.; Marchand, S.; Martin, N.; Reverdin, G. Mitigation of systematic errors in SMOS sea surface salinity. *Remote Sens. Environ.* **2016**, *180*, 164–177. [CrossRef]
26. Boutin, J.; Vergely, J.L.; Marchand, S.; D'Amico, F.; Hasson, A.; Kolodziejczyk, N.; Reul, N.; Reverdin, G.; Vialard, J. New SMOS Sea Surface Salinity with reduced systematic errors and improved variability. *Remote Sens. Environ.* **2018**, *214*, 115–134. [CrossRef]
27. Olmedo, E.; Martínez, J.; Turiel, A.; Ballabrera-Poy, J.; Portabella, M. Debiased non-Bayesian retrieval: A novel approach to SMOS Sea Surface Salinity. *Remote Sens. Environ.* **2017**, *193*, 103–126. [CrossRef]
28. Carval, T.; Keeley, R.; Takatsuki, Y.; Yoshida, T.; Loch, S.; Schmid, C.; Goldsmith, R.; Wong, A.; McCreadie, R.; Thresher, A.; et al. *Argo User 's Manual*; Ifremer: Brest, France, 2017.
29. Piepmeier, J.; Brown, S.; Gales, J.; Hong, L.; Lagerloef, G.; Le Vine, D.; de Matthaeis, P.; Meissner, T.; Bindlish, R.; Jackson, T.; et al. *Aquarius Radiometer Post-Launch Calibration for Product Version 2*; National Aeronautics and Space Administration: Washington, DC, USA, 2013.
30. Dinnat, E.P.; Le Vine, D.M.; Piepmeier, J.R.; Brown, S.T.; Hong, L. Aquarius L-band Radiometers Calibration Using Cold Sky Observations. *IEEE J. Sel. Top. Appl. Earth Obs. Remote Sens.* **2015**, *8*, 5433–5449. [CrossRef]
31. Martín-Neira, M.; Oliva, R.; Corbella, I.; Torres, F.; Duffo, N.; Durán, I.; Kainulainen, J.; Closa, J.; Zurita, A.; Cabot, F.; et al. SMOS instrument performance and calibration after six years in orbit. *Remote Sens. Environ.* **2016**, *180*, 19–39. [CrossRef]
32. Peng, J.; Misra, S.; Piepmeier, J.R.; Dinnat, E.P.; Hudson, D.; Le Vine, D.M.; De Amici, G.; Mohammed, P.N.; Bindlish, R.; Yueh, S.H.; et al. Soil Moisture Active/Passive L-Band Microwave Radiometer Postlaunch Calibration. *IEEE Trans. Geosci. Remote Sens.* **2017**, *55*, 5339–5354. [CrossRef]
33. Zine, S.; Boutin, J.; Font, J.; Reul, N.; Waldteufel, P.; Gabarro, C.; Tenerelli, J.; Petitcolin, F.; Vergely, J.L.; Talone, M.; et al. Overview of the SMOS sea surface salinity prototype processor. *IEEE Trans. Geosci. Remote Sens.* **2008**, *46*, 621–644. [CrossRef]
34. Meissner, T.; Wentz, F.J.; Ricciardulli, L. The emission and scattering of L-band microwave radiation from rough ocean surfaces and wind speed measurements from the Aquarius sensor. *J. Geophys. Res. C Ocean.* **2014**, *119*, 6499–6522. [CrossRef]
35. Piepmeier, J.R.; Hong, L.; Pellerano, F.A. Aquarius L-Band Microwave Radiometer: 3 Years of Radiometric Performance and Systematic Effects. *IEEE J. Sel. Top. Appl. Earth Obs. Remote Sens.* **2015**, *8*, 5416–5423. [CrossRef]
36. Lagerloef, G.; Kao, H.-Y.; Meissner, T.; Vazquez, J. *Aquarius Salinity Validation Analysis*; Data Version 4.0; National Aeronautics and Space Administration: Washington, DC, USA, 2015.
37. Yin, X.; Boutin, J.; Spurgeon, P. First assessment of SMOS data over open ocean: Part I-pacific ocean. *IEEE Trans. Geosci. Remote Sens.* **2012**, *50*, 1648–1661. [CrossRef]
38. Meissner, T.; Wentz, F.J.; Le Vine, D.M. The salinity retrieval algorithms for the NASA Aquarius version 5 and SMAP version 3 releases. *Remote Sens.* **2018**, *10*, 1121. [CrossRef]
39. Klein, L.A.; Swift, C.T. An improved model for the dielectric constant of sea water at microwave frequencies. *IEEE J. Ocean. Eng.* **1977**, *AP-25*, 104–111. [CrossRef]
40. Meissner, T.; Wentz, F.J. The complex dielectric constant of pure and sea water from microwave satellite observations. *IEEE Trans. Geosci. Remote Sens.* **2004**, *42*, 1836–1849. [CrossRef]

41. Meissner, T.; Wentz, F.J. The Emissivity of the Ocean Surface Between 6 and 90 GHz Over a Large Range of Wind Speeds and Earth Incidence Angles. *IEEE Trans. Geosci. Remote Sens.* **2012**, *50*, 3004–3026. [CrossRef]
42. Donlon, C.J.; Martin, M.; Stark, J.; Roberts-Jones, J.; Fiedler, E.; Wimmer, W. The Operational Sea Surface Temperature and Sea Ice Analysis (OSTIA) system. *Remote Sens. Environ.* **2012**, *116*, 140–158. [CrossRef]
43. Reynolds, R.W.; Smith, T.M.; Liu, C.; Chelton, D.B.; Casey, K.S.; Schlax, M.G. Daily high-resolution-blended analyses for sea surface temperature. *J. Clim.* **2007**, *20*, 5473–5496. [CrossRef]
44. Liebe, H.J.; Rosenkranz, P.W.; Hufford, G.A. Atmospheric 60-GHz oxygen spectrum: New laboratory measurements and line parameters. *J. Quant. Spectrosc. Radiat. Transf.* **1992**, *48*, 629–643. [CrossRef]
45. Wentz, F.J.; Meissner, T. Atmospheric absorption model for dry air and water vapor at microwave frequencies below 100GHz derived from spaceborne radiometer observations. *Radio Sci.* **2016**, *51*, 381–391. [CrossRef]
46. Meissner, T.; Wentz, F.J.; Scott, J.; Vazquez-Cuervo, J. Sensitivity of Ocean Surface Salinity Measurements From Spaceborne L-Band Radiometers to Ancillary Sea Surface Temperature. *IEEE Trans. Geosci. Remote Sens.* **2016**, *54*, 7105–7111. [CrossRef]
47. Dinnat, E.P.; Brucker, L. Improved Sea Ice Fraction Characterization for L-Band Observations by the Aquarius Radiometers. *IEEE Trans. Geosci. Remote Sens.* **2017**, *55*, 1285–1304. [CrossRef]
48. Tian-Kunze, X.; Kaleschke, L.; Maaß, N.; Mäkynen, M.; Serra, N.; Drusch, M.; Krumpen, T. SMOS-derived thin sea ice thickness: Algorithm baseline, product specifications and initial verification. *Cryosphere* **2014**, *8*, 997–1018. [CrossRef]
49. Maaß, N.; Kaleschke, L.; Tian-Kunze, X.; Drusch, M. Snow thickness retrieval over thick Arctic sea ice using SMOS satellite data. *Cryosphere* **2013**, *7*, 1971–1989. [CrossRef]
50. Boutin, J.; Chao, Y.; Asher, W.E.; Delcroix, T.; Drucker, R.; Drushka, K.; Kolodziejczyk, N.; Lee, T.; Reul, N.; Reverdin, G.; et al. Satellite and In Situ Salinity: Understanding Near-surface Stratification and Sub-footprint Variability. *Bull. Am. Meteorol. Soc.* **2016**, *97*, 1391–1408. [CrossRef]
51. Hasson, A.; Puy, M.; Boutin, J.; Guilyardi, E.; Morrow, R. Northward Pathway Across the Tropical North Pacific Ocean Revealed by Surface Salinity: How do El Niño Anomalies Reach Hawaii? *J. Geophys. Res. Ocean.* **2018**, *123*, 2697–2715. [CrossRef]
52. Chaigneau, A.; Morrow, R. Surface temperature and salinity variations between Tasmania and Antartica, 1993–1999. *J. Geophys. Res.* **2002**, *107*, 22–27. [CrossRef]
53. Yin, X.; Boutin, J.; Spurgeon, P. Biases between measured and simulated SMOS brightness temperatures over ocean: Influence of sun. *IEEE J. Sel. Top. Appl. Earth Obs. Remote Sens.* **2013**, *6*, 1341–1350. [CrossRef]
54. Zhou, Y.; Lang, R.H.; Dinnat, E.P.; Le Vine, D.M. L-Band Model Function of the Dielectric Constant of Seawater. *IEEE Trans. Geosci. Remote Sens.* **2017**, *55*, 6964–6974. [CrossRef]
55. Shibata, A. Description of GCOM-W1 AMSR2 Sea Surface Temperature Algorithm. In *Descriptions of GCOM-W1 AMSR2 Level 1R and Level 2 Algorithms*; Japan Aerospace Exploration Agency: Tokyo, Japan, 2013; p. 119.
56. Shibata, A. A change of microwave radiation from the ocean surface induced by air-sea temperature difference. *Radio Sci.* **2003**, *38*. [CrossRef]
57. Lang, R.; Zhou, Y.; Utku, C.; Le Vine, D. Accurate measurements of the dielectric constant of seawater at L band. *Radio Sci.* **2016**, *51*, 2–24. [CrossRef]

remote sensing

MDPI

Article

An Observational Perspective of Sea Surface Salinity in the Southwestern Indian Ocean and Its Role in the South Asia Summer Monsoon

Xu Yuan *, Mhd. Suhyb Salama and Zhongbo Su

Faculty of Geo-Information Science and Earth Observation (ITC), University of Twente, Hengelosestraat 99, 7514 AE Enschede, The Netherlands; s.salama@utwente.nl (M.S.S.); z.su@utwente.nl (Z.S.)

* Correspondence: x.yuan@utwente.nl

Received: 31 October 2018; Accepted: 27 November 2018; Published: 1 December 2018

Abstract: The seasonal variability of sea surface salinity anomalies (SSSAs) in the Indian Ocean is investigated for its role in the South Asian Summer Monsoon. We have observed an elongated spatial-feature of the positive SSSAs in the southwestern Indian Ocean before the onset of the South Asian Summer Monsoon (SASM) by using both the Aquarius satellite and the Argo float datasets. The maximum variable areas of SSSAs in the Indian Ocean are along (60° E–80° E) and symmetrical to the equator, divided into the southern and northern parts. Further, we have found that the annual variability of SSSAs changes earlier than that of sea surface temperature anomalies (SSTAs) in the corresponding areas, due to the change of wind stress and freshwater flux. The change of barrier layer thickness (BLT) anomalies is in phase with that of SSSAs in the southwestern Indian Ocean, which helps to sustain the warming water by prohibiting upwelling. Due to the time delay of SSSAs change between the northern and southern parts, SSSAs, therefore, take part in the seasonal process of the SASM via promoting the SSTAs gradient for the cross-equator currents.

Keywords: Aquarius; Argo; Sea Surface Salinity

---

1. Introduction

The South Asian Summer Monsoon (SASM) forms a vital source of water for one-sixth of the world's population. The onset and the intensity of the SASM control the occurrence of drought and flood events in South Asia, impacting on agricultural yields, water resources, infrastructure and humans. According to the monsoon's dynamic theory, the northward shift of the intertropical convergence zone (ITCZ) marks the onset of the SASM whereby the dynamic aspects of the ocean play critical roles in modulating the strength of the monsoon [1]. It has been found that Sea Surface Temperature (SST) Anomalies (SSTAs) over the Indian Ocean are good indicators of the differential heating between ocean and land and are correlated with both the onset and the intensity of the SASM [2–4].

As water density is controlled by both temperature and salinity, evidence suggests that Sea Surface Salinity (SSS) could be an indicator of abrupt changes in ocean dynamic and air-sea interaction [5]. Salinity affects many aspects of ocean stability [6,7], dynamic ocean variability [8–12] and complicates air-sea interactions [13–17]. In previous studies [6,18], the variability of SSS is attributed to a series of complex mechanisms, mainly including the freshwater flux, horizontal advection, vertical entrainment as well as some turbulence of mixing and diffusion but the latter in the mixed layer has little influence on the seasonal variability of salinity [19–21].

Sea Surface Salinity Anomalies (SSSAs) have noticeable changes during SASM. For example, during the SASM season, the extent of SSSAs in the Indian mini warm pool region [22] is minimized and the freshwater belt is formed along the west coast of India [23]. SSS along the equatorial Indian

Ocean can also reflect the imprints of Indian winter and summer monsoons via freshwater input and wind-induced mixing [24]. In 2004, minimum SSS was found before the onset of the monsoon [6]. Many scientific studies have shown that the inter-annual variability of SSSAs is connected to the local vortex of the monsoon onset [6,22,25,26] through the Arabian Sea mini warm pool.

Although the annual variance in SSS has been analysed in many studies [27–29], the annual variance in SSSAs is yet to be analysed. Neema et al. [22] revealed that the SSSAs in the Arabica Sea signal the onset of SASM by using only one-year simulated data, in this paper, we attempt to study the relationship between SSSA and SASM by providing observational and statistical evidence.

2. Data and Methods

SSS dataset. The SSS datasets comprised satellite products as well as in situ data. Satellite SSS products were obtained from the Version 5 Aquarius Combined Active-Passive (CAP) archive [30] for the period (09/2011–06/2015) on a grid of 0.5° × 0.5° (https://aquarius.umaine.edu/cgi/data_v5.htm). In situ measurements of SSS were obtained from the IPRC Array for Real-Time Geotropic Oceanography (Argo) product archive for the period (01/2005–12/2014) (http://apdrc.soest.hawaii.edu/projects/Argo).

Using in situ data obtained from Argo to study the Indian Ocean hold credibility because of the high density of data in the central Arabian Sea and central Bay of Bengal [31]. The Aquarius data, though limited in the period of coverage, are used for their higher spatial resolution to correct for discrepancies resulting from spatial interpolation of Argo data [32,33]. Both SSS data are mapped on a 1° × 1° grid on a monthly timescale.

Ancillary Datasets. The freshwater flux [Evaporation minus Precipitation (E-P)] is calculated using the monthly evaporation dataset obtained from the objectively analysed air-sea fluxes project (OAFLUX) [34] and monthly precipitation datasets from the Climate Prediction Centre (CPC) Merged Analysis of Precipitation (CMAP). Monthly MLD is also derived from Argo product provided by French Research Institute for Exploration of the Sea (Ifremer: http://www.ifremer.fr), with the MLD defined as the depth where the density has 0.03 kg/m^3 difference from that of the surface [35].

Monthly SST dataset is used from NOAA Optimum Interpolation (OI) version 2 [36]. The atmospheric circulation is obtained from the monthly wind of ERA-Interim reanalysis data produced by the European Centre for Medium-Range Weather Forecasts (ECMWF) at standard pressure levels (1000 hPa − 100 hPa). Monthly barrier layer thickness (BLT) is calculated from Argo products provided by Ifremer, by the equation $BLT = TTD_{DTm02} - MLD_{DReqDT02}$, where TTD_{DTm02} is the isothermal depth defined as the depth at which the surface temperature cools by 0.2 °C and $MLD_{DReqDT02}$ is the mixed layer depth [37,38].

All data used in this study are re-gridded into a 1° × 1° grid resolution from 2005 to 2014, except for wind stress which is from 2008 to 2014. The anomalies are calculated by the differences between the monthly data and climatological mean and the tendency of SSSAs and SSTAs are estimated by adopting the finite-difference method.

3. Observed Seasonal Variability in Sea Surface Salinity Anomalies

The SSS difference between the Arabian Sea and the Bay of Bengal, which is significant in climatological SSS distribution [28], was not found in monthly SSSAs. However, changes are apparent in several SSSAs maxima and minima centres shown by the Aquarius data (Figure 1a). For instance, in January, a positive SSSA centre appears in the eastern part (A_{EIO}), a negative SSSA centre in the central part (B_{CIO}) and a positive SSSA centre in the western part (C_{WIO}), depicted in Figure 1a, respectively. From boreal winter to spring, the B_{CIO} continuously intensifies and expands its area, whereas both the A_{EIO} and C_{WIO} diminish and disappear. In April, negative SSSAs distribute over almost the whole Indian Ocean. A notable change can be seen in May, with positive SSSAs slightly to the south of the equator, which may represent the suddenly intensifying A_{EIO}. This positive SSSA, emerging as an elongated spatial feature (ESF), divides B_{CIO} into two separated regions. Later, the positive SSSA ESF dominates almost the whole equatorial Indian Ocean and connects up with the intensifying C_{WIO} in the northern Indian Ocean. Subsequently, the B_{CIO} disappears during the boreal summer monsoon period with the positive SSSAs controlling the whole Indian Ocean, except for the equator and the western coast of the Arabian Sea. In the autumn, the positive SSSAs begin to dissipate gradually and the boreal winter mode is restored in December. This annual cycle of SSSAs also emerges when employing long-term SSS data obtained from Argo data for the period 2005–2014 (Figure 1b). Sparse subsampling of Argo data can reproduce the main features of the annual cycle defined by the satellite SSS from Aquarius.

Figure 2a shows the latitude-time plots of SSSAs averaged over 60°E–80°E, clearly depicting the positive SSSAs in May. The equator acts as a natural boundary, separating the changes in the SSSAs in the northern and southern Indian Ocean. There is an apparent time lag in these changes occurring across the southern to the northern Indian Ocean, which is consistent with the seasonal variability of the Indian Ocean induced by solar insolation. In this work, the SSSAs ESF area is defined as being the area of (60°E–80°E, 10°S–5°S) and its corresponding SSSAs in the northern hemisphere (60°E–80°E, 5°N–10°N) is also studied. Moreover, since the thermocline is shallower in the southwestern Indian Ocean and deeper in the eastern Indian Ocean and the easterlies are along the equator, the downwelling Rossby wave acts to suppress the continuous upwelling in the western Indian Ocean [39,40]. The symmetrical variability of SSSAs mode (Figure 2a) probably attributes to this downwelling Rossby wave [27,29,41]. However, there are no significant westward SSSAs motions along 10°S and 10°N in the hovmöller diagrams (Figure 2b). The maxima (minima) centres of SSSAs mainly locate in the SSSAs ESF areas. Consequently, the seasonal variability of SSSAs may be influenced by local atmospheric circulations and will be analysed in Section 5. But before doing so, we will examine next the relationship between SSSAs and SSTAs.

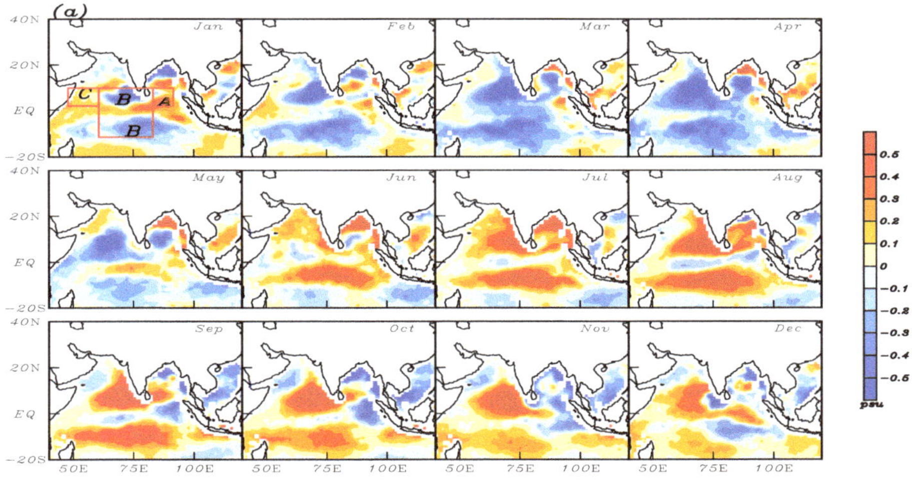

Figure 1. *Cont.*

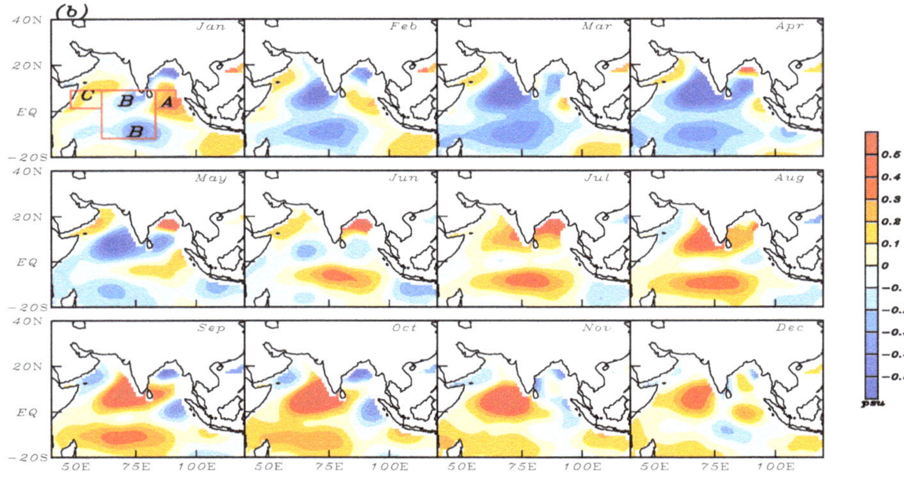

Figure 1. Seasonal variability of sea surface salinity in the Indian Ocean. The annual cycle of sea surface salinity anomalies by (**a**) Aquarius dataset during 2012–2014 and (**b**) Argo dataset during 2005–2014. A, B and C (see January) denotes the eastern part (A_{EIO}), the central part (B_{CIO}) and the western part (C_{WIO}) of the Indian Ocean respectively.

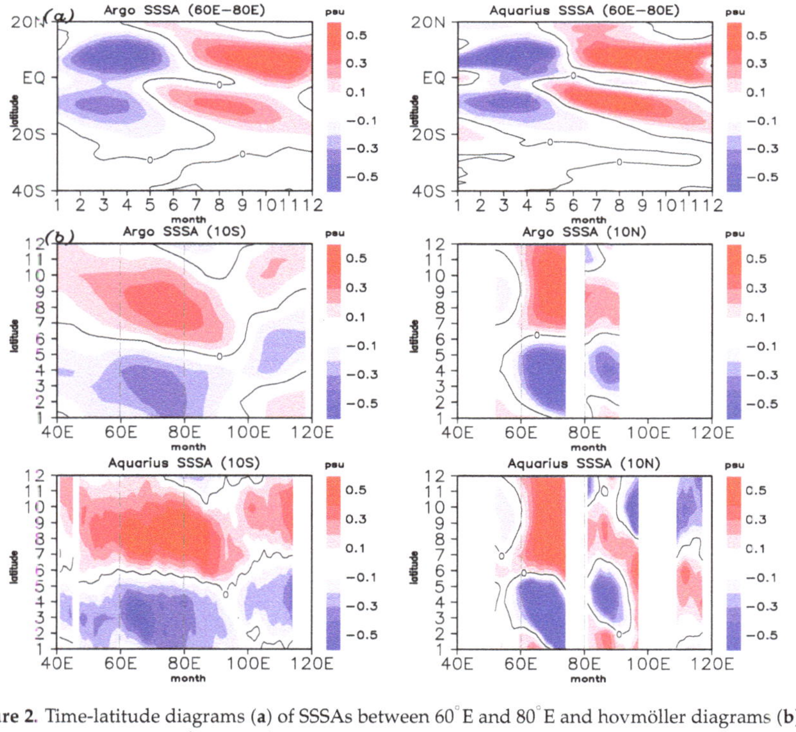

Figure 2. Time-latitude diagrams (**a**) of SSSAs between 60° E and 80° E and hovmöller diagrams (**b**) of SSSAs along the area of 10° S and 10° N in Argo and Aquarius (Unit: psu, the dashed lines enclose the SSSAs ESF areas).

4. The Relationship between SSSAs and SSTAs before the Onset of SASM

The SSSA ESF area corresponds roughly with the Seychelles Dome (SD) which is a remarkable oceanic thermal dome along (60°E–80°E, 5°S–10°S) [42,43]. Due to its unique location, which is under the control of both the monsoon wind and south-easterly trade wind, SD has a semi-annual cycle associated with upwelling in the boreal spring [42,43]. Above the SD, sea surface temperature anomalies (SSTAs) are sensitive to variability in the upwelling and this is especially so in its seasonal variation [44].

The seasonal variability of SSSAs is different from that of SSTAs, presenting not only in the trend of changing [when SSSAs increase, corresponding SSTAs decrease (Figure 3)] but also in the time of changing. Specifically, in the southern Indian Ocean, the tendency of SSSAs increases in the middle of March and that of SSTAs decreases in the early April. In the northern Indian Ocean, the tendency of SSSAs increases in the early April and that of SSTAs decreases in the early May. In other words, SSSAs have one and a half cycle/yr while SSTAs have one cycle/yr. Thus, SSSAs change faster than SSTAs. To give more convincing evidence of the quick change of SSSAs, we also calculated the seasonal variation in SSSAs and SSTAs for individual years with the Aquarius SSS dataset (Figure 4). Remarkably, the change of SSSAs tendency precedes that of SSTAs even in the individual years, except for 2013 (figure not shown), when the MJO (Madden Julian Oscillation) is very active during the onset of the SASM [45], resulting in heavy precipitation inhibiting SSSAs increase.

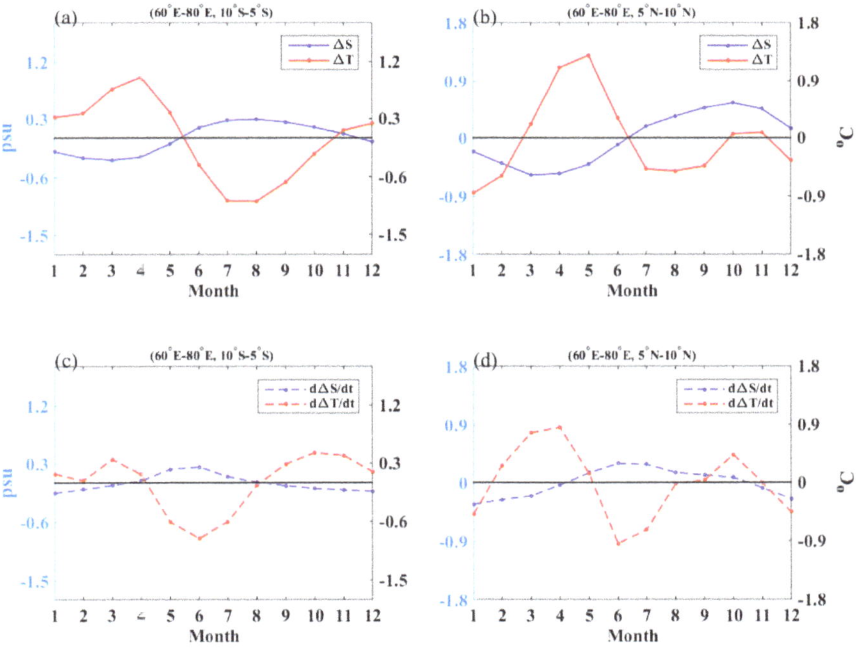

Figure 3. Seasonal variability and tendency for both SSSA (obtained from Argo, in blue) and SSTA (obtained from NOAA; in red) in the areas [(**a**,**c**); 60°E–80°E, 10°S–5°S] and [(**b**,**d**); 60°E–80°E, 5°N–10°N] for 2005 to 2014. Unit: psu; °C.

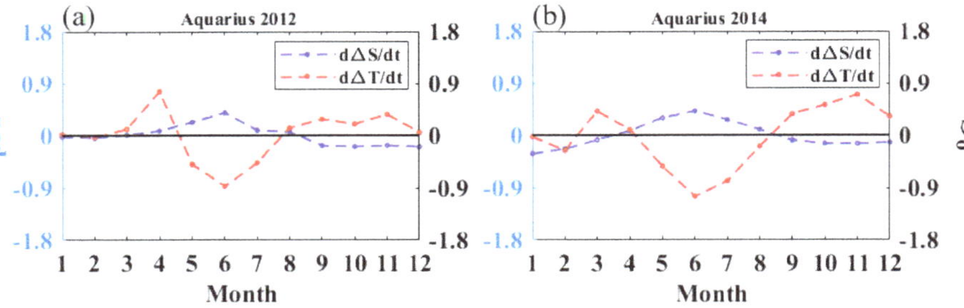

Figure 4. Same as Figure 2 but for SSS data obtained from Aquarius.

The earlier changed SSSAs may affect the oceanic stratification and, in turn, trigger a response in atmospheric circulation. To investigate this process, we show the seasonal variabilities of MLD and BLT anomalies in Figure 5. The top layer is the MLD which is defined as the depth that has a density change of 0.03 kg/m^3 from the surface. In the SD area, the MLD anomalies change in the middle of May and are associated with the change of SSSAs and SSTAs (Figure 3a). BLT is the difference between an isothermal layer and the isopycnal MLD, insulating the surface from the deeper layers. Thereby, BLT has influences on the SST by controlling the vertical mixing [46,47]. In the SD area (Figure 5a), the change of BLT anomalies is more sensitive to the tendency of SSSAs rather than the seasonal variability of SSSAs (Figure 3c). When the tendency of SSSAs increase to positive in early March, the BLD anomalies begin to decrease below zero and vice versa. According to the calculation of BLT (in Section 2) and given the slightly decreasing MLD anomalies (Figure 5a), in the southern Indian Ocean, decreasing BLT anomalies are attributed to upwelling. Then, upwelling eats away at the BLT leading to more cold and salty water entrained into the mixed layer. When MLD anomalies increase in April, SSTAs begin to be affected by this upwelling process (the tendency of the SSTAs decrease below zero as shown in Figure 3c). In the northern Indian Ocean, things respond a little differently (Figure 5b). Although BLT and MLD also change in March, the tendency of SSSAs changes in April. We assume that this is because of the strong salty surface water in the Arabian Sea, so it needs more time for upwelling to affect the SSSAs. With the increasing SSSAs, the significantly decreasing BLT brings more subsurface water into the mixed layer, which in turn, promotes the decrease of SSTAs. As such, SSSAs has the potential ability to affect SSTAs through shoaling the BLT.

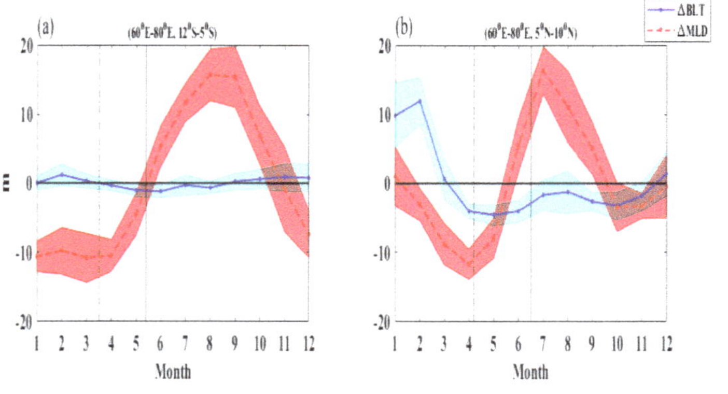

Figure 5. Seasonal variability for both BLT (in blue) and MLD anomalies (in red) in the areas [(**a**); 60° E–80° E, 10° S–5° S] and [(**b**); 60° E–80° E, 5° N–10° N] for 2005 to 2014 (The shaded areas are the standard deviation, the solid black line represents the time that SSSAs change and dotted black line represents the time that the tendency of SSSAs change). Unit: m.

5. The External Forcing for the SSSAs Change

We analysed next the atmospheric circulation anomalies before and after the SSSAs change to better understand the variability of SSSAs. On the one hand, freshwater flux is stressed to play a crucial role in the seasonal variability of SSS in the Indian Ocean [12,19]. Figure 6 shows the distributions of freshwater flux anomalies in February and April respectively. Before the SSSAs change (February), there is more precipitation over the SSSA ESF area, while less freshwater goes into the SSSA ESF area after the SSSAs change (April); instead, strong evaporation anomalies make this area much saltier. On the other hand, the upwelling effect on the variability of SSSAs can be found in the entrainment circulations before and after the increasing SSSAs (Figure 7). The wintertime north-westerly wind anomalies intensify in the southern Indian Ocean before SSSAs change and weaken after SSSAs (Figure 7a). In the southern Hemisphere, negative curl corresponds to cyclonic curl and relative upwelling. Thereby, before the SSSAs change, the positive wind stress curl anomalies over the SSSAs ESF area, represent Ekman downwelling. After the SSSAs change, the negative wind stress curl anomalies induce Ekman upwelling in the SSSAs ESF area (Figure 7b). Furtherly, due to the negative wind stress curl anomalies in the north of 10°S and the positive ones in the south of 10°S before the increasing SSSAs (Figure 7b), corresponding to anticyclone in the north and cyclone in the south, the easterlies is strengthened (Figure 7c), which in turn, provides a favourable environment for upwelling in the eastern Indian Ocean. The mode of wind stress curl change after the SSSAs increasing, with positive anomalies in the north and negative in the south, leading to anomaly westerlies (Figure 7c). In March, the SSSAs increase just as the atmospheric circulation pattern begins to change into the SASM pattern and the thermocline in the SD starts to shoal as a result of weakened downwelling [44].

Therefore, freshwater flux and entrainment circulation anomalies contribute to the increasing SSSAs in the southwestern Indian Ocean.

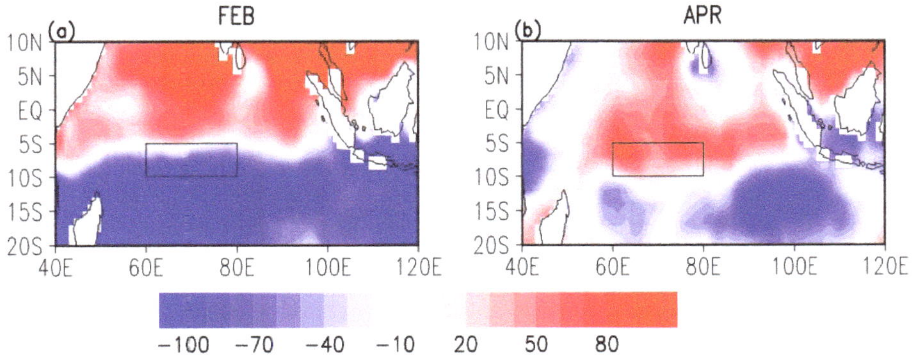

Figure 6. Freshwater flux anomalies. (**a**) Monthly mean freshwater flux in February (**a**) and April (**b**) from 2005 to 2014. (Unit: cm/yr). The box (in black line) denotes the SSSAs ESF area (60°E–80°E, 10°S–5°S).

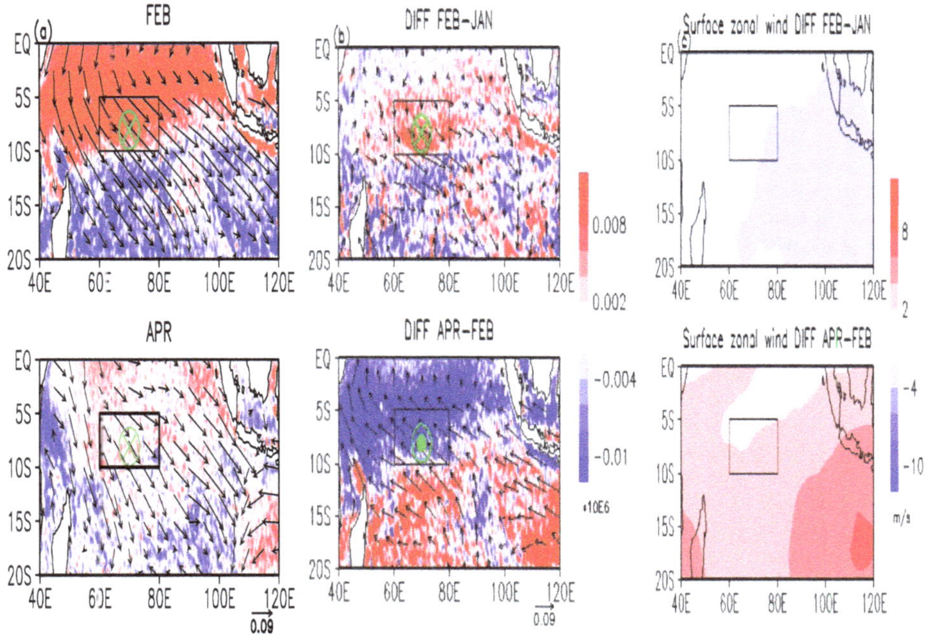

Figure 7. Wind stress and wind stress curl anomalies. (**a**) Monthly mean wind stress (vector) and wind stress curl anomalies (shaded) in February and April from 2008 to 2014. (**b**) Differences in wind stress and wind stress curl anomalies as February minus January and April minus February, respectively. (Unit: N/m^2; N/m^3). (**c**) Differences in the surface zonal wind (10m) obtained from ERA-interim difference as February minus January and for April minus February, respectively (Unit: m/s). [The boxes in black line are the SSSAs ESF area in the southern Indian Ocean (60°E–80°E, 5°S–10°S); the green plus mark represents downwelling and the green closed circle represents upwelling].

In summary (Figure 8), in the SD (5°S–10°S) from January to the early April, SSTAs stay in an increasing state (orange rectangle), while SSSAs start to increase in the middle of March (red filled rectangle), which is due to the less freshwater and Ekman upwelling caused by the wind stress. Thus increasing SSSAs, accompanied by the decreasing BLT anomalies, help to create a favourable environment for the SSTAs to decrease in the coming month via strengthening the upwelling. In other words, SSSAs, as one component in the ocean, are more sensitive in responding to the atmospheric influence than SSTAs. This phenomenon can also be found in the symmetrical area of the northern Hemisphere (5°N–10°N). One should mention here that there exist time-lags between the southern and northern Hemisphere, which leads to significant SST gradient in May (black rectangle). Thus, this heat contrast promotes the ITCZ to move northward and contributes to the change of vertical circulation.

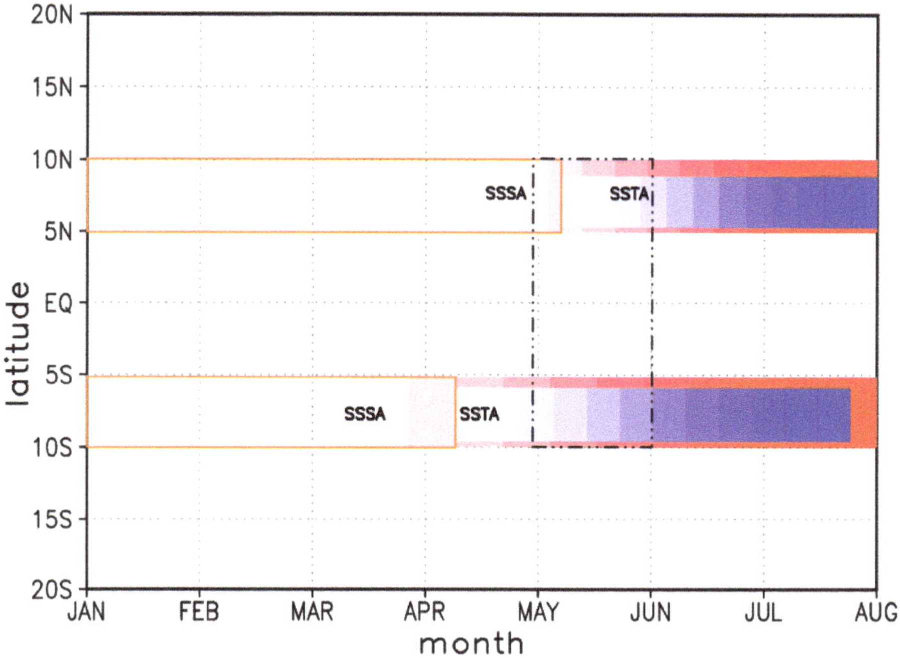

Figure 8. Diagram of the potential mechanism between SSSA and the onset of SASM along 60° E–80° E (Gradually increasing red rectangles represent the increasing SSSA and the blue ones represent the decreasing SSTA; In addition, the orange hollow rectangles represent the increasing SSTA and the black dot-dashed rectangle indicates the period with large SSTA gradient).

6. Conclusions

In this study, we analysed the annual variability in sea surface salinity anomalies (SSSAs) by employing Aquarius satellite datasets from 2012 to 2014 and Argo datasets from 2005 to 2014, as well as the corresponding freshwater fluxes, sea surface temperature anomalies (SSTAs) and also ocean dynamic and atmospheric circulations.

We found that a positive SSSAs as an elongated spatial feature (ESF) along the equator in May with its area roughly coinciding with the Seychelles Dome (SD). The seasonal variability of SSSAs in the SD change from negative to positive and, in contrast, SSTAs change from positive to negative. Although their changes happen in spring, SSSAs change earlier than that of SSTAs, associated with the thinner barrier layer thickness (BLT) anomalies in the corresponding area, which in turn, help to provide favourable circumstance for the decreasing SSTAs. The change of SSSAs is closely related to the atmospheric circulation anomalies, mainly attributed to the freshwater flux anomalies and entrainment anomalies. Moreover, there exists a time delay of SSSAs change between the northern and the southern Indian Ocean, resulting in cross-equatorial current by forming the meridional SSTAs gradient. Therefore, SSSAs contribute to the onset of SASM by affecting the SSTAs.

Author Contributions: Conceptualization, Z.S. and X.Y.; Methodology, X.Y.; Software, X.Y.; Formal analysis, X.Y.; Writing—Original Draft Preparation, X.Y.; Writing—Review & Editing, Z.S., M.S.S. and X.Y.

Funding: This research received no external funding.

Acknowledgments: We thank three anonymous reviewers for their valuable comments. The authors also would like to thank the following data providers: The gridded ocean parameter datasets are available at the Asian-Pacific Data-Research Centre (http://apdrc.soest.hawaii.edu/data/data.php), NASA Jet Propulsion Laboratory (https://podaac.jpl.nasa.gov/announcements), Woods Hole Oceanographic Institution (http://oaflux.whoi.edu/eva

p.html), Climate Prediction Centre (http://www.esrl.noaa.gov/psd/data/gridded/data.cmap.html), National Oceanic and Atmospheric Administration (https://www.esrl.noaa.gov/psd/data/gridded/data.noaa.oisst.v2.html), the Centre de Recherche et d'Exploitation Satellitaire (http://cersat.ifremer.fr/data/products/latest-products) and Centre National D'études Spatiales (https://igsac-cnes.cls.fr/html/products.html). In addition, the reanalysis datasets are from the European Centre for Medium-Range Weather Forecasts (http://www.ecmwf.int/en/research/climate-reanalysis/era-interim).

Conflicts of Interest: The authors declare no conflict of interest. The founding sponsors had no role in the design of the study; in the collection, analyses, or interpretation of data; in the writing of the manuscript and in the decision to publish the results.

References

1. Webster, P.; Fasullo, J. Monsoon: Dynamical theory. *Encycl. Atmos. Sci.* **2003**, *3*, 1370–1391.
2. Jang, Y.; Straus, D.M. Tropical stationary wave response to ENSO: Diabatic heating influence on the Indian summer monsoon. *J. Atmos. Sci.* **2013**, *70*, 193–222. [CrossRef]
3. Minoura, D.; Kawamura, R.; Matsuura, T. A mechanism of the onset of the South Asian summer monsoon. *J. Meteorol. Soc. Jpn.* **2003**, *81*, 563–580. [CrossRef]
4. Shenoi, S.; Shankar, D.; Shetye, S. On the sea surface temperature high in the Lakshadweep Sea before the onset of the southwest monsoon. *J. Geophys. Res.* **1999**, *104*, 15703–15712. [CrossRef]
5. Hénin, C.; Du Penhoat, Y.; Ioualalen, M. Observations of sea surface salinity in the western Pacific fresh pool: Large-scale changes in 1992–1995. *J. Geophys. Res. Ocean.* **1998**, *103*, 7523–7536. [CrossRef]
6. Nyadjro. E.S.; Subrahmanyam, B.; Murty, V.; Shriver, J.F. The role of salinity on the dynamics of the Arabian Sea mini warm pool. *J. Geophys. Res. Ocean.* **2012**, *117*. [CrossRef]
7. Schiller, A.; Mikolajewicz, U.; Voss, R. The stability of the North Atlantic thermohaline circulation in a coupled ocean-atmosphere general circulation model. *Oceanogr. Lit. Rev.* **1998**, *1*, 41–42. [CrossRef]
8. Du, Y.; Zhang, Y. Satellite and Argo observed surface salinity variations in the tropical Indian Ocean and their association with the Indian Ocean Dipole mode. *J. Clim.* **2015**, *28*, 695–713. [CrossRef]
9. Stammer, D. Global characteristics of ocean variability estimated from regional TOPEX/POSEIDON altimeter measurements. *J. Phys. Oceanogr.* **1997**, *27*, 1743–1769. [CrossRef]
10. Sun, S.; Lan, J.; Fang, Y.; Gao, X. A triggering mechanism for the Indian Ocean Dipoles independent of ENSO. *J. Clim.* **2015**, *28*, 5063–5076. [CrossRef]
11. Thompson, B.; Gnanaseelan, C.; Salvekar, P. Variability in the Indian Ocean circulation and salinity and its impact on SST anomalies during dipole events. *J Mar. Res.* **2006**, *64*, 853–880. [CrossRef]
12. Vinayachandran, P.N.; Nanjundiah, R.S. Indian Ocean sea surface salinity variations in a coupled model. *Clim. Dyn.* **2009**, *33*, 245–263. [CrossRef]
13. Grunseich, G.; Subrahmanyam, B.; Wang, B. The Madden-Julian oscillation detected in Aquarius salinity observations. *Geophys. Res. Lett.* **2013**, *40*, 5461–5466. [CrossRef]
14. Guan, B.; Lee, T.; Halkides, D.J.; Waliser, D.E. Aquarius surface salinity and the Madden-Julian Oscillation: The role of salinity in surface layer density and potential energy. *Geophys. Res. Lett.* **2014**, *41*, 2858–2869. [CrossRef]
15. Horii, T.; Ueki, I.; Ando, K.; Hasegawa, T.; Mizuno, K.; Seiki, A. Impact of intraseasonal salinity variations on sea surface temperature in the eastern equatorial Indian Ocean. *J. Oceanogr.* **2015**, *72*, 313–326. [CrossRef]
16. Simon, B.; Rahman, S.; Joshi, P. Conditions leading to the onset of the Indian monsoon: A satellite perspective. *Meteorol. Atmos. Phys.* **2006**, *93*, 201–210. [CrossRef]
17. Williams, P.D.; Guilyardi, E.; Madec, G.; Gualdi, S.; Scoccimarro, E. The role of mean ocean salinity in climate. *Dyn. Atmos.Ocean.* **2010**, *49*, 108–123. [CrossRef]
18. Rao, R.R. Seasonal variability of sea surface salinity and salt budget of the mixed layer of the north Indian Ocean. *J. Geophys. Res.* **2003**, *108*, 3009. [CrossRef]
19. Zhang, Y.; Du, Y. Seasonal variability of salinity budget and water exchange in the northern Indian Ocean from HYCOM assimilation. *Chin. J. Oceanol. Limn.* **2012**, *30*, 1082–1092. [CrossRef]
20. Dong, S.; Garzoli, S.L.; Baringer, M. An assessment of the seasonal mixed layer salinity budget in the Southern Ocean. *J. Geophys. Res. Ocean.* **2009**, *114*. [CrossRef]

21. Da-Allada, C.; Alory, G.; Du Penhoat, Y.; Kestenare, E.; Durand, F.; Hounkonnou, N. Seasonal mixed-layer salinity balance in the tropical Atlantic Ocean: Mean state and seasonal cycle. *J. Geophys. Res. Ocean.* **2013**, *118*, 332–345. [CrossRef]
22. Neema, C.; Hareeshkumar, P.; Babu, C. Characteristics of Arabian Sea mini warm pool and Indian summer monsoon. *Clim. Dyn.* **2012**, *38*, 2073–2087. [CrossRef]
23. Deshpande, R.; Muraleedharan, P.; Singh, R.L.; Kumar, B.; Rao, M.S.; Dave, M.; Sivakumar, K.; Gupta, S. Spatio-temporal distributions of δ 18 O, δD and salinity in the Arabian Sea: Identifying processes and controls. *Mar. Chem.* **2013**, *157*, 144–161. [CrossRef]
24. Saraswat, R.; Nigam, R.; Mackensen, A.; Weldeab, S. Linkage between seasonal insolation gradient in the tropical northern hemisphere and the sea surface salinity of the equatorial Indian Ocean during the last glacial period. *Acta Geol. Sin.* **2012**, *86*, 1265–1275.
25. Sijikumar, S.; Rajeev, K. Role of the Arabian Sea warm pool on the precipitation characteristics during the monsoon onset period. *J. Clim.* **2012**, *25*, 1890–1899. [CrossRef]
26. Maes, C.; O'Kane, T.J. Seasonal variations of the upper ocean salinity stratification in the Tropics. *J. Geophys. Res. Ocean.* **2014**, *119*, 1706–1722. [CrossRef]
27. Donguy, J.-R.; Meyers, G. Seasonal variations of sea-surface salinity and temperature in the tropical Indian Ocean. *Deep Sea Res. Oceanogr. Res. Pap.* **1996**, *43*, 117–138. [CrossRef]
28. Rao, R.; Sivakumar, R. Seasonal variability of sea surface salinity and salt budget of the mixed layer of the north Indian Ocean. *J. Geophys. Res. Ocean.* **2003**, *108*, 3009. [CrossRef]
29. Zhang, Y.H.; Du, Y. Seasonal variability of salinity budget and water exchange in the northern Indian Ocean from HYCOM assimilation. *Chin. J. Oceanol. Limn.* **2012**, *30*, 1082–1092. [CrossRef]
30. Kao, H.; Lagerloef, G.; Lee, T.; Melnichenko, O.; Hacker, P. Aquarius Salinity Validation Analysis; Data Version 5.0. 2017. Available online: http://podaac-ftp.jpl.nasa.gov/allData/aquarius/docs/v5;AQ-014-PS-0016_AquariusSalinityDataValidationAnalysis_DatasetVersion5.0.pdf (accessed on 21 August 2018).
31. Bhaskar, T.V.; Jayaram, C. Evaluation of Aquarius sea surface salinity with Argo sea surface salinity in the Tropical Indian Ocean. *Geosci. Remote Sens. Lett. IEEE* **2015**, *12*, 1292–1296. [CrossRef]
32. Ebuchi, N.; Abe, H. Evaluation of sea surface salinity observed by Aquarius. In Proceedings of the 2014 IEEE International Geoscience and Remote Sensing Symposium (IGARSS 2014), Quebec City, QC, Canada, 13–18 July 2014; pp. 4427–4430.
33. Qu, T.; Song, Y.T.; Maes, C. Sea surface salinity and barrier layer variability in the equatorial Pacific as seen from Aquarius and Argo. *J. Geophys. Res. Ocean.* **2014**, *119*, 15–29. [CrossRef]
34. Yu, L.; Jin, X.; Weller, R. *Multidecade Global Flux Datasets from the Objectively Analyzed Air-sea Fluxes (OAFlux) Project: Latent and Sensible Heat Fluxes, Ocean Evaporation, and Related Surface Meteorological Variables*; OAFlux Project Technical Report, OA-2008-01; Woods Hole Oceanographic Institution: Woods Hole, MA, USA, 2008; 64p.
35. De Boyer Montégut, C.; Madec, G.; Fischer, A.S.; Lazar, A.; Iudicone, D. Mixed layer depth over the global ocean: An examination of profile data and a profile-based climatology. *J. Geophys. Res. Ocean.* **2004**, *109*. [CrossRef]
36. Reynolds, R.W.; Rayner, N.A.; Smith, T.M.; Stokes, D.C.; Wang, W. An improved in situ and satellite SST analysis for climate. *J. Clim.* **2002**, *15*, 1609–1625. [CrossRef]
37. De Boyer Montégut, C.; Mignot, J.; Lazar, A.; Cravatte, S. Control of salinity on the mixed layer depth in the world ocean: 1. General description. *J. Geophys. Res. Ocean.* **2007**, *112*. [CrossRef]
38. Mignot, J.; de Boyer Montégut, C.; Lazar, A.; Cravatte, S. Control of salinity on the mixed layer depth in the world ocean: 2. Tropical areas. *J. Geophys. Res. Ocean.* **2007**, *112*. [CrossRef]
39. Shinoda, T.; Hendon, H.H.; Alexander, M.A. Surface and subsurface dipole variability in the Indian Ocean and its relation with ENSO. *Deep Sea Res. Oceanogr. Res. Pap.* **2004**, *51*, 619–635. [CrossRef]
40. Potemra, J.T. Contribution of equatorial Pacific winds to southern tropical Indian Ocean Rossby waves. *J. Geophys. Res. Ocean.* **2001**, *106*, 2407–2422. [CrossRef]
41. Yuhong, Z.; Yan, D.; Shaojun, Z.; Yali, Y.; Xuhua, C. Impact of Indian Ocean Dipole on the salinity budget in the equatorial Indian Ocean. *J. Geophys. Res. Ocean.* **2013**, *118*, 4911–4923. [CrossRef]
42. Izumo, T.; Montégut, C.B.; Luo, J.-J.; Behera, S.K.; Masson, S.; Yamagata, T. The role of the western Arabian Sea upwelling in Indian monsoon rainfall variability. *J. Clim.* **2008**, *21*, 5603–5623. [CrossRef]

43. Yokoi, T.; Tozuka, T.; Yamagata, T. Seasonal variation of the Seychelles Dome. *J. Clim.* **2008**, *21*, 3740–3754. [CrossRef]
44. Yokoi, T.; Tozuka, T.; Yamagata, T. Seasonal and interannual variations of the SST above the Seychelles Dome. *J. Clim.* **2012**, *25*, 800–814. [CrossRef]
45. Pai, D.; Bhan, S. *Monsoon 2013: A report (IMD Met. Monograph no: ESSO/IMD/SYNOPTIC MET/01-2014/15)*; India Meteorological Department, National Climate Center: Pune, India, 2014.
46. Maes, C.; Picaut, J.; Belamari, S. Importance of the salinity barrier layer for the buildup of El Niño. *J. Clim.* **2005**, *18*, 104–118. [CrossRef]
47. Masson, S.; Luo, J.J.; Madec, G.; Vialard, J.; Durand, F.; Gualdi, S.; Guilyardi, E.; Behera, S.; Delécluse, P.; Navarra, A. Impact of barrier layer on winter-spring variability of the southeastern Arabian Sea. *Geophys. Res. Lett.* **2005**, *32*, 193–222. [CrossRef]

remote sensing

MDPI

Article

The Potential and Challenges of Using Soil Moisture Active Passive (SMAP) Sea Surface Salinity to Monitor Arctic Ocean Freshwater Changes

Wenqing Tang [1,*], Simon Yueh [1], Daqing Yang [2], Alexander Fore [1], Akiko Hayashi [1], Tong Lee [1], Severine Fournier [1] and Benjamin Holt [1]

1. Jet Propulsion Laboratory, California Institute of Technology, Pasadena, CA 91109, USA; Simon.H.Yueh@jpl.nasa.gov (S.Y.); Alexander.Fore@jpl.nasa.gov (A.F.); Akiko.K.Hayashi@jpl.nasa.gov (A.H.); tlee@jpl.nasa.gov (T.L.); Severine.Fournier@jpl.nasa.gov (S.F.); Benjamin.M.Holt@jpl.nasa.gov (B.H.)
2. Environment and Climate Change Canada, Water and Climate Impacts Research Centre, Victoria, BC V8P 5C2, Canada; daqing.yang@canada.ca

* Correspondence: Wenqing.Tang@jpl.nasa.gov; Tel.: +1-818-354-8199

Received: 5 May 2018; Accepted: 28 May 2018; Published: 4 June 2018

Abstract: Sea surface salinity (SSS) links various components of the Arctic freshwater system. SSS responds to freshwater inputs from river discharge, sea ice change, precipitation and evaporation, and oceanic transport through the open straits of the Pacific and Atlantic oceans. However, in situ SSS data in the Arctic Ocean are very sparse and insufficient to depict the large-scale variability to address the critical question of how climate variability and change affect the Arctic Ocean freshwater. The L-band microwave radiometer on board the NASA Soil Moisture Active Passive (SMAP) mission has been providing SSS measurements since April 2015, at approximately 60 km resolution with Arctic Ocean coverage in 1–2 days. With improved land/ice correction, the SMAP SSS algorithm that was developed at the Jet Propulsion Laboratory (JPL) is able to retrieve SSS in ice-free regions 35 km of the coast. SMAP observes a large-scale contrast in salinity between the Atlantic and Pacific sides of the Arctic Ocean, while retrievals within the Arctic Circle vary over time, depending on the sea ice coverage and river runoff. We assess the accuracy of SMAP SSS through comparative analysis with in situ salinity data collected by Argo floats, ships, gliders, and in field campaigns. Results derived from nearly 20,000 pairs of SMAP and in situ data North of 50°N collocated within a 12.5-km radius and daily time window indicate a Root Mean Square Difference (RMSD) less than ~1 psu with a correlation coefficient of 0.82 and a near unity regression slope over the entire range of salinity. In contrast, the Hybrid Coordinate Ocean Model (HYCOM) has a smaller RMSD with Argo. However, there are clear systematic biases in the HYCOM for salinity in the range of 25–30 psu, leading to a regression slope of about 0.5. In the region North of 65°N, the number of collocated samples drops more than 70%, resulting in an RMSD of about 1.2 psu. SMAP SSS in the Kara Sea shows a consistent response to discharge anomalies from the Ob' and Yenisei rivers between 2015 and 2016, providing an assessment of runoff impact in a region where no in situ salinity data are available for validation. The Kara Sea SSS anomaly observed by SMAP is missing in the HYCOM SSS, which assimilates climatological runoffs without interannual changes. We explored the feasibility of using SMAP SSS to monitor the sea surface salinity variability at the major Arctic Ocean gateways. Results show that although the SMAP SSS is limited to about 1 psu accuracy, many large salinity changes are observable. This may lead to the potential application of satellite SSS in the Arctic monitoring system as a proxy of the upper ocean layer freshwater exchanges with subarctic oceans.

Keywords: SMAP; sea surface salinity; Arctic Ocean; sea ice; river discharge; Arctic Gateways

1. Introduction

The Arctic Ocean is experiencing unprecedented changes: the loss of sea ice [1–3], becoming fresher [4–6] and warmer [7,8], and the intensification of the regional freshwater cycle [9]. The urgency to closely monitor changes in the Arctic freshwater system and its exchange with subarctic oceans has been widely recognized by the scientific communities [10–16] Among the key observables, ocean salinity plays a critical role of linking the freshwater components within the atmospheric, terrestrial, and cryospheric water cycles to ocean circulation [17–23]. In particular, salinity responds to terrestrial runoff (river discharge), sea ice melt and growth, surface freshwater forcing (precipitation and evaporation), and exchanges with subarctic oceans via oceanic transports through Arctic gateways [24]. The Arctic freshwater changes that are associated with salinity alter the horizontal and vertical density structure of seawater, thereby influencing regional oceanic processes with global consequences [25–29].

In situ salinity observations in the Arctic Ocean have been historically very sparse. Spaceborne observation of sea surface salinity (SSS) is anticipated to advance our monitoring capability for the Arctic Ocean. SSS is a good tracer of the freshwater movement in the upper Arctic Ocean. The synoptic large-scale spatial variability of the SSS field that was obtained from the vantage point of space will not only monitor freshwater changes due to regional processes in the Arctic Ocean, but it will also provide a means to improve our understanding of the interaction between the Arctic and subarctic oceans.

Global SSS observations have been available from three satellite missions. The first one is the European Space Agency (ESA) Soil Moisture and Ocean Salinity (SMOS) mission launched on 2 November 2009 [30,31], carrying a passive L-band microwave system. The second one is the Aquarius/SAC-D, a joint venture of National Aeronautics and Space Administration (NASA) and Argentina's Comision Nacional de Actividades Espaciales (CONAE) [32–34], which was in operation from 10 June 2011 to 7 June 2015. The NASA Soil Moisture Active Passive (SMAP) observatory, which was launched in 31 January 2015, designed to measure the soil moisture over land [35], is also observing global ocean with an L-band radiometer and is currently being used to estimate SSS [36].

Satellite SSS data have enabled many new scientific results for tropical and subtropical oceans. However, its applications in the polar and sub-polar oceans did not progress as well due to two reasons [37,38]. First, satellite SSS retrieval in high latitude seawater is complicated by several factors. L-band instrument sensitivity to SSS is greatly reduced in cold seawater [39,40]. Leakage of emissivity from sea ice into the satellite antenna's main lobe or through the antenna side-lobes can contaminate salinity signals if undetected [41]. The larger uncertainty in the ocean surface roughness correction in the rough seas frequently encountered under the conditions of high winds and the strong waves of polar oceans influences the accuracy of the SSS retrievals [42,43]. The relatively large uncertainties of the ancillary satellite sea surface temperature (SST) data in the Arctic Ocean used in the SSS retrievals also affect the SSS accuracy [44]. Second, the paucity of in situ measurements hinders the validation of satellite SSS [45–47]. The validation of satellite SSS using in situ data can also be complicated by the difference in spatial scale between the satellite measurements over satellite footprints, and the point-wise in situ measurements in regions with strong sub-footprint variability [48–50]. As a result, quite large gaps near land and ice are masked out (~110 km for Aquarius and ~50 km for previously released SMAP SSS) in order to avoid contaminated retrieval, which could be misinterpreted or interfere with global validation results.

The recent release of SMAP SSS (V4.0) produced at the Jet Propulsion Laboratory (JPL) [51] was based on a more advanced correction strategy, aiming to deliver the valuable information in the critical regions near land and sea ice as much as possible. One of the main improvements is land contamination correction on measured brightness temperature (TB) before SSS retrievals. Land contamination is the leakage of energy from land surface into the radiometer receiver through side lobes or partially through the main lobe. In land-contaminated TB, non-oceanic signatures mixed with oceanic signatures may result in differences between two categories that exceed 100 kelvin (K). If undetected or uncorrected, the non-oceanic signature would cause large bias in the satellite salinity values. The JPL SMAP SSS algorithm has implemented the land contamination correction that is based on two look-up-tables (LUT), both being derived from SMAP data itself. One LUT is the beam-integrated land fraction as a function of footprint position and orientation, while the other is a monthly climatology of land TB extending out to 1000 km from coast, which represents effective contributions of TB from side-lobes over land. Combining the two LUTs, the land contamination contribution for each SMAP footprint is estimated and removed from TB. With improved land correction, the JPL algorithm retrieves SSS within 35 km from the land wherever the sea ice concentration values are less than 3% (based on NCEP SIC analysis [52]).

This study investigates the utility and accuracy of SMAP SSS retrieved in the Arctic Ocean via two complementary approaches: (1) comparing with collocated in situ salinity data, and (2) analyzing the co-variability of SMAP SSS with changes in sea ice and river discharges. We also examined the seasonal and interannual variability of SMAP SSS over major Arctic Ocean gateways to explore the potential of satellite SSS as a proxy to infer the freshwater exchange between the Arctic and sub-oceans. The datasets that were used for the analysis are described in Section 2. The results are presented in Section 3. Section 4 provides further discussion, and Section 5 summarizes the findings.

2. Data

2.1. SMAP SSS

SMAP SSS are produced at Level 2 (L2) and Level 3 (L3). L2 data represents the spatial average of the instantaneous measurements in the swath footprint, delivered for each of the salinity-wind-cell (SWC) along the satellite swath with 1624 × 76 cells along/cross track per satellite revolution. L2 data covers the global ocean in eight days with a spatial resolution of ~60 km, posted on a 25 km grid. There are two Level 3 (L3) products, the first with monthly temporal averaging and the second with eight-day averaging, both on a global 0.25° × 0.25° grid. The eight-day product is created daily by averaging eight days of L2 data centered at noon UTC (Coordinated Universal Time) of the day with a search radius of 45 km and a Gaussian weighting half-power distance of 30 km. Also, provided along with L2 and L3 products are the surface salinity output from the Hybrid Coordinate Ocean Model (HYCOM) [53] interpolated to SMAP measurement time and location.

Currently, there are two SMAP SSS products. One is the JPL product [36,51] that is mentioned in Section 1, and the other is based on algorithm independently developed at the Remote Sensing System (RSS) [54]. Both datasets are available at NASA Physical Oceanography Distributed Active Archive Center (PO.DAAC) (http://podaac.jpl.nasa.gov). The RSS product leaves large gaps in the Arctic Ocean near the ice edge, and is therefore not suitable for this study. We will focus on analyzing the JPL SMAP SSS V4.0 product (hereafter named SMAP SSS).

2.2. In Situ Salinity Data

Individual Argo float profiles are obtained from USGODAE (The United States (US) Global Ocean Data Assimilation Experiment, http://www.usgodae.org/argo/argo.html). The Argo array consists of approximately 3700 floats in the global ocean measuring salinity and temperature [55,56]. Argo data were collected and were made freely available by the International Argo Program and the national programs that contribute to it, as part of the Global Ocean Observing System (GOOS).

Salinity data from ships and gliders in the Arctic Ocean are obtained from the Copernicus Marine Environment Monitoring Service (CMEMS), which is an earth observing data center under the European Commission [57]. We use the product of Arctic Ocean-In situ Near Real Time Observations [58], which contains data collected from the Arctic ROOS (Regional Ocean Observing System) members, and quality controlled using automated procedures that are available from the In Situ Thematic Assembly Centre (INS TAC).

We also used data sets that were collected in recent field campaigns. During the NASA Ocean Melting Greenland (OMG) mission [10,11] field campaigns that took place in September/October

2016 and October 2017, NASA's G-III research plane, flying along the Greenland coast, deployed Airborne expandable CTD (AXCTD) sensors to measure temperature and salinity profiles from the surface to about 1000 m depth [58]. Quality controlled OMG data is made publicly available at (https://omg.jpl.nasa.gov/portal/). University of East Anglia, UK provided us with the salinity data from the seaglider missions MASSMO4, AlterEco and Oman579 (personal communication with Bastien Queste, PI) (http://ueaglider.uea.ac.uk).

2.3. Sea Ice Concentration

Sea ice concentration (SIC) data not only identifies ice-free areas where satellite SSS retrieval is possible, but also may influence the SSS values through freshwater exchanges that were associated with sea ice changes [59]. We obtain the daily and monthly SIC data from the NASA DAAC at the National Snow and Ice Data Center (NSIDC). The SIC data set [NSIDC-0051] is generated from multi-frequency brightness temperature data and is designed to provide a consistent time series of SIC combining the coverage of several passive microwave instruments [60]. The data are provided in the polar stereographic projection at a grid cell size of 25 × 25 km^2. The SIC data are also available in the Near-Real-Time (NRT) with a daily update [61]. The JPL SMAP SSS retrieval algorithm uses the high resolution (1/12°) SIC analyses from NCEP [52] to identify ice-free regions (ftp://ftpprd.ncep.noaa.gov/pub/data/nccf/com/omb/prod/). The SMAP L1B TB footprints falling into bins with SIC < 3% are flagged as ice-free.

2.4. River Discharge

Many large rivers in the northern regions produce huge amounts of flow to the Arctic Ocean. River discharge is therefore a dominant component in the Arctic freshwater system. Land hydrology in the polar regions are closely related with climate change and permafrost conditions [62,63]. The daily discharge data from six largest Arctic rivers are available from Arctic Great River Observatory (Arctic-GRO) (http://arcticgreatrivers.org), which is a component of the National Science Foundation's (NSF) Arctic Observing Network (AON). Arctic-GRO implemented identical sampling and analytical protocols that were developed by the PARTNERS Project (Pan-Arctic River Transport of Nutrients, Organic Matter, and Suspended Sediments), a coordinated international effort to collect and analyze a time-series of water samples [64,65]. We use discharge data from Ob' and Yenisey River during May 2015 to April 2017 for this study [66]. Other studies examined the flow regimes and changes for these two rivers [67,68]. We focus on the discharge at the downstream stations for basin freshwater input to the Arctic Ocean.

2.5. Moorings at Arctic Gateways

The main Arctic gateways are well monitored by in situ mooring arrays [24,69,70]. The mooring data over the Bering Strait [69,70] was obtained from the Applied Physics Laboratory (APL) at the University of Washington (http://psc.apl.washington.edu/HLD/Bstrait/bstrait.html). Although the measuring depth of moorings (~50 m) is largely different from the penetration depth (~1–2 cm) of L-band radiometer, and is therefore is not included in the in situ data for validation (Section 3.1), it should provide an indication of the dynamical range of the seasonal variation and interannual trend of SSS. Historical mooring data also exists at the Davis Strait and the Fram Strait. Unfortunately, we were not able to find up-to-date mooring data at either the Davis Strait or the Fram Strait.

3. Results

3.1. Validation with In Situ Salinity

We aggregated in situ salinity data collected by Argo floats, ships, gliders, and in OMG field campaigns to create a gridded in situ data set on 0.25° × 0.25° grids daily from 1 April 2015 to 31 March 2018. Measurements of different instruments, such as CTD (Conductivity, Temperature and

Depth), XCTD (Expandable CTD), and TSG (Thermosalinograph) taken within 5 m of the sea surface are used. Figure 1 shows the distribution of in situ measurements North of 50°N on 0.25° grids, as long as there is as least one daily record in three years on the grid point. Areas South of 60°N are mostly covered by Argo floats, complemented by a few ship tracks. Areas between 60°N and 80°N in Greenland area are well covered combining ships, gliders, and OMG data. However, above the Arctic Circle, particularly North of the Bering Strait, there are almost no in situ salinity data.

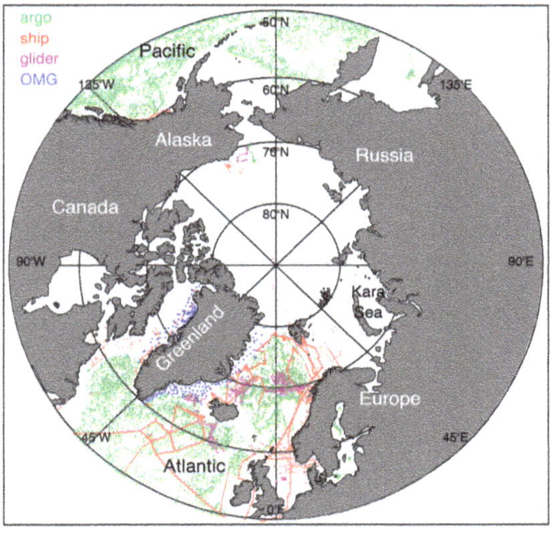

Figure 1. Location where in situ salinity are collected within 5 m from surface by Argo float (green), ship (red), glider (purple), and OMG AXCTD (blue) during the three years from April 2015 to March 2018.

The three-year mean of SSS averaged over collocated grid points from April 2015 to March 2018 are illustrated in Figure 2. In situ data confirms the large-scale salinity feature that was observed in SMAP SSS, i.e., the much saltier seawater on the Atlantic side than on the Pacific side of the Arctic Ocean. The biases with respect to in situ salinity are less than 1 psu in a majority of the areas, which are masked out in white so areas with large biases stand out more clearly (Figure 2, right column). It can be seen that SMAP SSS shows positive biases in a few places near the coast, and negative biases along the east coast of Greenland and near Davis Strait (Figure 2d). On the other hand, HYCOM SSS shows mostly positive biases east of Greenland (Figure 2e). The contrast between SMAP and HYCOM east of Greenland is particularly interesting. It suggests that SMAP SSS retrieval may contain un-detected sea ice effect, while HYCOM is not able to catch the freshening signal due to ice melting where no Argo floats are available for HYCOM's operational data assimilation.

The statistical results of comparing in situ salinity with collocated SMAP SSS and HYCOM SSS are presented in Table 1 and Figure 3. All data are gridded on 0.25° grids daily, covering the period from April 2015 to March 2018. North of 50°N, the bias, standard deviation, and Root Mean Square Difference (RMSD) between 19,738 pairs of SMAP SSS and in situ are 0.442, 2.391 and 2.431 psu with correlation of 0.81. The statistics between HYCOM SSS and in situ is slightly better. The open circles in the scatter plot indicate outliers defined as those pairs where the absolute difference exceeds three times the standard deviation. The number of such defined outliers is less than 1% of the total collocated data set. The entire set of collocated data can be visually divided into two groups: one group of high SSS (>25 psu) and the other of low SSS (<10 psu). It is interesting to note that in the low SSS group, in situ salinity are distinct at three sub-groups that are centered around 2, 5, and 7 psu, with collocated SMAP SSS spread from 0 to 10 psu, while HYCOM SSS almost perfectly aligned with the in situ data.

Again, this is expected because most of the in-situ data being used for the evaluation here are from Argo profiling floats, and HYCOM assimilates the Argo float data operationally.

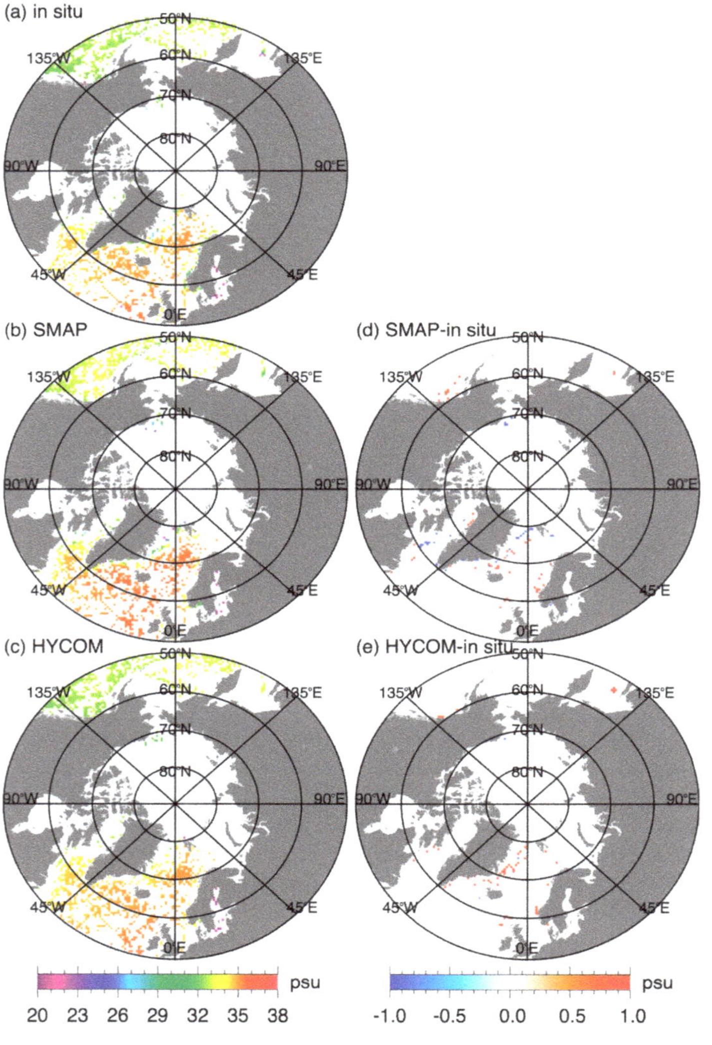

Figure 2. Salinity maps North of 50°N averaged from 1 April 2015 to 31 March 2018 from (**a**) in situ data, (**b**) Soil Moisture Active Passive Sea Surface Salinity (SMAP SSS), and (**c**) Hybrid Coordinate Ocean Model (HYCOM) SSS, and the difference of (**d**) SMAP SSS minus in situ, and (**e**) HYCOM SSS minus in situ, wherever absolute differences less than 1 psu are masked out in white.

Table 1. Results of statistical comparison between in situ salinity and SMAP SSS (JPL V4.0) and HYCOM SSS, for North of 50°N or North of 65°N, respectively. All data are gridded on 0.25° grids averaged daily from April 2015 to March 2018. N is the number of collocated pairs; bias (data minus in situ), standard deviation (Std.) and Root Mean Square Difference (RMSD) are in psu; Corr. stands for the correlation coefficients.

	North of 50°N					North of 65°N				
	N	Bias	Std.	RMSD	Corr.	N	Bias	Std.	RMSD	Corr.
SMAP	19738	0.442	2.391	2.431	0.805	5785	0.342	2.829	2.849	0.509
HYCOM	19738	0.270	2.110	2.128	0.934	5785	0.285	2.337	2.354	0.892
After excluding outliers										
SMAP	19543	0.385	0.987	1.060	0.817	5712	0.339	1.179	1.227	0.518
HYCOM	19617	0.149	0.661	0.678	0.942	5749	0.182	0.840	0.860	0.900

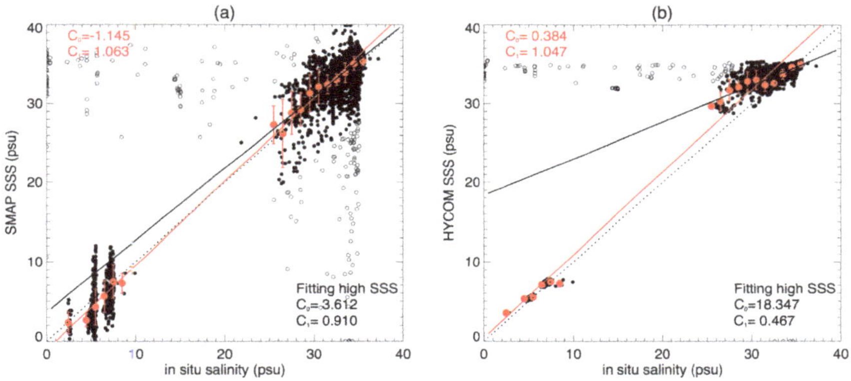

Figure 3. Scatterplot of in situ salinity collected within 5 m from surface and collocated data of (**a**) SMAP SSS and (**b**) HYCOM SSS. Collocations are based on gridded data on 0.25° grids averaged daily from April 2015 to March 2018. The three clusters associated with low in situ salinity (2–7 psu) are located in the Gulf of Bothania and the Baltic Sea. Open circles indicate where the absolute difference between the pair exceeds three times of the standard deviation. Linear regression of all data (excluding outliers) bin-averaged in terms of in situ SSS (bin size of 1 psu) is illustrated in red with zero intercepted denote by C_0 and slope by C_1, while the regression of high SSS (>25 psu, excluding outliers and the low salinity clusters.) is illustrated in black. Red dots indicate bin-average and vertical error bars indicate ±1 standard deviation in the bin. Dotted line indicates the diagonal for reference.

Significant outliers of SMAP SSS are seen as being associated with the high SSS group in two branches, with positive and negative biases, respectively, where positive bias may indicate an over estimation of the surface roughness correction, and the negative bias may suggest un-detected ice contamination. This will be considered for the next version of SMAP geophysical model function to improve the SSS retrieval algorithm. Note that there is no negative bias in HYCOM for the high SSS group. However, there is similar positive bias in the HYCOM versus in situ. After excluding the outliers, the bias, standard deviation, and RMSD between 19,543 pairs of SMAP SSS and in situ are 0.385, 0.987, and 1.060 psu with correlation of 0.82.

North of the Arctic Circle, the number of collocated points drops more than 70%. Table 1 also provides of statistical results North of 65°N. The bias and RMSD between 5712 pairs of SMAP and in situ after removing outliers are 0.339 and 1.227 psu, respectively.

After excluding the outliers, we performed a linear regression of the SMAP (or HYCOM) vs. in situ data. The regression slope for SMAP is close to 0.9 if only the data for in situ greater than 25 psu are considered (Figure 3). If all data are included, then the regression slope is 1.06. Interestingly, HYCOM

SSS seems to have a systematic positive bias for in situ SSS in the range of 25 to 30 psu, resulting in a reduced regression slope of 0.467 for the high SSS group of data. The SMAP SSS conditioned on the in situ within this range has a smaller bias, although being more noisy.

3.2. The Arctic Ocean SSS and Sea Ice

Figure 4 illustrates the monthly mean SSS from 2015 to 2017 in August, when the Arctic Ocean ice has the minimum extent with maximum ice-free surface areas for satellite salinity retrieval. The large-scale salinity feature, i.e., saltier Atlantic and fresher Pacific is consistently observed in three consecutive years and it agrees with the known contrast between the two ocean basins. North of the Arctic Circle (~66°N), however, SMAP SSS shows large interannual variations, not only in magnitudes of SSS retrieved, but also in areas with a valid retrieval. For example, North of the Bering Strait, in August 2015, SSS is retrieved in the large areas of the Chukchi Sea and the East Siberian seas; in August 2016 in the Beaufort Sea and the East Siberian Sea; and, in August 2017 areas with valid retrieval extended further north in all three marginal seas. This is due to the differing sea ice extent for the same months among three years, because SMAP SSS retrieval is only possible in open water. For the regions with SSS retrieval in all three years, e.g., in the Hudson Bay or Kara Sea, SMAP SSS shows large interannual differences. In contrast, HYCOM SSS shows a much small variation in the marginal seas within the Arctic Circle. It is known that HYCOM uses climatology river discharge and its SSS is relaxed to a seasonal climatology; both may suppress the magnitude of interannual variations of SSS in HYCOM [71].

SMAP SSS in marginal seas near the ice edge often show very low values. These fresh signatures should be examined and validated very carefully because it could be a mixture of real signal and ice contamination. It is known that in seawater near the ice edge, sea ice formation, and melt have significant contributions to the mixed layer salt budget of the ambient waters, with growing importance toward the ice edge [59]. Sea ice concentration (SIC) averaged in the areas North of 65°N changes dramatically through the seasonal cycle, from 20% in October or November to more than 70% in winter (January–May) (blue curve in Figure 5). It appears that the seasonal cycle of SSS averaged North of 65°N closely follows that of SIC, with the lowest SSS being observed within one or two months of the minimum SIC. This covariance should be taken with a grain of salt though, since the low SSS in summer also includes the effect of river runoff, and the high SSS in winter largely comes from the Atlantic side when most marginal seas are frozen. However, it is evident that SMAP SSS (black curve in Figure 5) reveals large interannual variations that are consistent with the sea ice conditions. Note that while averaging of SIC was over the entire region including zero or non-zero SIC values, the averaging of SSS is obtained over ice-free areas, which itself changes with time. We found the ice-free area during the ice melting seasonal peak of 2017 is 5% and 10% more than the two previous seasons of 2015 and 2016 (red curves in Figure 5). Correspondingly, the Arctic Ocean averaged SMAP SSS is the lowest in August 2017 despite HYCOM SSS showing essentially no changes. To alienate the possibility that summer SSS minimum was dominated by a few extremely low SSS retrievals or river discharges, Figure 5 also shows the time series of SSS after excluding outliers, which is identified as where the absolute differences from the daily mean exceeding three times of the standard deviation in the averaging domain. After removing the outliers (about 1 to 5% of valid retrievals of the day), the domain averaged SSS increases by 1–2 psu in its summer low for both SMAP and HYCOM, indicating that most outliers are in the low end of SSS. It is important to note that with outliers removed, SMAP SSS still shows clear interannual variability with minimum of 2017 season more than 2 psu lower than that of 2016, in contrast to HYCOM SSS. The consistent seasonal variation of SSS and SIC suggests that SMAP SSS retrieval correctly characterized the effect of sea ice changes.

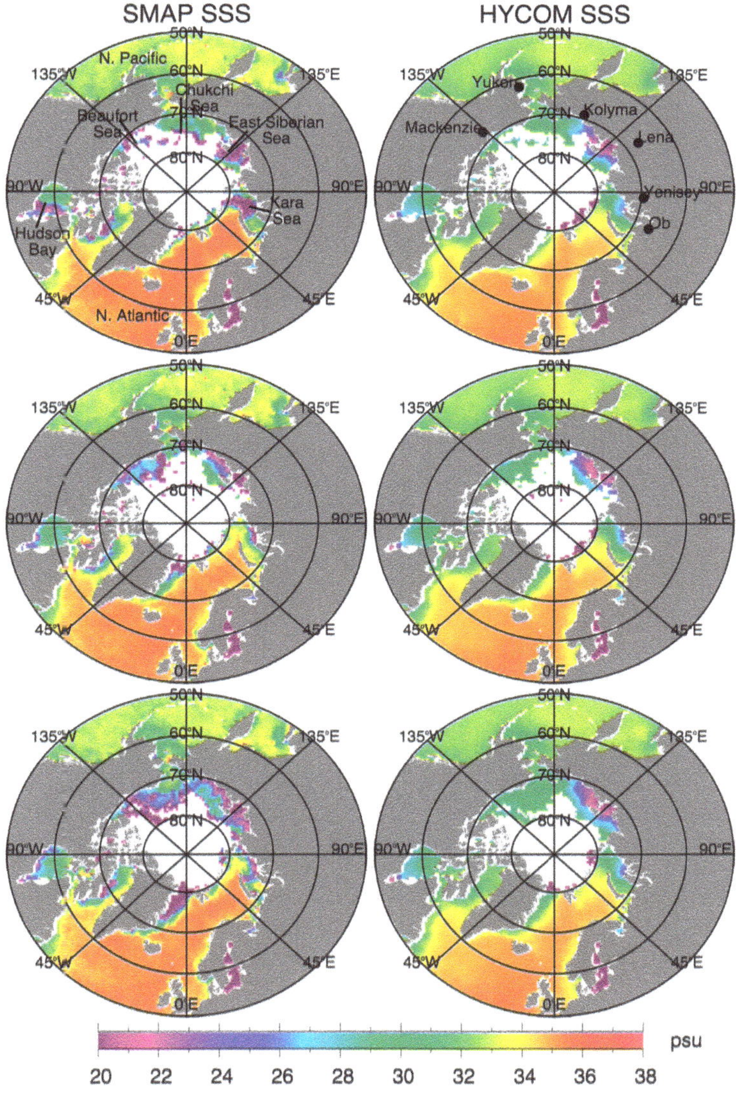

Figure 4. Salinity maps North of 50°N from SMAP (left) and HYCOM (right) for the months of August 2015, 2016, and 2017 (top to bottom). Marginal seas are indicated in the top left panel. Black dots in the top right panel indicate locations of river discharge data are collected.

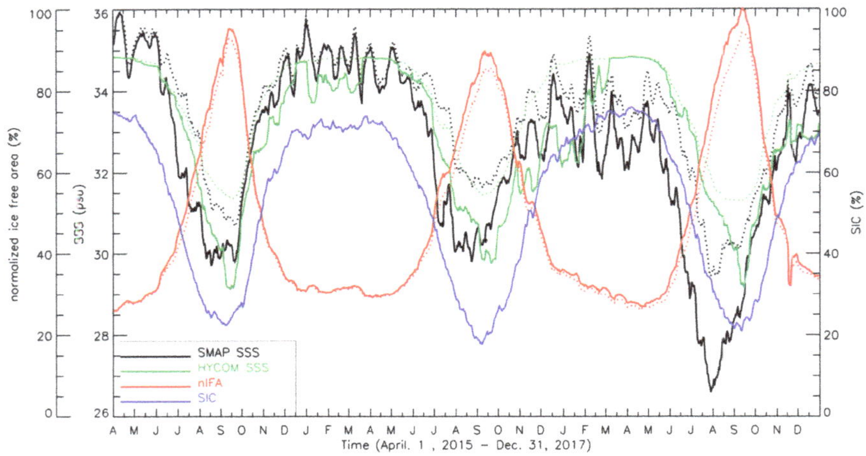

Figure 5. The time series of SMAP SSS (black) and HYCOM SSS (green) averaged in the domain North of 65°N in the ice-free-area (red) over all retrievals (solid line), or with outliers removed (dotted line). Outliers are identified as where the absolute differences from the daily mean exceeding three times of the standard deviation (~1 to 5%). Ice-free-area is normalized (nIFA) by the record maximum (observed in September 2017). Sea ice concentration (SIC) (blue) is averaged daily North of 65°N including zero and non-zero values.

However, this agreement is encouraging but insufficient to completely exclude the effect of un-detected ice-contamination impacting SMAP SSS. The appearance of frazil-pancake ice maybe increasing in the margins of the Arctic during autumn, as identified in a recent field campaign [16]. This type of new sea ice forms in the presence of incoming surface waves and it commonly occurs in the wave-dominated marginal ice zone of the Southern Ocean. With the retreat of the Arctic summer ice edge, the increased fetch has led to the increase of surface waves in the summer and fall, particularly in the Beaufort-Chukchi seas. Frazil-pancake ice may present a challenge to identify, say with passive microwave, as the sea ice concentrations may be quite variable over short time scales, which may lead to contamination in the SSS fields. Frazil-pancake ice has long been identifiable in SAR imagery. One recent study [72] describes the validation of a method to estimate the thickness of this young ice type based on wave dispersion. Being able to identify this form of new ice in the Arctic along with other sea ice conditions (e.g., melt, thin ice thickness, eddy formation, presence of waves) and the extent within the marginal ice zone will improve the accurate detection of ice and reduce the uncertainty in the satellite SSS retrieval.

3.3. SSS and River Discharge

As indicated in Section 3.1, there are very few in situ salinity data in the marginal seas within the Arctic Circle: for example, in the Hudson Bay and Kara Sea, no in situ data during the three years of SMAP were identified. In this section, we explore an alternative validation by examining the change of SMAP SSS in response to independently measured river discharge data. Massive northern rivers transport huge quantities of water from the continents to the Arctic Ocean. The freshwater inputs that are associated with river runoff should be reflected in the SSS field, particularly near the river mouths and over the shallow shelf areas. Due to its proximity to the two major Arctic rivers (Ob' and Yenisey), the Kara Sea provides an ideal case to examine SSS response to river discharge.

Figure 6 illustrates the evolution of SMAP SSS over the Kara Sea for the warm season (May to October) of 2015 and 2016. No SSS were retrieved in the Kara Sea before June and after October because it was completely covered by sea ice. The first valid SSS retrievals appeared in June, but at different locations in 2015 and 2016 due to the different spatial coverage of sea ice. The differences between these two years became more dramatic throughout the season. In the summer of 2015, the freshwater patch first appeared in June, in the area east of the northern tip of the Novaya Zemlya archipelago. It grew and spread to cover almost half of the Kara Sea in August. In contrast, in the season of 2016, the freshwater signature was limited to areas near Ob' and Yenisey Bay. Because the whole region has become almost ice-free after July, the impact of new freshwater inputs from sea ice melt is likely to be minimal. Therefore, the dramatic freshening signature spreading through middle of the Kara Sea from July to September in 2015 and along the Siberia coast in 2016, must have originated from the other freshwater source—river discharge. Below, we show that the contrast in SMAP SSS clearly reflects the differences in river runoff in those two years.

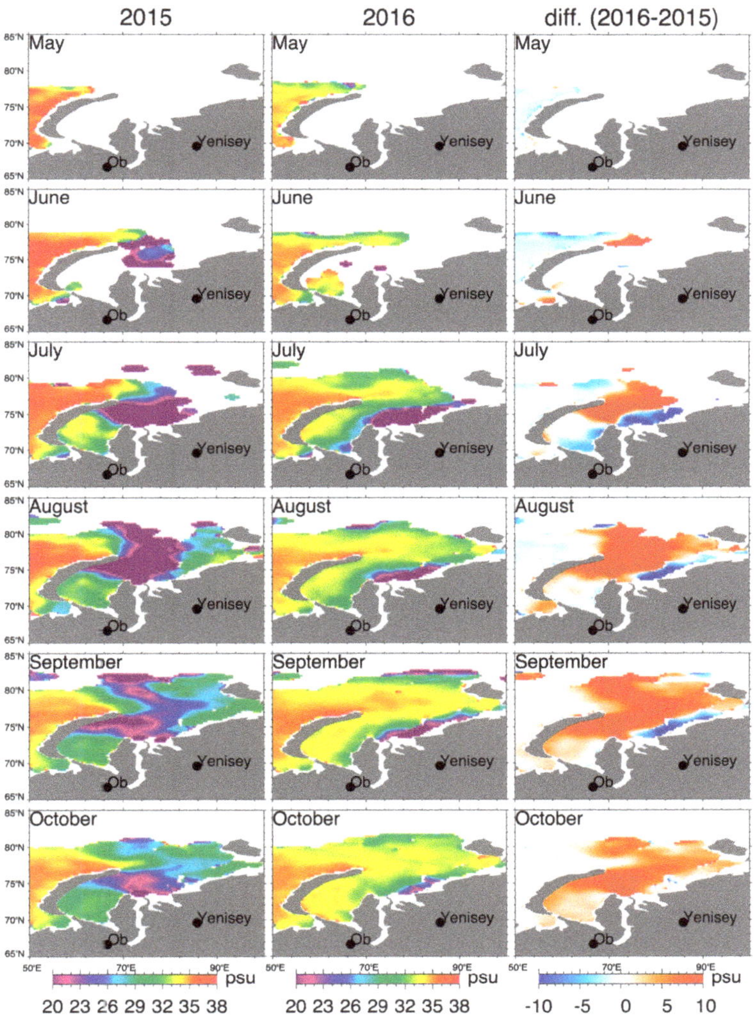

Figure 6. SSS in Kara Sea for the months from May to October (top to bottom) for the year 2015 (left) and 2016 (middle). The corresponding differences (2016 minus 2015) are shown in the right column. Sampling locations of the Ob and Yenisey River are indicated by black dots.

As a reference, Figure 7 illustrates the SSS evolving patterns in the Kara Sea from HYCOM SSS. The anomaly in HYCOM SSS is an order of magnitude smaller. As mentioned before, HYCOM is forced by climatological river discharges and its SSS is relaxed to a seasonal climatology [71]. These two climatological forcing would suppress the magnitude of interannual variations of SSS in HYCOM for regions without in-situ data or where in-situ data are insufficient to constraint the model SSS, such as the Kara Sea.

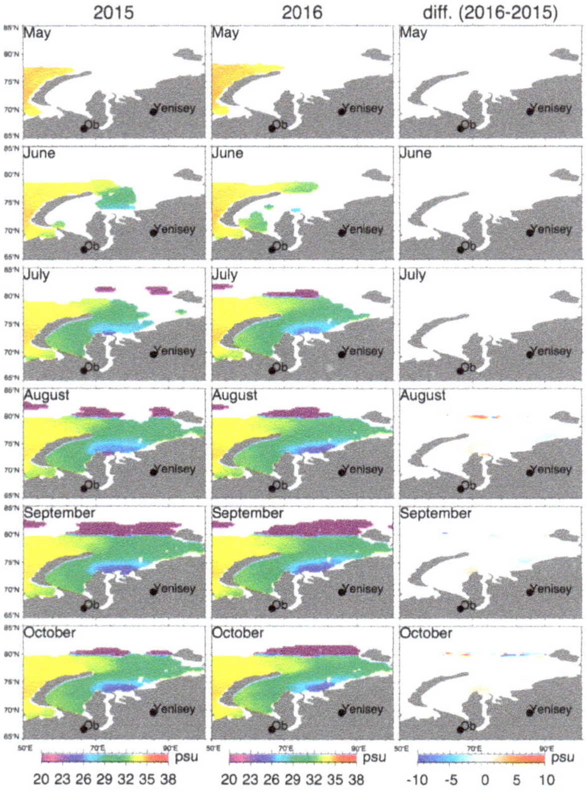

Figure 7. Similar as Figure 6 for HYCOM SSS.

In the surface layer of Kara Sea ocean currents carry waters from the River Ob′ and River Yenisey to the north and northeast. Based on the SMAP SSS observations in the Kara Sea (Figure 6), we find that river discharge adjacent to the Kara Sea must be much larger in 2015 than in 2016. The additional freshwater in 2015 was transported northward and further spread along the path, while the spreading of freshwater discharge in 2016 was limited along the coast.

Indeed, the daily discharge data from the Ob′ River and the Yenisey River (Figure 8) show differences that are consistent with SMAP observation. In the first two months of the warm season (May and June), discharge from Ob′ shows a similar magnitude for the two years, while Yenisey discharge peaked early and injected 93.4 km^3 more water in 2015 than 2016. In the following four months from July to October, Ob′ became the main player, putting 90.9 km^3 more freshwater into the Kara Sea, while Yenisey added another extra 31.5 km^3 (Table 2). Combining the discharges from the Ob′ and Yesiney during May to October together, the Kara Sea received more than 210 km^3 extra freshwater in 2015 relative to 2016.

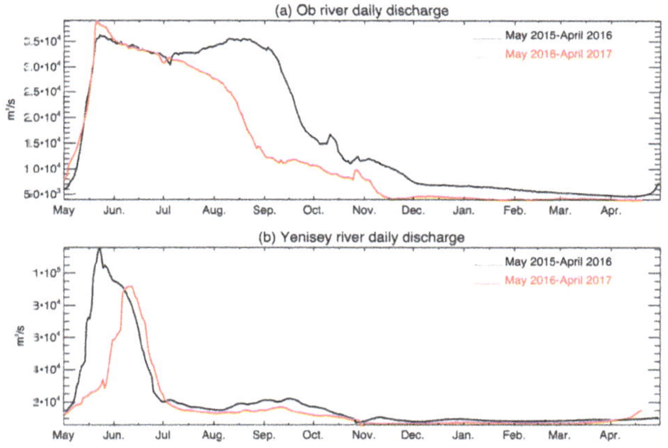

Figure 8. Daily river discharge from (**a**) Ob′ and (**b**) Yenisey Rivers during May 2015 to April 2016 (black) and from May 2016 to April 2017 (red).

Table 2. Comparison of Ob′ and Yenisey River integrated discharge for the warm seasons of 2015 and 2016.

Unit: km^3	May–June		July–October	
	Ob′	Yenisey	Ob′	Yenisey
2015	148.89	335.17	285.03	184.17
2016	154.11	241.78	194.12	152.70
Δ (2015 minus 2016)	−5.22	93.39	90.90	31.47
Δ (Ob′ & Yenisey)	88.17		122.37	

It might be useful to roughly estimate the effect of this extra amount of freshwater on the SSS anomaly. About 3 cm of freshwater are needed to dilute 1 m of seawater with a change of salinity by 1 psu. Assuming the extra ~210 km^3 freshwater spread over half of the Kara Sea (total surface area 926,000 km^2), it may produce 15 psu salinity changes within top 1 m surface water layer, or 7.5 psu within top 2 m. As seen in Figure 6 (right column), the areas with positive SSS differences that are exceeding 10 psu covering about half of the Kara Sea areas, which is in the same order of magnitude as the freshening effect that was possibly produced by the river discharge difference in the two years. It is understood that large differences in SSS anomaly may depend on how the discharge freshwater transported horizontally and vertically, in terms of depth and spread of the diluted water body. Nevertheless, the agreement in order of magnitude is encouraging.

3.4. SSS Variability at Arctic Ocean Gateways

We have demonstrated that SMAP SSS retrieves reasonably good quality data in the Arctic Ocean in terms of comparisons with in situ salinity data (Section 3.1) and in response to river discharge (Section 3.3). Here, we further explore the feasibility of using satellite SSS to monitor the surface salinity variability at Arctic Ocean Gateways. We examine if the SSS variability at the gateways are significantly greater than the retrieval accuracy of 1 psu.

The Arctic Ocean exchanges freshwater with the sub-oceans through four major gateways, as shown schematically in the currents map (Figure 9a from Figure 2a in [14]): the Bering Strait inflows of relative fresh Pacific waters; the Barents Sea Opening (BSO) and part of the Fram Strait inflow of the salty Atlantic water; and, through the Davis Strait and the part of the Fram Strait between Greenland and Svalbard, which comprise the major outflow locations of water modified by the Arctic Ocean in addition to freshwater that is associated with sea ice flux and subsequent melt [12,14]. Because the general directions of the ocean currents at gateways are known, the surface salinity variability observed at the gateways carries the information of the freshwater passing through the regions in the upper ocean.

We first describe the procedure to extract SSS information at each of the Arctic gateways. The exact locations of the gateways defined for this study are given in Table 3 and are indicated in Figure 9b. To ensure currents flow through the passage in a roughly unique direction, we divided the Fram Strait into two parts: one between east Greenland coast and 0° longitude (named EG), the other from 0° longitude to the west coast of Svalbard (named Fram Strait). We also added a section to extend the narrow Bering Strait from 62°N to 68°N so sufficient grid points can be included in the calculation. According to this definition, a maximum of around 250 to 500 grid points can be extracted from SMAP L3 daily maps with 0.25° spacing, as listed in Table 3. The actual number of grid points with valid SSS retrieval changes with time depending on the ice situation, which varies with the gateway. This is illustrated in terms of the percentage of the ice-free area at each gateway (Figure 10, right column). We note that BSO and the Fram Strait are mostly ice-free year round, while EG and the Davis Strait are only open for about three months in summer, with more than 50% ice cover the rest of the year. The Bering Strait, which is the only gateway connecting the Arctic Ocean and Pacific Ocean, opens about half of the year from June to December. The time series of SSS are obtained by daily averaging over all grid points with valid retrievals at each gateway. In the extremely challenging environment of the Arctic Ocean, we designed two criteria to avoid possible false retrievals. First, we include in the time series calculation only if >30% of the grid points are ice-free. The daily mean and standard deviation were obtained from 30% or more grid points at the gateway with valid SSS data, named $SSS_{Gateway}(t)$ and $\delta_{Gateway}(t)$. Examining the time series, we found extremely large values of $\delta_{Gateway}(t)$ often occurred at seasonal transitions of ice melting or freezing, suggesting the possibility of undetected ice contamination. Therefore, we defined the second criteria to exclude those outliers. We calculate the standard deviation of $SSS_{Gateway}(t)$ over the whole period ($\sigma_{Gateway}$), and defined outliers as those with $\delta_{Gateway}(t)$ exceeding twice of $\sigma_{Gateway}$. After filtering with these two criteria, we apply a 30 days moving average on $SSS_{Gateway}(t)$.

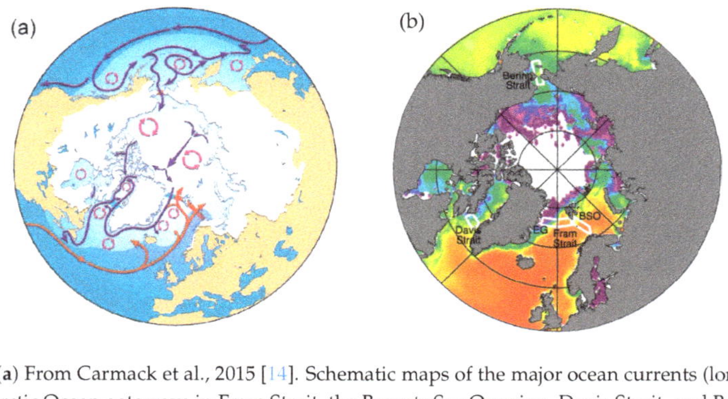

Figure 9. (**a**) From Carmack et al., 2015 [14]. Schematic maps of the major ocean currents (long arrows), the four Arctic Ocean gateways in Fram Strait, the Barents Sea Opening, Davis Strait, and Bering Strait (thick bars with red denoting inflow and blue denoting outflow), the gyral circulation patterns (circular arrows), the salt-stratified ocean domains are shown in light blue, and the All Arctic Regions definition of the terrestrial contributing areas shown in white. (**b**) Locations of the Arctic Gateways indicated on SMAP SSS gridded fields (three years mean), which were used in the calculation of mean SSS and the percentage of ice-free area (Figure 10). In this study, the Fram Strait is divided into two parts where currents flow in opposite directions: one between east Greenland coast and 0° longitude, the other from 0° longitude to the west coast of Svalbard. Exact coordinates of the four corners of the polygon used for each gateway are given in Table 3.

Table 3. List of the mean, standard deviation, minimum and maximum values of SMAP SSS retrieved at the Arctic Gateways from 1 May 2015 to 30 April 2018.

		BSO	Fram Strait	E Greenland	Davis Strait	Bering Strait
* Gateway location (lon, lat)	UL	(17°E, 75°N)	(0°E, 77°N)	(344°E, 77°N)	(62°W, 66°N)	(171°W, 68°N)
	UR	(19°E, 77°N)	(15°E, 77°N)	(360°E, 77°N)	(54°W, 66°N)	(167°W, 68°N)
	LL	(27°E, 73°N)	(0°E, 75°N)	(344°E, 75°N)	(62°W, 64°N)	(171°W, 62°N)
	LR	(29°E, 71°N)	(15°E, 75°N)	(360°E, 75°N)	(54°W, 64°N)	(167°W, 62°N)
Number of grid points over the gateway		512	480	512	256	384
Mean SSS (psu)		35.1255	34.8499	31.2173	32.3675	31.6195
SSS Std. (psu)		0.3145	0.5222	2.3011	1.0619	1.5067
Min. SSS (psu)		33.8045	32.6700	22.7833	29.4338	23.4985
Max. SSS (psu)		35.7872	35.6629	35.0302	34.2970	33.3319
Number of valid SSS		871	1023	959	660	621

* Gateway locations are given in (longitude, latitude) for upper-left (UL), upper-right (UR), low-left (LL), and low-right (LR) corners.

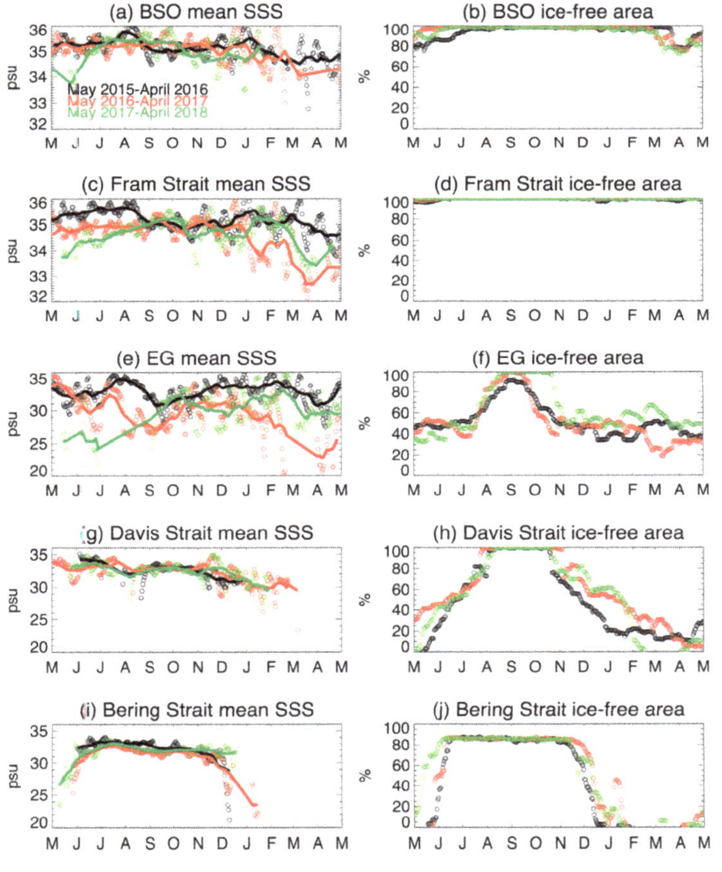

Figure 10. Time series at Arctic Ocean gateways of 1 May 2015–30 April 2016 (black), 1 May 2016–30 April 2017 (red), and 1 May 2017–30 April 2018 (green) of (left) SMAP SSS averaged and (right) percentage of the ice-free area over Arctic gateways defined in Figure 9b. (**a**,**b**) The Barents Sea Opening (BSO); (**c**,**d**) the Fram Strait east of 0° longitude; (**e**,**f**) the Fram Strait opening between east coast of Greenland and Svalbard (EG); (**g**,**h**) the Davis Strait; and, (**i**,**j**) the Bering Strait. 30 days moving average is applied on SSS time series (left column, thick lines).

The time series of SSS at Arctic gateways that were observed by SMAP during the past three summers reveal very rich information (Figure 10, left column). First, it correctly reflects the typical characteristics of the freshwater sources from different ocean basins: relative salty water from the Atlantic and fresh water from the Pacific, with more than 3 psu differences between mean SSS_{BSO} and SSS_{Bering} (Table 3). SSS_{Fram} is similar to SSS_{BSO} in terms of mean and variability; both bring warm and salty Atlantic water into the Arctic Ocean. Also note, SSS_{Fram} is slightly fresher than SSS_{BSO}, possibility due to its vicinity to Greenland. On the other hand, SSS_{EG} and SSS_{Fram}, although being geographically adjacent to each other, bear dramatic contrast in the properties of water transport out from (SSS_{EG}, mean 31.2 psu) and into (SSS_{Fram}, mean 34.8 psu) the Arctic Ocean. The much fresher signature and larger dynamical range of SSS_{EG} suggest the impact of Arctic ice. We note the exceptionally large drop in SSS_{Fram} (~3 psu) and SSS_{EG} (~6 psu) from late 2016 to early 2017. It is unlikely that this large interannual change is caused by the SMAP algorithm or long-term calibration changes because the changes at BSO are much smaller. At this moment, we do not have enough data to pin point the exact mechanism causing such freshening. Also, the magnitude of SSS_{EG} may need further calibration due to less coverage (~40% ice-free area). We speculate that this freshening and recovering in the later part of 2017 is influenced by the Greenland sea ice melting and freezing. This hypothesis may be supported by the observation that the Davis Strait, another passage like EG where outflows of Arctic Ocean modified water mixed with sea ice drift to the northern Atlantic. Different from SSS_{EG}, SSS_{Davis} shows similar seasonal cycles for the three summers that were captured by SMAP.

Consistent seasonal cycles are also observed over the Bering Strait, with a peak-to-peak range of about 4 psu. It is also interesting to note the inter-annual difference. SSS_{Bering} in June–December 2016 (red curve in Figure 10i) is exceptionally low, about 2 psu less than the same period in 2015 (black curve in Figure 10i); while it recovered partially one year later, as seen in Figure 10i, the green curve (June–December 2017) falls between the previous two years. Year 2016 was the strongest *El Niño* in recent two decades. One well-known characteristic of an *El Niño* event is the zonal displacement of the western equatorial Pacific warm/fresh pool. The edge of the pool extends eastward during *El Niño*, retreats westward during *La Niña* [73–75].

Moorings are installed at all major Arctic Ocean gateways, measuring salinity at depth ~50 m year round At the close of this study, we only obtained mooring data at the Bering Strait up to the summer of 2016 [24,69], with about one year overlapping with SMAP data. Figure 11 illustrates the time series of SMAP SSS_{Bering} (from Figure 10i) along with salinity form June 2015 to June 2016 at three mooring sites: A2 (66°19′N, 168°57′W) and A3 (65°46′N, 168°34′W) in the US channel of the strait, and A4 (65°44′N, 168°15′W) close to the Alaskan coast. All three mooring sites are within the white polygon, defined as the Bering Strait (Figure 9b). They are within 90 km from each other. During August–November, the salinities from the three moorings were mostly within 1 psu from each other, indicating a relatively small spatial variability. However, the A4 mooring data reduced by about 5 psu from November to December, whereas the other two moorings had smaller changes. The SMAP SSS_{Bering} at 1–2 cm depth depicted a change of about 5 psu, well resolved by the 1 psu accuracy (Section 3.1), and appeared to have a consistent trend with the A4 mooring on the amplitude and timing of the freshening peak in December 2015. Nevertheless, the differences with the other two moorings were clearly larger. A2 and A3 moorings revealed less variability relative to SMAP, which is likely due to the vertical and horizontal dilution from the surface to 50 m depth. A longer time series of data is required to bring a more definite conclusion on the consistency or discrepancy, possibly resulting from differences in sensing depth or spatial variability.

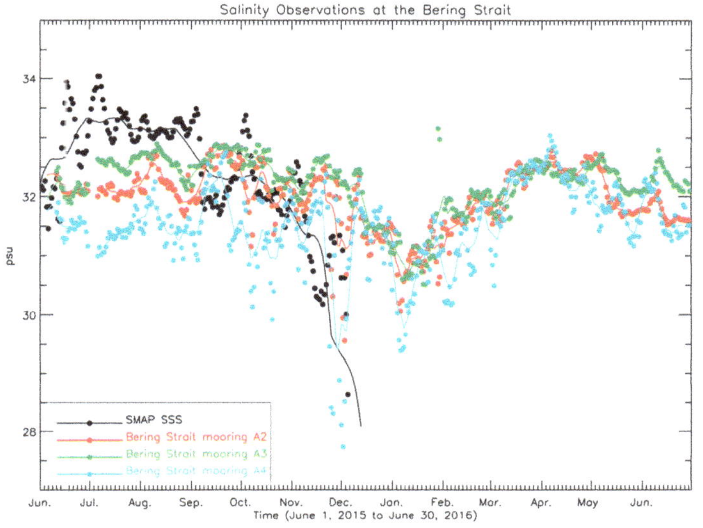

Figure 11. Daily time series of salinity at the Bering Strait retrieved from SMAP (black), and mooring measurement at mooring sites A2 (red), A3 (greed) and A4 (blue). Daily data are indicated by solid dots and lines illustrate the 8 days moving average.

4. Discussion

This study explores the utility of SMAP SSS in monitoring the Arctic freshwater system. We emphasize the importance to assess the uncertainty and understanding of the limitations of satellite SSS at this explorative stage in the extremely challenging environment of the Arctic Ocean. We discuss several related issues, including the complementarity of satellite SSS with other satellite and in-situ observations.

First, we point out the urgent need for in situ measurements of salinity in the polar regions to enhance satellite salinity products. Currently, there are very few in situ salinity data available in the areas North of the Arctic Circle (Section 3.1). A sufficiently large database of in situ salinity data (in conjunction with simultaneous geophysical information, such as SST, wind speed and direction, air temperature, and ocean current) will be invaluable. It will not only provide a robust assessment of satellite SSS uncertainties in the Arctic Ocean, but also support the retrieval algorithm refinement, including the SST effect on the roughness correction in cold water.

Second, improved retrievals of sea ice concentration would be of considerable value to reduce possible ice-contaminated SMAP SSS retrievals. It is known that the quality of sea ice concentration product degrades in thin ice, especially at the ice margins and during summer melt. The method of identifying Frazil-pancake ice based on SAR imagery [72] with the wave dispersion technique may have an advantage over passive microwave, as the sea ice concentrations may be quite variable over short time scales [76]. The added information of thickness of this type of young ice [72] will allow quantitative analyses of the SSS response to ice formation and related salt rejection during ice growth on the mixed layer budget.

Third, integrated data analysis is important and necessary to realize the benefit of satellite SSS. One example is using satellite SSS to trace the river discharge. Because satellite observes salinity in the top 1–2 cm of sea surface, SMAP SSS only reveals freshwater variability in the top layer. In stratified regions, SSS alone cannot give the volume of the freshwater distribution, which is traced by satellite from the surface, such as in the Kara Sea (Section 3.3). Quantitative estimation of river discharge impact and distribution over the Arctic Ocean can only be achieved by combining SSS with the salinity stratification structure of the upper ocean layer, based on in situ measurements of salinity profiles and ocean circulation models. Another example is the need of ocean currents in estimating freshwater transport. We demonstrated that the SMAP SSS variability could be used to discern the freshwater transport through Arctic Ocean gateways. However, to quantify the freshwater transport that is associated with those SSS variability, one also needs information of the magnitude and direction of ocean currents in addition to salinity profiles.

Finally, we want to emphasize that the value of SMAP SSS relative to data assimilation products, such as HYCOM, cannot be judged from the comparison with co-located in-situ measurements alone. We have shown (e.g., Section 3.3) that SMAP data reveal much stronger interannual variation of SSS than HYCOM. Given HYCOM's relaxation of SSS to seasonal climatology and the use of seasonal climatology of river discharges, SMAP SSS may provide a more reliable dataset to study interannual freshwater changes of the Arctic Ocean in regions where no or few in-situ data are available to constrain HYCOM.

5. Conclusions

Here, we demonstrated the potential utility of SMAP SSS in studying Arctic Ocean freshwater changes. The JPL SMAP SSS team adopts an advanced retrieval algorithm near the land and ice edge, aiming to provide the scientific community potentially useful information. We assessed the accuracy of SMAP SSS retrieved North of 50°N through comparative analysis with in situ salinity data collected by Argo floats, ships, gliders, and in field campaigns. Results that were derived from near 20,000 pairs collocated within a 12.5-km radius and daily time window indicate a RMSD less than ~1 psu and a correlation coefficient ~0.82. We have demonstrated the feasibility of using SMAP SSS to study Arctic Ocean salinity response to changes in sea ice concentration and river discharge. In the Arctic Ocean, SMAP SSS reveals seasonal and interannual variations that are consistent with sea ice concentration. In the Kara Sea, the large SSS contrast between 2015 and 2016 observed by SMAP is corroborated by the anomalous freshwater inputs from the Ob' and Yenisey Rivers. We also demonstrated the feasibility to use SSS as a proxy in monitoring the upper layer freshwater exchanges between the Arctic and sub-Arctic Oceans and through the four major Arctic Ocean gateways.

Results presented in this paper may lead to interesting future studies. Particularly, to understand the surface salinity variability that is observed by SMAP at Arctic Gateways, it will be interesting to examine its relation with other variables, such as SST, wind and current, locally and beyond. In conjunction with in situ data, satellite SSS can provide complementary and critical surface information to advance our knowledge on the linkage between the climate changes in polar regions and sub-oceans.

Author Contributions: Data curation, W.T., A.F. and A.H.; Formal analysis, W.T. and S.Y.; Investigation, W.T. and S.Y.; Visualization, W.T.; Writing—original draft, W.T.; Writing—review & editing, W.T., S.Y., D.Y., A.F., A.H., T.L., S.F. and B.H.

Acknowledgments: This research was carried out at the Jet Propulsion Laboratory, California Institute of Technology, under a contract with the National Aeronautics and Space Administration. Environment and Climate Change Canada provided necessary support to this research collaboration. The SMAP SSS products were provided by the NASA Physical Oceanography Distributed Active and Archive enter (PO.DAAC) at the Jet Propulsion Laboratory. The in situ salinity data were obtained from USGODAE (The US Global Ocean Data Assimilation Experiment), CMEMS (the Copernicus Marine Environment Monitoring Service), and NASA OMG (Ocean Melting Greenland) mission. We also thank Bastien Queste of University of East Anglia, UK for providing the salinity data from seaglider missions. The sea ice concentration data was provided by the National Snow and Ice Data Center (NSIDC). The Arctic river discharge data was provided by the Arctic Great Rivers Observatory at Woods Hole Research Center. The Bering Strait mooring data was provided by the Applied Physics Laboratory at University of Washington. Authors thank the reviewers for their constructive comments and suggestions.

Conflicts of Interest: The authors declare no conflict of interest.

Abbreviations

AON	Arctic Observing Network
APL	Applied Physics Laboratory
Arctic-GRO	Arctic Great River Observatory
AXCTD	Airborne eXpandable CTD
BSO	The Barents Sea Opening
CMEMS	Copernicus Marine Environment Monitoring Service
CTD	Conductivity Temperature Depth
CONAE	Comision Nacional de Actividades Espaciales
ESA	The European Space Agency
GOOS	Global Ocean Observing System
HYCOM	Hybrid Coordinate Ocean Model
INS TAC	In Situ Thematic Assembly Centre
JPL	Jet Propulsion Laboratory
LUT	look-up-table
NASA	The National Aeronautics and Space Administration
NCEP	National Centers for Environmental Prediction
NRT	near-real-time
NSF	The National Science Foundation
NSIDC	The National Snow and Ice Data Center
OMG	Ocean Melting Greenland
PARTNERS	Pan-Arctic River Transport of Nutrients, Organic Matter, and Suspended Sediments
PO.DAAC	Physical Oceanography Distributed Active Archive Center
RMSD	Root Mean Square Difference
ROOS	Regional Ocean Observing System
RSS	Remote Sensing System
SIC	sea ice concentration
SMAP	Soil Moisture Active Passive
SMOS	Soil Moisture and Ocean Salinity
SSS	sea surface salinity
SST	sea surface temperature
SWC	salinity-wind-cell
TB	brightness temperature
TSG	Thermosalinograph
USGODAE	The US Global Ocean Data Assimilation Experiment
UTC	The Coordinated Universal Time
XCTD	eXpendable CTD

References

1. Kwok, R.; Cunningham, G.F.; Wensnahan, M.; Rigor, I.; Zwally, H.J.; Yi, D. Thinning and volume loss of Arctic sea ice: 2003–2008. *J. Geophys. Res.* **2009**, *114*, C07005. [CrossRef]
2. Comiso, J.C. A rapidly decline perennial sea ice cover in the Arctic. *Geophys. Res. Lett.* **2002**, *29*, 1956. [CrossRef]
3. Cavalieri, J.D.; Parkinson, C.L. Arctic sea ice variability and trends. 1979–2010. *Cryosphere* **2012**, *6*, 881–889. [CrossRef]
4. Proshutinsky, A.; Krishfield, R.; Timmemans, M.-L.; Toole, J.; Carmack, E.; McLaughlin, F.; Williams, W.J.; Zimmermann, S.; Itoh, M.; Shimada, K. The Beaufort Gyre Fresh Water Reservior: State and variability from observations. *J. Geophys. Res.* **2009**, *114*, C00A10. [CrossRef]
5. Haine, T.W.N.; Curry, B.; Gerdes, R.; Hansen, E.; Karcher, M.; Lee, C.; Rudels, B.; Spreen, G.; Steur, L.; Stewart, K.D.; et al. Arctic freshwater export: Status, mechanisms, and prospects. *Glob. Planet Chang.* **2015**, *125*, 13–35. [CrossRef]

6. Rage, B.; Karcher, M.; Schauer, U.; Toole, J.; Krishfield, R.; Pisarev, S.; Kauker, F.; Gerdes, R.; Kikuchi, T. An assessment of Arctic Ocean freshwater content changes from the 1990s to 2006–2008. *Deep Sea Res.* **2011**, *58*, 173–185. [CrossRef]
7. McLaughlin, F.A.; Carmack, E.C.; Williams, W.J.; Zimmerman, S.; Shimada, K.; Itoh, M. Joint effects of boundary currents and thermo-haline intrusions on the warming of Atlantic water in the Canada Basin, 1993–2007. *J. Geophys. Res.* **2009**, *114*, C00A12. [CrossRef]
8. Polyakov, I.V.; Pnyushkov, A.V.; Timokhov, T.A. Warming of the intermediate Atlantic Water of the Arctic Ocean in the 2000s. *J. Clim.* **2012**, *25*, 8362–8370. [CrossRef]
9. Rawlins, M.A.; Steele, M.; Holland, M.; Adam, J.; Cherry, J.; Francis, J.; Groisman, P.; Hinzman, L.; Huntington, T.; Kane, D.; et al. Analysis of the Arctic System for Freshwater Cycle Intensification: Observations and Expectations. *J. Clim.* **2010**. [CrossRef]
10. Willis, J.K.; Rignot, E.; Nerem, R.S.; Lindstrom, E. Introduction to the special issue on ocean-ice interaction. *Oceanography* **2016**, *29*, 19–21. [CrossRef]
11. Fenty, I.; Willis, J.K.; Khazendar, A.; Dinardo, S.; Forsberg, R.; Fukumori, I.; Holland, D.; Jakobsson, M.; Moller, D.; Morison, J.; et al. Oceans Melting Greenland: Early results from NASA's ocean-ice mission in Greenland. *Oceanography* **2016**, *29*, 72–83. [CrossRef]
12. Prowse, T.; Bring, A.; Mård, J.; Carmack, E. Arctic Freshwater Synthesis: Introduction. *J. Geophys. Res. Biogeosci.* **2015**, *120*, 2121–2131. [CrossRef]
13. Prowse, T.; Bring, A.; Mård, J.; Carmack, E.; Holland, M.; Instanes, A.; Vihma, T.; Wrona, F.J. Arctic Freshwater Synthesis: Summary of key emerging issues. *J. Geophys. Res. Biogeosci.* **2015**, *120*, 1887–1893. [CrossRef]
14. Carmack, C.E.; Yamamoto-Kawai, M.; Haine, T.W.N.; Bacon, S.; Bluhm, B.A.; Lique, C.; Melling, H.; Polyakov, I.V.; Straneo, F.; Timmermans, M.-L. Freshwater and its role in the Arctic Marine System: Sources, disposition, storage, export, and physical and biogeochemical consequences in the Arctic and global oceans. *J. Geophys. Res. Biogeosci.* **2016**, *121*. [CrossRef]
15. Lique, C.; Holland, M.M.; Dibike, Y.B.; Lawrence, D.M.; Screen, J.A. Modeling the Arctic freshwater system and its integration in the global system: Lessons learned and future challenges. *J. Geophys. Res. Biogeosci.* **2016**, *121*, 540–566. [CrossRef]
16. Thomson, J.; Ackley, S.; Girard-Ardhuin, F.; Ardhuin, F.; Babanin, A.; Boutin, G.; Brozena, J.; Cheng, S.; Collins, C.; Doble, M. Overview of the Arctic Sea State and Boundary Layer Physics Program. *J. Geophys. Res. Oceans* **2018**. [CrossRef]
17. Durack, P.J. Ocean salinity and the global water cycle. *Oceanography* **2015**, *28*, 20–31. [CrossRef]
18. Durack, P.J.; Wijffels, S.E.; Matear, R.J. Ocean Salinities Reveal Strong Global Water Cycle Intensification During 1950 to 2000. *Science* **2012**, *336*, 455–458. [CrossRef] [PubMed]
19. Schmitt, R.W. Salinity and the global water cycle. *Oceanography* **2008**, *21*, 12–19. [CrossRef]
20. Schmitt, R.W. The ocean component of the global water cycle: US National Report to International Union of Geodesy and Geophysics, 1991–1994. *Rev. Geophys.* **1995**, *33* (Suppl. 1), 1395–1409. [CrossRef]
21. Wijffels, S.E.; Schmitt, R.W.; Bryden, H.L.; Stigebrandt, A. On the transport of fresh water by the oceans. *J. Phys. Oceanography* **1992**, *22*, 155–162. [CrossRef]
22. Timmermans, M.-L.; Proshutinsky, A.; Golubeva, E.; Jackson, J.M.; Krishfield, R.; McCall, M.; Platov, G.; Toole, J.; Williams, W.; Kikuchi, T.; et al. Mechanisms of Pacific Summer Water variability in the Arctic's Central Canada Basin. *J. Geophys. Res. Oceans* **2014**, *119*, 7523–7548. [CrossRef]
23. Timmermans, M.-L.; Proshutinsky, A.; Krishfield, R.A.; Perovich, D.K.; Richter-Menge, J.A.; Stanton, T.P.; Toole, J.M. Surface freshening in the Arctic Ocean's Eurasian Basin: An apparent consequence of recent change in the wind-driven circulation. *J. Geophys. Res.* **2011**, *116*, C00D03. [CrossRef]
24. Woodgate, R.A. Increases in the Pacific inflow to the Arctic from 1990 to 2015, and insights into seasonal trends and driving mechanisms from year-round Bering Strait mooring data. *Prog. Oceanogr.* **2018**, *160*, 124–154. [CrossRef]
25. Frajka-Williams, E.; Bamber, J.L.; Våge, K. Greenland melt and the Atlantic meridional overturning circulation. *Oceanography* **2016**, *29*, 22–33. [CrossRef]
26. Yang, Q.; Dixon, T.H.; Myers, P.G.; Bonin, J.; Chambers, D.; van den Broeke, M.R.; Ribergaard, M.H.; Mortensen, J. Recent increases in Arctic freshwater flux affects Labrador Sea convection and Atlantic overturning circulation. *Nat. Commun.* **2016**, *7*, 10525. [CrossRef] [PubMed]

27. Jackson, L.C.; Kahana, R.; Graham, T.; Ringer, M.A.; Woollings, T.; Mecking, J.V.; Wood, R.A. Global and European climate impacts of a slowdown of the AMOC in a high resolution GCM. *Clim. Dyn.* **2015**, *45*, 299–316. [CrossRef]
28. Leuliette, E.W.; Nerem, R.S. Contributions of Greenland and Antarctica to global and regional sea level change. *Oceanography* **2016**, *29*, 154–159. [CrossRef]
29. Agarwal, N.; Köhl, A.; Mechoso, C.R.; Stammer, D. On the early response of the climate system to a meltwater input from Greenland. *J. Clim.* **2014**, *27*, 276–296. [CrossRef]
30. Kerr, Y.H.; Waldteufel, P.; Wigneron, J.E.; Delwart, S.; Cabot, F.; Boutin, J.; Escorihuela, M.A.; Font, J.; Reul, N.; Gruhier, C. The SMOS mission: New tool for monitoring key elements of the global water cycle. *Proc. IEEE* **2010**, *98*, 666–687. [CrossRef]
31. Font, J.; Camps, A.; Borges, A.; Martin-Neira, M.; Boutin, J.; Reul, N.; Kerr, Y.H.; Hahne, A.; Mecklenburg, S. SMOS: The challenging sea surface salinity measurement from space. *Proc. IEEE* **2010**, *98*, 649–665. [CrossRef]
32. Le Vine, D.M.; Lagerloef, G.S.E.; Colomb, F.R.; Yeh, S.H.; Pellerano, F.A. Aquarius: An instrument to monitor sea surface salinity from space. *IEEE Trans. Geosci. Remote Sens.* **2007**, *45*, 2040–2050. [CrossRef]
33. Lagerloef, G.; Colomb, F.R.; le Vine, D.; Wentz, F.; Yueh, S.; Ruf, C.; Lilly, J.; Gunn, J.; Chao, Y.; deCharon, A.; et al. The Aquarius/Sac-D Mission: Designed to Meet the Salinity Remote-Sensing Challenge. *Oceanography* **2008**, *21*, 68–81. [CrossRef]
34. Yueh, S.H.; Tang, W.; Fore, A.; Neumann, G.; Hayashi, A.; Freedman, A.; Chaubell, J.; Lagerloef, G. L-band Passive and Active Microwave Geophysical Model Functions of Ocean Surface Winds and Applications to Aquarius Retrieval. *IEEE Trans. Geosci. Remote Sens.* **2013**, *51*, 4619–4632. [CrossRef]
35. Entekhabi, D.; Njoku, E.G.; O'Neill, P.E.; Kellogg, K.H.; Crow, W.T.; Edelstein, W.N.; Entin, J.K.; Goodman, S.D.; Jackson, T.J.; Johnson, J.; et al. The Soil Moisture Active Passive (SMAP) Mission. *Proc. IEEE* **2010**, *98*, 704–716. [CrossRef]
36. Fore, A.; Yueh, S.; Tang, W.; Stiles, B.; Hayashi, A. Combined Active/Passive Retrievals of Ocean Vector Wind and Sea Surface Salinity with SMAP. *IEEE Trans. Geosci. Remote Sens.* **2016**, *54*. [CrossRef]
37. Linkages of Salinity with Ocean Circulation, Water Cycle, and Climate Variability. Community White Paper in Response to Request for Information #1 by the US National Research Council Decadal Survey for Earth Science and Applications from Space 2017–2027. Available online: http://surveygizmoresponseuploads.s3.amazonaws.com/fileuploads/15647/2289356/66-d5c509554e258d30eb31a63804edbf70_LeeTong.docx (accessed on 5 May 2018).
38. Linkages of Salinity with Ocean Circulation, Water Cycle, and Climate Variability. Community White Paper in Response to Request for Information #2 by the US National Research Council Decadal Survey for Earth Science and Applications from Space 2017–2027. Available online: http://surveygizmoresponseuploads.s3.amazonaws.com/fileuploads/15647/2604456/107-1abc9aa1a37ab7e77d91d86598954a50_LeeTong.pdf (accessed on 5 May 2018).
39. Lang, R.; Zhou, Y.; Utku, C.; le Vine, D. Accurate measurements of the dielectric constant of seawater at L band. *Radio Sci.* **2016**, *51*, 2–24. [CrossRef]
40. Klein, L.; Swift, C. An improved model for the dielectric constant of seawater at microwave frequencies. *IEEE Trans. Antennas Propag.* **1977**, *25*, 104–111. [CrossRef]
41. Dinnat, E.P.; Brucker, L. Improved sea ice fraction characterization for L-band observations by the aquarius radiometers. *IEEE Trans. Geosci. Remote Sens.* **2017**, *55*, 1285–1304. [CrossRef]
42. Brucker, L.; Dinnat, E.P.; Koenig, L.S. Weekly gridded Aquarius L-band radiometer/scatterometer observations and salinity retrievals over the polar regions—Part 1: Product description. *Cryosphere* **2014**, *8*, 905–913. [CrossRef]
43. Brucker, L.; Dinnat, E.P.; Koenig, L.S. Weekly gridded Aquarius L-band radiometer/scatterometer observations and salinity retrievals over the polar regions—Part 2: Initial product analysis. *Cryosphere* **2014**, *8*, 915–930. [CrossRef]
44. Castro, S.L.; Wick, G.A.; Steele, M. Validation of satellite sea surface temperature analyses in the Beaufort Sea using UpTempO buoys. *Remote Sens. Environ.* **2016**, *187*, 458–475. [CrossRef]
45. Garcia-Eidell, C.; Comiso, J.C.; Dinnat, E.; Brucker, L. Satellite observed salinity distributions at high latitudes in the Northern Hemisphere: A comparison of four products. *J. Geophys. Res. Oceans* **2017**, *122*, 7717–7736. [CrossRef]

46. Tang, W.; Yueh, S.H.; Fore, A.G.; Hayashi, A. Validation of Aquarius sea surface salinity with in situ measurements from Argo floats and moored buoys. *J. Geophys. Res. Oceans* **2014**, *119*, 6171–6189. [CrossRef]
47. Tang, W.; Fore, A.; Yueh, S.; Lee, T.; Hayashi, A.; Sanchez-Franks, A.; Martinez, J.; King, B.; Baranowski, D. Validating SMAP SSS with in situ measurements. *Remote Sens. Environ.* **2017**, 326–340. [CrossRef]
48. Lee, T. Consistency of Aquarius sea surface salinity with Argo products on various spatial and temporal scales. *Geophys. Res. Lett.* **2016**, *43*, 3857–3864. [CrossRef]
49. Boutin, J.; Chao, Y.; Asher, W.E.; Delcroix, T.; Drucker, R.; Drushka, K.; Kolodziejczyk, N.; Lee, T.; Reul, N.; Reverdin, G.; et al. Satellite and in situ Salinity: Understanding Stratification and Sub-Footprint Variability. *Bull. Am. Met. Soc.* **2016**, *97*, 1391–1407. [CrossRef]
50. Vinogradova, T.N.; Ponte, R.M. Small-scale variability in sea surface salinity and implications for satellite-derived measurements. *J. Atmos. Ocean. Technol.* **2018**, *30*, 2689–2694. [CrossRef]
51. JPL Climate Oceans and Solid Earth Group. *JPL SMAP Level 3 CAP Sea Surface Salinity Standard Mapped Image Monthly or 8-Day Running Mean*; V4.0 Validated Dataset; PO.DAAC: Pasadena, CA, USA, 2018.
52. NCEP Sea Ice Concentration Analyses. Available online: http://polar.ncep.noaa.gov/seaice/Analyses.shtml (accessed on 5 May 2018).
53. Chassignet, E.P.; Hurlburt, H.E.; Metzger, E.J.; Smedstad, O.M.; Cummings, J.; Halliwell, G.R.; Bleck, R.; Baraille, R.; Wallcraft, A.J.; Lozano, C.; et al. U.S. GODAE: Global Ocean Prediction with the HYbrid Coordinate Ocean Model (HYCOM). *Oceanography* **2009**, *22*, 64–75. [CrossRef]
54. Meissner, T.; Wentz, F.J. *Remote Sensing Systems SMAP Ocean Surface Salinities [Level 2C, Level 3 Running 8-day, Level 3 Monthly]*; Version 2.0 validated release; Remote Sensing Systems: Santa Rosa, CA, USA, 2016.
55. Roemmich, D.; the Argo Steering Team. Argo: The challenge of continuing 10 years of progress. *Oceanography* **2009**, *22*, 46–55. [CrossRef]
56. Argo. Argo float data and metadata from Global Data Assembly Centre (Argo GDAC). *SEANOE* **2000**. [CrossRef]
57. European Union Copernicus Marine Environment Monitoring Service (CMEMS). The Arctic Ocean In-Situ Near-Real-Time Observations; Product Identifier INSITU_ARC_NRT_OBSERVATIONS_013_031. Available online: http://copernicus.eu/situ-thematic-centre-ins-tac/ (accessed on 9 April 2018).
58. OMG Mission. *Conductivity, Temperature and Depth (CTD) Data from the Ocean Survey*; Version 0.1; OMG SDS: Needham, MA, USA, 2016.
59. Ren, L.; Speer, K.; Chassignet, E.P. The mixed layer salinity budget and sea ice in the Southern Ocean. *J. Geophys. Res.* **2011**, *116*, C08031. [CrossRef]
60. Cavalieri, D.J.; Parkinson, C.L.; Gloersen, P.; Zwally, H.J. *Sea Ice Concentrations from Nimbus-7 SMMR and DMSP SSM/I-SSMIS Passive Microwave Data, Version 1*; [NSIDC-0051]; NASA National Snow and Ice Data Center Distributed Active Archive Center: Boulder, CO, USA, 1996. [CrossRef]
61. Maslanik, J.; Stroeve, J. *Near-Real-Time DMSP SSMIS Daily Polar Gridded Sea Ice Concentrations, Version 1*; [NSIDC-0081]; NASA National Snow and Ice Data Center Distributed Active Archive Center: Boulder, CO, USA, 1999. [CrossRef]
62. Ye, B.; Yang, D.; Zhang, Z.; Kane, D.L. Variation of hydrological regime with permafrost coverage over Lena Basin in Siberia. *J. Geophys. Res.* **2009**, *114*, D07102. [CrossRef]
63. Woo, K.; Kane, D.; Carey, S.; Yang, D. Progress in Permafrost Hydrology in the New Millennium. *Permafr. Periglac. Process.* **2008**, *19*, 237–254. [CrossRef]
64. Peterson, B.J.; Holmes, R.M.; McClelland, J.W.; Vorosmarty, C.J.; Lammers, R.B.; Shiklomanov, A.I.; Rahmstorf, S. Increasing river discharge to the Arctic Ocean. *Science* **2002**, *298*, 2171–2173. [CrossRef] [PubMed]
65. McClelland, J.W.; Dery, S.J.; Peterson, B.J.; Holmes, R.M.; Wood, E.F. A pan-arctic evaluation of changes in river discharge during the latter half of the 20th century. *Geophys. Res. Lett.* **2006**, *33*, L06715. [CrossRef]
66. Arctic Great Rivers Observatory. Discharge Dataset, Version 20180319. 2018. Available online: https://www.arcticrivers.org/data (accessed on 19 March 2018).
67. Yang, D.; Ye, B.; Shiklomanov, A. Discharge characteristics and changes over the Ob river watershed in Siberia. *J. Hydrometeorol.* **2004**, *5*, 595–610. [CrossRef]
68. Yang, D.; Ye, B.; Kane, D. Streamflow changes over Siberian Yenisei river basin. *J. Hydrol.* **2004**, *296*, 59–80. [CrossRef]

69. Woodgate, R.A., Stafford, K.M.; Prahl, F.G. A Synthesis of Year-Round Interdisciplinary Mooring Measurements in the Bering Strait (1990–2014) and the RUSALCA Years (2004–2011). *Oceanography* **2015**, *28*, 46–67. [CrossRef]
70. De Steur, L.; Hansen, E.; Gerdes, R.; Karcher, M.; Fahrbach, E.; Holfort, J. Freshwater fuxes in the East Greenland Current: A decade of observations. *Geophys. Res. Lett.* **2009**, *36*, L23611. [CrossRef]
71. Fournier, S.; Lee, T.; Gierach, M. Seasonal and interannual variations of sea surface salinity associated with the Mississippi River plume observed by SMOS and Aquarius. *Remote Sens. Environ.* **2016**, *180*, 431–439. [CrossRef]
72. Wadhams, P.; Aulicino, G.; Parmiggiani, F.; Persson, P.O.G.; Holt, B. Pancake ice thickness mapping in the Beaufort Sea from wave dispersion observed in SAR imagery. *J. Geophys. Res. Oceans* **2018**, *123*. [CrossRef]
73. Jin, F.F. An equatorial ocean recharge paradigm for ENSO. Part I: Conceptual model. *J. Atmos. Sci.* **1997**, *54*, 811–829. [CrossRef]
74. Picaut, J.; Masia, F.; Penhoat, Y.D. An advective-reflective conceptual model for the oscillatory nature of the ENSO. *J. Geophys. Res.* **1997**, *103*, 14261–14290. [CrossRef]
75. Qu, T.; Yu, J.Y. ENSO indices from sea surface salinity observed by Aquarius and Argo. *J. Oceanogr.* **2014**, *70*, 367–375. [CrossRef]
76. Steele, M.; Ermold, W. Loitering of the retreating sea ice edge in thearctic seas. *J. Geophys. Res. Oceans* **2015**, *120*, 7699–7721. [CrossRef] [PubMed]

remote sensing

MDPI

Article

Assessing Coastal SMAP Surface Salinity Accuracy and Its Application to Monitoring Gulf of Maine Circulation Dynamics

Semyon A. Grodsky [1,*], Douglas Vandemark [2] and Hui Feng [2]

[1] Department of Atmospheric and Oceanic Science, University of Maryland, College Park, MD 20742, USA
[2] Ocean Process Analysis Laboratory, Institute for the Study of Earth, Oceans, and Space, University of New Hampshire, Durham, NH 03824, USA; doug.vandemark@unh.edu (D.V.); hui.feng@unh.edu (H.F.)
* Correspondence: senya@atmos.umd.edu; Tel.: +1-301-405-5330

Received: 22 June 2018; Accepted: 3 August 2018; Published: 6 August 2018

Abstract: Monitoring the cold and productive waters of the Gulf of Maine and their interactions with the nearby northwestern (NW) Atlantic shelf is important but challenging. Although remotely sensed sea surface temperature (SST), ocean color, and sea level have become routine, much of the water exchange physics is reflected in salinity fields. The recent invention of satellite salinity sensors, including the Soil Moisture Active Passive (SMAP) radiometer, opens new prospects in regional shelf studies. However, local sea surface salinity (SSS) retrieval is challenging due to both cold SST limiting salinity sensor sensitivity and proximity to land. For the NW Atlantic, our analysis shows that SMAP SSS is subject to an SST-dependent bias that is negative and amplifies in winter and early spring due to the SST-related drop in SMAP sensor sensitivity. On top of that, SMAP SSS is subject to a land contamination bias. The latter bias becomes noticeable and negative when the antenna land contamination factor (LC) exceeds 0.2%, and attains maximum negative values at LC = 0.4%. Coastward of LC = 0.5%, a significant positive land contamination bias in absolute SMAP SSS is evident. SST and land contamination bias components are seasonally dependent due to seasonal changes in SST/winds and terrestrial microwave properties. Fortunately, it is shown that SSS anomalies computed relative to a satellite SSS climatology can effectively remove such seasonal biases along with the real seasonal cycle. SMAP monthly SSS anomalies have sufficient accuracy and applicability to extend nearer to the coasts. They are used to examine the Gulf of Maine water inflow, which displayed important water intrusions in between Georges Banks and Nova Scotia in the winters of 2016/17 and 2017/18. Water intrusion patterns observed by SMAP are generally consistent with independent measurements from the European Soil Moisture Ocean Salinity (SMOS) mission. Circulation dynamics related to the 2016/2017 period and enhanced wind-driven Scotian Shelf transport into the Gulf of Maine are discussed.

Keywords: satellite salinity; Gulf of Maine; bias characteristics; Scotian Shelf

---

1. Introduction

Monitoring interactions between the cold and productive waters of the Gulf of Maine (GoM) and the adjoining northwestern Atlantic shelf is an important yet challenging regional need, with seasonal and longer timescale change impacts that affect ecosystem, fishery, and coastal management efforts. Key water mass inflows influencing this marginal sea include the upstream fresh/cold Scotian Shelf and Labrador Sea currents, and the adjoining salty/warm northwestern (NW) Atlantic shelf (Mountain [1]; Townsend et al. [2]; Feng et al. [3]; Peterson et al. [4]). The primary inflow to the GoM is through deep, typically salty, water along the Northeast Channel (NEC) that lies between Browns and Georges

Banks (see Figure 1). Separately, fresher Scotian Shelf water enters through the relatively shallow (100 m) Northern Channel between Cape Sable and Browns Bank (Ramp et al. [5]; Feng et al. [3]). The variability and relative strength of these transports are thought to control interior Gulf of Maine nutrient and density levels (Townsend et al. [2]).

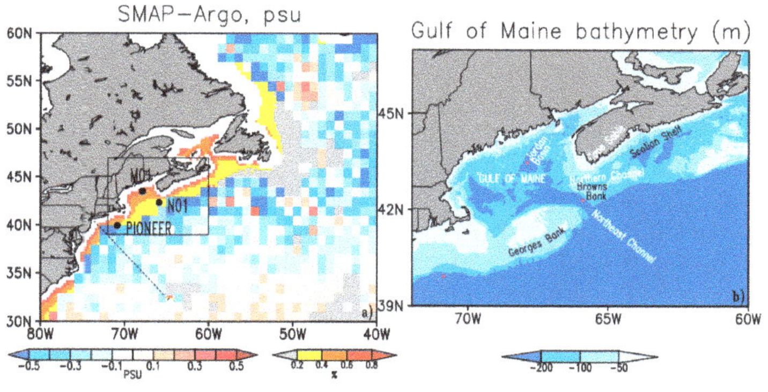

Figure 1. (**a**) Time mean Soil Moisture Active Passive (SMAP) minus Array for Real-Time Geostrophic Oceanography (Argo) salinity differences (PSU) over the northwestern (NW) Atlantic shelf. SMAP antenna land contamination (%) is shown in coastal areas. Data gaps are shaded in gray. The Oleander ship transect (dashed), PIONEER buoy, M01 and N01 NERACOOS buoy sites are overlain. (**b**) Gulf of Maine (GoM) and adjacent area bathymetry. Buoy locations are shown in magenta. The geographic area of panel (**b**) is marked by the box in panel (**a**).

The dominance of subsurface GoM exchange pathways has, so far, limited the utility of surface remote sensing data for monitoring GoM dynamics. However, Feng et al. [3] recently used altimeter data to demonstrate that seasonal and interannual variability in near-surface freshwater advection from the Scotian Shelf into the coastal GoM can be related to local wind-induced changes in the remote upstream coastal geostrophic currents. That study, amongst others cited, highlighted the need to resolve spatial information on surface salinity fields just offshore of and inside the GoM. While focused on open ocean waters, two recent NW Atlantic sea surface salinity (SSS) studies showed new satellite capabilities to resolve along shelf dynamics and Gulf Stream perturbations (Grodsky et al. [6] and Reul et al. [7]). The present study explores SSS data capabilities nearer to shore and in colder waters, including a focus on wintertime interactions between the GoM and NW Atlantic shelf slope region. For this, we use Soil Moisture Active Passive (SMAP) microwave SSS data at a spatial resolution near 40 km. This is a courser resolution than that used for visible or infrared sea surface temperature (SST) data (1–10 km), but it provides weekly-to-monthly data that is unaffected by winter period rain and clouds. The choice of microwave remote sensing data is important, because stronger NEC transport events into the GoM tend to occur in winter (Ramp et al. [5]) when SST is cold and its gradients are at their weakest due to strong surface cooling.

The application of microwave data for SSS observation in marginal seas is complicated by proximity to land (e.g., Dinnat and LeVine [8]). A few successful applications of satellite SSS to study salinity variation near to shore have been in warmer tropical seas where the measurement sensitivity is higher (e.g., Gierach et al. [9]). However, salinity retrievals over cold coastal seas must contend with both known major and competing sources of bias, water temperature, and land effects. Satellite SSS is subject to an SST-dependent bias that is, at present, negative and amplified in winter and early spring due to the SST-related drop in microwave sensor sensitivity (e.g., Köhler et al. [10]; Meissner et al. [11]). On top of that, satellite SSS is subject to a land contamination bias (e.g., Tang et al. [12]). The two bias components have pronounced seasonal signatures due to seasonal changes in SST-related sensor sensitivity and land emissivity. Global ocean SSS analyses have shown that if one is interested in true SSS change (e.g., seasonal anomalies) with respect to these consistently repeating biases, then one can reference SSS data to the sensor-specific SSS climatology to obtain the anomaly. This effectively removes or strongly attenuates all such bias effects. Thus, we expect that the monthly SSS anomalies used for dynamical analyses in this paper have sufficient accuracy and applicability to extend beyond the tropics to mid-latitude oceans and nearer to the coast (Boutin et al. [13]; Lee [14]). This expectation is aided by the relatively large, ~1 psu, spatial and temporal salinity changes that are often observed across the NW Atlantic shelf, which are gradients that are known to be detectable by remote sensing salinity (e.g., Kubryakov et al. [15]; Garcia-Eidell et al. [16]).

In this paper, we first address SMAP SSS data accuracy using concurrent buoy, Array for Real-Time Geostrophic Oceanography (Argo) float, glider, and thermosalinograph datasets with a focus on the land contamination (LC) and SST biases. We also seek to demonstrate that one can successfully remove the observed seasonal SSS cycle to retrieve an accurate estimate of the monthly SMAP SSS anomaly across this marginal sea region. We then compare apparent water mass changes within the GoM and between Georges Banks and the Scotian Shelf in the winters of 2016/17 and 2017/18. Regional wind speed data suggest that the 2016/17 GoM freshening event is likely tied to enhanced Scotian Shelf water inflow in fall 2016.

2. Materials and Methods

Satellite SSS: SMAP SSS data are provided globally and daily using an eight-day running mean on a 0.25° spatial grid (Version 2 as produced by the Remote Sensing Systems (RSS, www.remss.com/missions/smap, Meissner and Wentz [17]) with the effective spatial resolution of ~40 km. Besides SSS retrieval, this dataset provides reference sea surface temperature (SST) as well as antenna land contamination (LC or GLAND) and ice contamination (GICE) fractions (both weighted by antenna gain). To independently confirm and support SMAP results, we also use the debiased 0.25° SSS dataset obtained using the European Soil Moisture Ocean Salinity sensor (SMOS version 2.0, www.catds.fr/sipad/; Boutin et al. [18]).

Mooring data: As noted earlier, satellite SSS data in colder waters (<5–7 °C) and approaching coastlines are still subject to several limitations that may lead to regional and seasonally dependent SSS biases (e.g., Meissner et al. [11]). Fortunately, a number of observational programs are providing valuable salinity ground truth measurements to aid in the characterization of measurement noise and bias. The NorthEastern Regional Association of Coastal and Ocean Observing Systems (NERACOOS) operates a number of GoM coastal and deeper basin moorings (gyre.umeoce.maine.edu/buoyhome.php). We focus on the mid-basin GoM mooring (M01, 2003-present, Figure 1) located in Jordan Basin and the most seaward mooring located in the Northeast Channel (NEC) to monitor GoM inflows (N01, 2004–2017, Figure 1). These buoys monitor temperature and salinity (S), including measurements at depths of $z = 1$ m, 20 m, and 50 m. A more recent addition (2014–present) to this regional observation system is provided by the Ocean Observing Initiative (OOI, https://ooinet.oceanobservatories.org/). Their coastal PIONEER array includes an offshore surface mooring (ASIMET Bulk Meteorology Instrument Package) that monitors 1-m depth salinity. Daily buoy S anomalies at each buoy were computed by calculating monthly seasonal cycles (based on the full buoy data temporal coverage) and daily resampling using cubic spline interpolation. All of the buoy data have been averaged daily, in part to reduce the significant variability associated with tides.

Thermosalinograph and glider: Another valuable data source that was used here specifically for SMAP SSS land contamination assessment is ship-based thermosalinograph (TSG) observations collected near coastlines. Study data come from the long-term M/V Oleander transect program (http://po.msrc.sunysb.edu/Oleander). These along transect salinity measurements (3-m depth) cover from the (United States) US east coast to Bermuda with near monthly sampling. Salinity data quality was confirmed against Argo float datasets from 2012–2016. We also use a second higher-latitude Atlantic TSG dataset collected by the M/V NUKA-ARCTICA, which are data that are quality controlled, flagged, and distributed by the Laboratoire d'Etudes en Géophysique et Océanographie (LEGOS) (www.legos.obs-mip.fr/observations/sss/datadelivery/dmdata/). These data include Europe–Greenland transects and episodic summer period coastal Labrador Sea and Greenland coastline visits. Only NUKA-ARCTICA salinity observations that have been flagged as good are used. In the RSS SMAP v.2.0 gridded product, SSS is set to an undefined value for sea ice concentration, GICE > 0.1%. There are no valid SMAP/TSG collocations for GICE > 0.1%. Even after this ice-filtering criterion, GICE > 0.1, some residual sea-ice contamination could still persist. Their possible remaining effect on the SMAP/TSG matchups is not considered herein, but may lead to a fresh SMAP SSS bias (see Section 3.2, Tang et al. [19]). For additional in situ context, Slocum glider T/S data (near-surface data are from 3–5 m) on the Scotian Shelf in fall 2016 and 2017 were obtained via the Marine Environmental Observation Prediction and Response Network (MEOPAR) at gliders.oceantrack.org/data/slocum. Glider near-surface salinity data are averaged across a given day to provide one daily estimate at a daily center position.

Argo data: Our chosen global near-surface in situ salinity database uses the daily $1^\circ \times 1^\circ$ binned version of the original Argo profiles, which are available from ftp://usgodae.org/. The shallowest depth Argo level (typically 5–10 m) is used as a proxy for SSS.

Other satellite data: Finally, the AVISO multi-satellite merged absolute dynamic topography and its related geostrophic currents are used to identify regional ocean circulation dynamics. These satellite altimeter data products are now distributed through the Copernicus Marine Environment Monitoring Service (http://marine.copernicus.eu). Gridded 10-m near-surface winds from the Advanced Scatterometer (ASCAT) scatterometer onboard the European Meteorological Satellite Organization MetOp (ftp.ifremer.fr/ifremer/cersat/products/gridded/MWF/L3/ASCAT) are produced by Bentamy and Croize-Fillon [20].

All of the anomalies used in this paper are calculated as deviations from corresponding monthly seasonal cycles.

3. Results

3.1. SMAP SSS Bias Characterization

3.1.1. Global Distribution of SMAP Bias from Comparisons with Argo

The general characterization of SMAP versus in situ salinity differences (bias hereafter) is illustrated by comparison to quasi-synchronous (same day) Argo salinity measurements, as shown in Figure 2. The time mean SMAP bias (Figure 2a) has a clear spatial distribution (resembling SST, Melnichenko personal communication) with a salty bias in the warmer tropics and a fresh bias in the colder mid-latitude ocean. Noting the strong seasonal cycle of SST, one may anticipate a corresponding seasonal variation in the bias. This time variability is indeed present, but mostly for the higher latitude fresh bias regions where the magnitude increases during cold seasons (Figure 2b). In contrast, the tropical positive bias has only minor seasonal changes, in turn suggesting it may not be related to SST (Tang et al. [12]). Also, note these bias estimates implicitly include any haline stratification that is present between the surface and the shallowest Argo measurement level ~5 m, which is a factor that at present is poorly characterized (Grodsky et al. [21]; Liu et al. [22]; Song et al. [23]).

Besides the meridional structure, the SMAP SSS biases in Figure 2a also depict clear negative patterns in coastal areas. This fresh along coast bias feature is present at all latitudes, including the tropics, reflecting an underestimation of salinity due to the SMAP antenna land contamination (LC, leakage of land surface radiation through antenna side lobes). These coastal features have significant impacts on the accuracy of SMAP SSS over the northwestern (NW) Atlantic shelf, which is our area of focus. Most of the NW shelf shows a negative SSS bias (Figure 1a) that can be examined with further coastal and independent in situ measurements.

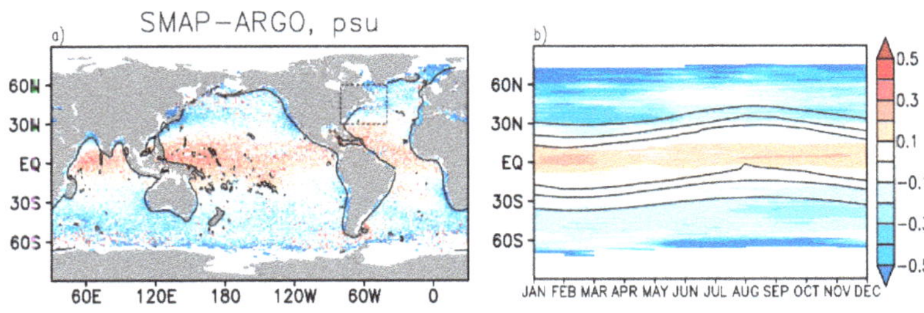

Figure 2. (**a**) Time mean (April 2015–April 2018) of a daily salinity difference of one degree, SMAP-Argo SMAP antenna land contamination 0.2% contour is shown. (**b**) Time-latitude diagram of zonally averaged SMAP-Argo. Zonally averaged 20C, 25C, and 27C SST contours are overlain.

3.1.2. SMAP Comparison with TSG Data

Figure 3 data indicate that the Oleander TSG salinity data are remarkably consistent with collocated (same day, 25-km search radius) SMAP SSS over the open ocean south of the Gulf Stream (latitude, or LAT < 37°N). This consistency partly contradicts the fresh bias versus Argo salinity (Figure 2a), which is a point to be addressed later. The negative SMAP bias becomes noticeable only approaching the continent (LAT > 37.5°N) as well as approaching Bermuda (LAT < 33.5°N). There is also a large salty bias very close to the continent (for $LC \geq 0.5\%$, LAT > 39.5°N), which may be interpreted as a result of an overestimation of the land contamination impact. Note that coastal areas, where SMAP land contamination levels likely require attention, cover much of the GoM (Figure 3a).

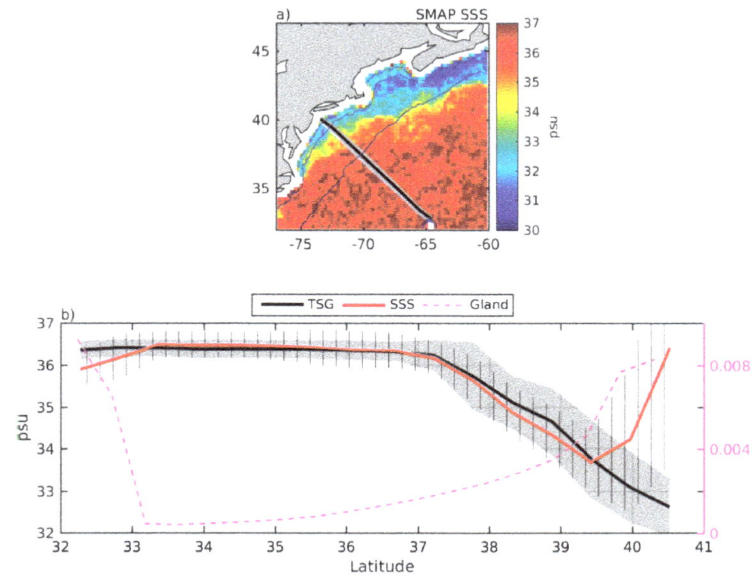

Figure 3. (**a**) Sample regional SMAP sea surface salinity (SSS) data and mapping of the M/V OLEANDER transect line. Also shown are SMAP data product estimates for land contamination given as contours at 0.2%, 0.5%, and 0.8%. (**b**) Time mean SMAP SSS, thermosalinograph (TSG) salinity, and antenna land contamination (Gland). Gray shading and vertical bars are standard deviations of TSG and SMAP salinity, respectively.

Conditionally, the total SMAP bias can be partitioned into two major components. We define an inherent SMAP retrieval bias (SRB) as being the SSS bias for open ocean conditions; i.e., away from coastal impacts. A second land contamination bias (LCB) is solely due to land-based emissivity that enters through the side lobes of SMAP antenna. Binning SMAP-TSG S differences versus LC reveals that time mean LCB becomes noticeable in the M/V OLEANDER dataset at $LC \geq 0.2\%$, and reaches maximum negative values at $LC \approx 0.4\%$ (Figure 4). Closer to the coast ($LC \geq 0.5\%$), the LCB switches to positive values due to an overestimation of the LC effect. Although this behavior should be geographically dependent due to relative changes in sea and land microwave properties, it is generally confirmed by the separate TSG data from the M/V NUKA-ARCTICA collected in the colder North Atlantic region (Figure 4). The SMAP biases along these more northern NUKA-ARCTICA transects have slightly larger negative values than SMAP biases along OLEANDER transects, perhaps due to the contribution from the time mean SRB, which is more negative over cold SST.

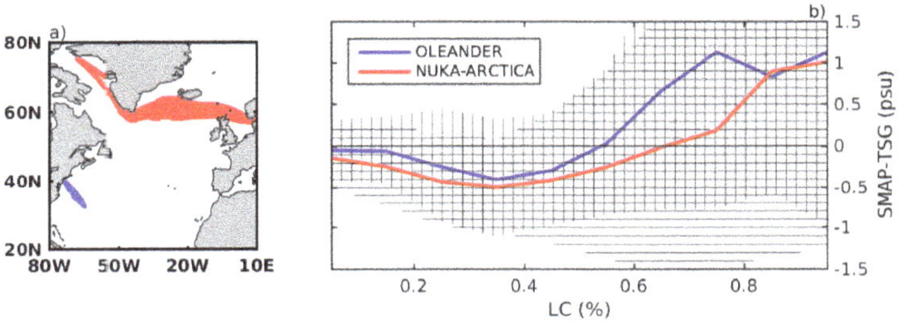

Figure 4. (**a**) Geographical map of M/V OLEANDER and NUKA-ARCTICA transects. (**b**) SMAP-TSG salinity difference binned versus land contamination factor (LC) for the two ship datasets. Data scatter (STD) in each $\Delta LC = 0.1\%$ bin is shown by vertical and horizontal lines for M/V OLEANDER and NUKA-ARCTICA, respectively.

We predict that both SRB and LCB will be seasonally dependent due to seasonal changes in ocean conditions (SST, wind) and in terrestrial microwave properties (soil moisture, in particular), respectively. The more extensive Oleander transect time coverage allows the best TSG-based seasonal characterization of SMAP bias, but only locally along the mean transect line (Figure 5). Note that the Figure 5 endpoint data also need to consider varying distance to nearby land (hence the LC factor, see also Figure 3a). Close to Bermuda (LAT ≤ 33°N) and approaching the continent (37°N ≤ LAT ≤ 39.5°N), the bias is mostly negative due to land contamination and has stronger negative values in colder seasons. In line with Figure 3b, the LCB becomes positive (>1 psu) closer to the coast (LAT ≥ 39.5°N), reflecting an overestimation of the LC effect. This overestimation has clear seasonal dependence. It is weaker or even reversed in boreal fall. This may reflect uncorrected seasonal changes in sea–land temperature contrasts and land emissivity (moisture).

In an offshore segment of the OLEANDER transects (33°N ≤ LAT ≤ 37°N), where SRB should dominate, SMAP bias results show a slight overestimation during warm months and a clear underestimation in January-March (Figure 5). When averaged over the entire year, Figure 3b data suggest that these seasonal changes effectively cancel and lead to nearly unbiased SMAP SSS along this segment of the OLEANDER track. As noted before, this somewhat contradicts the time mean fresh bias versus Argo that is present in this area (Figure 2a). One possible reason for this disagreement could be the difference between TSG (~3 m) and Argo (5–10 m) measurement depths in combination with the presence of barrier layers and the associated shallow haline stratification known to prevail in this area (e.g., Liu et al. [22]). In fact, collocating TSG and Argo along the Oleander transects reveals that Argo is consistently saltier than TSG (Figure 6a) due to the stable near-surface halocline. The difference between Argo salinity and skin SSS (seen by SMAP and SMOS) would be expectably larger if there is a systematic salinity gradient above the Argo (5–10 m) and/or TSG measurement depths (~3 m). Binning Argo-TSG salinity difference versus wind speed (using only data from salinity homogeneous area south of the Gulf Stream and excluding Bermuda region, 33°N ≤ LAT ≤ 36°N), suggests that a fresh near-surface salinity contrast exists, and is largest (up to 0.5 psu) under calm winds (Figure 6b). This stratification will weaken when winds strengthen above $U_{10} = 5$ m/s, but not completely. The mean Argo-TSG salinity difference remains ~ 0.15 psu ($U_{10} > 5$ m/s), suggesting that at least a portion of the fresh SMAP bias seen in Figure 2a may be attributed to shallow stable haline stratification in rainy and river-impacted zones such as the NW Atlantic shelf. Figure 6a,c also indicate that near-surface halocline may be stronger in the high gradient area north of the Gulf Stream wall, where the lateral intrusion processes may contribute (e.g., Grodsky et al. [21]; Gawarkiewicz et al. [24]).

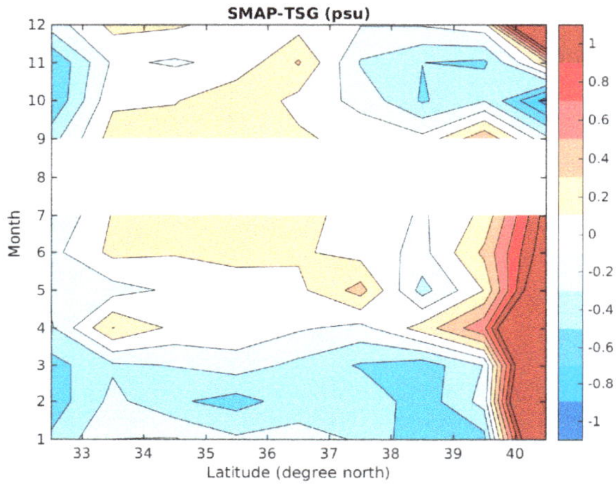

Figure 5. Latitude calendar month diagram of SMAP-TSG salinity difference along the OLEANDER line. So far, there are no August OLEANDER TSG data during the SMAP period.

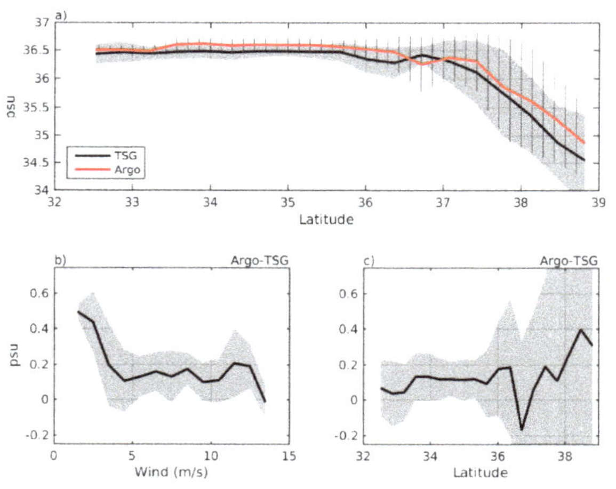

Figure 6. (**a**) Time mean salinity difference between Argo (5–10 m) and TSG (~3 m) along the OLEANDER line. Data scatter (STD) in each 0.5° latitude bin is shown by gray shading and vertical lines for TSG and Argo, respectively. (**b**) Argo-TSG south of 36N as a function of wind speed. (**c**) Argo-TSG as a function of latitude. Data scatter (STD) is shown by gray shading in (**b**,**d**).

3.1.3. SMAP SSS Comparison with Mooring Data

Daily time series of near-surface salinity from buoys allow for a more complete temporal evaluation of SMAP SSS biases (Figure 7) at three regional sites. Haline stratification impacts on biases referenced to buoy measurements are expected to be weak, especially given that the seasonal cycle in regional haline impacts has been shown to be small (see Figure 5 in Li et al. [25]). The general physics of haline stratification above the shallowest buoy level, $z = 1$ m, are discussed by Boutin et al. [26], but there is insufficient data to evaluate such shallow haline impacts quantitatively.

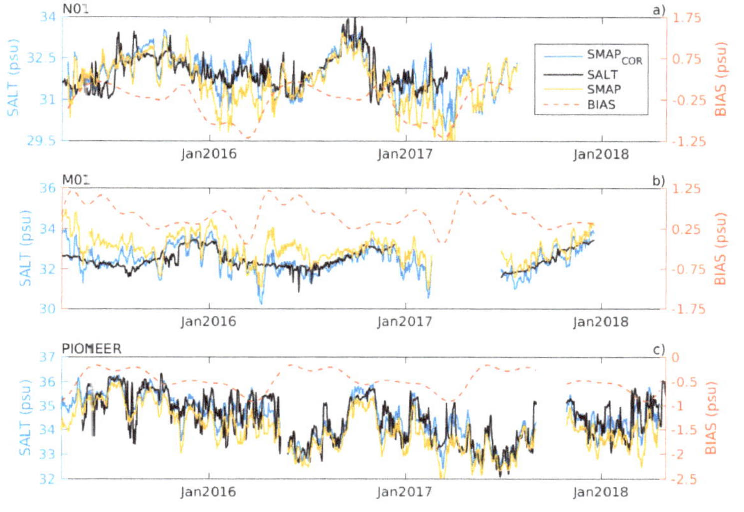

Figure 7. Buoy 1-m salinity (SALT), SMAP salinity, and SMAP salinity corrected for the seasonal bias ($SMAP_{COR}$). Seasonally dependent BIAS=SMAP-BUOY is shown against the right hand vertical axis. (**a**,**b**) show data from NorthEastern Regional Association of Coastal and Ocean Observing Systems (NERACOOS) moorings N01 and M01. (**c**) shows data from OOI PIONEER mooring.

The SMAP biases referenced to 1-m buoy salinity, $SMAP - S_{1m}$, exhibit clear seasonal variation around the annual mean bias at each node. The most negative bias values are in phase with the coldest SST typically occurring in February and March. This seasonal behavior is consistent among all of the buoys, and qualitatively agrees with the TSG-based results of Figure 5. NERACOOS buoy N01 (see Figure 1) is located near the entrance gate to the GoM where the land contamination is relatively low ($LC \approx 0.25\%$). At N01, the seasonal SST varies from ~20 °C in summer down to ~5 °C in winter. During warm months, SMAP SSS closely follows the buoy data in Figure 7 to within some high frequency scatter. This contrasts with the cold season SMAP SSS, which is observed to be 1 psu fresher than in-situ S_{1m}. This appears to reflect an SST-dependent bias for coldest SST. The bias seasonal cycle is dominated by the first annual harmonic. Its removal largely corrects deviations between SMAP and buoy salinities, both decreasing the overall RMS difference and increasing temporal correlation between $SMAP_{cor} = SMAP - BIAS$ and S_{1m} (see Table 1). However, this seasonal debiasing doesn't affect the high frequency (sampling) SMAP variability that is reflected in the high STD = 0.57 psu. The strong sampling variability of SMAP data results in the low correlation of daily SMAP and buoy salinity anomaly data (CORR = 0.25). This correlation between anomalies improves after monthly averaging of the data (CORR = 0.49). This indicates that SMAP SSS estimation noise is significantly reduced in monthly averages.

Table 1. Comparison of the daily mean buoy (z = 1 m) and collocated SMAP salinity (black/red is for original/corrected SMAP SSS, respectively). Correlation (CORR) of buoy vs. SMAP salinity anomalies (SSSA) is included in the last two columns for monthly and daily anomalies. See Figure 2 for buoy locations.

BUOY	BIAS (psu) SMAP-BUOY	RMS (psu)	STD (psu)	CORR	Monthly Anomaly CORR	Daily Anomaly CORR
N01	−0.34	0.80/0.57	0.72/0.57	0.56/0.63	0.49	0.25
M01	0.56	0.68/0.29	0.43/0.29	0.62/0.84	0.58	0.27
PIONEER	−0.48	0.77/0.57	0.61/0.57	0.77/0.80	0.97	0.76

A similar error budget assessment at NERACOOS buoy M01 is more complex. In contrast to N01, where the SMAP retrieval bias (SRB) dominates, buoy M01 is located inside the GoM and closer to land. At M01, the LC bias component is stronger because the LC factor is near 0.8%. Surprisingly, at this location, the strongest bias (~+1 psu) is observed in early summer, when SMAP SST impacts on SRB should weaken (Figure 7, M01). This seasonal behavior may reflect a combination of the SRB (negative in winter) and the LCB (positive except in boreal fall). Overall, the elevated positive bias at the M01 buoy is likely dominated by the positive LC component that arises due to overestimated land contamination effect at $LC > 0.5\%$. During spring through fall, the SRB is small, and the total bias is close to the seasonal cycle of the local LCB, which is positive throughout the year and decreases in fall (in comparison with Figure 5). In winter to early spring, the positive LCB may be compensated by the negative SRB (Figure 7, M01). This possible cancellation effect could explain why SMAP SSS at the M01 location is more accurate in winter, just when SSS retrieval is most challenging.

The LC factor at the PIONEER offshore mooring site, $LC \approx 0.4\%$ (Figure 2a) predicts that SMAP SSS will be affected by both SRB and LCB (Figure 4b). Since the LC contamination overestimation dominates at larger $LC > 0.5\%$, the LCB is expected to be negative here. A combination of the two bias components results in the negative total salinity bias (Figure 7, PIONEER). In summer, the SRB is expected to be small (similar to at N01 location, Figure 7, N01). Thus, the summer bias of ≈ -0.3 psu is attributable to the LCB. Both bias components have stronger negative values during colder seasons. At the PIONEER offshore mooring, this leads to the seasonal peak in negative values of the total SMAP bias in winter through early spring (-1 psu).

Very recently, the SMAP project released a new data version (2.1) intended for technical evaluation purposes only. This new version includes corrections that are meant to improve the SRB versus the SST. For completeness, we provide V2.0 versus V2.1 results in Figure 8. The data in Figures 7 and 8 illustrate that these new SMAP SSS data at buoy N01 show very little SRB change in cold water conditions. One can observe that version 2.1 SMAP SSS is smoother (in time) in comparison with version 2.0 SSS. This difference is related to an effective 70-km spatial smoothing in version 2.1 (which is the only currently available). Since cold water SRB appears to be similar for the two versions, this study focuses on the publicly available SMAP version 2.0 datasets.

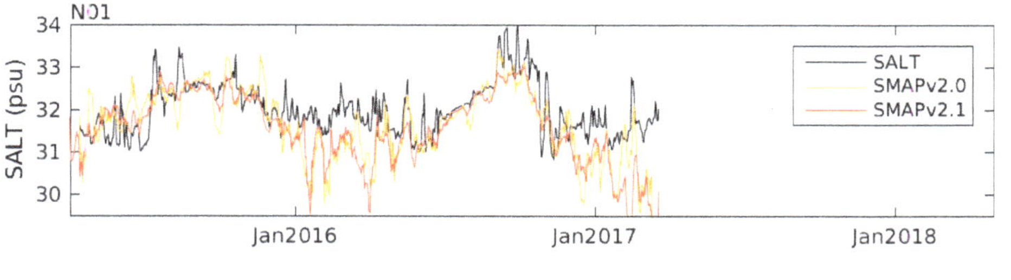

Figure 8. Buoy N01 1-m salinity (SALT), SMAP version 2.0 and version 2.1 collocated salinity.

To further evaluate SRB characteristics, Figure 9 presents salinity differences binned in wind speed/SST space. The intent is to discern the remaining SRB signal due to either: (a) SST-dependent salinity sensitivity or (b) sea surface roughness (wave) effects on microwave emission and reflectance. The SRB structure is not easy to assess at the more coastal buoys (PIONEER and M01), because both LCB and SRB factor into the total SMAP bias, and both bias components have correlated seasonal variations. The bias component combination is evident at the M01 (Figure 9b), where, due to high $LC \approx 0.8\%$, the total bias is dominated by the positive LCB.

At buoy N01, where LCB is weaker ($LC \approx 0.2\%$), the total bias is close to the negative SRB. Its magnitude increases for cold SST and for stronger winds. Since these two factors are seasonally correlated, their relative impact is unclear, but their combination (cold SST and strong winds) leads to the most negative SRB levels (Figure 9a). A similar codependence is present at the PIONEER location ($LC \approx 0.4\%$) where negative values increase at low SST and strong winds. However, even during warm seasons, the total bias remains negative due to the contribution of negative LCB, which peaks at $LC \approx 0.4\%$ (Figure 4b).

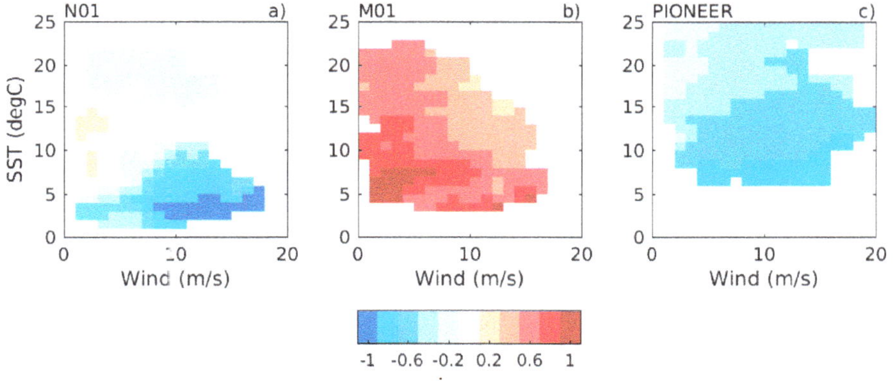

Figure 9. SMAP-BUOY salinity bias binned versus SST and wind speed for (**a**,**b**) NERACOOS moorings N01 and M01, and (**c**) OOI PIONEER mooring.

Annual and semiannual harmonics are the highest peaks in the spectrum of SMAP-BUOY salinity deviations (Figure 10). As expected, the removal of these annual peaks corrects for the most of the apparent deviations between SMAP and buoy, decreasing RMS salinity deviation and increasing the temporal correlation of collocated SMAP and buoy records (Table 1). However, the seasonal debiasing doesn't affect the high frequency (sampling) SMAP variability that is reflected in the still high STD = 0.57 psu. The high frequency sampling variability present in the SMAP data results in the low correlation of daily anomalous salinity from SMAP and buoy (CORR $\approx$ 0.25, Table 1) except at the PIONEER, where interannual anomalies have larger amplitude than in the N01 and M01 records (Figure 7). However, the salinity anomaly skill improves after monthly averaging (CORR $\approx$ 0.5 at N01 and M01) due to the corresponding reduction in SMAP sampling variability. In the data discussion to follow, the anomalous SSS from SMAP will be used to monitor water exchanges between the GoM and the adjacent Atlantic shelf. In the data discussion to follow, these SMAP SSS anomaly estimates are used to evaluate the water exchange between the GoM and the adjacent Atlantic shelf.

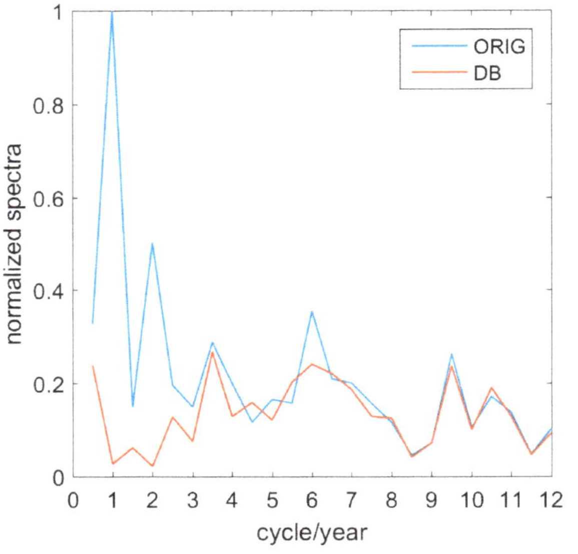

Figure 10. Spectrum of SMAP-S (z = 1 m) difference for original (ORIG) and debiased (DB) SMAP SSS. Spectra for N01, M01, and PIONEER data are averaged after normalization on their spectrum peak values. Spectra based on debiased data are normalized on the same peak values of original spectra.

3.1.4. SMAP SSS Comparison with Glider Data

There are limited coincident glider and SMAP data matchups to augment SMAP LCB assessment. The available results are provided in Figure 11. The data are focused on two regions, as shown in Figure 11a. The results of SMAP-glider S differences are stratified into ad hoc seaward (LC < 0.5%) and coastal (LC > 0.7%) cases based on the LC impacts shown earlier using the TSG datasets. Seaward data in Figure 11b exhibit a seasonal cycle in SMAP bias that resembles the SRB shown earlier. It is weak year round, but strong (−1 psu) negative values are present in the phase with cold SST during winter to early spring. This is consistent with buoy (Figures 7 and 9) and TSG (Figure 5) comparisons. Coastal data, as shown in Figure 11c, exhibit strong positive LCB values (associated with an overestimation of the LC effect) in the first half of the year. This bias component is significantly attenuated during the second half of the year, which is in contrast with the OLEANDER data (Figure 5) that exhibit weaker positive LCB values only in the fall. This seasonal behavior is mostly dictated by the Bedford Institute of Oceanography (BIO) glider data on the more northern Scotian Shelf region. These data have better seasonal coverage than the Rutgers University glider data collected to the southwest (Figure 11c). It may be not surprising to see the difference in LCB seasonality between the Nova Scotia shelf (Figure 11b) and Mid-Atlantic Bight (Figure 5) due to differing ocean brightness temperatures and terrestrial properties.

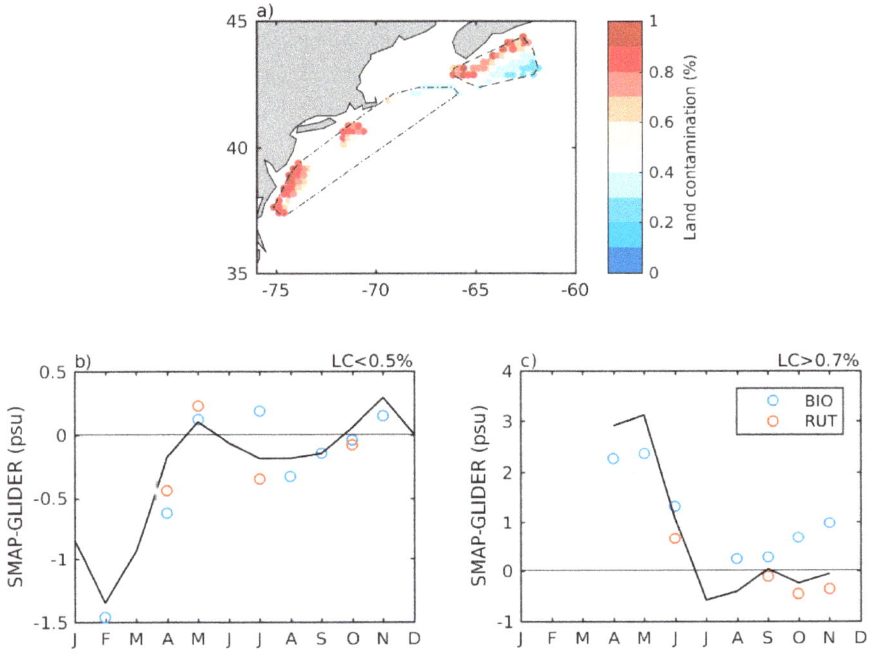

Figure 11. (**a**) Glider coverage for Bedford Institute of Oceanography (BIO) data near Nova Scotia and Rutgers University (RUT) data in the Mid-Atlantic Bight. Colors represent SMAP antenna land contamination (LC, %). (**b**) SMAP-GLIDER bias binned each calendar month for LC < 0.5%. (**c**) SMAP-GLIDER bias in the coastal regions with strong land contamination (LC > 0.7%). The black curves in (**b**,**c**) are the seasonal fit with the first three annual harmonics. Only calendar months with at least 10 collocated data are used.

3.2. Applying SMAP Data to Gulf of Maine SSS Monitoring

Spatial SST gradients in the Gulf of Maine region are significantly reduced during the winter due to strong near-surface mixing and destratification. This, in part, helps to open a gateway for winter period water exchanges with the adjacent NW Atlantic shelf. During the winters of 2016/17 and 2017/18, monthly SMAP SSS anomaly (SSSA) data indicate two differing episodes of Gulf of Maine water inflow (Figures 12 and 13). In both winters, there is the localized intensity of SSSA anomalies fields near the inflow pathways in between Georges Banks and Nova Scotia (i.e., near buoy station N01, see also Figure 2b). Figures 12 and 13 also show that the SSSA patterns observed by SMAP are remarkably consistent with the independent measurements produced using the European Soil Moisture Ocean Salinity (SMOS) mission sensor. This lends confidence to the SMAP-observed phenomena. The dynamics and details tied to the high salinity feature observed in January and February 2018 (Figure 13) are the subject of a separate study, and are thus not discussed further here. Here, we focus on the apparent freshwater inflow to the GoM in winter 2016/17.

The initial stage of the fresh GoM inflow is evident in October 2016 as an anomalous SSS pattern that shows a tongue of low SSS entering the GoM along the southwestern Nova Scotia coast (Figure 12). The pattern is consistent in SMAP and SMOS data. Notably, the SSSA (salinity anomaly) front lies northeast of buoy N01, meaning upstream of the Northeast Channel (NEC), which is the main gateway for winter shelf and slope water inflows that normally penetrate the GoM as deeper inflow (Ramp et al. [5]). In this case, the October 2016 fresh anomaly does not involve NEC transport, at least at this stage. By November 2016 (Figure 12), the core of the low SSSA feature near the NEC still lies northeast of buoy N01 and the NEC. Moreover, the fresh pattern is shifted closer to the Cape Sable coast, which is consistent with inflow through the Northern Channel (see Figure 2b) and inflow through the interior of the GoM.

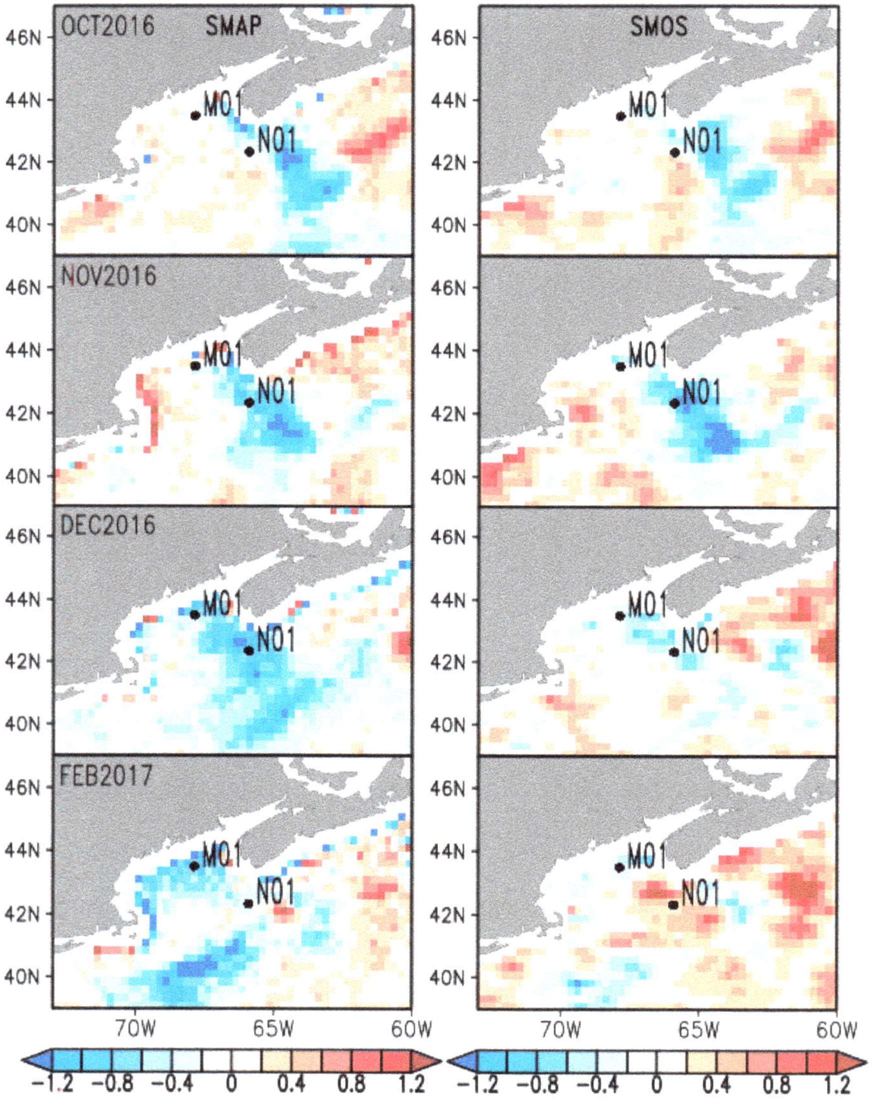

Figure 12. Anomalous SSS from (**left**) SMAP and (**right**) Soil Moisture Ocean Salinity (SMOS) during the fall and winter of 2016–2017.

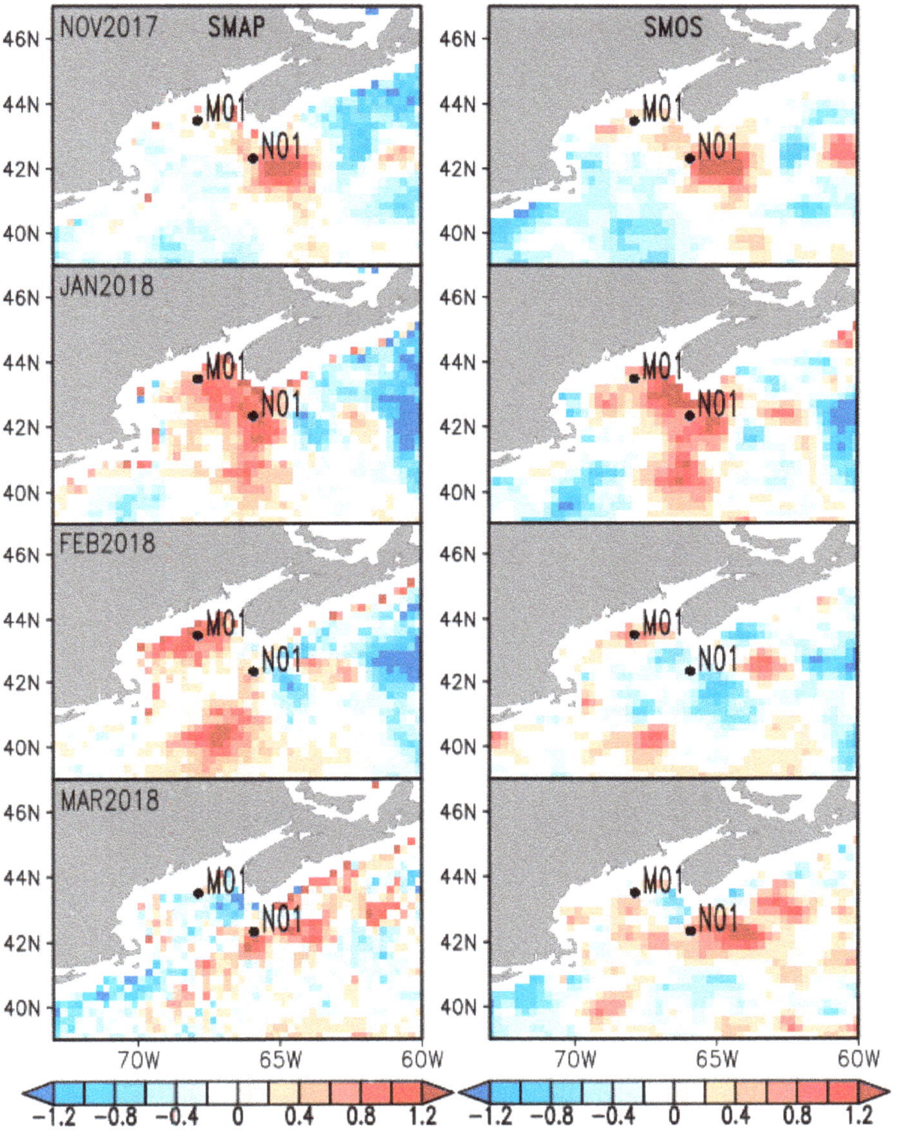

Figure 13. Anomalous SSS from (**left**) SMAP and (**right**) SMOS during the fall and winter of 2017–2018.

In situ observations from buoy N01 show that first evidence of near-surface freshening at this site is observed in late October 2016 (Figure 14a). The fresh intrusion (more than −1 psu) is evident to at least 20-m depth in the water column, but doesn't penetrate down to 50 m. This anomalously freshwater is present in the surface layers of the NEC through January 2017. The spatial SSSA SMAP data suggest this freshwater intrusion is not advected into the NEC from the Atlantic to the southeast, but is probably a lateral diffusion of the upstream (poleward) main fresh feature located to the northeast of the NEC. This hypothesis is consistent with the upper ocean currents measured at N01 (Figure 14b).

The upper 20-m velocity data show a generally southwest (SW) (down channel) NEC flow (negative velocity components, in line with Brooks [27]) during this fresh feature entrance period (October 2016 to January 2017).

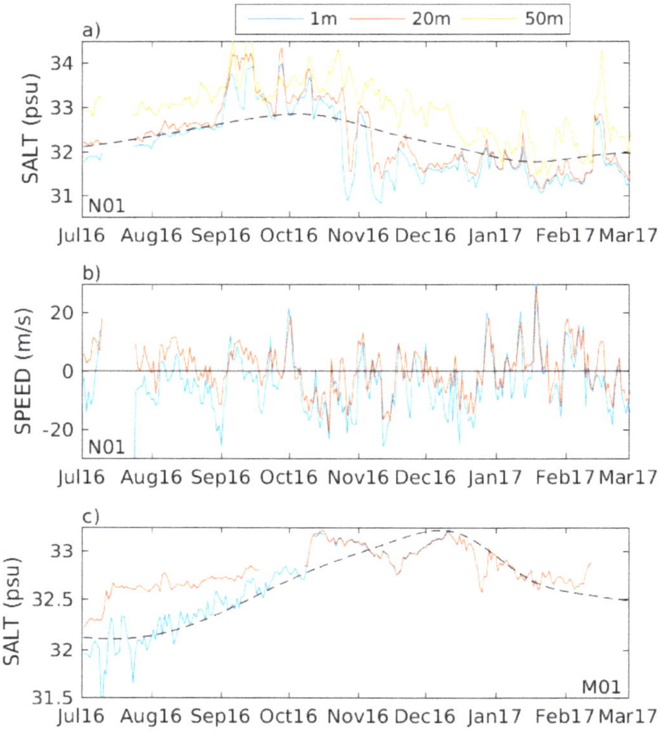

Figure 14. Time series of (**a**) salinity and (**b**) currents at buoy N01 located in the Northeastern Channel. Currents are projected on the direction perpendicular to the general orientation of the coast (30 degrees counterclockwise from the true north). Positive currents correspond to Gulf of Maine inflow along the Northeastern Channel. (**c**) Time series of salinity at buoy M01. Climatological salinity at $z = 1$ m is shown as black dashed line. The x-axis labels correspond to the first day of the month.

From December 2016 through February 2017, the Figure 12 correspondence between SMAP and SMOS SSS anomalies is not as clear. We posit that this is probably due to SMOS debiasing issues. During this period, SMAP data show that anomalously freshwater occupies a significant portion of the northeastern GoM (December 2016). In February 2017, it starts circulating cyclonically along the northwestern Gulf coast in the Maine Coastal Current, while the open ocean part of the fresh anomaly travels further down shelf along the shelf–slope current. In situ buoy data from station M01, located inside the GoM, confirm that transient fresh SSS anomalies of ~−0.5 psu are observed in the upper 20-m water column during November and December 2016 (Figure 14c), which is consistent with SMAP data. However, the 20-m depth M01 data in January and February 2017 indicate that salinity at the M01 is close to the normal, which is at odds with the strong negative SSSA in the last SMAP panel of Figure 12 at site M01. Unfortunately, the M01 surface (1-m) salinity data are unavailable after mid-December 2016. Overall, the N01 and M01 buoy data certainly confirm the timing and general magnitude of a near-surface freshening in the GoM from at least October into December 2016.

Next, we consider a possible scenario of dynamical events leading to this local 2016/17 winter fresh inflow. The freshest water in the region is known to be present along the southeastern Nova Scotia coast (Figure 15). It is transported southwestward down the coast within the Scotian Shelf Current (SSC, inshore branch). The variability of this coastal transport is an important factor controlling freshwater inflows into the GoM (Feng et al. [3]). To assess this as a possible explanation for the observed SMAP SSSA data near buoys N01 and M01, we present climatological and anomalous ocean current and wind data for October and November 2016 in Figure 15. The climatological geostrophic currents along the SSC (~65° W, 43.5° N in Figure 15a,c) are apparently weaker than the SSC current in fall of 2016 (Figure 15b,d). These anomalously strong currents result in stronger coastal volume transport and thus lower salinity around the southern tip of Nova Scotia (Figure 15). Moreover, the dynamical forcing relevant to this SSC acceleration may be related to wind forcing. During October and November 2016, the satellite scatterometer wind data depict a southwestward wind anomaly over the NW Atlantic (Figure 15b,d). Such a wind anomaly reflects a weakening of prevailing alongshore westerly winds on the shelf near southeastern Nova Scotia. The associated coastal sea level set up enhances the cross-coastal sea level gradient, increasing the SSC following the valve mechanism suggested by Li et al. [25].

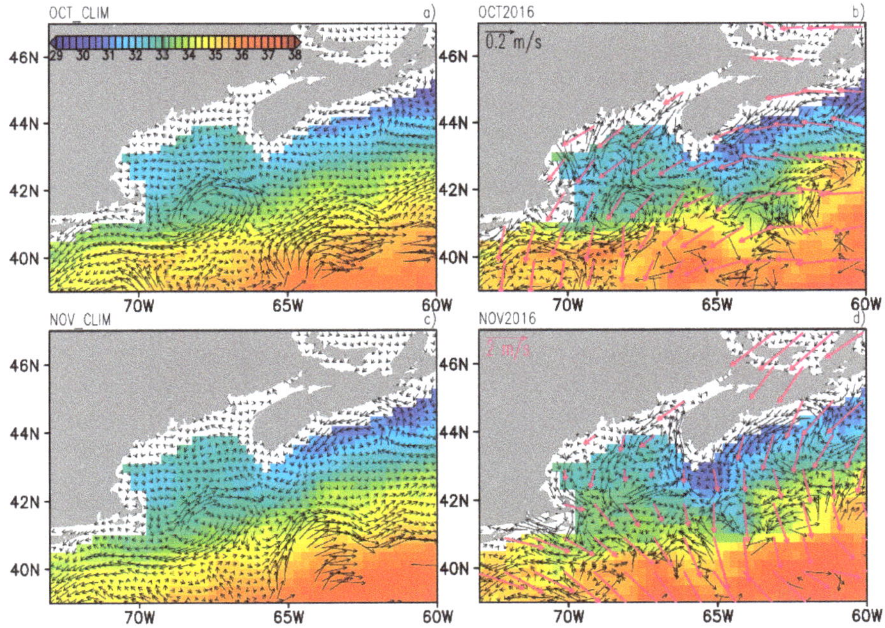

Figure 15. SMOS SSS (shaded) and surface geostrophic currents (black arrows) during the fresh feature entrance to the Gulf of Maine in October and November 2016 (see Figure 12). The left column (**a**,**c**) shows climatological conditions for October and November 2016. The right column (**b**,**d**) shows conditions for October and November 2016. Advanced Scatterometer (ASCAT) anomalous wind for October and November 2016 is overlaid in panels (**b**,**d**), respectively. For simplicity, debiased SMOS (rather than SMAP) is used for absolute SSS, because all of the near coastal grid points, which are subject to strong bias, are filtered out from this product.

The positive wind-induced sea level anomaly evolves southwestward along the Nova Scotia coast as a coastal western boundary Kelvin wave that distributes the sea level high further south and thus accelerates the SSC around the entire southern part of Nova Scotia. In November 2016, the accelerated SSC encompassed Nova Scotia and branched its freshwaters into the GoM (Figure 15d). This is also coincident with cyclonic anomalous upwelling wise winds over the GoM (Figure 15d) that depress in-Gulf sea levels and additionally contribute to coastal currents acceleration and the resulting anomalously strong freshwater inflow.

4. Conclusions

This paper addresses several challenging issues related to satellite ocean salinity data accuracy and application within the challenging salinity remote sensing domains of both cold and coastal waters. The focus region is the Gulf of Maine and the adjoining Scotian Shelf and NW Atlantic shelf break, where upstream fresh coastal subarctic waters are advected into a region that also includes high salinity Gulf Stream eddies. Both can lead to strong salinity gradients and variable water mass exchanges that are often best identified using salinity fields rather than temperature. An extensive regional approach to SMAP SSS data accuracy assessment is provided using not only Argo data (which do not collect data in shelf and coastal areas) but also all of the available in situ measurements. The paper also examines anomalous salinity signatures propagating into the productive waters of the Gulf of Maine (GoM) to assess their interactions with the nearby NW Atlantic shelf.

A regional SMAP SSS error assessment exposes remaining SSS bias using a suite of near-surface salinity measurements that includes buoy, thermosalinograph (TSG), and glider data. For the NW Atlantic, our analysis shows that SMAP SSS is subject to an SST-dependent bias that is negative and at its maximum in winter and early spring (up to −1 psu). Our speculation is that this is due to the SST-related decrease in microwave sensor sensitivity to SSS when SST falls below 5–7 °C. The data also indicate that regional SMAP SSS is subject to a land contamination bias associated with the contribution of land radiation via side lobes of SMAP antenna. The latter bias is identified by comparing SMAP SSS with salinity data from repeating M/V OLEANDER TSG ship transects. SSS differences are found to be a function of the antenna land contamination factor (LC), becoming significant and negative for $LC \geq 0.2\%$. Maximum negative SSS biases near –0.5 psu occur for $LC = 0.4\%$. Coastward, a very different SMAP SSS bias arises for $LC \geq 0.5\%$, where a larger and positive land contamination bias grows to reach $\geq +1$ psu. This near coastal positive land contamination bias is probably related to a local overestimation of land contamination impact along the M/V OLEANDER line. However, while this land contamination bias is expected to be regionally dependent due to geographic changes in the difference between land and sea microwave radiation, a qualitatively similar land contamination bias is also observed along the distant M/V NUKA-ARCTICA Europe-to-Greenland repeating transects, which are further north near 60°N.

Data also indicate that SMAP SSS bias components related to SST correction and land contamination are seasonally dependent. We attribute this to seasonal changes in SST/winds and coastal terrestrial microwave emissivity. Seasonal harmonics are the highest peaks in spectra of SMAP-TSG salinity difference. As for the other open ocean satellite SSS studies, the results show that a pragmatic remedy to work around these SSS bias issues is the computation and use of SMAP SSS anomalies; anomalies are formed taking the difference relative to a satellite SSS climatology. The approach effectively removes seasonal biases along with the real seasonal cycle, but provides access to spatial and temporal information on interannual SSS change. The NW Atlantic SMAP results shown in Table 1 and Figures 12 and 13 suggest that SMAP monthly SSS anomalies have sufficient accuracy and applicability to allow new insight into shelf and coastal dynamics at the monthly scale. Satellite data for 2015–2018 were used to examine Gulf of Maine water inflow, with data suggesting distinct and important water intrusions localized between Georges Banks and Nova Scotia in the winters of 2016/17 and 2017/18. Water intrusion patterns observed by SMAP are consistent with independent measurements from the European SMOS mission.

A cursory evaluation of the dynamics of the winter 2016/17 freshening anomaly inflow used satellite-based ocean current and wind vector data (Figure 15) to suggest the change is controlled by an anomalously strong increase in the Scotian Shelf Current (SSC) that transports Scotian Shelf Water (the freshest water in the region) southwestward down the coast. This results in lower salinity around the southern tip of Nova Scotia, and stronger coastal freshwater flux into the GoM that agrees with interior GoM buoy salinity data at station M01. The forcing behind this SSC acceleration is likely wind-driven in that wind anomaly data in October and November 2016 are consistent with a relaxation in the prevailing Scotian Shelf alongshore westerlies. This finding generally agrees with the circulation perturbation mechanism discussed recently in Li et al. [25].

In summary, study results indicate favorable possibilities for both the immediate use and future improvement of SMAP surface salinity measurements along the slope and sea regions of the NW Atlantic. The new regional in situ datasets compiled to help evaluate coastal and cold-water SMAP data accuracy issues clearly add new information to support improvements of the data processing algorithms, as well as validation data to support the use of salinity anomaly datasets for scientific applications using the presently available V2.0 SMAP data.

Author Contributions: Conceptualization and Methodology, S.A.G. and D.V.; Writing-Original Draft Preparation, S.A.G.; Glider data processing, H.F.

Funding: This research is supported by the NASA Earth Science Division via Ocean Salinity Science Team funding (OSST grants NNX17AK08G and NNX17AK02G).

Acknowledgments: SMAP salinity is produced by Remote Sensing Systems and sponsored by the NASA OSST. NERACOOS and OOI projects are acknowledged for providing the buoy hydrographic data. The glider operations are funded by the Ocean Tracking Network (OTN) and the Marine Environmental Observation Prediction and Response Network (MEOPAR). MAB Slocum glider datasets are provided by the Rutgers University COOL group. Argo data were collected and made freely available by the International Argo Program and the national programs that contribute to it (http://www.argo.ucsd.edu, http://argo.jcommops.org). The Argo Program is part of the Global Ocean Observing System. TSG salinity data are provided by the LEGOS SSS observation service (http://www.legos.obs-mip.fr/observations/sss/) and by the OLEANDER Project (http://po.msrc.sunysb.edu/Oleander/TSG/TSG.html).

Conflicts of Interest: The authors declare no conflict of interest.

References

1. Mountain, D.G. Labrador slope water entering the Gulf of Maine-response to the North Atlantic Oscillation. *Cont. Shelf Res.* **2012**, *47*. [CrossRef]
2. Townsend, D.W.; Pettigrew, N.R.; Thomas, M.A.; Neary, M.G.; Mcgillicuddy, D.J.; Donnell, J.O. Water masses and nutrient sources to the Gulf of Maine. *J. Mar. Res.* **2015**, *73*. [CrossRef] [PubMed]
3. Feng, H.; Vandemark, D.; Wilkin, J. Gulf of Maine salinity variation and its correlation with upstream Scotian Shelf currents at seasonal and interannual time scales. *J. Geophys. Res. Ocean.* **2016**, *121*. [CrossRef]
4. Peterson, I.; Greenan, B.; Gilbert, D.; Hebert, D. Variability and wind forcing of ocean temperature and thermal fronts in the Slope Water region of the Northwest Atlantic. *J. Geophys. Res. Oceans* **2017**, *122*. [CrossRef]
5. Ramp, S.R.; Schlitz, R.J.; Wright, W.R. The Deep Flow through the Northeast Channel, Gulf of Maine. *J. Phys. Oceanogr.* **1985**, *15*. [CrossRef]
6. Grodsky, S.A.; Reul, N.; Chapron, B.; Carton, J.A.; Bryan, F.O. Interannual surface salinity on Northwest Atlantic shelf. *J. Geophys. Res. Oceans* **2017**, *122*. [CrossRef]
7. Reul, N.; Chapron, B.; Lee, T.; Donlon, C.; Boutin, J.; Alory, G. Sea surface salinity structure of the meandering Gulf Stream revealed by SMOS sensor. *Geophys. Res. Lett.* **2014**, *41*. [CrossRef]
8. Dinnat, E.P.; Le Vine, D.M. Effects of the Antenna Aperture on Remote Sensing of Sea Surface Salinity at L-Band. *IEEE Trans. Geosci. Remote Sens.* **2007**, *45*, 2051–2060. [CrossRef]
9. Gierach, M.M.; Vazquez-Cuervo, J.; Lee, T.; Tsontos, V.M. Aquarius and SMOS detect effects of an extreme Mississippi River flooding event in the Gulf of Mexico. *Geophys. Res. Lett.* **2013**, *40*, 5188–5193. [CrossRef]
10. Köhler, J.; Sena Martins, M.; Serra, N.; Stammer, D. Quality assessment of spaceborne sea surface salinity observations over the northern North Atlantic. *J. Geophys. Res. Oceans* **2015**, *120*, 94–112. [CrossRef]

11. Meissner, T.; Wentz, F.J.; Scott, J.; Vazquez-Cuervo, J. Sensitivity of Ocean Surface Salinity Measurements From Spaceborne L-Band Radiometers to Ancillary Sea Surface Temperature. *IEEE Trans. Geosci. Remote Sens.* **2016**, *54*, 7105–7111. [CrossRef]
12. Tang, W.; Fore, A.; Yueh, S.; Lee, T.; Hayashi, A.; Sanchez-Franks, A.; Martinez, J.; King, B.; Baranowski, D. Validating SMAP SSS with in situ measurements. *Remote Sens. Environ.* **2017**, *200*, 326–340. [CrossRef]
13. Boutin, J.; Martin, N.; Kolodziejczyk, N.; Reverdin, G. Interannual anomalies of SMOS sea surface salinity. *Remote Sens. Environ.* **2016**, 128–136. [CrossRef]
14. Lee, T. Consistency of Aquarius sea surface salinity with Argo products on various spatial and temporal scales. *Geophys. Res. Lett.* **2016**, *43*, 3857–3864. [CrossRef]
15. Kubryakov, A.; Stanichny, S.; Zatsepin, A. River plume dynamics in the Kara Sea from altimetry-based lagrangian model, satellite salinity and chlorophyll data. *Remote Sens. Environ.* **2016**, *176*, 177–187. [CrossRef]
16. Garcia-Eidell, C.; Comiso, J.C.; Dinnat, E.; Brucker, L. Satellite observed salinity distributions at high latitudes in the Northern Hemisphere: A comparison of four products. *J. Geophys. Res. Ocean.* **2017**, *122*, 7717–7736. [CrossRef]
17. Meissner, T.; Wentz, F.J. *Remote Sensing Systems SMAP Ocean Surface Salinities [Level 2C, Level 3 Running 8-day, Level 3 Monthly], Version 2.0 Validated Release*; Remote Sensing Systems: Santa Rosa, CA, USA, 2016.
18. Boutin, J.; Vergely, J.L.; Marchand, S.; D'Amico, F.; Hasson, A.; Kolodziejczyk, N.; Reul, N.; Reverdin, G.; Vialard, J. New SMOS Sea Surface Salinity with reduced systematic errors and improved variability. *Remote Sens. Environ.* **2018**, *214*, 115–134. [CrossRef]
19. Tang, W.; Yueh, S.; Yang, D.; Fore, A.; Hayashi, A.; Lee, T.; Fournier, S.; Holt, B. The Potential and Challenges of Using Soil Moisture Active Passive (SMAP) Sea Surface Salinity to Monitor Arctic Ocean Freshwater Changes. *Remote Sens.* **2018**, *10*, 869. [CrossRef]
20. Bentamy, A.; Fillon, D.C. Gridded surface wind fields from Metop/ASCAT measurements. *Int. J. Remote Sens.* **2012**, *33*, 1729–1754. [CrossRef]
21. Grodsky, S.A.; Carton, J.A.; Liu, H. Comparison of bulk sea surface and mixed layer temperatures. *J. Geophys. Res. Oceans* **2008**, *113*. [CrossRef]
22. Liu, H.; Grodsky, S.A.; Carton, J.A. Observed subseasonal variability of oceanic barrier and compensated layers. *J. Clim.* **2009**, *22*, 6104–6119. [CrossRef]
23. Song, Y.T.; Lee, T.; Moon, J.H.; Qu, T.; Yueh, S. Modeling skin-layer salinity with an extended surface-salinity layer. *J. Geophys. Res. Oceans* **2015**, *120*, 1079–1095. [CrossRef]
24. Gawarkiewicz, G.; Todd, R.; Zhang, W.; Partida, J.; Gangopadhyay, A.; Monim, M.-U.-H.; Fratantoni, P.; Malek Mercer, A.; Dent, M. The Changing Nature of Shelf-Break Exchange Revealed by the OOI Pioneer Array. *Oceanography* **2018**, *31*, 60–70. [CrossRef]
25. Li, Y.; Ji, R.; Fratantoni, P.S.; Chen, C.; Hare, J.A.; Davis, C.S.; Beardsley, R.C. Wind-induced interannual variability of sea level slope, along-shelf flow, and surface salinity on the Northwest Atlantic shelf. *J. Geophys. Res. Oceans* **2014**, *119*, 2462–2479. [CrossRef]
26. Boutin, J.; Chao, Y.; Asher, W.E.; Delcroix, T.; Drucker, R.; Drushka, K.; Kolodziejczyk, N.; Lee, T.; Reul, N.; Reverdin, G.; et al. Satellite and in situ salinity understanding near-surface stratification and subfootprint variability. *Bull. Am. Meteorol. Soc.* **2016**, *97*, 1391–1407. [CrossRef]
27. Brooks, D.A. The influence of warm-core rings on slope water entering the Gulf of Maine. *J. Geophys. Res.* **1987**, *92*, 8183. [CrossRef]

remote sensing

MDPI

Article

SMAP and CalCOFI Observe Freshening during the 2014–2016 Northeast Pacific Warm Anomaly

Jorge Vazquez-Cuervo [1,*] and Jose Gomez-Valdes [2]

1 Jet Propulsion Laboratory/California Institute of Technology, Pasadena, CA 91109, USA
2 Physical Oceanography Department, Center for Scientific Research and Higher Education at Ensenada, 22860 Ensenada, Mexico; jgomez@cicese.mx
* Correspondence: jorge.vazquez@jpl.nasa.gov; Tel.: +1-818-354-6980

Received: 5 October 2018; Accepted: 26 October 2018; Published: 31 October 2018

Abstract: Data from NASA's Soil Moisture Active Passive Mission (SMAP) and from the California Cooperative Oceanic Fisheries Investigations (CalCOFI) were used to examine the freshening that occurred during 2015–2016 in the Southern California Current System. Overall, the freshening was found to be related to the 2014–2016 Northeast Pacific Warm Anomaly. The primary goal was to determine the feasibility of using SMAP data to observe the surface salinity signal associated with the warming and its coastal impact. As a first step, direct comparisons were done with salinity from the CalCOFI data at one-meter depth. During 2015, SMAP was saltier than CalCOFI by 0.5 Practical Salinity Units (PSU), but biases were reduced to <0.1 PSU during 2016. South of 33°N, and nearer to the coast where upwelling dominates, SMAP was fresher in 2015 by almost 0.2 PSU. CalCOFI showed freshening of 0.1 PSU. North of 33°N, SMAP and CalCOFI saw significant freshening in 2016, SMAP by 0.4 PSU and CalCOFI by 0.2 PSU. Differences between SMAP and CalCOFI are consistent with the increased stratification in 2015 and changes in the mixed layer depth. SMAP observed freshening that reached the Baja California Coast.

Keywords: coastal; salinity; upwelling

1. Introduction

A well-documented marine heat wave occurred in the Northeast Pacific during the period 2014–2016 (the 2014–2016 warm anomaly). Based on an analysis using NASA's Multiscale Ultra-high-Resolution Sea Surface Temperature (MUR SST), [1] reported that SST anomalies along the West Coast of the United States were warmer than usual during the period 2014–2016, relative to the 2002–2012 climatology. They found that during the upwelling season, the anomalies were abated along the coast. As evidence of the regional effect of upwelling on the warming event, hydrographic data were used by [2] in finding that a strong upwelling event during the spring transition of 2015 abated the warm anomaly in Monterey Bay, California. The authors of [3], in their study of the pelagic ecosystem response of the 2014–2016 warm anomaly off Oregon, found the same effect of the strong upwelling event. Based on underwater gliders' observations and ancillary data products, [4] found that during 2014–2015, the warming of the Southern California Current System (Southern CCS) was a combination of high downward heat flux and weak winds, conditions that also lead to the stratification of the water column. At the same time as the 2014–2016 warm anomaly, a strong El Niño event occurred. The work in [5] analyzed the influence of the 2015–2016 El Niño years on the CCS. They found that the effects of the equatorial Kelvin wave activity and the upwelling-favorable winds on the CCS were weaker than expected, especially for a strong El Niño. At the present time, a study of the salinity variability from remote sensing during the 2014–2016 warm anomaly is lacking.

The salinity signal associated with the 2014–2016 warm anomaly has been documented for some locations in the California Current domain by several authors. The work in [3] showed evidence of negative salinity anomalies for both surface and subsurface waters off the slope/shelf of Oregon. The works in [4] and [2] also found negative salinity anomalies off California. Recent results [6] have shown that salinity derived from the Soil Moisture Active Passive (SMAP) Mission can be used to detect changes in coastal waters associated with river discharge. In a study in the Gulf of Mexico, large signals in freshening were detected. The flooding event of 2015 in Texas caused an increase in river discharge, which led to a freshwater plume in the Gulf of Mexico and significant freshening of greater than five Practical Salinity Units (PSU). Changes in salinity of this magnitude could potentially have impacts on the biology [7]. The work outlined here seeks to determine, using the SMAP data, the observed surface salinity signal during the 2014–2016 warm anomaly. At the same time, direct comparisons will be made with the California Cooperative Oceanic Fisheries Investigations (CalCOFI) data.

A key issue, because of the comparison of remote sensing-derived surface salinity with in situ data, is how near-surface stratification can affect differences between the measurements. The issues were addressed by [8] where they examined possible differences between the two measurements due to evaporation, rain and specific types of wind conditions. Under low wind speeds and high evaporation, differences between one meter and the surface can reach at least 0.2 PSU. Thus, the interpretation of differences between remote sensing-derived salinity and in situ data must be interpreted with respect to the known stratification. This is critical in interpreting the validation of the results.

CalCOFI is a successful and enduring monitoring program, which includes a collection of hydrographic and biological data off the West Coast of North America. Since 1997, CalCOFI scientists have reported annually the state of the California Current. Their reports of 2015, 2016 and 2017 [9–11] have shown that the CalCOFI dataset is a very useful tool to study the hydrography of the 2014–2016 warm anomaly. The CalCOFI data have also been used to validate satellite products [12]. In particular, the CalCOFI hydrographic dataset has been used to evaluate the importance of in situ salinity observations in heat storage estimation from satellite altimetry [13]. To evaluate the performance of SMAP Sea Surface Salinity (SSS) during an open-ocean freshening event, this study compares SSS from SMAP with salinity (retrieved at one meter) from the CalCOFI array during the period 2015–2016. The variability of the freshening is examined along with explanations for differences between SMAP SSS and CalCOFI.

2. Materials and Methods

2.1. Methods

The work was divided into several steps:

(1) Examination of the space-time variability of SSS from SMAP using Hovmöller diagrams.
(2) Validation of SMAP SSS by direct comparisons with CalCOFI-derived salinity at a one-meter depth.
(3) Evaluation of the space-time variability of salinity and stratification parameters associated with the warming event using CalCOFI data.
(4) Evaluation of SMAP SSS versus CalCOFI salinity differences based on air-sea interaction.

For Step 1, Hovmöller diagrams based on latitude versus time were created from the SMAP SSS. The resolution of the SMAP SSS (40 km) allowed for an examination of the time-space evolution of the signal along the two longitudes: 118°W (see Figure 1), close to the coast, and at 120°W, further from the coast. The rationale was to see if there were differences in the variability associated with the coastal upwelling signal compared to further offshore. Additionally, the goal was to use the SMAP data to examine the propagation of the freshening signal from north to south. The two longitudes were chosen to determine if there was any spatial variability in the propagation of the signal as one got close to shore.

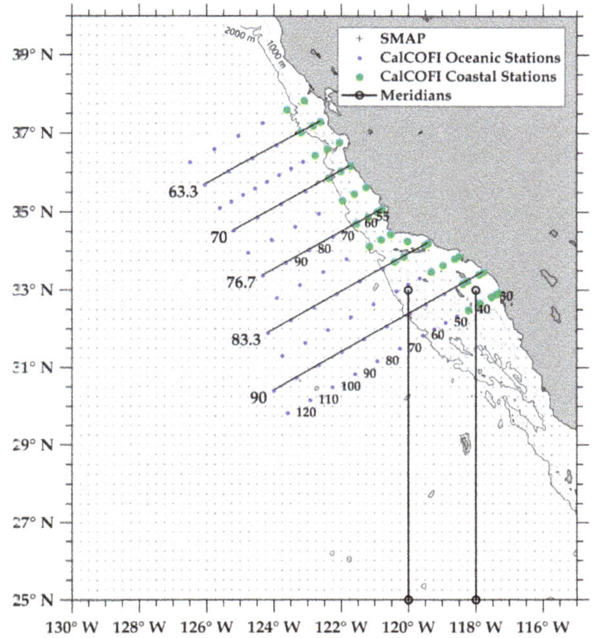

Figure 1. Map of the study area. Soil Moisture Active Passive Mission (SMAP) data domain (+) and location of the California Cooperative Oceanic Fisheries Investigations (CalCOFI) lines and stations. Coastal stations (<100 km) are denoted by green circles, and oceanic stations (>100 km) are denoted by blue circles. The isobaths for 1000 and 2000 m are contoured. Longitude lines 118°W and 120°W are highlighted.

For Step 2, comparisons were done directly between the SMAP SSS and the CalCOFI salinity at a 1-m depth from April 2015–November 2016. The data were co-located in space-time using a simple nearest neighbor approach. Because SMAP is in an 8-day repeat orbit, the co-location window used was ±3 days and 25 km. As with Step 1, comparisons were divided into coastal and open ocean components. A simple criterion was used to divide coastal and open water stations (see Figure 1). For the sake of comparisons with the SMAP data, all CalCOFI stations within 100 km of the coast were considered coastal stations. For Step 3, Hovmöller diagrams were also derived directly using the CalCOFI data.

For the third step, we examined the space-time evolution of the freshening associated with the 2014–2016 warm anomaly. CalCOFI data were used to determine the full space-time evolution of the salinity signal associated with the warming event.

We based our analysis on using the CalCOFI data during the period 2003–2016. The CalCOFI data will be discussed more in Section 2.2.2. We used both basic statistics and analysis of anomalies. Basic statistics were obtained using the four CalCOFI surveys of each year. Temperature and salinity anomalies at the surface (1-m) depth were examined at CalCOFI Lines 76.7 (north), 83.3 (middle) and 90 (south) (see Figure 1), following the methodology of [14] and [15], which consists of the calculation of the mean seasonal variation based on the annual and semiannual harmonics for all stations, where the harmonics were obtained by a least-squares regression of the data. For each station, every anomaly was obtained relative to the mean seasonal variation and the 2003–2012 average. For mapping, Kriging interpolation was applied to the set of anomalies of each line of the CalCOFI sampling grid. Finally, Hovmöller diagrams were elaborated for each line.

For each line, the mixed layer depth was calculated following the method of [16]. The mixed layer depth anomalies were obtained following the same methodology as the one used with temperature and salinity anomalies.

As a final step, an evaluation was done of the air-sea interaction effect on the salinity variations. The rationale was to determine if differences between SMAP SSS and CalCOFI could be explained by the air-sea interaction. Following [17], a simple model was applied to evaporation and precipitation satellite-derived data. Assuming a balance between the salinity changes at the sea surface and the precipitation minus evaporation rates, the salinity budget equation is in the form:

$$\int_{-h}^{0} \frac{\partial S}{\partial t}\, dz = -S_0(P - E) \quad (1)$$

where h is the depth of the layer, S_0 is the mean surface salinity and P and E are the precipitation and evaporation rates, respectively. In this work, h = 1 m.

2.2. Data

2.2.1. JPL SMAP Data

The SMAP SSS data used in the study were from the JPL Version 3 product (doi:10.5067/SMP30-3TMCS).

The data were based on the Level 3 8-day running means gridded at a 0.25-degree resolution. The SMAP product, at the time of the study, was available from April 2015 through the end of 2016. The dates used overlapped with the CalCOFI data from spring 2015–fall 2016. More information on the data and its use may be found in the user's guide [18]. Because of the proximity of the study area to land, a brief review of the data and the correction for land contamination set are given below.

Level 3 products are generated from the Level 2 products using a Gaussian weighting. Weights are assigned based on the distance to the center of the given grid cell. The Level 3 product also contains the Hybrid Coordinate Ocean Model (HYCOM) as an ancillary field, which is shown in the study for comparison purposes. More information on the HYCOM model may be found at https://www.hycom.org. The HYCOM model does not assimilate CalCOFI data. Information on the assimilated parameters used may be found at: https://hycom.org/attachments/084_5_Lozano.pdf. Briefly, parameters assimilated include sea surface temperature from the Advanced Very High Resolution Radiometer (AVHRR), geostationary GOES satellites and sea surface height from the Jason satellite. Low frequency boundary conditions are relaxed to climatologies, including temperature and salinity. HYCOM is used for comparison purposes with the JPL SMAP product to determine the influence of land contamination on the SMAP SSS retrieval. This is critical in a study that attempts to use SMAP SSS data to examine coastal dynamics such as upwelling. The land contamination issue is explained as follows. Radiometers measure energy from the entire visible disk seen on the Earth's surface. If part of that disk lies over land, then the signal will partially include the brightness temperature generated by the land emissivity. In this case, the goal becomes to remove the signal due to land, leaving only the part due to the ocean. Corrections for land contamination are extremely important and need to be considered carefully when applying SMAP SSS to coastal studies.

To remove possible land contamination, the JPL SMAP product calculates a land climatology for the brightness temperature Tb at both the horizontal and vertical polarizations on a monthly time scale [18]. The climatology is then used to estimate the land contamination within 500 km of the coast. This is critical for deriving accurate SSS in the upwelling region off Southern California, which lies within areas of land contamination close to the coast. With the land correction, biases of SSS are reduced to −0.2 PSU with SSS being too fresh [18]. Beyond 50 km, biases are reduced to near zero. Root mean squared (RMS) differences are reduced to 0.5 PSU. Thus, the application of the land correction is critical for resolving the coastal SSS signal, where upwelling is expected to bring saltier deep water to the surface. Negative biases (fresher) near the coast would indicate possible land contamination.

2.2.2. CalCOFI Data

CalCOFI has made observations of the hydrography and biology off the West Coast of North America since 1949. Quarterly cruises have been conducted off Southern California since 1984. The CalCOFI's standard sampling grid is based on parallel lines oriented perpendicular to the coast; the separation of lines and stations are both 74 km [14]. Data collected at depths down to 515 m include a Seabird 911 CTD (Conductivity, Temperature, Depth) mounted on a 24-bottle rosette with continuous measurements of pressure, temperature and conductivity. CalCOFI CTD data are computed by Seasoft based on (equation of state for seawater, 1980) EOS-80. Salinity is (practical salinity scale 1978) PSS-78. Salinity values in the CalCOFI array are typically between 32 and 36 PSU, with of course the equivalency of PSS and PSU. More information on CalCOFI CTD general practices and algorithms may be found at http://calcofi.org/. The modern CalCOFI salinity measurements provide a unique coastal dataset that spans decades. Thus, the data will provide a validation of SSS during the overlap period of the SMAP mission. Presented here is the time evolution of the seawater properties using data obtained during the 2003–2017 period in the CalCOFI 104 stations plan (see Figure 1). Details of the sampling protocol of this stations plan can be found in [14].

2.2.3. Evaporation Data

Evaporation data were retrieved from the Woods Hole Oceanographic Institution Objectively Analyzed Air-Sea Flux dataset [19]. More information on the dataset may be retrieved from http://oaflux.whoi.edu. The evaporation data use an advanced optimal interpolation technique to combine satellite and model from the numerical weather prediction. This is necessary because satellite data alone are not sufficient to derive parameters needed for the air-sea flux determination.

2.2.4. Precipitation Data

Rain rate was extracted from the Special Sensor Microwave Imager (SSM/I) onboard the Defense Meteorological Satellite Program (DMSP) satellites since 1987. DMSP satellites are polar orbiting and listed by satellite number, F08–F18. Rainfall data were downloaded from Remote Sensing Systems (http://www.remss.com/missions/ssmi/). More information on the processing and latest processing algorithms may be found in [20].

3. Results

3.1. Evaluation of SMAP Data

The JPL SMAP data were directly compared/co-located with all the CalCOFI data to determine the overall quality and validate the data in the region. The data were also divided into sub-regions for statistical comparisons to evaluate possible differences based on the proximity to the coast. For the SMAP and CalCOFI co-locations, we used the criterion of 100 km from the coast to delineate open ocean and coastal comparisons (see Figure 1). Comparisons were also done with the HYCOM model as a third independent dataset. The HYCOM model-derived salinity is readily available as an ancillary field in the SMAP SSS. Comparisons are only shown for the JPL product.

Figure 1 shows the location of the CalCOFI lines used in the study. Along the coast, the CalCOFI monitoring program spans the California Coastline. The stations' plan also extends several hundred kilometers offshore. We analyzed all the lines shown in Figure 1. The lines 76.7, 83.3 and 90 had the highest occupations during the period 2003–2016.

Figure 2a–f shows the monthly averaged values for SMAP for (a) April 2015, (b) April 2016, (c) July 2015, (d) July 2016, (e) November 2015 and (f) November 2016. The most prominent differences are clearly seen in July and November. July of 2015 clearly shows the saltier water along the coast associated with the seasonal upwelling. The work in [15] outlined three periods of upwelling off the Central to Northern California; April–June is associated with the intensification of upwelling due to alongshore Equatorward Winds. For July–September, one sees a relaxation of the upwelling associated with the weakening of the alongshore winds. November through March is considered the storm season and a transitional period. These patterns are consistent with the variability seen in Figure 2. Thus, the observed seasonal cycle in the SMAP SSS is consistent with the historical analysis. November of 2016 shows fresher waters along the coast, specifically between 30°N and 35°N. Overall, the SMAP SSS shows freshening from 2015–2016, with a clear impact along the coast and the upwelling signal seen in July. To our knowledge, this is the first time salinity data from SMAP have been used to observe the coastal upwelling signal off California that is associated with the 2014–2016 warm anomaly.

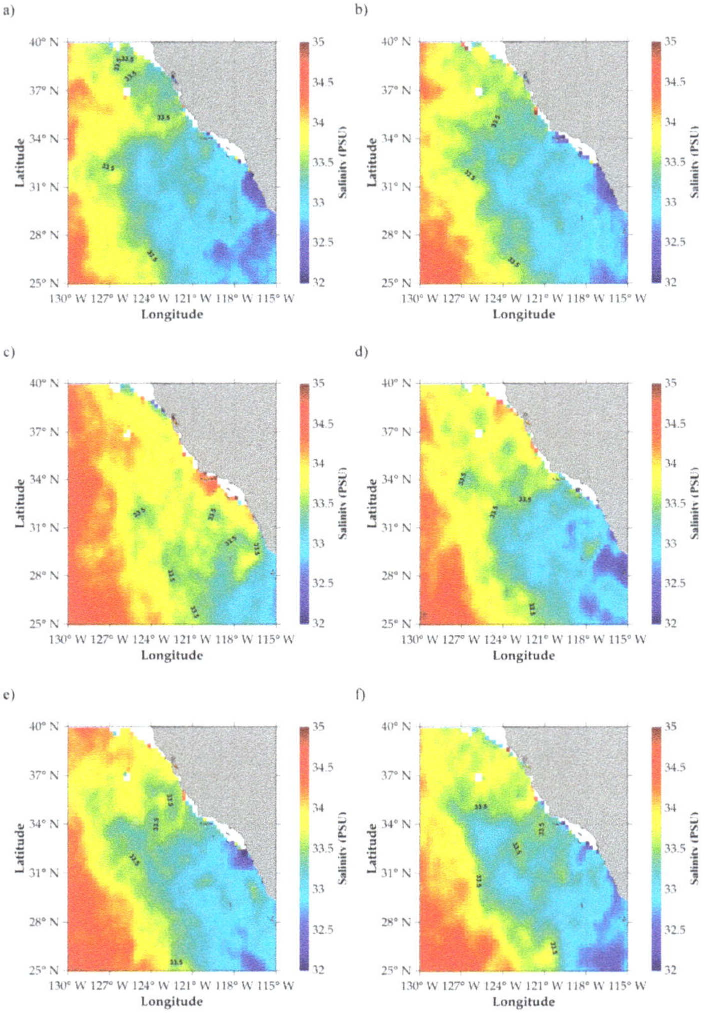

Figure 2. SMAP monthly. (**a**) Map of Sea Surface Salinity (SSS) averaged from SMAP for April 2015; (**b**) map of SSS averaged from SMAP for April 2016; (**c**) map of SSS averaged from SMAP for July 2015; (**d**) map of SSS averaged from SMAP for July 2016; (**e**) map of SSS averaged from SMAP for November 2015; (**f**) map of SSS averaged from SMAP for November 2016.

Figure 3a,b illustrates further the space-time variability of SSS from SMAP using Hovmöller diagrams at 120°W (a) and 118°W (b). Upwelling scales in this region can be variable, and the two longitudes were chosen based on work by [21]. The work in [21], for similar latitudes, chose 119°W to be representative of the coastal upwelling. Thus, the two longitudes were simply chosen to determine if there were changes in the propagation of the north-south freshening based on closeness to shore. At 120°W (further offshore), one clearly sees the propagation of freshwater from north to south, except during the summer and the seasonal upwelling season. From April 2015–December 2016, freshwater had propagated from 33°N–25°N. The dashed line in Figure 3a indicates a north-south propagation of approximately 2 cm/s. This is consistent with the observed effects of the warming off the Baja California Coast [22] The work in [23] observed the propagation of the SST signal reaching the Baja California Coast. Freshening is also observed closer to the coast at 118°W, but the signal in 2015 is dominated by an annual/semiannual component, most likely due to the coastal upwelling signal. One still sees a semiannual component closer to the coast, but localized further south at approximately 25°N. The freshening has impacted the coastal waters and likely weakened the upwelling signal. The observed gap in data at approximately 29°N in Figure 3b is due to the island of Guadalupe. The results are consistent with [1] and [2] in showing that the warming along the coast abated during the spring upwelling season. Overall, the results are consistent in showing that although the impacts of the warming were felt along Southern California, and the maxima impact of SST and SSS were seen north of 35°N. Additionally, the saltier water seen near the coast for July 2015 is consistent with the 2015 El Niño [10]. To quantify and validate these results, direct comparisons were first made with the salinity at a one-meter depth from the CalCOFI monitoring program.

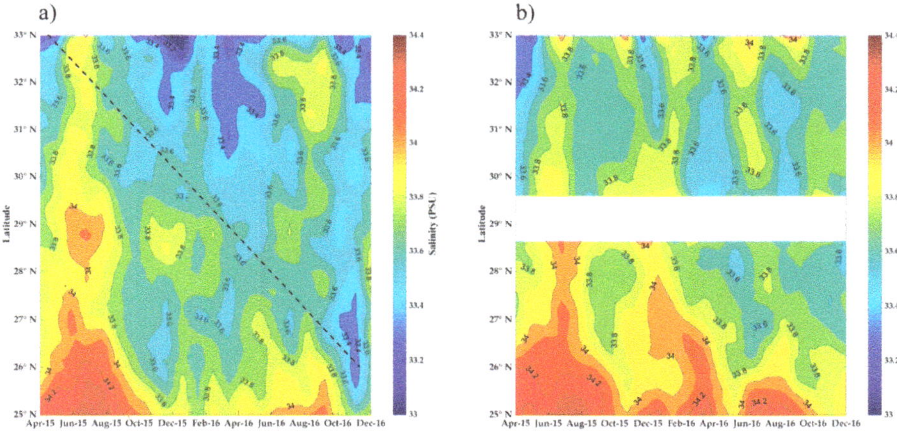

Figure 3. (**a**,**b**) Hovmöller diagrams using SMAP-derived SSS of the latitudinal variation at two ((**a**) 120°W and (**b**) 118°W) longitudes. The dashed line shows the approximate propagation of the freshening from north to south.

Figure 4 was generated using values of SMAP co-located with the CalCOFI lines from Figure 1. All the CalCOFI data indicated in Figure 1 were used. Results were separated into coastal and open ocean stations (see Figure 1). Figure 4a shows the mean SSS values over the entire CalCOFI region for April 2015, July 2015, November 2015, January 2016, April 2016 and November 2016. Over the entire region for 2015, values for SMAP SSS (red) were saltier than either HYCOM (blue) or CalCOFI (black). Values show SMAP SSS saltier by almost 0.4 PSU for 2015. In 2016, SMAP SSS and CalCOFI agreed to less than 0.1 PSU. Over the entire CalCOFI region, all three, SMAP, CalCOFI and HYCOM, showed a freshening from 2015–2016, but the freshening was magnified in SMAP. The freshening in SMAP was approximately 0.3 PSU, while in CalCOFI and HYCOM, the freshening was around 0.1

PSU. Figure 4b shows the bias between SMAP-CalCOFI (black) and HYCOM-CalCOFI (blue). Biases between HYCOM-CalCOFI were consistently smaller than 0.1 PSU over the entire region. Biases defined as SMAP-CalCOFI showed the saltier SMAP bias of 0.4 PSU for 2015 decreasing to less than 0.1 PSU in 2016. Thus, over the entire region, SMAP SSS was saltier than CalCOFI for 2015, almost reaching zero in 2016. Figure 4c is the same as Figure 4a, but for the area between 30°N and 33°N and using only the coastal stations (see Figure 1). The intent here was to focus on the upwelling region where also saltier SSS values were observed in July 2016 (see Figure 2). The latitudes 30°N and 33°N were chosen to capture the upwelling region (saltier) identified in Figure 2. Close to the coast (Figure 4c) in the Southern CCS region, all three: SMAP, CalCOFI, and HYCOM, showed fresher values in 2015, becoming slightly saltier in 2016. This is consistent with upwelling conditions returning to normal in mid–late 2016. SMAP shows an increase in saltiness by about 0.3 PSU, while CalCOFI and HYCOM show an increase of about 0.1 PSU. Figure 4e shows the SMAP, CalCOFI and HYCOM values for the region between 34°N and 38°N for the coastal stations only. All three datasets show an increase in freshening in 2016 with minimum values seen in April of 2016. However, the freshening in SMAP is even more pronounced, greater than 0.4 PSU. After April of 2016, all three datasets showed trends of increasing saltiness. Biases (Figure 4f) are reflective of the saltiness of SMAP in 2016. The change in the SSS trends between 30°N and 38°N would be consistent with the difference upwelling conditions associated with the California Coast. Error bars in Figure 4b,d,f show consistent RMS differences that are reduced to less than 0.3 PSU for 2016. Overall biases between the HYCOM SSS and CalCOFI SSS were <0.1 PSU. Based on Figure 4, the large salty bias in the SMAP SSS when compared with the CalCOFI in situ salinity will be addressed.

The saltier values of SMAP in 2015 may be understood in terms of the stratification of the water column. Previous studies comparing SMAP with ARGO data have shown that there are no significant changes in the biases between 2015 and 2016 [24]. There is also no indication that such biases exist in the CalCOFI data. Biases in satellite-derived SSS have other possible causes. Other possible explanations for the biases could be due to sea surface temperature [23]. However, the sea surface temperature biases dominate at cold temperatures [25]. This is due to the decreased sensitivity of salinity to changes in brightness temperatures. This bias would be minimized along the California Coast. Additionally, the bias would have a dominant seasonal cycle. As the largest biases occurred during the warmer waters of 2015, warm SST-induced biases effecting SSS are inconsistent with cold temperatures and a seasonal component. Thus, the sea surface temperature bias is an unlikely explanation for the differences seen between SMAP and CalCOFI. Another possible cause of biases is due to land contamination. However, maxima biases occurred at distances >100 km from the coast (open ocean), a distance that would not be associated with possible land contamination. Thus, another explanation must be found for the observed differences between 2015 and 2016. This will be discussed further in the next section with respect to the stratification of the water column and air-sea coupling. Overall results using the CalCOFI data showed increased stratification of the water column in 2015, with saltier values.

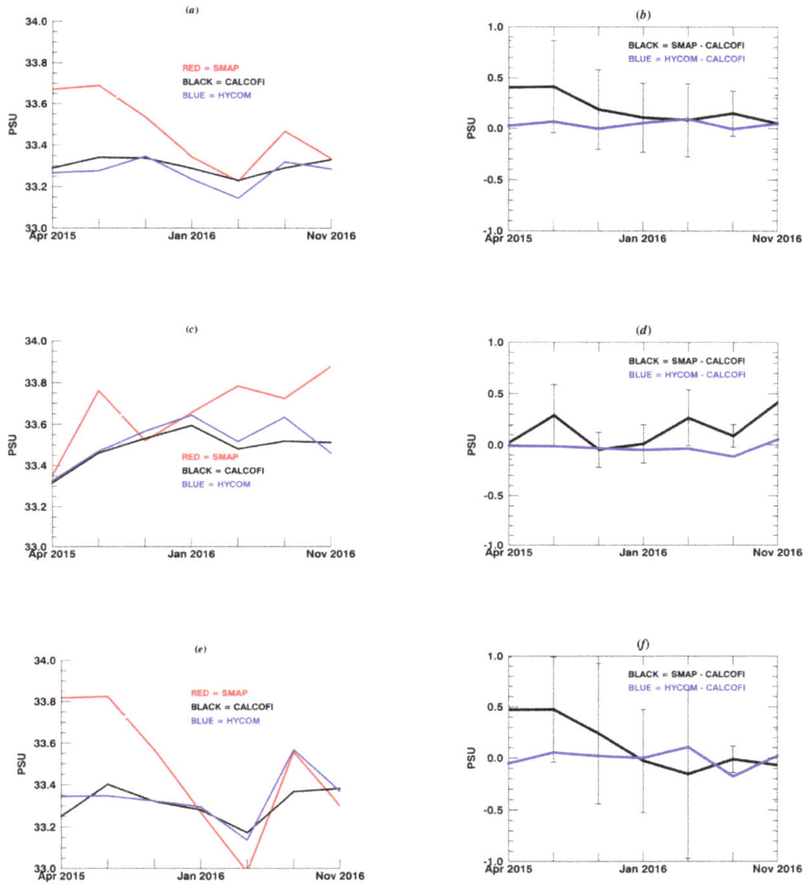

Figure 4. (**a**) shows SSS averaged for the entire CalCOFI area defined in Figure 1 for coastal and open ocean stations. Salinity for all CalCOFI stations as measured by SMAP = red, CalCOFI = black and Hybrid Coordinate Ocean Model (HYCOM) = blue. (**b**) shows bias and RMS over the CalCOFI area for SMAP-CalCOFI (black) and HYCOM-CalCOFI (blue). (**c**) The same as (a), except for coastal stations between 30 and 33°N. (**d**) The same as (b), except for coastal stations between 30°N and 33°N. (**e**) The same as (a), except for coastal stations between 34°N and 38°N. (**f**) The same as (b), except for coastal stations between 34°N and 38°N.

3.2. Salinity Variations Induced by Precipitation and Evaporation

Table 1 shows the year-to-year salinity changes associated with the rate of change of precipitation minus the evaporation factor. The change in salinity is based on using Equation 1. The maximum salinity change occurred in the fall and the minimum in summer. Because the evaporation rate was greater than precipitation rate, the salinity changes are mainly associated with the former. Table 1 reveals that the salinity changes in 2015 induced by the evaporation reached −0.6 PSU in April of 2015 and were greater than −2.2 PSU in November. The results are consistent with the saltier values derived from SMAP in 2015. The increased evaporation is also consistent with the increased stratification in 2015. This will be discussed more in later sections.

Table 1. Year-to-year salinity changes. PSU, Practical Salinity Units.

Month	2015–2016 (Units Are PSU)
January	−1.3
April	−0.6
July	0.1
November	−2.2

3.3. CalCOFI Observations

Figure 5 shows the histograms of the salinity distribution for the coastal zone at a 1-m depth for the four-year period 2013–2016. Station 60 of each line was used as the delimiter between the coastal and the oceanic zone. Because the arithmetic mean is sensitive to outliers, we used the median as the central tendency and higher central moments such as kurtosis and skewness to evaluate the dispersion. A positive kurtosis corresponds to a sharper peak, while the skewness is a measure of the asymmetry of the distribution. Whenever the distribution is extended more to the left of the mean value, it will always be a sign of a negative skew.

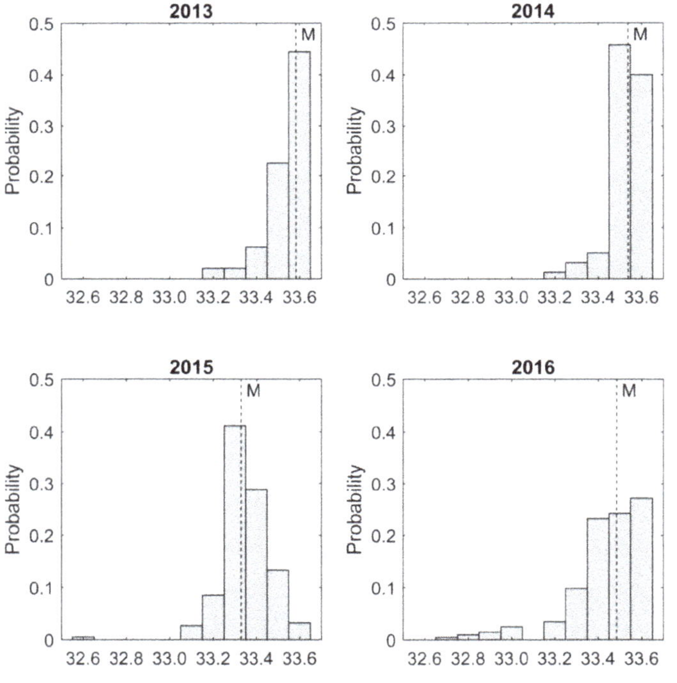

Figure 5. Histograms of salinity at a 1-m depth for the coastal zone of the CalCOFI sampling region: 2013, 2014, 2015 and 2016. The dashed line indicates the median of the distribution.

In 2013, the range of salinity was between 33.2 and 33.9 PSU with a skewness of −0.50 and a kurtosis of 4.9, while the median stood at 33.6 PSU (Table 2). In 2014, the range of the salinity was the same as in 2103, the skewness was −1.12, and the kurtosis was 5.1. In 2015, the range changed; it moved toward low values. The skewness value decreased, and the kurtosis increased, with the median decreasing. Finally, in 2016, the measures of central tendency and dispersion followed the same pattern as in 2015. The 2013–2014 period was saltier than the 2015–2016 period. It is also worth mentioning that the kurtosis was higher in the second period (2015–2016).

Table 2. Year-to-year measures of central tendency and dispersion of the salinity distribution at a 1-m depth from the CalCOFI data for the coastal zone. Units are PSU.

Year	Range	Median	Skewness	Kurtosis
2013	33.2–33.9	33.6	−0.5	4.9
2014	33.2–33.6	33.5	−1.1	5.1
2015	32.6–33.6	33.3	−1.3	10.7
2016	32.3–33.7	33.5	−2.3	10.0

Figure 6 shows the histograms of the salinity distribution for the oceanic zone at a 1-m depth for the same four-year period. For the oceanic zone, these histograms show clearly the freshening from 2014 to the 2015–2016 period. In 2013, the range of salinity was between 32.8 and 33.7 PSU with a skewness of −0.30 and a kurtosis of 3.4, while the median stood at 33.3 PSU (Table 3). In 2014, the range of the salinity was the same as in 2013; the kurtosis was positive, and the median was larger than in 2013. In 2015, the range changed to 32.7–33.7 PSU; the skewness was −0.12, and the kurtosis increased to 2.3. In this case, the median decreased to 33.2 PSU. Finally, in 2016, the range of salinity was between 32.5 and 33.6 PSU; the skewness was −0.25; the kurtosis was 2.8; and the median decreased to 33.1.

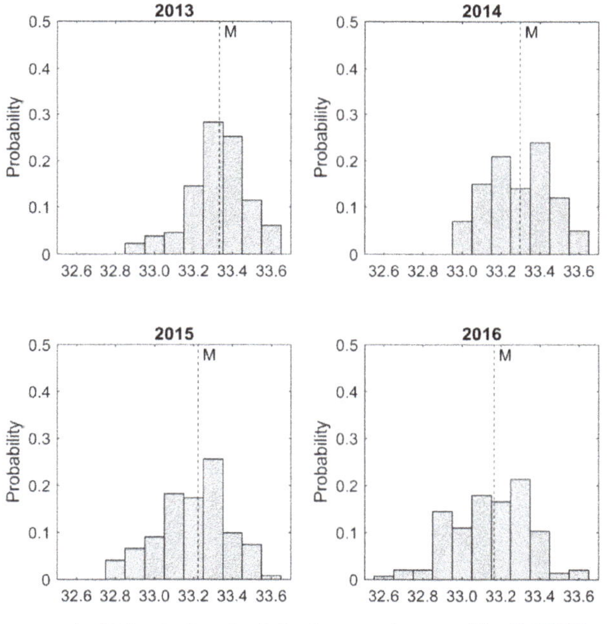

Figure 6. Histograms of salinity at a 1-m depth for the oceanic zone of the CalCOFI sampling region: 2013, 2014, 2015 and 2016. The dashed line indicates the median of the distribution.

Table 3. Same as Table 2, except for the oceanic zone.

Year	Range	Median	Skewness	Kurtosis
2013	32.8–33.7	33.33	−0.30	3.48
2014	32.9–33.7	33.73	−0.18	2.41
2015	32.7–33.7	33.22	−0.12	2.67
2016	32.5–33.6	33.16	−0.25	2.78

CalCOFI data at a 1-m depth registered the 2014–2016 warm anomaly (Figure 7a,c,e). In the CalCOFI region, the warming started in 2014 and was most pronounced in 2015 when the positive anomalies were around 4 °C.

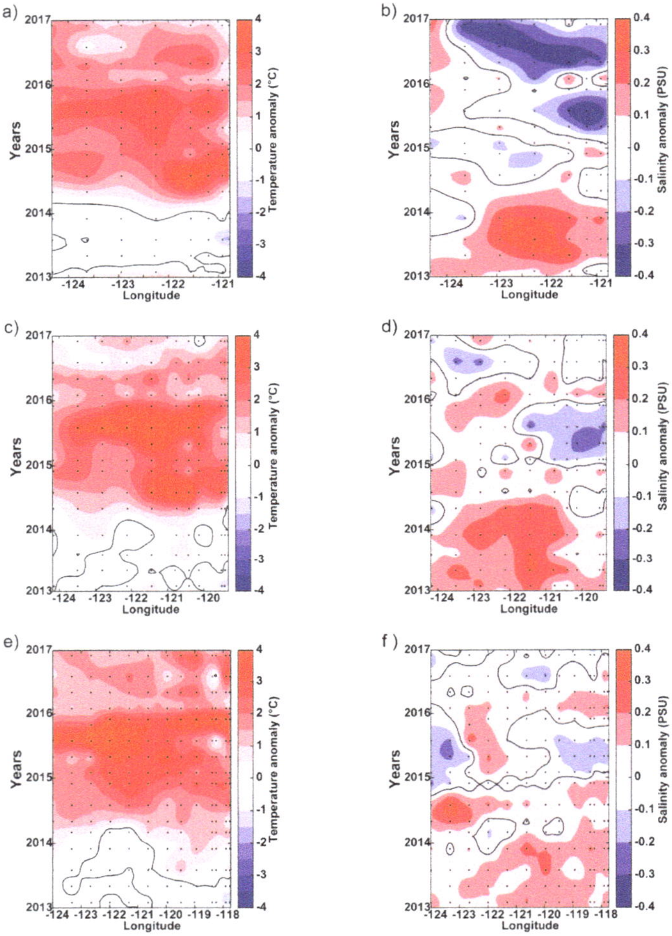

Figure 7. (**a**–**f**) Hovmöller diagrams of temperature anomalies (°C) at a 1-m depth for Lines 76.7 (**a**), 83.3 (**c**) and 90 (**d**) and salinity anomalies (PSU) at a 1-m depth for Lines 76.7 (**b**), 83.3 (**e**) and 90 (**f**). The anomalies are relative to the 2003–2012 climatology.

The geographic distribution of SSS anomalies in the CalCOFI region was irregular during the period 2014–2016 (Figure 7b,d,f)), which is an indication that SSS was responsive to local dynamics and air-sea fluxes. Near the coast, negative SSS anomalies prevailed in 2015 at the three CalCOFI lines. In the northern part of the CalCOFI region (Line 76.7), negative SSS anomalies (less than −0.2 PSU) occurred in 2016 almost over the entire line, while for the rest of the CalCOFI lines' negative SSS anomalies were of lesser magnitude. The minimum anomaly (−0.4 PSU) occurred in 2016 along Line 76.7. The freshening associated with the 2014–2016 warm anomaly was evident in the SSS anomalies.

The Mixed Layer Depth (MLD) is an indication of the degree of stratification; the smaller the MLD, the greater the stratification. During the 2014–2016 warm anomaly, MLD was overall anomalously negative in the CalCOFI region (Figure 8a–c). At Line 76.7, the negative MLD anomalies prevailed during 2014–2015, when the highest number of negative anomalies occurred further from the coast. In exchange, the negative anomalies occurred near the coast in 2016. As for Line 83.3 the negative MLD anomalies also prevailed during 2014–2015, but not in the winter season. Meanwhile, at Line 90, the negative MLD anomalies were less abundant than the other lines. This would also be consistent with results from Figure 4, which showed that between 30°N and 33°N, the large biases between SMAP SSS and CalCOFI salinity were minimized, with both showing a trend to increasing saltiness in 2016.

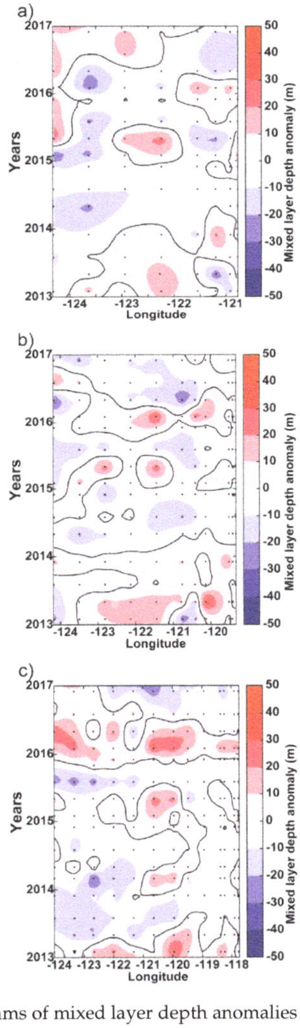

Figure 8. (**a–c**) Hovmöller diagrams of mixed layer depth anomalies (m) for Lines 76.7 (**a**), 83.3 (**b**) and 90 (**c**). The anomalies are relative to the 2003–2012 climatology.

4. Discussion

Results illustrate and compare both the surface salinity signal as measured by SMAP and the 1-m depth salinity signal as determined by CalCOFI during the 2014–2016 warm anomaly. Unique results are presented using SMAP SSS to observe the variability of salinity in a coastal region. Results are consistent with the CalCOFI 1-m depth salinity and the known stratification during that period of time.

Freshening was observed in the CalCOFI region throughout 2015 and 2016. The freshening was observed to be greater in the open ocean in both the CalCOFI and SMAP data. In the SMAP SSS, freshening offshore appears in 2015, spreading throughout the entire region in 2016. Offshore SMAP SSS shows freshening of 0.5 PSU starting in the north in 2015, but then extending south through 2016. The magnitude of the freshening, approximately 0.5 PSU, is most likely driven by the increased stratification in 2015. The stratification is consistent with MLD derivations for the same time period. Freshening is also observed in the CalCOFI lines, but at a decreased magnitude. Maximum freshening appears at Line 76.7 (north) in 2016 with anomalies, based on the CalCOFI climatology at one-meter depth, reaching 0.4 PSU. Further south, freshening is reduced with anomalies of 0.2 PSU appearing along the coast in 2015. The differences in SMAP SSS and CalCOFI salinity between 2015 and 2016 can be explained by the stratification of the water column. Stratification occurred in 2015 and 2016, but had a maximum in 2015. Biases between SMAP SSS and CalCOFI were larger than 0.4 PSU during 2015, but reduced to less than 0.1 PSU in 2016. The difference is consistent with the increased stratification in 2015. Results are consistent with [4], who showed that during the 2014–2015 period, local anomalous heat flux and wind strength led to the warming in the region. During the period of greater mixing and decreased stratification, differences between the SMAP SSS and CalCOFI were minimal. The stratification is consistent with increased evaporation rates in 2015 and changes in surface salinity.

Close to shore, both SMAP and CalCOFI show a trend towards saltier values between 2015 and 2016. This is consistent with the increasing strength of the upwelling signal in 2016. Results are consistent with [1] and [2].

Results are extremely important in showing that both satellite-derived SSS and in situ salinity are needed to monitor changes in the stratification of the water column associated with air-sea interactions. There are predominately two reasons for this. First, satellites measure a response at the surface of the ocean, while in situ measurements are at depth. Thus, differences between the two measurements would reflect the stratification of the water column, as they are related to the air-sea coupling. A second validation using in situ data is important in assessing the quality of the satellite data. Once validated, the satellites give enhanced spatial and temporal coverage not possible with in situ data. Off the California Coast, upwelling scales, along with fronts, can change on weekly to monthly time scales. The agreement between SMAP SSS and CalCOFI in the region is very encouraging for using SMAP SSS in coastal regions. Biases of less than 0.1 PSU would allow for monitoring of changes in salinity in critical ocean regions associated with coastal upwelling. SMAP is consistent in showing that near the coast, freshening was minimized. In 2015, the observed signal in freshening is consistent with a weakening of coastal upwelling [1,3].

Summarizing, direct comparisons between the SMAP SSS and salinity derived from the CalCOFI array showed that biases were observed with the SMAP SSS saltier than CalCOFI for 2015. During 2016, biases between SMAP SSS and CalCOFI were reduced to less than 0.1 PSU. One of the goals of the work was to determine the primary reason behind the observed differences between 2015 and 2016 and relate them to salinity differences between SMAP and CalCOFI caused by possible ocean conditions. Biases between the SMAP SSS and CalCOFI salinity could not be attributable to sea surface temperature and/or possible land contamination.

The 2015 differences between SMAP SSS and CalCOFI were found to be associated with the stratification of the water column. Based on the derivation, Figure 8 indicates a stronger stratification in 2015, consistent with an increase in salinity at the surface observed by the SMAP satellite. Differences in salinity between the two measurements are consistent with the results in [8].

5. Conclusions

Several key results were found from using the SMAP SSS. Overall, this should be considered an important step forward in applying satellite-derived salinity to coastal processes and their connection to basin-scale changes. We identified four major conclusions in the study. First, the SMAP SSS can be used to monitor the freshening in a coastal region associated with a major warming event. SMAP observed the freshening extending to the Baja California Coast. Second, using the SMAP SSS, freshening due to the Northeast Pacific heat wave was observed to propagate and reach the Baja California Coast. Third, the SMAP SSS identifies changes in salinity associated with a coastal upwelling system. Fourth, coastal station-based programs, such as CalCOFI, are critical for fully understanding and validating satellite-derived SSS. In periods of low stratification, biases between the SMAP-derived SSS and CalCOFI were less than 0.1 PSU, increasing to greater than 0.4 during periods of high stratification

Parameters that identify the air-sea coupling are critical for a comprehensive understanding of the differences between satellite-derived and in situ surface salinities. This study shows that in situ data are not only needed for validation purposes, but also to understand fully the issues of stratification and mixing that may lead to differences between satellite-derived SSS and buoy salinity. Using both SMAP and CalCOFI leads to further evidence of the increased stratification during the 2014–2016 Northeast Pacific Warm Anomaly.

Future research should focus on validation efforts of SMAP in other coastal regions where upwelling is pronounced. This includes the Peru/Chile Coast, the Canarias, Benguela and Western Australia.

Author Contributions: J.V.-C. carried out the work at the Jet Propulsion Laboratory as part of the Salinity Calibration/Validation effort. J.G.-V. carried out the work while on sabbatical at the Jet Propulsion Laboratory/ California Institute of Technology from the Centro de Investigación Científica y de Educación Superior de Ensenada, Baja California. CONACyT supported the preparation of this paper through Contract No. 257125 to J.G.-V.

Funding: The lead author was funded through a contract with the National Aeronautics and Space Administration at the Jet Propulsion Laboratory/California Institute of Technology.

Acknowledgments: The research was carried out at the Jet Propulsion Laboratory, California Institute of Technology, under a contract with the National Aeronautics and Space Administration. The SMAP data are accessible through the Physical Oceanography Distributed Active Archive Center (https://podaac.jpl.nasa.gov/dataset/SMAP_JPL_L3_SSS_CAP_8DAY-RUNNINGMEAN_V3). CalCOFI data are available through http://calcofi.org/data.html. Luis E. Miranda, at the Centro de Investigación Científica y de Educación Superior de Ensenada (CICESE), gave computing and technical support to the project.

Conflicts of Interest: The authors declare no conflict of interest.

References

1. Gentemann, C.L.; Fewings, M.L.; Garcia-Reyes, M. Satellite sea surface temperature along the West Coast of the United States during the 2014–2016 northeast Pacific marine heat wave. *Geophys. Res. Lett.* **2016**, *44*, 312–319. [CrossRef]
2. Ryan, J.P.; Kudela, R.M.; Birch, J.M.; Blum, M.; Bowers, H.A.; Chavez, F.P.; Doucette, G.J.; Hayashi, K.; Marin, R., III; Mikulski, C.M.; Pennington, J.T.; et al. Causality of an extreme harmful algal bloom in Monterey Ba, California, during the 2014-2016 northeast Pacific warm anomaly. *Geophys. Res. Lett.* **2017**, *44*, 5571–5579. [CrossRef]
3. Peterson, W.T.; Fisher, J.L.; Strub, P.T.; Du, X.; Risien, C.; Peterson, J.; Shaw, C.T. The pelagic ecosystem in the Northern California Current off Oregon during the 2014-2016 warm anomalies within the context of the past 20 years. *J. Geophys. Res. Oceans* **2017**, *122*. [CrossRef]
4. Zaba, K.D.; Rudnick, D.L. The 2014–2015 warming anomaly in the Southern California Current System observed by underwater gliders. *Geophys. Res. Lett.* **2016**, *43*, 1241–1248. [CrossRef]
5. Jacox, M.G.; Hazen, E.L.; Zaba, K.D.; Rudnick, D.L.; Edwards, C.A.; Moore, A.M.; Bograd, S.J. Impacts of the 2015–2016 El Niño on the California Current System. Early assessment and comparison to past events. *Geophys. Res. Lett.* **2006**, *43*, 7072–7080. [CrossRef]

6. Fournier, S.; Reager, J.T.; Lee, T.; Vazquez-Cuervo, J.; David, C.H.; Gierach, M.M. SMAP observes flooding from land to sea: The Texas event of 2015. *Geophys. Res. Lett.* **2016**, *43*. [CrossRef]
7. Pares-Escobar, F.; Lavaniegos, B.E.; Ambriz-Arreola, I. Interannual variability in oceanic euphausiid communities off the Baja California western coast during 1998–2008. *Progr. Oceanogr.* **2018**, *160*, 53–67. [CrossRef]
8. Boutin, J.; Chao, Y.; Asher, W.E.; Delcroix, T.; Drucker, R.; Drushka, K.; Kolodziejczyk, N.; Lee, T.; Reul, N.; Reverdin, G.; et al. Satellite and in situ salinity: Understanding near-surface stratification and sub-footprint variability. *Bull. Am. Meteorol. Soc.* **2016**. [CrossRef]
9. Leising, A.W.; Schroeder, I.D.; Bograd, S.J.; Abell, J.; Durazo, R.; Gaxiola-Castro, G.; Bjorkstedt, E.P.; Field, J.; Sakuma, K.; Robertson, R.R.; et al. State of the California Current 2014–15: Impacts of the warm- water "blob". *Calif. Coop. Fish. Investig. Rep.* **2015**, *56*, 31–68.
10. McClatchie, S.; Goericke, R.; Leising, A.; Auth, T.D.; Bjorkstedt, E.; Robertson, R.R.; Brodeur, R.D.; Du, X.; Daly, E.A.; Morgan, C.A.; et al. State of the California Current 2015–2016: Comparisons with the 1997–98 El Niño. *Calif. Coop. Fish. Investig. Rep.* **2016**, *57*, 5–61.
11. Wells, B.K.; Schroeder, I.D.; Bograd, S.J.; Hazen, E.L.; Jacox, M.G.; Leising, A.; Mantua, N.; Santora, J.A.; Fisher, J.; Peterson, W.T.; et al. State of the California Current 2016–17: Still anything but "normal" in the north. *Calif. Coop. Fish. Investig. Rep.* **2017**, *58*, 1–55.
12. McClain, C.R. A decade of satellite ocean color observations. *Annu. Rev. Mar. Sci.* **2008**, *1*, 19–42. [CrossRef] [PubMed]
13. Sato, O.T.; Polito, P.S.; Liu, W.T. Importance of salinity measurement in the heat storage estimation from TOPEX/POSEIDON. *Geophys. Res. Lett.* **2000**, *27*, 549–551. [CrossRef]
14. Bograd, S.J.; Lynn, R.J. Long-term variability in the Southern California Current System. *Deep-Sea Res. II* **2003**, *50*, 2355–2370. [CrossRef]
15. Garcia-Reyes, M.; Largier, J.L. Seasonality of coastal upwelling off central and northern California: New insights, including temporal and spatial variability. *J. Geophys. Res.* **2012**, *117*, C03028. [CrossRef]
16. Torres, H.S.; Gomez-Valdes, J. Coastal Circulation driven by short-period upwelling-favorable winds in the northern Baja California region. *Deep-Sea Res. I* **2015**, *98*, 31–42. [CrossRef]
17. Feng, M.; Hacker, P.; Lukas, R. Upper ocean heat and salt balances in response to a westerly wind burst in the western equatorial Pacific during TOGA COARE. *J. Geophys. Res.* **1998**, *103*, 10289–10311. [CrossRef]
18. Fore, A.; Yueh, S.; Tang, W.; Hayashi, A. *SMAP Salinity and Wind Speed Data User's Guide*; version 3; Jet Propulsion Laboratory/California Institute of Technology: Pasadena, CA, USA, 2016.
19. Yu, L.; Jin, X.; Weller, R.A. *Multidecadal Global Flux Datasets from the. Objectively Analyzed Air-sea Fluxes (OAFlux) Project: Latent and Sensible Heat Fluxes,. Ocean Evaporation, and Related Surface Meteorological Variables (OAFlux Project Technical Report OA-2008-01)*; Woods Hole Oceanographic Institution: Woods Hole, MA, USA, 2008; p. 64.
20. Wentz, F.J. *SSM/I Version-7 Calibration Report*; RSS Technical Report 011012; Remote Sensing Systems: Santa Rosa, CA, USA, 2013; p. 46.
21. Schwing, F.B.; O'Farrell, M.; Steger, J.M.; Baltz, K. *Coastal Upwelling Indices West Coast of North America 1946-95*; NOAA Technical Memorandum, NMFS, NOAA-TM-NMFS-SWFSC-231; United States Department of Commerce: Washington, DC, USA, 1996.
22. Gómez-Ocampo, E.; Gaxiola-Castro, G.; Durazo, R.; Beier, E. Effects of the 2013–2016 warm anomalies on the California Current Phytoplankton. *Deep-Sea Res. II* **2017**. [CrossRef]
23. Di Lorenzo, E.; Mantua, N. Multi-year persistence of the 2014/15 North Pacific marine heat wave. *Nat. Clim. Chang.* **2016**, *6*. [CrossRef]
24. Tang, W.; Fore, A.; Yueh, S.; Lee, T.; Hayashi, A.; Sanchez-Franks, A.; Martinez, J.; King, B.; Baranowski, D. Validating SMAP SSS with in situ measurements. *Remote Sens. Environ.* **2017**, *200*, 326–340. [CrossRef]
25. Meissner, T.; Wentz, F.J.; Scott, J.; Vazquez-Cuervo, J. Sensitivity of Ocean Surface Salinity Measurements from Spaceborne L-Band Radiometers to Ancillary Sea Surface Temperature. *IEEE Trans. Geosci. Remote Sens.* **2016**, *54*, 7105–7111. [CrossRef]

remote sensing

MDPI

Article

Seasonal Variability of Retroflection Structures and Transports in the Atlantic Ocean as Inferred from Satellite-Derived Salinity Maps

Paola Castellanos [1,2,*], Estrella Olmedo [1,3], Josep Lluis Pelegrí [1,4], Antonio Turiel [1,3] and Edmo J. D. Campos [5,6]

1. Department d'Oceanografia Física i Tecnològica, Institut de Ciencias del Mar, Consejo Superior de Investigaciones Científicas, 08003 Barcelona, Spain; olmedo@icm.csic.es (E.O.); pelegri@icm.csic.es (J.L.P.); turiel@icm.csic.es (A.T.)
2. Marine and Environmental Sciences Centre, Faculdade de Ciências da Universidade de Lisboa, 1749-016 Lisboa, Portugal
3. SMOS Barcelona Expert Center, Pg. Marítim 37-49, 08003 Barcelona, Spain
4. Department d'Oceanografia Física i Tecnològica, Institut de Ciencias del Mar, Consejo Superior de Investigaciones Científicas, Unidad Asociada ULPGC-CSIC, 08003 Barcelona, Spain
5. Instituto Oceanográfico, Universidade de São Paulo, São Paulo 01000-000, Brazil; edmo@usp.br
6. American University of Sharjah, P.O. Box 26666, Sharjah, UAE

* Correspondence: castellanos@icm.csic.es or pcossa@fc.ul.pt; Tel.: +34-93-230-95-00

Received: 29 December 2018; Accepted: 28 March 2019; Published: 3 April 2019

Abstract: Three of the world's most energetic regions are in the tropical and South Atlantic: the North Brazil Current Retroflection, the Brazil-Malvinas Confluence, and the Agulhas Current Retroflection. All three regions display offshore diversions of major boundary currents, which define the intensity of the returning limb of the Atlantic meridional overturning circulation. In this work, we use a sea-surface salinity (SSS) satellite product, combined with a high-resolution numerical model and in situ measurements, in order to explore the seasonal variation of the surface currents and transports in these three regions. The analysis of the model output shows that the SSS patterns reflect the surface velocity structure, with the largest horizontal SSS gradients coinciding with those areas of highest velocity and the most predominant velocity vector being 90° anticlockwise (clockwise) from the horizontal SSS gradient in the northern (southern) hemisphere. This information is then applied to the SSS satellite product to obtain maps of water velocity and salt transports, leading to a quantitative tool to estimate both water and salt transports in key regions of the world ocean.

Keywords: sea surface salinity; SMOS; retroflections; surface velocity; water transport; salt transport

1. Introduction

The long-term variations of the Atlantic meridional overturning circulation (AMOC) can lead to regional changes in the distribution of sea surface temperature (SST) and salinity (SSS) (e.g., [1–3]). Conversely, some key regional-scale processes in the tropical and South Atlantic Oceans can largely influence the dynamics and variability of the returning limb of the AMOC. Of particular relevance to both the regional processes and the overall latitudinal heat and salt transports, are the pathways followed by the returning limb of the AMOC, from the Southern Ocean to the deep-water formation regions in the North Atlantic [4]. While the regions of formation of the North Atlantic Depth Water (NADW) are relatively well known, there are still large uncertainties regarding the origin and pathways followed by the water parcels crossing the South Atlantic into the North Atlantic.

One element that contributes to this high uncertainty is the existence of major regional diversions, commonly named retroflections, where variability is high. These retroflections play an important role on the interconnections among the ocean basins, along the returning AMOC's pathway. In the tropical regions and South Atlantic, there are three key locations where substantial retroflections take place: the Agulhas Retroflection, the Brazil-Malvinas Confluence and the North Brazil Retroflection (Figure 1; top panel).

In the subpolar region, the Malvinas Current (MC) flows towards the equator carrying waters from the Antarctic Circumpolar Current. At approximately 36–38°S, the MC encounters the Brazil Current (BC), forming the Brazil-Malvinas Confluence (BMC), where both subantarctic and subtropical waters collide frontally and are diverted offshore. On the eastern side of the basin, the Agulhas Current (AC) tips along the southern coast of Africa. As the AC surpasses the southern end of the continent, it curls back upon itself before turning East, in a process that leaks rings and filaments into the Atlantic Ocean (Agulhas Leakage, AL) [3,5]. Finally, the North Brazil Current (NBC) also experiences a major retroflection (NBCR) after surpassing the equator along the northeastern coast of South America. This retroflection changes seasonally as waters pile up in the upper interior ocean, and the North Equatorial Counter Current (NECC) eventually connects with the western boundary current [6,7].

Retroflection phenomena are also observed in other regions of the world ocean, usually associated with strong gradients of temperature and salinity. The variability of SSS in these retroflection regions provides useful regional descriptions and, even more importantly, gives insight into the predominant pathways connecting the adjacent ocean gyres. Hence, the continuous monitoring of variables such as SSS is essential to determine and predict the relationship between oceanic and climate variability from regional to global scales. This motivates us to examine the relationship between the SSS and velocity structures in these three western-boundary-retroflection regions. The objective is to examine if we can use the information in the SSS fields not only to describe the variability in regional patterns but also to estimate how the volume, heat, and salt transports change in time, thereby assessing the intensity of the returning limb of the AMOC.

For our analysis, we take advantage of recent spatially and temporally dense SSS measurements obtained using the Soil Moisture Ocean Salinity (SMOS) mission (Figure 1; middle panel), together with outputs from an eddy-resolving simulation with the Hybrid Coordinate Ocean Model (HYCOM) (Figure 1; top panel) and in situ data from the Argo float constellation (Figure 1; bottom panel). We focus on the seasonal variations in the three retroflection regions, investigating the SSS patterns and their coherence with the velocity structures. The paper is structured as follows. In Section 2, we present the data and methods. The variability of the SSS in the retroflection regions is examined in Section 3, the connection between the horizontal SSS-gradient and the velocity fields is developed in Section 3.2 In Section 3.3, we use the above results to calculate the seasonal variability in the water and salt transports associated with the three retroflection regions, and we close the article with some concluding remarks in Section 4.

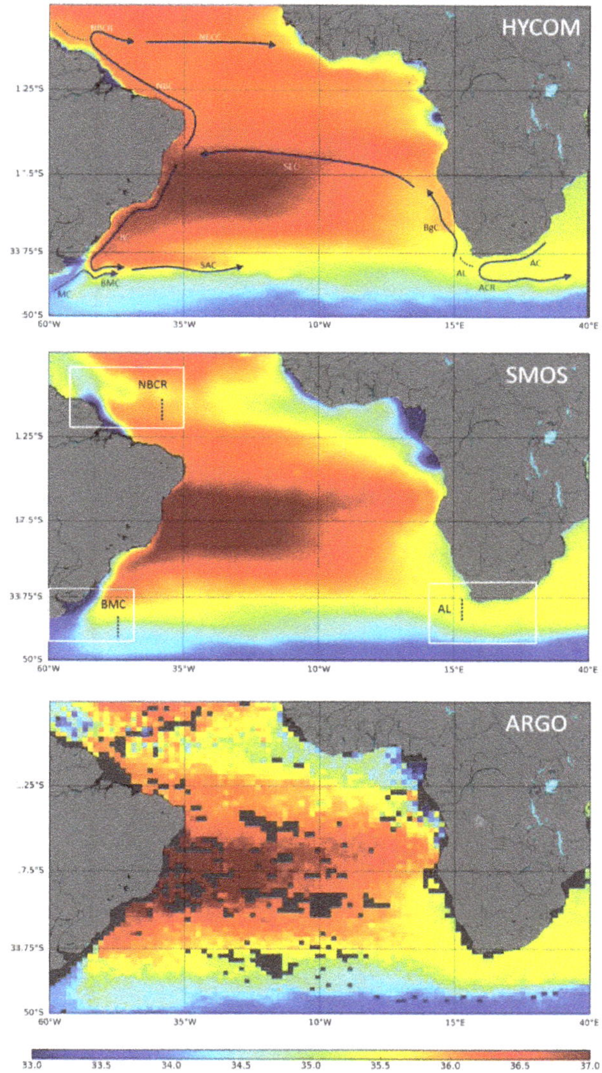

Figure 1. Mean sea-surface salinity (SSS) from January 2011 to December 2015 for (**top panel**) Hybrid Coordinate Ocean Model (HYCOM) at 1/12° resolution, (**middle panel**) Soil Moisture Ocean Salinity (SMOS) at 1/4° resolution, and (**bottom panel**) Argo at 1° resolution. The top panel shows a scheme of sea surface currents, highlighting those features relevant to our study: North Brazil Current (NBC), North Brazil Current Retroflection (NBCR), North Equatorial Countercurrent (NECC), South Equatorial Current (SEC), Brazil Current (BC), Malvinas Current (MC), Brazil-Malvinas Confluence (BMC), South Atlantic Current (SAC), Benguela Current (BgC), Agulhas Current (AC), Agulhas Current Retroflection (ACR), and Agulhas Leakage (AL). The middle panel includes rectangles that locate our three study areas, with dotted lines denoting the sections used for the zonal transport calculations.

2. Data and Methods

2.1. HYCOM Simulation

We have used the output from an eddy-resolving (1/12°) implementation of HYCOM for the Atlantic and Indian Oceans, in a domain extending from 98°W to 114°E and from 65°S to 65°N. The numerical experiment was run at the Ocean Modeling Laboratory (LABMON) of the Oceanographic Institute of the University of Sao Paulo (IOUSP). HYCOM is a primitive-equation hybrid-coordinate ocean general circulation model [8,9].

The model's products analyzed here correspond to the period from 2011 to 2015, extracted from the experiment forced with monthly-mean fields of long-wave radiation, short wave radiation, precipitation and specific humidity, as deduced from NCEP/NCAR reanalysis [10], (Figure 1). The model computes evaporation and sensible heat flux from the precipitation and air temperature data. In the experiment, the SST was relaxed to the climatology, but the SSS was allowed to evolve freely. The bathymetry was extracted from the ETOPO 5 (Data Announcement 88-MGG-02, NOAA, National Geophysical Data Center). During the past decade, HYCOM has become a widely used ocean circulation model, being validated in numerous ocean climate investigations. A complete list of references is available at the HYCOM consortium website, additional information about the experiment and its validation can be found in [11].

2.2. SMOS Data

The SMOS mission is an innovative Earth Observation satellite, launched on November 2009, to remotely sense soil moisture over land and sea surface salinity over the ocean [12,13]. The SMOS single payload is the Microwave Imaging Radiometer using Aperture Synthesis (MIRAS), an L-band 2D synthetic aperture radiometer, with multi-angular and full polarization capabilities. This new instrument suggested a technological challenge that required the development of dedicated calibration and image reconstruction algorithms [14].

The SMOS SSS maps are produced from five years (2011–2015) of brightness temperatures measured by SMOS and provided by the European Space Agency (ESA), following the methodology explained in [15] (middle panel of Figure 1). As a result of this novel technique, the SMOS SSS maps are devoid of land-sea contamination, recovering more measurements near the coast. A complete description of the methodology as well as an extensive validation of the product can be found in [14]. Daily SSS maps are computed from objective analysis of the SMOS data in a time window of 9 days at 0.25° resolution (more details in Olmedo et al., 2017 [15]).

2.3. Argo Data

We have also used in situ salinity data from close-to-surface acquisitions by Argo floats between January 2011 and December 2015. Argo salinities deeper than 10 m and shallower than 0.5 m are removed, the latter because of the possible presence of air bubbles that increase the error of the conductivity measurements.

Every available Argo surface salinity measurement is compared with the corresponding SMOS/HYCOM SSS monthly values. Figure S1 shows the spatial distribution of the root-mean-square differences or errors (RMSEs) between the SMOS/HYCOM and Argo data; given the different resolutions of the HYCOM and SMOS data, we have uniformized this comparison employing for both cases a 1° grid. Figure S2 presents the temporal evolution of the region-averaged RMSEs between SMOS/HYCOM and Argo for each study area.

The comparison between Argo and SMOS/HYCOM salinity data was satisfactory, with region-averaged RMSEs in the ranges of 0.3–0.4, 0.3–0.6 and 0.2–0.3 for the NBCR, BMC and AL, respectively; these errors, when contrasted with the range of SMOS SSS variability, represent relative variations of 5–7%, 3.5–7.5% and 10.5–15.5% for the NBCR, BMC, and AL, respectively. The largest RMSE occurs for HYCOM in the NBCR and BMC regions, undoubtedly because the model uses climatological forcing that does not incorporate individual events of high river discharge. In contrast, the SMOS values remain moderate or low except in some particular instances in the NBCR, possibly when the temperature of the Amazon River runoff matches the temperature of the boundary current.

3. Results and Discussion

3.1. Sea Surface Salinity Variability

In this section we explore the seasonal variability of the SSS, as provided by the monthly SMOS data, and its relation with the surface velocity fields, as derived from the HYCOM model outputs. For this purpose, we first compute the SSS and surface velocity climatologies. All 2011–2015 SMOS data are used to generate the SSS monthly fields at 0.25° resolution. Similarly, the 2011–2015 HYCOM surface velocity fields are used to produce monthly SSS and sea-surface velocities at 1/12° resolution, which are then averaged in order to generate the fields at 0.25° resolution over the same SMOS grid points.

3.1.1. North Brazil Current Retroflection

In the Atlantic Ocean, the transport of heat and salt away from the equatorial and tropical regions takes place largely thanks to its western boundary current, the NBC. However, this transport is largely blocked on a seasonal basis, as the NBCR diverts waters offshore in what becomes the origin of the NECC. To understand this poleward transport of properties, we must assess the seasonal cycle of the NBCR.

The Amazon and the Orinoco Rivers discharge almost 20% of the global freshwater river outflow directly to the surface waters of the western equatorial Atlantic [16]. Particularly, the Amazon flushes its waters close to the equator, which are then transported along-slope by the NBC until at least near 6–8°N. At these latitudes, the waters may either continue along-slope as the NBC or be diverted offshore as the NBCR. Hence, we may use the near-surface freshwater plume in order to track the seasonal changes in SSS and its relation to the NBCR [17].

Figure 2 shows the monthly variability of SSS in the NBC and NBCR. The seasonal pattern is clear in the SSS from both SMOS and HYCOM, with the plume of low salinity waters stretching offshore between July and December, reaching its maximum expression in September. These changes in the eastward extension of the low-salinity plume are not directly related to the temporal variation in Amazon River discharge—minimum in October-November and maximum, about 3.5 times greater, in May–June [18]—but rather to the seasonal appearance of the interior NECC [6,7]. For our purposes, the low-salinity plume behaves only as a tracer of the subjacent dynamics. The velocity fields from HYCOM also display the high seasonality of the NBC and NBCR: between January and June, most of the flow continues along-slope while from July to December a large fraction of the flow retroflects offshore between about 5 and 10°N.

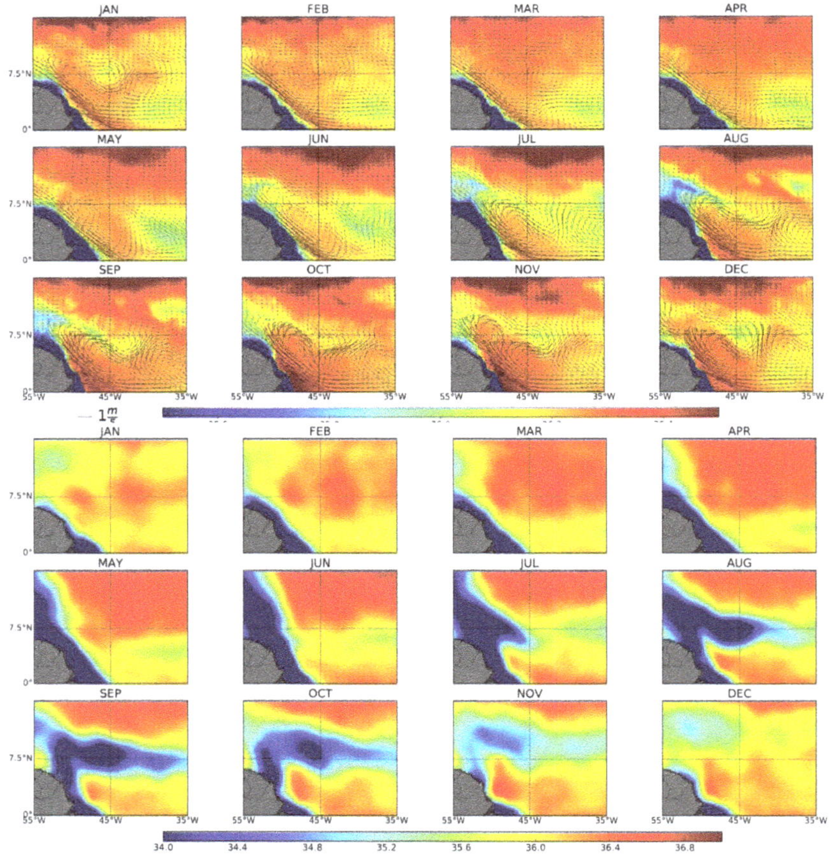

Figure 2. (**Top panels**) Monthly mean SSS (colored) and surface velocity field (velocity vectors) from January 2011 to December 2015 for HYCOM at 1/12° resolution in the NBCR. (**Bottom panels**) The same using SMOS data but without velocity vectors.

3.1.2. Brazil-Malvinas Confluence

The convergence of the BC and MC leads to the BMC, an intense frontal system between warm-salty subtropical and cold-fresh subantarctic waters [19,20]. The subtropical and subantarctic waters are diverted eastward along the frontal system, in a process that sheds numerous eddies of both signs [21]. This may be viewed as a process that enhances eddy-like latitudinal diffusion of both heat and salt.

Despite the averaging over five years, the HYCOM monthly distributions of SSS and velocity illustrate how the encounter between the BC and the MC gives rise to the sharp turn of the less-salty subantarctic waters (the MC retroflection, MCR) and the southward penetration of the salty subtropical waters (the BC overshoot, BCO) [22] (Figure 3, upper panels). These regional patterns also appear, although less clear, in the SMOS SSS fields, certainly because of their more limited spatial resolution.

The SMOS salinity fields show the outflow of La Plata River stretching along the continental margin all year long, although the low-salinity values extend further south between April and August (Figure 3, lower panels); this agrees with the moderate seasonal changes in river discharge (variations of about 50% from the mean, with minimum values in September and December–January, and maximum values in April–May) [23].

According to both HYCOM and SMOS, the frontal system displays moderate variations in latitudinal position. These changes are small near the shelf break but become more visible offshore (east of 52.5°W), with the high subtropical SSS reaching further south during the austral summer (November through March) in both outputs. Despite the different resolutions, both HYCOM and SMOS display similar seasonal patterns of intensification of the BCO (high SSS between about 40 and 44°S), being enhanced in January–April and September–December.

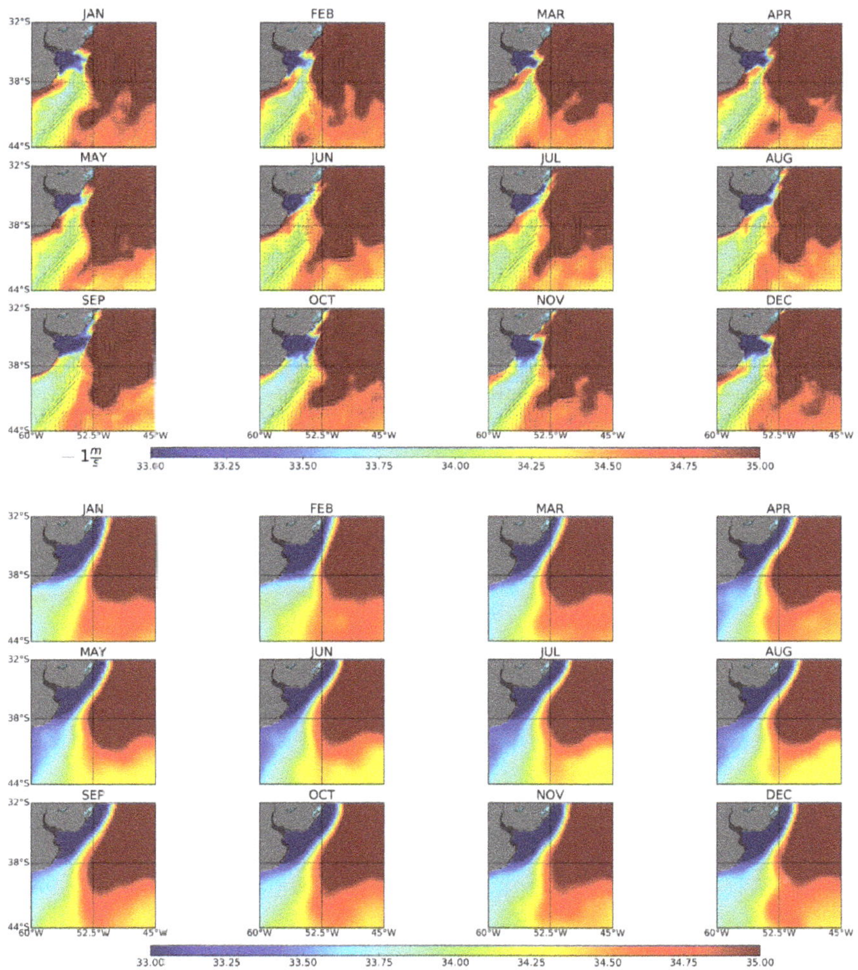

Figure 3. As in Figure 2 but for the BMC.

3.1.3. Agulhas Leakage

Possibly the most energetic among all open-ocean diversions is the abrupt turning of the Agulhas Current (AC) as this western boundary current enters the Atlantic Ocean, in what has been named the AC retroflection (ACR) [5]. The counterpart of the ACR is the AL, either as rings or filaments, whereby some Indian Ocean waters become incorporated into the South Atlantic. The stronger the AL is, the larger transports of salt and heat between the Indian and Atlantic Oceans, as part of the returning AMOC limb [24].

Maps of SSS and surface velocity provide information on the regional dynamics in the AL region (Figure 4). Despite the ubiquity of the ACR, with some mesoscale variability probably related to the relatively small number of years used for calculating the monthly averages, both the SMOS and the model display a seasonal cycle in SSS for a zonal band southwest of Africa—from the continent to about 40°S and stretching between 10°E and 20°E—where the AL is to occur. The SMOS data show maximum SSS values between December and April, and minimum values between June and September. However, the seasonal appearance of high SSS values is more related to the oscillation in the longitude of retroflection than to the intensity of the AL [25–27].

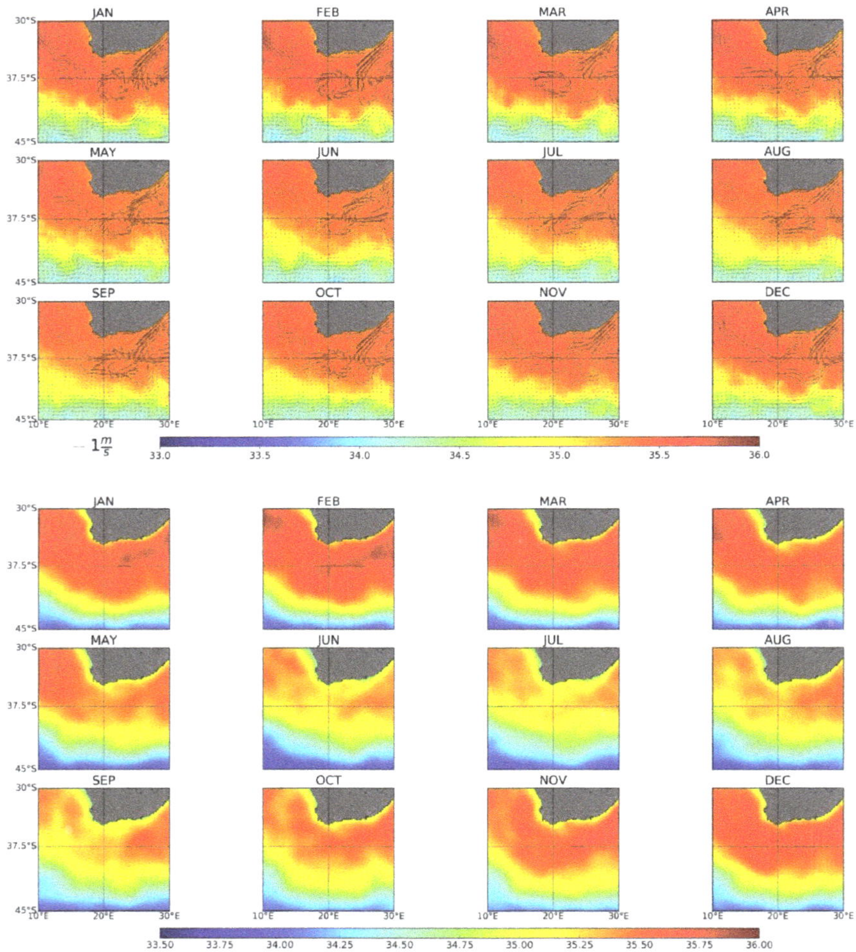

Figure 4. As in Figure 2 but for the AL.

3.2. *Relation Between the Surface Salinity and Velocity Fields*

The salinity fields are related to the velocity fields in two different ways. First, if the water parcels conserve salinity, then the contours of constant salinity will coincide with trajectories; if the field is steady, these constant-salinity contours will also coincide with the streamlines. Second, if we consider the deep ocean to be in geostrophic balance then the velocity field will be normal to the pressure gradients; in a baroclinic ocean this pressure gradient is related to both the temperature and salinity gradients. Both considerations point at the relevance of calculating the horizontal SSS gradient and comparing it with the velocity field.

Regarding the above second point, it is worth emphasizing that both salinity and temperature increase towards the sea surface for most of the Atlantic Ocean. Consequently, both the SSS and SST gradient vectors will usually be anti-parallel to the pressure-gradient force, so that, for sea-surface geostrophic velocities, the angle between the SSS gradient vector and the velocity vector will be 90° (measured from the former to the latter) in the northern hemisphere and 270° in the southern hemisphere. Another related issue is that temperature and salinity variations oppose each other in setting the density changes; in the upper ocean, temperature has always a greater effect than salinity in setting the density but the relation between both may change from one region to another. Hence, the SSS gradient vector is indeed an indicator of the intensity and direction of the velocity field, although the precise relation will change from place to place, to the point that in some areas the speed may be directly proportional to the amplitude of the salinity gradient, while in other places there may be an inverse relation.

In this section, we use the HYCOM model salinity and velocity outputs to infer, for each of the three retroflection regions, the functional relationships that best relate both fields. These dependences will then be applied to the SSS SMOS data in order to infer the surface velocities and transports (Section 3.3). The procedure, done separately for each region and month, has three steps. First, we produce the monthly climatologies of surface horizontal velocity and SSS, and, from the latter, we compute the SSS horizontal gradients using center differences (Figures S3–S5, Supplementary Materials; hereafter we will always refer to the horizontal components of the velocity and the SSS gradient). Secondly, we produce a probability density function of surface water speed V for each absolute value of the SSS gradient $|\nabla S|$. The functional relation $V = f(|\nabla S|)$ is then set as the maximum probability value of the speed V for each value of $|\nabla S|$ (Figures 5–7, top panels). Finally, we compute the angle between the SSS-gradient and velocity vectors (measured anticlockwise from ∇S) and produce a frequency distribution of the occurrence of each angle; the angle that sets the relationship is then selected to be the most frequent one, θ_{max} (Figures 5–7, bottom panels), and the angle between ∇S and the eastward direction is given by $\theta = \theta_{gs} + \theta_{max}$, where θ_{gs} correspond to the orientation of ∇S.

Hence, the zonal and meridional components of the velocity field, may be written as follows:

$$u = f(|\nabla S|) \cos\theta \quad (1)$$

$$v = f(|\nabla S|) \sin\theta \quad (2)$$

In the NBCR, the most frequent argumental difference between the SSS gradient and velocity vectors is 90° (Figure 5); in this region, the functional relation between the speed and the amplitude of the salinity gradient is not lineal. Regarding the BMC, the argument histogram is bi-modal, though the most probable value is 270° (Figure 6), so we set this value as the constant difference between the SSS gradient and velocity vectors. The existence of a second peak at about 90° is caused by the presence of relatively fresh Subantarctic Shelf Waters along the Patagonian shelf [28], which causes the SSS gradient vector to be locally parallel to the pressure gradient force (Figure S6, Supplementary Materials). For both the BMC and NBCR regions, the relationships between speed and the amplitude of the salinity gradient are quite lineal.

In the AL region (Figure 7), we have also considered the angle between the SSS gradient and velocity vectors as constant and equal to 270°. An interesting situation happens, we observe a decreasing relation between the SSS gradient and the velocity (see Figure 7, top). This decreasing relation happens because the SSS gradient is compensated by the temperature gradient. The compensation between the temperature and salinity gradient typically occurs in the regions where the horizontal mixing dominates the dynamics [29]. The AL region is a region where eddies are generated, which it is increases the eddy diffusivity. Therefore, we could say that we are in the previous situation.

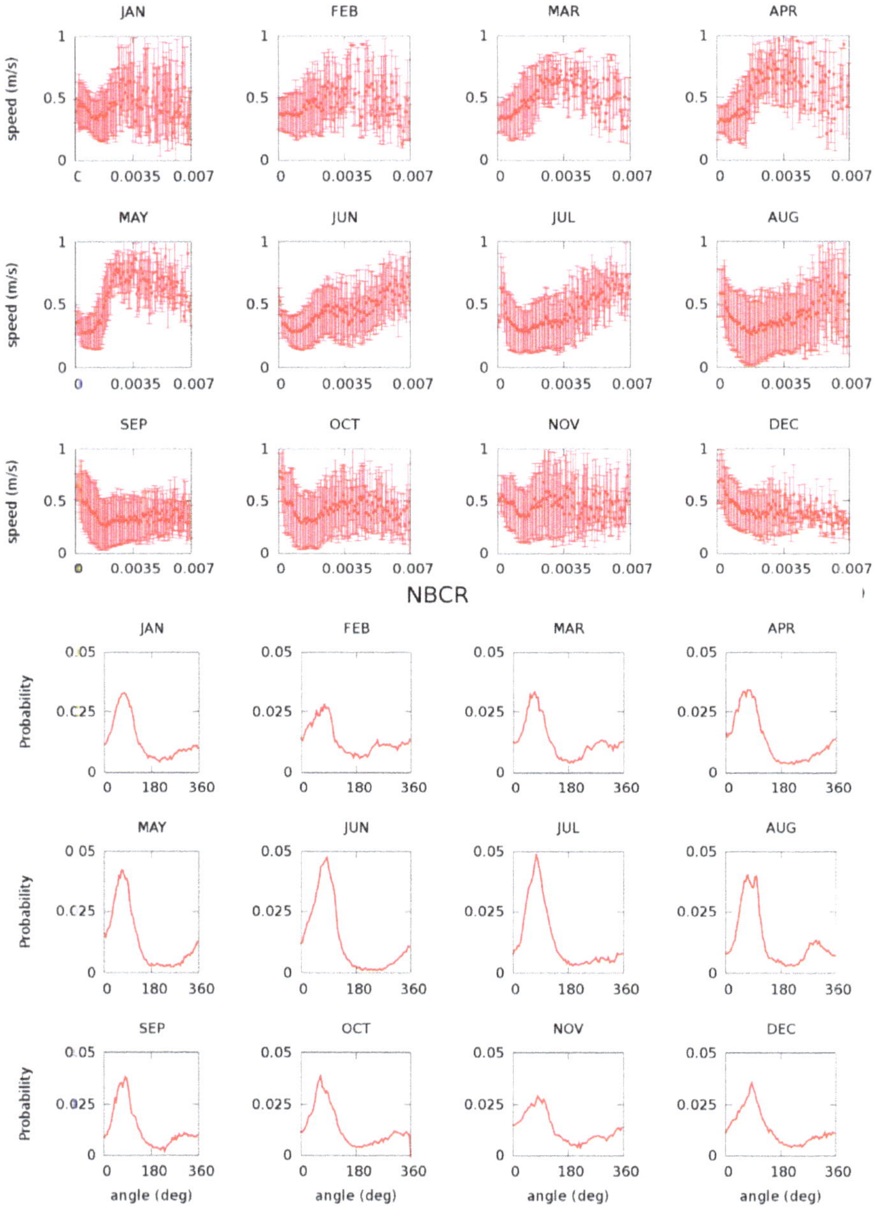

Figure 5. Monthly dependence of the horizontal speed of water and the SSS horizontal gradient in the NBCR, as inferred from HYCOM. (**Top panels**) Maximum probability value and standard deviations for water speed as a function of the absolute value of the SSS gradient. (**Bottom panels**) Frequency distribution of the angle between SSS gradient and surface water velocity vectors (measured from the former to the latter).

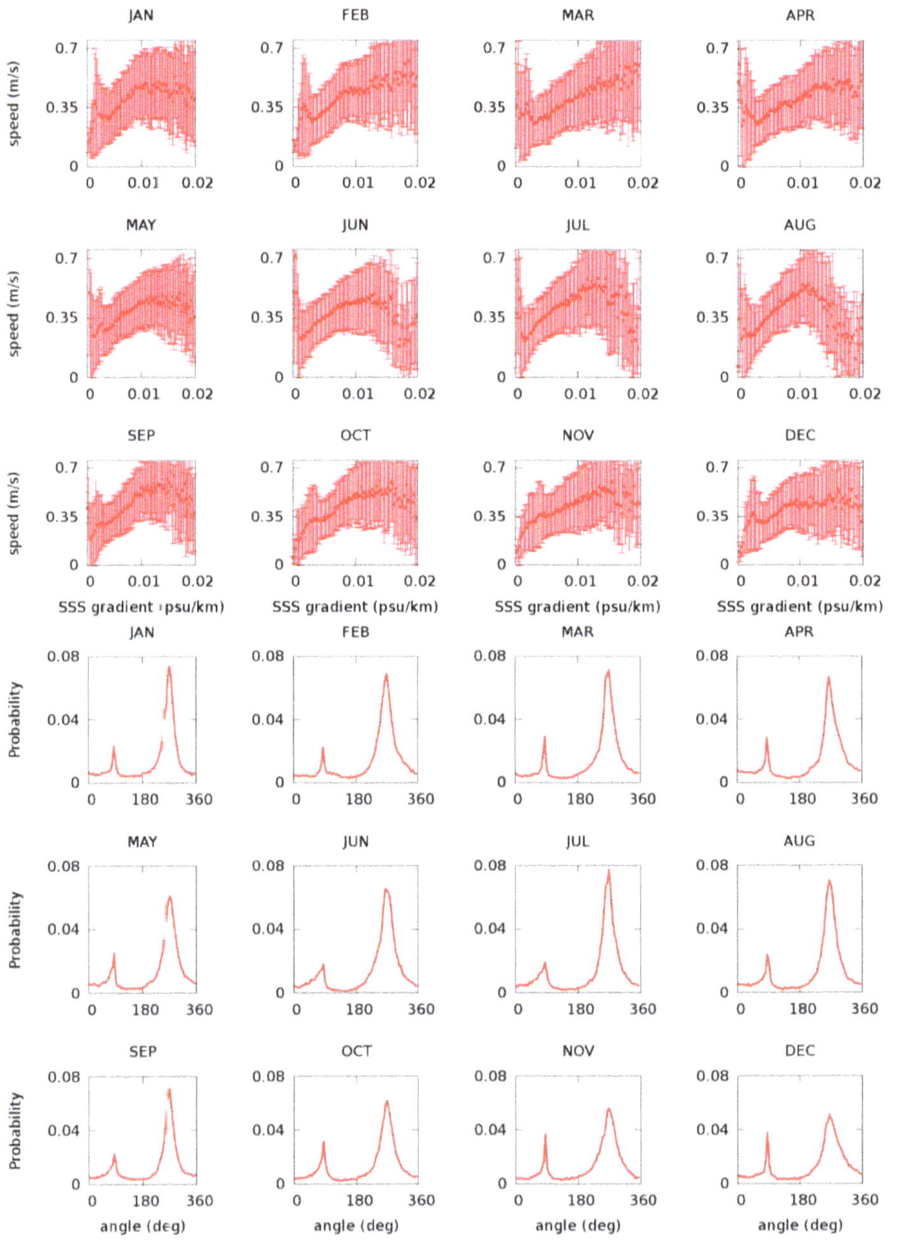

Figure 6. As in Figure 5, but for the BMC.

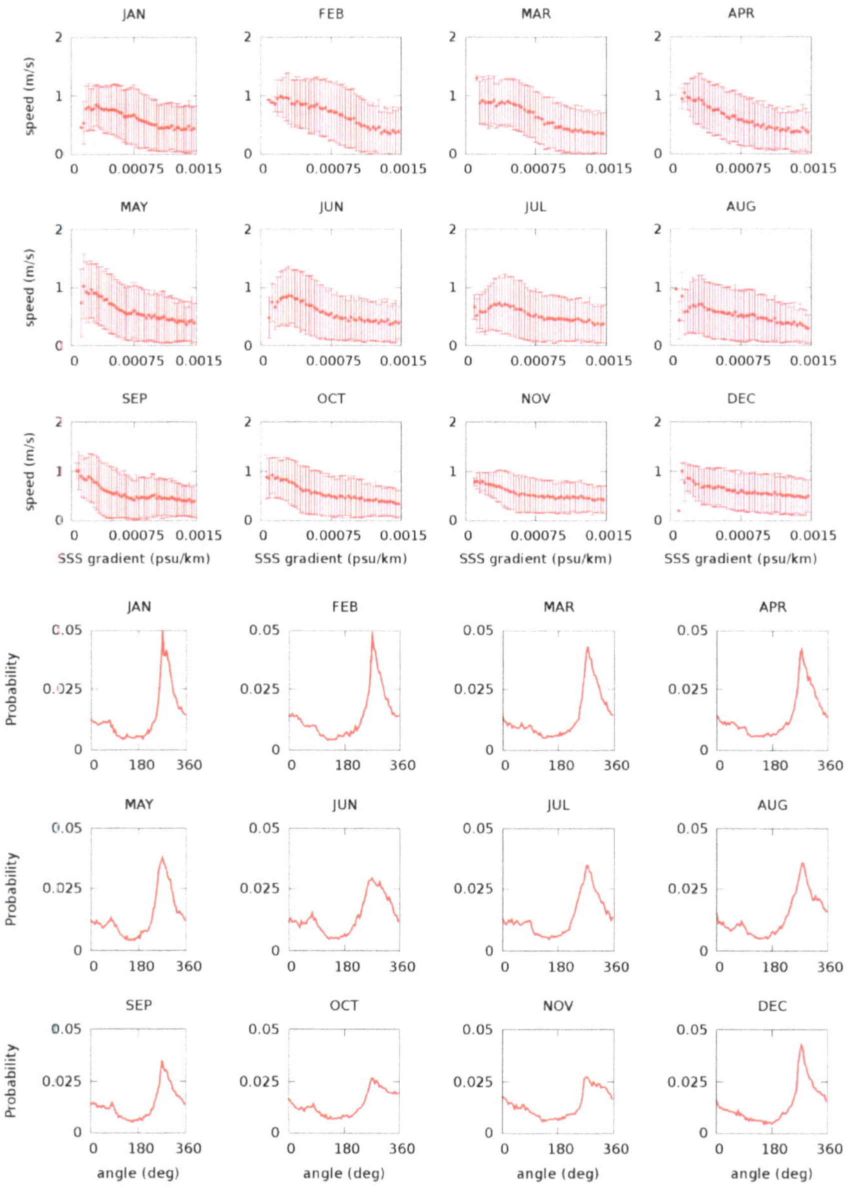

Figure 7. As in Figure 5, but for the AL.

Finally, the procedure presents some limitations that are worth mentioning. One drawback is that the amplitude of the SSS gradients is smaller in the model than as deduced from SMOS, which will likely lead to an underestimation of the velocities. Another limitation is that we may sometimes find very different adjacent regimes, thereby constraining the application of the method to one selected region. One such example is the southern edge of the ACR, which is characterized by very large salinity gradients because of the influence of the circumpolar current; hence, in this work we will restrict the method to the northern and core regions of the ACR. Nevertheless, the results are very encouraging, both in terms of the direct relation between speed and the amplitude of the SSS gradients as well as the agreement of the most frequent angles (either 90° or 270°) with what we would expect for geostrophic currents.

3.3. Salt Transport Variability

In last section we developed a simple methodology to infer the surface velocities from the SSS fields, which we now apply in order to obtain the seasonal cycles of water and salt surface transports associated with the three retroflections. This calculation is entirely done using the SMOS SSS fields, both to infer the velocities and integrated water transports and also to obtain the salt fluxes and integrated salt transports. The salt fluxes are simply calculated as the product of velocity, SSS and an average surface density (here taken as 1025 kg m^{-3}).

The velocity and salt fluxes characterize the surface layers so that the integration is done in one horizontal direction, thereby obtaining transports per unit depth over selected sections. The changes in SSS are much smaller than the changes in surface velocity so the monthly patterns of velocity and salt flux (not shown) are very much alike. Similarly, the seasonal variations of the water and salt integrated transports follow analogous variations (Figure 8). We must keep in mind that the subsurface velocity and salinity fields often bear similarities with the surface ones. Hence, our results may likely reflect the character of the depth-integrated transports—i.e., a water transport of 10^5 m^2 s^{-1} over a depth of 100 m would represent a water transport of 10 Sv—but this calculation falls beyond the objectives of our study.

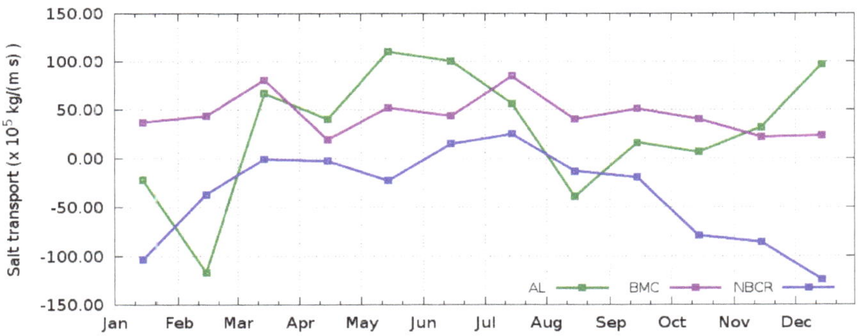

Figure 8. Salt zonal transport per unit depth for the North Brazil Current Retroflection (NBCR), the Brazil-Malvinas Confluence (BMC), and the Agulhas Leakage (AL). In all cases, the integration is over 4° of latitude, as explained in the text, with positive/negative values denoting eastward/westward transports.

3.3.1. North Brazil Current Retroflection

When considering the NBC, we are interested in assessing how much water and salt are diverted zonally by the NBCR, towards the interior tropical North Atlantic Ocean (Figure 1). The stronger this diversion, the weaker will be the meridional transfer of heat and salt from the tropical Atlantic Ocean towards the temperate North Atlantic waters.

In order to assess the monthly strength of the NBCR, we integrate the zonal velocity and salt flux along 40°W between 2°N and 6°N (Figure 1). The resultant water and salt zonal transports show a clear seasonal pattern, with intense westward transports between October and January that revert in June and July (Figure 8). This pattern is characteristic of the year-long dominance of the westward SEC except during the summer months, when the NBCR develops and connects to the downstream eastward NECC [6,7,30].

3.3.2. Brazil-Malvinas Confluence

The encountering of the BC and MC sets up the BMC, a remarkable frontal system that behaves as a zonal guide from the western boundary to the interior South Atlantic Ocean. The flow along this front represents the origin of the South Atlantic Current (SAC), carrying both warm-salty subtropical and fresh-cold subantarctic waters (Figure 1). This current is characterized by the presence of numerous eddies of both signs, with subtropical (warm and salty) eddies moving southwards and subantarctic (cold and fresh) eddies following north. The more intense the BMC and SAC, the larger the eddy activity will be, thereby increasing the meridional heat exchange.

In order to evaluate the monthly intensity of the flow emerging zonally at the BMC, we integrate the zonal velocity and salt flux along 47°W between 40°S and 44°S (Figure 1). The latitudinally-integrated water and salt zonal transports show fairly constant eastward values for all months. These values are indicative of an intense BMC throughout the year, with latitudinal BC and MC transports that do not display a prominent seasonal cycle (Figure 8) [31]. These currents converge onto the frontal system and are diverted eastwards to set up the origin of the SAC [32–35].

3.3.3. Agulhas Leakage

The returning limb of the AMOC depends largely on the ACR and AL, as the relative strength of both currents controls the transfer of warm and salty Indian Ocean waters into the South Atlantic Ocean. The AC cannot overtake the southern tip of Africa and returns east as the ACR. The actual penetration proceeds intermittently as rings and filaments, in what is known as the AL, connecting the AC with the equatorward Benguela Current (BgC).

In order to assess the seasonal variation of the Agulhas Leakage (AL), we integrate the zonal velocity and salt flux along 17°E between 34°S and 38°S (Figure 1). The subsequent water and salt zonal transports display substantial intermittency (Figure 8). This agrees fairly well with observations that show no prominent seasonal pattern in the AL, with the actual monthly transports changing as a function of the shedding of rings from the AC [36,37].

4. Conclusions

For conditions of weak vertical and horizontal mixing, and in the absence of intense evaporation or precipitation, surface water parcels approximately conserve their salinity and SSS serves as a good indicator of the surface flow field. This result is the case for relatively short time scales, shorter than the time scale that characterizes both the local rate of change and advection through the area of study. In these circumstances, the flow is nearly stationary and the constant salinity contours coincide with both trajectories and streamlines. Hence, the sources of high (e.g., Indian Ocean) or low salinity (e.g., Amazon River discharge) serve to delineate the flow field, with the velocity vector tangent to the constant SSS contours.

There is still a second justification for using the SSS as an indicator of the flow field. The subinertial open-ocean flow (with time scales longer than one day) is largely in geostrophic balance: the velocity field is normal to the contours of constant pressure, which often closely parallel the temperature and salinity contours. Further, the greater the pressure gradient, i.e., the closer the constant pressure contours are, the faster the velocity fields.

Both above reasons point at the possibility of inferring the velocity field from the constant-salinity contours, as long as we do our analysis on the proper time scale: longer than inertial and shorter than the advective and local rate-of-change time scales. The local rate-of-change time scale is often dominated by the seasonal cycles, so a reasonable time scale is one month. Using one month as the advective time scale, and considering a swift ocean flowing at 0.2–0.4 m s^{-1}, leads to considering regions no larger than 1000–2000 km approximately.

We have used these ideas to explore if we can use the SSS fields to characterize the flow in three retroflection regions with typical lengths of 1000–2000 km. Our approach has consisted, first, in using a well-contrasted ocean model to infer monthly functional relations between SSS and surface velocity and, second, in applying these relationships to the SMOS data, an independent and novel SSS data set. Our results are encouraging, as we have obtained rather robust functional dependences between the model SSS and surface velocity, and because the most-frequent angle between both vectors is consistent with what we expect for geostrophic currents. These relations, when translated to the SMOS data, lead to consistent patterns of seasonal variability for the water and salt transports associated with the retroflections.

Future research should involve using long model outputs for analyzing the relationships between SSS gradients and surface velocities at different temporal scales, from shorter time scales in more local sites to longer time scales over much larger areas. This will allow a better understanding of the dynamics behind each functional relation, thereby providing a better assessment of the advantages and limitations of this approach. Once the relationships are well established, and longer SMOS time series become available, we will have the capacity to identify the inter-annual changes in water and salt transport for different ocean domains, including the critical western-boundary retroflection regions.

Supplementary Materials: The following are available online at http://www.mdpi.com/2072-4292/11/7/802/s1, Figure S1. (Left panels) Root Mean Square Error (RMSE) in SSS between HYCOM and Argo in a 1° × 1° grid, in (top) the North Brazil Current Retroflection, (middle) the Brazil-Malvinas Confluence and (bottom) the Agulhas Leakage. (Right panels) As in the left panels but using SMOS and Argo data. Figure S2: Time evolution of the RMSE in SSS between HYCOM and Argo data (dashed lines) and between SMOS and Argo data (solid lines), for (top panel) the North Brazil Current Retroflection, (middle panel) the Brazil-Malvinas Confluence and (bottom panel) the Agulhas Leakage. Figure S3: Monthly climatology in the North Brazil Current Retroflection from HYCOM data. (Top panels) Speed of the surface water. (Bottom panels) Absolute value of the SSS horizontal gradient. Figure S4: Monthly climatology in the Brazil-Malvinas Confluence from HYCOM data. (Top panels) Speed of the surface water. (Bottom panels) Absolute value of the SSS horizontal gradient. Figure S5: Monthly climatology in the Agulhas Leakage from HYCOM data. (Top panels) Speed of the surface water. (Bottom panels) Absolute value of the SSS horizontal gradient.

Author Contributions: Conceptualization, P.C., E.O., J.L.P. and A.T.; Investigation, P.C. and E.O.; Methodology, P.C., E.O., A.T. and E.J.D.C.; Writing – original draft, P.C. and E.O.; Writing – review & editing, J.L.P. and E.J.D.C.

Funding: This work has been funded by the Spanish government through the National R+D Plan through projects VA-DE-RETRO (reference number CTM2014-56987-P), Promises (reference number ESP2015-67549-C3) and L-Band (reference number ESP2017-89463-C3-1-R). The HYCOM numerical experiment was supported by the São Paulo State Foundation for Research Support (FAPESP, Grants: 2008/58101-9, 2010/01943-8, and 2011/50552-4). E. Campos acknowledges the Brazilian National Council for Scientific and Technological Development (CNPq) for a Research Fellowship (Grant 302018/2014-0) and FAPESP (Grants, 2017/09659-6).

Acknowledgments: The Argo data were collected and made freely available by the International Argo Program that contributes to it (http://www.argo.ucsd.edu, http://argo.jcommops.org). The Argo Program is part of the Global Ocean Observing System. Finally, we are very grateful to our two reviewers for a number of useful comments and suggestions.

Conflicts of Interest: The authors declare no conflict of interest.

References

1. Haarsma, R.J.; Campos, E.; Hazeleger, W. Influence of the meridional overturning circulation on tropical Atlantic climate and variability. *J. Clim.* **2008**, *21*, 1403–1416. [CrossRef]
2. Huisman, S.E.; Toom, M.; Dijkstra, H.A.; Drijfhout, S. An indicator of the multiple equilibria regime of the Atlantic Meridional Overturning Circulation. *J. Phys. Oceanogr.* **2010**, *40*, 551–567. [CrossRef]
3. Garzoli, S.L.; Matano, R. The South Atlantic and the Atlantic Meridional Overturning Circulation. *Deep Sea Res. II* **2011**, *58*, 1337–1847. [CrossRef]
4. Talley, L.D. Shallow, intermediate and deep overturning components of the global heat budget. *J. Phys. Oceanogr.* **2003**, *33*, 530–560. [CrossRef]
5. Gordon, A.L. The brawniest retroflection. *Nature* **2003**, *421*, 904–905. [CrossRef] [PubMed]
6. Castellanos, P.; Pelegrí, J.L.; Campos, E.D.; Rosell-Fieschi, M.; Gasser, M. Response of the surface tropical Atlantic Ocean to wind forcing. *Progr. Oceanogr.* **2015**, *134*, 271–292. [CrossRef]
7. Rosell-Fieschi, M.; Pelegrí, J.L.; Gourrion, J. Zonal jets in the equatorial Atlantic Ocean. *Progr. Oceanogr.* **2015**, *130*, 1–18. [CrossRef]

8. Halliwell, G.R.J. Simulation of North Atlantic decadal/multidecadal winter SST anomalies driven by basin-scale atmospheric circulation anomalies. *J. Phys. Oceanogr.* **1998**, *28*, 5–21. [CrossRef]
9. Bleck, R. An oceanic general circulation model framed in hybrid isopycnic-Cartesian coordinates. *Ocean Model.* **2002**, *4*, 55–88. [CrossRef]
10. Kalnay, E.; Kanamitsu, M.; Kistler, R.; Collins, W.; Deaven, D.; Gandin, L.; Iredell, M.; Saha, S.; White, G.; Woollen, J.; et al. The NCEP/NCAR 40-year reanalysis project. *Bull. Amer. Meteor. Soc* **1996**, *77*, 437–470. [CrossRef]
11. Castellanos, P.; Campos, E.J.D.; Giddy, I.; Santis, W. Inter-comparison studies between high- resolution HYCOM simulation and observational data: The South Atlantic and the Agulhas leakage system. *J. Mar. Syst.* **2016**, *159*, 76–88. [CrossRef]
12. Font, J.; Camps, A.; Borges, A.; Martin-Neira, M.; Boutin, J.; Reul, N.; Kerr, Y.; Hahne, A.; Mechlenburg, S. SMOS: The challenging sea surface salinity measurement from space. *Proc. IEEE* **2010**, *98*, 649–654. [CrossRef]
13. Kerr, Y.; Waldteufel, P.; Wigneron, J.-P.; Delwart, S.; Cabot, F.; Boutin, J.; Escorihuela, M.-J.; Font, J.; Reul, N.; Gruhier, C.; et al. The SMOS mission: New tool for monitoring key elements of the global water cycle. *Proc. IEEE* **2010**, *98*, 666–687. [CrossRef]
14. McMullan, K.D.; Brown, M.; Martin-Neira, M.; Rits, W.; Ekholm, S.; Marti, J.; Lemanczyk, J. SMOS: The payload. *IEEE Trans. Geosci. Remote Sens.* **2008**, *46*, 594–605. [CrossRef]
15. Olmedo, E.; Martínez, J.; Turiel, A.; Ballabrera-Poy, J.; Portabella, M. Debiased non-Bayesian retrieval: A novel approach to SMOS sea surface salinity. *Remote Sens. Environ.* **2017**, *193*, 103–126. [CrossRef]
16. Masson, S.; Delecluse, P. Amazon River runoff on the tropical Atlantic. *Phys. Chem. Earth B* **2001**, *26*, 137–142. [CrossRef]
17. Grodsky, S.A.; Carton, J.A.; Bryan, F.O. A curious local surface salinity maximum in the northwestern tropical Atlantic. *J. Geophys. Res. Ocean* **2014**, *119*, 1–12. [CrossRef]
18. Oltman, R.E. *Reconnaissance Investigations of the Discharge and Water Quality of the Amazon River*; Geological Survey Circular 552; U.S. Geological Survey: Washington, DC, USA, 1968; 16p.
19. Campos, E.J.D.; Olson, D.B. Stationary Rossby Waves in western boundary current extensions. *J. Phys. Oceanogr.* **1991**, *21*, 1202–1224. [CrossRef]
20. Matano, R. On the separation of the Brazil Current from the coast. *J. Phys. Oceanogr.* **1992**, *23*, 79–90. [CrossRef]
21. Mason, E.; Pascual, A.; Gaube, P.; Ruiz, S.; Pelegrí, J.L.; Delepoulle, A. Subregional characterization of mesoscale eddies across the Brazil-Malvinas Confluence. *J. Geophys. Res. Oceans* **2017**, *122*, 3329–3357. [CrossRef]
22. Saraceno, M.; Provost, C.; Piola, A.R.; Bava, J.; Gagliardini, A. Brazil Malvinas Frontal System as seen from 9 years of advanced very high resolution radiometer data. *J. Geophys. Res. Oceans* **2004**, *109*, C05027. [CrossRef]
23. Borús, J.; Giacosa, J. *Evaluación de caudales diarios descargados por los grandes ríos del Sistema del Plata al Río de la Plata*; Dirección de Sistemas de Información y Alerta Hidrológico Instituto Nacional del Agua: Ezeiza, Argentina, 2014.
24. Beal, L.M.; De Ruijter, W.P.M.; Biastoch, A.; Zhan, R. On the role of the Agulhas system in ocean circulation and climate. *Nature* **2011**, *472*, 429–436. [CrossRef] [PubMed]
25. Lutjeharms, J.R.E.; Van Ballegooyen, R.C. The retroflection of the Agulhas Current. *J. Phys. Oceanogr.* **1988**, *18*, 1570–1583. [CrossRef]
26. Boebel, O.; Lutjeharms, J.; Schmid, C.; Zenk, W.; Rossby, T.; Barron, C. The Cape Cauldron: A regime of turbulent inter-ocean exchange. *Deep Sea Res II* **2003**, *50*, 57–86. [CrossRef]
27. Piola, A.R.; Campos, E.J.; Möller, O.O.; Charo, M.; Martínez, C. Subtropical Shelf Front off eastern South America. *J. Geophys. Res. Oceans* **2000**, *105*, 6565–6578. [CrossRef]
28. Dencausse, G.; Arhan, M.; Speich, S. Spatio-temporal characteristics of the Agulhas Current retroflection. *Deep Sea Res. I* **2010**, *57*, 1392–1405. [CrossRef]
29. Rudnick, D.L.; Ferrari, R. Compensation of horizontal temperature and salinity gradient in the ocean mixed layer. *Science* **1999**, *283*, 526–529. [CrossRef] [PubMed]
30. Fonseca, C.A.; Goni, G.J.; Johns, W.E.; Campos, E.J.D. Investigation of the North Brazil Current retroflection and North Equatorial Countercurrent variability. *Geophys. Res. Lett.* **2004**, *31*, L21304. [CrossRef]

31. Artana, C.; Ferrari, R.; Koenig, Z.; Sennéchael, N.; Saraceno, M.; Piola, A.R.; Provost, C. Malvinas Current volume transport at 41°S: A 24 yearlong time series consistent with mooring data from 3 decades and satellite altimetry. *J. Geophys. Res. Oceans* **2018**, *123*, 378–398. [CrossRef]
32. Peterson, R.; Stramma, L. Upper-level circulation in the South Atlantic Ocean. *Progr. Oceanogr.* **1991**, *26*, 1–73. [CrossRef]
33. Matano, R.; Schlax, M.; Chelton, D. Seasonal variability in the Southwestern Atlantic. *J. Geophys. Res.* **1993**, *98*, 027–035. [CrossRef]
34. Valla, D.; Piola, A.R.; Meinen, C.S.; Campos, E. Strong mixing and recirculation in the northwestern Argentine Basin. *J. Geophys. Res. Oceans* **2018**, *123*, 4624–4648. [CrossRef]
35. Orúe-Echevarría, D.; Pelegrí, J.L.; Machín, F.; Hernández-Guerra, A.; Emelianov, M. Inverse modeling the Brazil-Malvinas Confluence. *J. Geophys. Res. Oceans* **2019**, *124*. [CrossRef]
36. Dijkstra, H.; de Ruijter, W.P.M. On the physics of the Agulhas Current: Steady retroflection regimes. *J. Phys. Oceanogr.* **2001**, *31*, 2971–2985. [CrossRef]
37. Cheng, Y.; Putrasahan, D.; Beal, L.; Kirtman, B. Quantifying Agulhas Leakage in a high-resolution climate model. *J. Clim.* **2016**, *29*, 6881–6892. [CrossRef]

remote sensing

MDPI

Article

Comparison of the Retrieval of Sea Surface Salinity Using Different Instrument Configurations of MICAP

Lanjie Zhang [1,2], Zhenzhan Wang [1,*] and Xiaobin Yin [3]

1 Key Laboratory of Microwave Remote Sensing, National Space Science Center, Chinese Academy of Sciences, Beijing 100190, China; zhanglanjie2012@163.com
2 University of Chinese Academy of Sciences, Beijing 100094, China
3 Beijing Piesat Information Technology Co. Ltd., Beijing 100195, China; yinxiaobin@piesat.cn
* Correspondence: wangzhenzhan@mirslab.cn; Tel.: +86-10-6258-6454

Received: 23 January 2018; Accepted: 29 March 2018; Published: 4 April 2018

Abstract: The Microwave Imager Combined Active/Passive (MICAP) has been designed to simultaneously retrieve sea surface salinity (SSS), sea surface temperature (SST) and wind speed (WS), and its performance has also been preliminarily analyzed. To determine the influence of the first guess values uncertainties on the retrieved parameters of MICAP, the retrieval accuracies of SSS, SST, and WS are estimated at various noise levels. The results suggest that the errors on the retrieved SSS have not increased dues poorly known initial values of SST and WS, since the MICAP can simultaneously acquire SST information and correct ocean surface roughness. The main objective of this paper is to obtain the simplified instrument configuration of MICAP without loss of the SSS, SST, and WS retrieval accuracies. Comparisons are conducted between three different instrument configurations in retrieval mode, based on the simulation measurements of MICAP. The retrieval results tend to prove that, without the 23.8 GHz channel, the errors on the retrieved SSS, SST, and WS for MICAP could also satisfy the accuracy requirements well globally during only one satellite pass. By contrast, without the 1.26 GHz scatterometer, there are relatively large increases in the SSS, SST, and WS errors at middle/low latitudes.

Keywords: MICAP; forward model; combined active/passive SSS retrieval algorithm; different instrument configurations; retrieval errors

---

1. Introduction

At a given pressure, the salinity and temperature determine the density of sea water so that the salinity plays an important role in the formation and circulation of water masses [1,2]. The SSS is also a key variable for understanding the role of the fresh water input and output in ocean dynamics [3]. The variations of the water cycle can be directly monitored by tracking SSS [4,5]. Knowledge of the distribution of SSS is key to understanding how the water cycle affects the ocean circulation and the climate variability [6]. In addition, knowledge of the distribution of SSS is also important for providing valuable estimations of rainfall over the oceans through its link to the evaporation–precipitation balance [6,7]. Although the full deployment of the Agro profiling float array expands the in situ observing system of SSS, the in situ sample density of SSS is still sparse in both time and space [8,9]. For the past ten years, with the successful launch of the Soil Moisture and Ocean Salinity (SMOS) mission, the Aquarius/SAC-D mission and the Soil Moisture Active-Passive (SMAP) mission, satellite SSS has become a reality. These three missions can systematically map SSS over all areas of the open ocean except near land and ice boundaries [8], and offer complementary information to existing in situ measurements [8,10].

The sensitivity of brightness temperature (TB) to SSS remains high only in the low frequency (~1 GHz, L-band), and the L-band is also protected against human-made emissions [11,12]. Thus,

L-band frequency range has been chosen for SMOS, Aquarius and SMAP missions. The unique payload of SMOS, L-band Microwave Imaging Radiometer using two-dimensional Aperture Synthesis, is a fully polarimetric radiometer (with an optional dual polarized mode), and it has multi-angular imaging capability [11,13]. Different from SMOS, Aquarius includes an L-band radiometer with a larger size real aperture antenna and an integrated L-band scatterometer to measure simultaneous oceanic backscatter [14]. The SMAP mission also originally utilized combined passive/active microwave instruments. However, the radar of SMAP ceased transmission on July 2015, and only the radiometer of SMAP still operates continuously over the ocean to provide SSS observations [15]. The SSS maps at 150–200 km, one-month resolution with an accuracy of ~0.2 psu or better at an open ocean are basically achieved by the above three missions [10,16,17].

Although SMOS, Aquarius, and SMAP have provided abundant global SSS measurements with adequate resolution and coverage, it is still a challenge to achieve higher retrieval accuracy of SSS from the observations of these satellites. Moreover, research on seasonal and interannual variations of SSS in near-coastal and middle/high latitudes is relatively limited due to the low-quality satellite-derived SSS measurements in these regions [18], reinforcing the requirement for accurate SSS measurements to possibly capture the seasonal and interannual variations of SSS. Developing a new payload to obtain high-quality global observations of SSS is necessary. Thus, the payload MICAP of the Ocean Salinity Satellite mission is designed. MICAP is dedicated to "all-weather" estimate the SSS at high spatial resolution. At L-band, TB not only depends on SSS, but also depends on SST, and sea roughness, which is the main geophysical error source in the retrieval of SSS from L-band TB [19]. Yueh et al. found that the impact of wind on the retrieved SSS is equivalent to several psu [20]. Thus, an L-band scatterometer is included in the MICAP to correct sea surface roughness.

The goal of MICAP is to produce SSS with an accuracy of 1 psu during one satellite pass, and provide the science community with monthly averaged SSS maps at a 200 km spatial scale over the open ocean to an accuracy of ~0.1 psu. In our previous studies, we have analyzed the preliminary performance (potentials and limitations) of MICAP using a combined active/passive SSS retrieval algorithm [21]. Result indicate that the SSS, SST, and WS errors with the original configuration of MICAP can satisfy the accuracy requirements. In this study, the influence of the first guess values uncertainties on the retrieved parameters is estimated at various noise levels. In addition, to obtain the simplified configuration of MICAP, the retrieval accuracies of SSS, SST, and WS with different instrument configurations are mainly studied.

Datasets and forward models are presented in Section 2. The instrument characteristics and simulation method of measurements are described in Section 3. Then, the combined active/passive SSS retrieval algorithm is introduced. To obtain the simplified instrument configuration of MICAP for SSS, SST, and WS retrieval, different channel combinations are given in Section 4. Finally, the retrieval accuracies of SSS, SST, and WS are evaluated using a simulated database (Section 5). Conclusions are provided in Section 6.

2. Datasets and Model

2.1. Datasets

The initialized oceanic and atmospheric inputs, which are used to simulate the sea surface TBs and backscatter coefficient in the following section, include three datasets: the monthly averaged grid data of the In Situ Analysis System (ISAS), the Advanced Microwave Scanning Radiometer 2 (AMSR-2) and the WindSat.

Specifically, the monthly reanalysis of SSS maps of the ISAS product by Laboratoire de Physique des Océans is used. These SSS maps preserve as much as possible the time and space sampling capabilities of the Argo network of profiling floats [22]. In general, the first layer of the monthly salinity data of ISAS is used to represent the SSS measurements [5]. These data are downloaded from the website (http://www.seanoe.org/data/00348/45945/). From 77° S to 66.5° N, the latitude grid of SSS of ISAS is 0.5 degree Mercator, where it is thus isotropic with a resolution that increases with latitude. From 66.5° to North Pole, the latitude step is fixed. Different from the latitude, from 180° W to 180° E, the longitude grid of SSS of ISAS is fixed (0.5 degree). To unify the spatial resolution of the above three datasets, the SSS map of ISAS in August 2012 is interpolated to 0.25° × 0.25° grid (approximately 25 km grid) using a bilinear interpolation method.

To simulate sea surface TBs, the monthly averaged SST, WS, and atmospheric products also need to be used apart from SSS field. Thus, the monthly averaged SST, WS, vapor (V), and cloud liquid (L) products of AMSR-2 in August 2012 provided by Remote Sensing Systems (RSS) [23] are used, which all are organized on the 0.25 × 0.25 degree grid. Because the monthly averaged products of AMSR-2 are produced by averaging of all data within the calendar month, the specific time information is not contained in these data products. To simulate the backscatter, the monthly averaged wind direction (WD) product of WindSat in August 2012 obtained from RSS [24] is also used. This WD product available on RSS are created by simple bin resampling to a 0.25 degree Earth grid, and it also does not contain specific time information. All these data from RSS are downloaded from the website (http://www.remss.com/). Note that the data values of SST, WS, V, L, and WD from RSS fall between 0 and 255 in their respective data file. These data values between 0 and 250 need to be scaled to obtain meaningful geophysical units before being used. To scale the data, multiply by the scale factors (and add the offsets) specifically listed in the website of RSS. In addition, the first guess values for those variables (SSS, SST, WS, WD, V, and L) used in the inversion also come from the above datasets.

2.2. Forward Models of L/C/K Band Radiometers

The forward models are used to calculate the TB that reaches radiometer antennas from the emitting top ocean layer [25]. The TBs comprise of the flat sea emission plus the rough sea surface component, atmospheric upwelling radiation and downwelling radiation scattered by the rough sea surface as well as atmospheric absorption, and the galactic noise being scattered on the roughened surface [25,26]. At polarization p, sea surface TB at the top of the atmosphere can be calculated by [27]:

$$TB_p = Tb_{up} + e^{-\tau_{atm}}\left(Tb_{p,flat} + Tb_{p,wind} + Tb_{down}\Gamma_p + T_{galref}\right), \quad (1)$$

where p represents the vertical polarization (V-pol.) and horizontal polarization (H-pol.), Tb_{up} represents the upwelling atmospheric self-emitted direct to the antenna, $e^{-\tau_{atm}}$ represents the equivalent optical thickness of the atmosphere, $Tb_{p,flat}$ denotes the TB of a flat sea surface, $Tb_{p,wind}$ is the wind-induced contribution sea surface TB, Tb_{down} denotes the self-emitted of TB by the atmospheric downward and attenuated along the downward path, Γ_p is the reflection coefficient of the sea surface, which is computed as $1-(Tb_{p,flat} + \mathrm{Tb}_{p,wind})/SST$, and Tb_{galref} is the contribution of cosmic and galactic radiation which is already scattered by the sea surface. In the following, we show the specific models which are used in this study.

Meissner et al. indicated that the uncertainty between TB (in Kelvin) and sea surface salinity (in psu) is approximately 1:2 [28]. Thus, the accuracy of TB calculated models directly affects the retrieval accuracy of SSS [29]. Here, to better understand the influence of the dielectric constant model on the TB of a flat sea surface, Figure 1 shows the differences of L-band TBs in August 2012 calculated by Klein and Swift (KS) model (1977) [30], Meissner and Wentz (MW) model (2012) [31] and Yiwen Zhou et al. (GW) model (2017) [32] based on the datasets in Section 2.1. The incident angle is chosen as 30° which is approximately the incident angle for the inner-most beam on the Aquarius. The mean bias (standard deviation) between the TB with GW model and TB with KS model are −0.10 K (1.22 K) at H-pol., and −0.12 K (1.47 K) at V-pol.. The mean bias (standard deviation) between the TB with GW model and TB with MW model are −0.26 K (1.22 K) at H-pol., and −0.31 K (1.47 K) at V-pol.. It is worth noting that the bias between GW and MW is higher than that between GW and KS at middle/high latitudes. The results indicate that the TB will be affected by the dielectric constant model, and further the error on the retrieved SSS will also be affected.

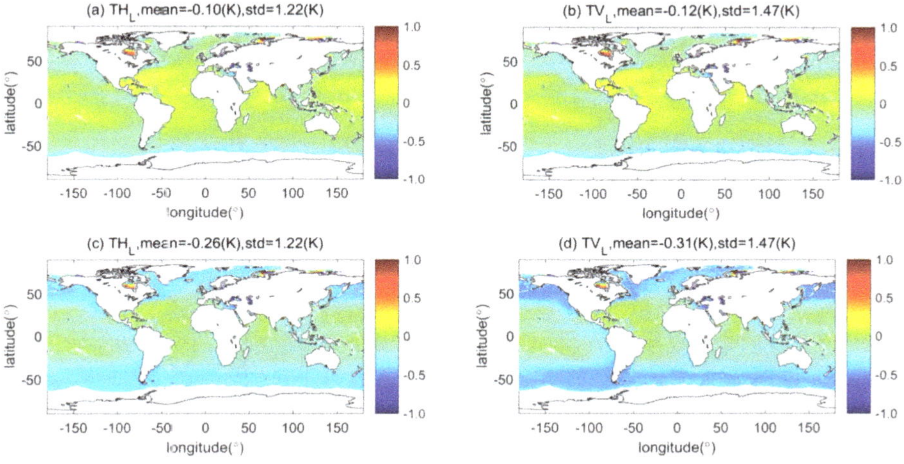

Figure 1. The difference of the TB of flat sea surface with KS, MW and GW dielectric constant models. This is an example of the differences are averaged over one month (August 2012) and in 0.25° bin in longitude and latitude: (**a**,**b**) the difference between GW and KS models at 30° incident angle for H-pol. and V-pol., respectively; and (**c**,**d**) the difference between GW and MW models at 30° incident angle for H-pol. and V-pol., respectively.

In this study, the choice of dielectric constant model is not critical since we are working only with simulated data. Thus, we chose the KS model to calculate the TBs of a flat sea surface. The two-scale model of SMOS from ESA Level 2 Ocean Salinity Processor (L2OSP) is used as the TB model of L-band radiometer [33,34]. The model function from Meissner and Wentz (2012) is used as TB model of C and K band radiometers [31]. Other contributions from atmospheric emission and absorption of L-band and C/K band are calculated by Liebe et al. (1993) [35] and Wentz et al. (2000) atmosphere models [36], respectively.

2.3. Backscatter Coefficient Model of L-Band Scatterometer

The backscatter coefficient is usually calculated by the theoretical model and the empirical model (also known as the geophysical model function). Currently, for L band, the major domestic and international geophysical model function (GMF) is derived from the scatterometer data of Aquarius and the synthetic aperture radar data of Phased Array type L-band Synthetic Aperture Radar (PALSAR). The GMF relates the microwave backscatter to the wind speed and direction, and is generally expressed as:

$$\sigma_0 = A_0(w,\theta) + A_1(w,\theta)\cos\varphi + A_2(w,\theta)\cos 2\varphi, \tag{2}$$

where σ_0 denotes the backscatter coefficient, $A_n(w,\varphi)(n = 0,1,2)$ is a function of the WS w and incident angle θ, and φdenotes the relative WD. Based on the GMF of PALSAR [37] and Aquarius (Meissner et al.) [28], the backscatter coefficients versus WD and 10 m/s wind speed for HH and VV polarizations are calculated in Figure 2. The incident angles are chosen as 29.36°, 39.44° and 46.29°, which are approximately the incident angles for the three beams on the Aquarius.

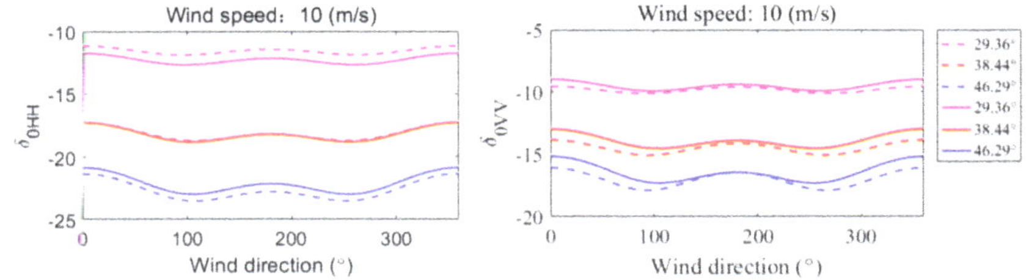

(a) HH polarization (b) VV polarization

Figure 2. Comparison of WD dependence of the backscatter coefficient for: HH polarization (**a**); and VV polarization (**b**), using GMF of Isoguchi and Shimada for PALSAR (solid line) and Meissner et al. (2014) for Aquarius (dashed line) at 10 m/s. The incident angles are chosen as 29.36° (mauve), 39.44° (red) and 46.29° (blue).

Figure 2 shows that the WD dependence of backscatter coefficient for the GMFs of PALSAR and Aquarius are almost the same, especially for HH polarization. The amplitude changes are less than 1 dB for each incident angle. Particularly, for the 46.29° incident angle, in which the backscatter coefficient is calculated with the GMF of PALSAR using the extrapolate method, there is a similar trend for WD dependence of backscatter coefficient to other incident angles. In this study, the GMF from PALSAR is used as the GMF of L-band scatterometer, since the incident angle range of MICAP (30°–55°) is approximate to that of PALSAR (17°–43°) compared with Aquarius (29.36°, 39.44° and 46.29°). For the large incident angles (43°–55°), the extrapolation method is used to calculate the backscatter coefficient of MICAP.

3. Description of MICAP

3.1. Instrument Characteristics

The MICAP will be one of the payloads for Ocean Salinity Satellite mission. It uses one-dimensional microwave interferometric radiometers that operates at 1.4 GHz, 6.9 GHz, 18.7 GHz and 23.8 GHz (abbreviated as MIR in the following) combined with a digital beam forming scatterometer at 1.26 GHz (abbreviated as DBFS in the following) to monitor SSS and reduce geophysical errors caused by surface roughness and SST in the original design. Compared with the two-dimensional microwave interferometric radiometer of SMOS, the one-dimensional MIR of MICAP has lower complexity. Compared with the real aperture radiometer of Aquarius, the MICAP can achieve better spatial resolution and wider swath using one-dimensional MIR [38]. In addition, combined the MIR with DBFS, MICAP can achieve simultaneous remote sensing of the SSS, SST, and WS. The detailed system architectures of MICAP are given in [38]. The main specifications of MICAP are summarized in Table 1 [38].

Table 1. The main system specifications of MICAP.

Instruments	L-Band Radiometer	C-Band Radiometer	K-Band Radiometer	L-Band Scatterometer
frequency	1.4 GHz	6.9 GHz	18.7 GHz, 23.8 [1] GHz	1.26 GHz
sensitivity	0.1 K	0.3 K	0.3 K	0.1 dB
polarization	H, V, T3	H, V,	H, V,	HH, HV, VV, VH
incident angle	30–55°			
spatial resolution	along-track: 65/15/15 km; cross-track: 50–100 km			
antenna size	Reflector: 3.0 m × 5.5 m (after deployment), Feed array: 4 m × 0.5 m			
FOV	>1000 km			
revisit	3 days			
SSS accuracy	<1 psu, over one satellite pass; < 0.1 psu, over 200 km and 30 days			

[1] The 23.8 GHz of K-band radiometer of MICAP is a candidate channel, and whether to keep this frequency is now under consideration.

3.2. Simulate TBs and Backscatter of MICAP

Because the MICAP is not launched yet, it is necessary to simulate TBs and backscatter as would be measured by MICAP using forward models and GMF as well as the initialized oceanic and atmospheric inputs (SSS, SST, WS, WD, V and L) in Section 2. An illustration of the simplified flow chart of the simulation process is displayed in Figure 3.

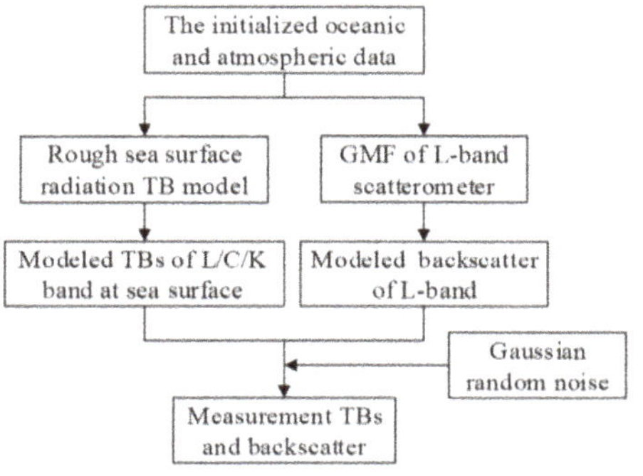

Figure 3. The simplified flow chart of the simulation process for measurements of MICAP.

Based on Figure 3 and the following three assumptions, TBs and backscatter of the MICAP can be produced: (1) in all directions, the ocean–atmosphere parameters of each grid point are homogeneous at 30° to 55° incident angles (±500 km swath); (2) all of the differences between measurements (TB/backscatter) and modeling values provided by model functions in Section 2 are regarded as measurement noises, and these noises are Gaussian; and (3) the measurement data (TB/backscatter) are generated by adding Gaussian random noises, which are related to the instrument sensitivities and incident angles of MICAP in the Table I, to TB and backscatter that are predicted by the above mentioned forward models (Section 2.2) and GMF (Section 2.3), respectively, and then repeated 2000 times.

4. Algorithm and Configurations

4.1. Combined Active/Passive SSS Retrieval Algorithm

Based on the SSS retrieval algorithm of SMOS and JPLCAP v4.0 of Aquarius, the active/passive SSS retrieval algorithm of MICAP is proposed. It retrieves SSS, SST, and WS simultaneously by finding the best fit solution to minimize a cost function (weighted sum of square differences between MICAP measurements and forward models). In addition, we add five additional terms in the cost function to constrain the SSS, SST, and WS solutions. The cost function of MICAP can be expressed as the following form:

$$\chi^2 = \sum_{p=V,H} \frac{\left(T_{Bpf} - T_{Bpmf}\right)^2}{\Delta T_{pf}^2} + \sum_{p=V,H} \frac{\left(\sigma_{0p} - \sigma_{0pm}\right)^2}{\left(\gamma_p \sigma_{0p}\right)^2} + \sum_{i=1}^{M} \frac{\left[P_i - P_{ia}\right]^2}{\Delta P^2}, \quad (3)$$

where V and H represent the V-pol. and H-pol.; the TB_{pf} represent the measured TBs of multi-incident angles at 1.4, 6.9, 18.7 and 23.8 GHz; TB_{pmf} is the simulated TB of the forward model at the above four frequencies; σ_{0p} and σ_{0pm} are the measured and simulated backscatter at 1.26 GHz, respectively; M is the number of the parameters; P_i consists of the parameters that may influence the modeled TBs, including SSS, SST, WS, WD, V, and L; the subscript a denote the initial fields of variables; the value of Δ is the weighting factors, which relate to the respective variance of expected error initial fields of variables; and ΔT_{pf} and γ_p relate to the instrument sensitivities in Table 1.

Once the TBs and backscatter are simulated, as described in Section 3.2, which correspond the MICAP measurements, the SSS, SST, and WS can be simultaneously retrieved using the combined active/passive SSS retrieval algorithm. During the inversion, the first guess geophysical inputs, i.e., SSS, SST, WS, WD, V, and L, are adjusted to minimize the cost function (Equation (3)).

4.2. Different Instrument Configurations of MICAP

As shown before, the 1.4 GHz provides a good sensitivity to the SSS over a large range. However, even in this frequency, the sensitivity of TB to SSS remains low due to the uncertainty on the TB variation related to the surface roughness and SST. Thus, the L-band scattrometer and C/K band radiometers are also included into the original instrument design of MICAP except L-band radiometer (default configuration). In this study, to obtain the simplified instrument configuration of MICAP that can provide SSS, SST, and WS retrieval with the required characteristics using the least frequency bands, three different instrument configuration selections (Table 2) in retrieval mode are compared.

Table 2. The different instrument configuration selections.

Number	Configurations
No. 1	1.26 GHz, 1.4 GHz, 6.9 GHz, 18.7 GHz and 23.8 GHz (default)
No. 2	without 1.26 GHz scatterometer
No. 3	without 23.8 GHz

Because the retrieval accuracies of SSS, SST, and WS may be affected by the polarization modes of instruments, except for the different instrument configurations listed in Table 2, polarization selections in retrieval mode are also tested, such as with all MICAP frequencies except the H-pol. of 23.8 GHz. In this study, we ignore the effect of different polarization selections on the SSS, SST, and WS retrieval accuracies, and focus only on the effect of different instrument configurations, as shown in Table 2.

5. Results

5.1. The Retrieval Results at Single Pixel

In this section, the root mean square (RMS) errors on the retrieved SSS, SST, and WS at the single pixel with No. 1 configuration in Table 2 are estimated. The average of their RMS errors for seven homogeneous scenes specified in Table 3 is shown in Table 4. As can be seen, the smallest RMS errors are obtained for the scene having high SST and low WS, where the sensitivity of TB to SSS and WS is high. The RMS errors for low SST scenes are higher than those for high SST scenes due to the lower sensitivity of TB to SSS, SST, and WS at low SST. The RMS errors are smaller at 3 m/s than those obtained at 7 m/s. The smaller error is explained by the greater sensitivity of TB to SSS, SST and WS at low WS. In addition, the greatest SSS and WS RMS errors (1.22 psu) are obtained for low SST and low SSS scenes. The greatest SST RMS error is obtained for low SST scenes (1.54 °C).

Table 3. The seven homogeneous scenes used in the retrievals.

Scene	SSS (psu)	SST (°C)	WS (m/s)
Reference	35	15	7
High SST	35	25	7
High SST and SSS	38	25	7
High SST, Low WS	35	25	3
Low SST	35	5	7
Low SST and SSS	33	5	7
Low SST and WS	35	5	3

Table 4. The average errors of SSS, SST, and WS from 30° to 55° incident angles for the seven scenes in Table 3.

Scene	SSS (psu)	SST (°C)	WS (m/s)
Reference	0.64	1.19	0.68
High SST	0.52	1.12	0.68
High SST and SSS	0.57	1.13	0.76
High SST, Low WS	0.49	1.00	0.46
Low SST	1.17	1.54	0.78
Low SST and SSS	1.22	1.50	0.79
Low SST and WS	1.10	1.44	0.50

Further, the SSS, SST, and WS RMS errors are tested for various noise levels on first guess values (Table 5), under the reference scene (Table 3). It is obvious that SSS RMS error is almost invariable regardless of whether there is noise on the first guess values for SST and WS. The greatest SSS RMS error (0.82 psu) is obtained only for poorly known SSS, increasing about 0.2 psu which is lower than the standard deviation (std) of the noise on the SSS first guess values (0.5 psu). It indicates that the SSS RMS error of MICAP is only weakly sensitive to the uncertainties of the first guess values for SST and WS. In addition, the greatest SST (2.24 °C) and WS (2.03 m/s) RMS errors are also obtained for poorly known SST and WS, respectively. The abovementioned results indicate that the errors on the retrieved SSS, SST, and WS of MICAP depend primarily upon the noise added to each first guess values at 15 °C, which may be because the MICAP can achieve simultaneous remote sensing of SSS, SST, and WS.

Table 5. The average RMS errors of SSS, SST, and WS from 30° to 55° incident angles for the various levels of noise. σ_{sss}, σ_{sst}, and σ_{ws} represent the std of the noise on the first guess values for SSS, SST, and WS, respectively.

Scene	SSS (psu)	SST (°C)	WS (m/s)
Reference	0.64	1.19	0.68
$\sigma_{sss} = 0.5$	0.82	1.19	0.68
$\sigma_{sst} = 1$	0.64	1.50	0.68
$\sigma_{sst} = 2$	0.64	2.24	0.68
$\sigma_{ws} = 1$	0.64	1.19	1.10
$\sigma_{ws} = 2$	0.64	1.19	2.03

5.2. Comparison of Retrieval Results with Different Configurations Over the Globe

5.2.1. Default Configuration and without 23.8 GHz

To assess whether 23.8 GHz channel is necessary to provide SSS retrieval with the required characteristics of MICAP, the SSS RMS errors during only one satellite pass with No. 1 and No. 3 configurations are shown in Figure 4a,b, respectively. It shows that the SSS RMS error is below 1 psu and varies around 0.61 psu at most of the middle/low latitudes (from 50° S to 45° N). The smallest SSS RMS error is obtained in regions of low wind speed and high temperature because the sensitivities of TB to WS at low WS and TB to SSS at high SST are greater. The SSS RMS errors increase with latitude in the Southern/ Northern Hemisphere is primarily due to the increase of WS and the decrease of SST. The greatest SSS RMS error (achieve or exceed 1 psu) is obtained at high latitudes for which the sensitivity of TB to SSS is lower at cold SST values. Compared to No. 1 configuration, the SSS RMS errors with No. 3 increase only slightly for most of the middle/low latitudes. Especially near the equator and in the Southern Hemisphere, the SSS RMS errors with No. 3 are almost consistent with No. 1, which indicates that 23.8 GHz channels (H-pol. and V-pol.) of MICAP may not bring much information in terms of SSS.

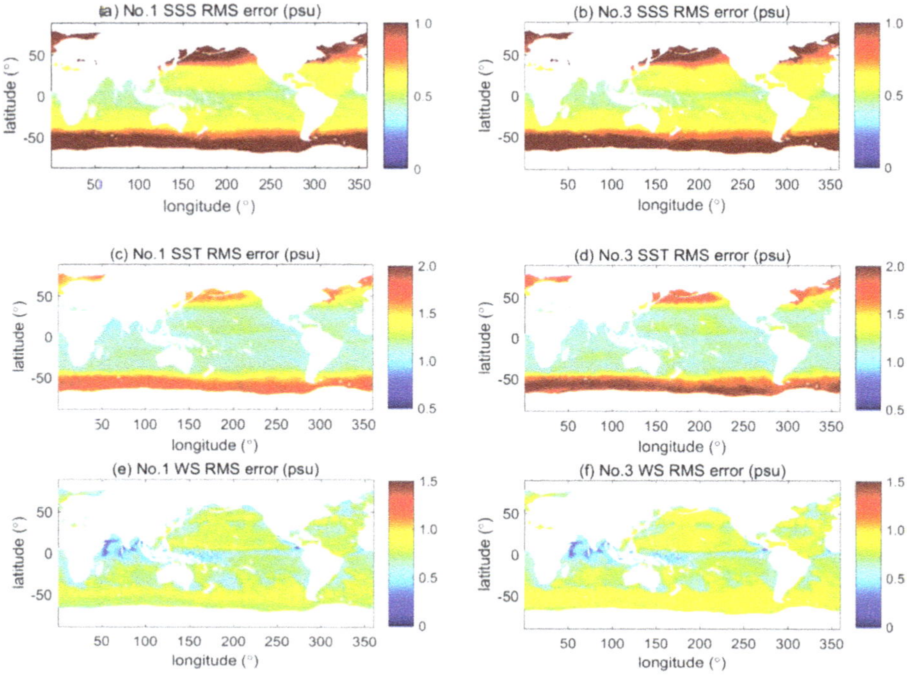

Figure 4. The errors on the retrieved SSS, SST, and WS with default configuration and with default configuration except 23.8 GHz of MICAP during only one satellite pass: SSS (**a**); SST (**c**); and WS (**e**) errors with No. 1 configuration; and SSS (**b**); SST (**d**); and WS (**f**) errors with No. 3 configuration.

Figure 4c,d shows the SST RMS error with No. 1 and No. 3 configurations, respectively. The SST RMS errors vary around 1.15 °C at most of the middle/low latitudes, and they slightly increase at high WS. At high latitude, a significant change of the SST RMS errors is observed because of the low sensitivity of TB to cold SST values. The smallest SST RMS error is obtained near the equator because of the greater sensitivity of TB to high SST. In the Northern Hemisphere where SST is higher, the SST RMS errors are slightly lower than those in the Southern Hemisphere at equivalent latitude. Compared with No. 1 configuration, the SST RMS errors with No. 3 also increase slightly for most of the middle/low latitudes.

The WS RMS errors with No. 1 and No. 3 configurations are presented in Figure 4e,f, respectively. At middle/low latitudes, the WS RMS errors vary around 0.75 m/s. The smallest WS RMS errors are observed in regions of low wind speed because the sensitivity of TB to WS is higher than at moderate wind speeds. Near the Arabian Sea, the Bay of Bengal and the equator (the wind speed is around 3 m/s), the WS RMS error is lower than 0.5 m/s. A greater increase in WS RMS error is observed from the equator to 10° N/S because the WS increases quickly with the increasing latitude. The differences of WS RMS errors between No. 3 and No. 1 configurations are greater for most of the middle/low latitudes compared with SSS and SST.

Figure 4 indicates that 23.8 GHz channels (H-pol. and V-pol.) do not bring much information for the SSS, SST, and WS retrieval accuracy of MICAP at the pixel level. This is explained by the weak sensitivity of SSS, SST, and WS RMS errors to the atmospheric parameters (V and L) except for low SST. It is proven that a simple instrument configuration of MICAP without 23.8 GHz could satisfy the accuracy requirements well in terms of SSS over the globe during only one satellite pass.

5.2.2. Default Configuration and Without Scatterometer

To estimate the impact of the 1.26 GHz scatterometer, which is used to simultaneously correct surface roughness, on the MICAP retrieved parameters, the SSS, SST, and WS RMS errors with No. 1 and No. 2 configurations are shown in Figure 5. For No. 2 configuration, the SSS RMS errors (Figure 5b) are below 1 psu at most of the middle/low latitudes (from 49° S to 40° N), and they vary around 0.7 psu. Compared with No. 1 configuration, the SSS RMS errors with No. 2 increase significantly at the middle/low latitudes, especially for high WS. The result emphasizes the larger dependence of the SSS error to roughness parameter. In the high latitude, the SSS RMS errors increase slightly with the increasing WS. It is explained that the impact of the roughness parameter on the SSS RMS error will decrease at low SST (below about 7 °C). In the Northern Hemispheres, SST is warmer where the sensitivity of TB to SSS is higher. Thus, the SSS RMS errors at equivalent latitude are slightly smaller than those in the Southern Hemisphere.

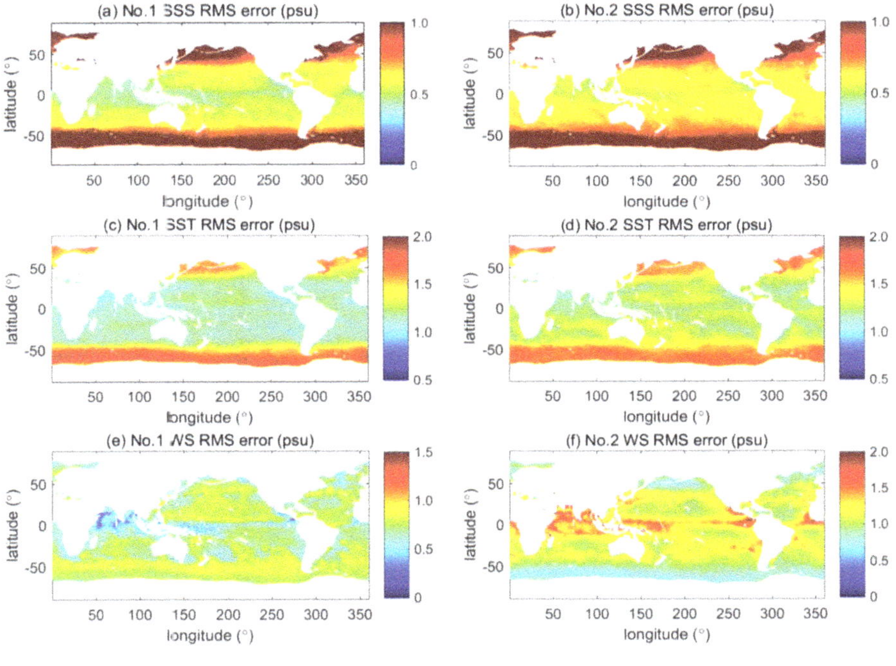

Figure 5. The errors on the retrieved SSS, SST, and WS with default configuration and with default configuration except the 1.26 GHz scatterometer of MICAP during only one satellite pass: SSS (**a**); SST (**c**); and WS (**e**) errors with No. 1 configuration; and SSS (**b**); SST (**d**); and WS (**f**) errors with No. 2 configuration.

The SST RMS errors with No. 2 configuration vary around 1.29 °C at the middle/low latitudes (from 50° S to 40° N) in Figure 5d, and they may exceed 1.5 °C in some regions. At most of the middle/low latitudes and high latitudes of the Northern Hemisphere, the SST RMS errors with No. 2 configuration are greater than those with No. 1. Conversely, the SST RMS errors are smaller at high latitudes of the Southern Hemisphere (above 50° S).

The WS RMS error with No. 2 configuration is shown in Figure 5f. Several interesting results can be seen. The WS RMS errors are greater at middle/low latitudes (around 1.17 m/s) than those at high latitudes, and they even exceed 2 m/s in low WS regions (below 3 m/s). Conversely, the WS RMS errors are relatively smaller in high WS regions (above 12 m/s). Compared to No. 1 and

No. 3 configurations, the WS RMS errors with No. 2 increase significantly over the globe except the south of 52° S. Especially near the equator, the WS RMS errors increase by about 0.78 m/s.

Figure 5 shows that the SSS, SST, and WS RMS errors with No. 2 configuration are greater than those with No. 1 and No. 3. Near the equator, an opposite variation tendency for WS RMS error with No. 2 configuration is observed compared with No. 1 and No. 3. Because the signal to noise ratio of high WS is higher than that of the low WS under the same noise level, the effect of noise on high WS is relatively smaller than low WS. Accordingly, the error on the retrieved WS is small at high WS.

To further analyze the influence of different instrument configurations on MICAP retrieved parameters at the middle/low latitudes, the latitudinal profiles (from 50° S to 40° N) of SSS, SST, and WS RMS errors are shown in Figure 6. It clearly shows that the differences of the latitudinal profiles of SSS, SST, and WS RMS errors between No. 1 and No. 3 configurations are the smallest near the equator. From 10° N to 20° N, the differences between No. 1 and No. 3 are the greatest due to the rapidly increasing wind speed, but still less than 0.02 psu, 0.03 °C and 0.05 m/s, respectively. It indicates that there is little impact on the SSS, SST and WS RMS errors without the 23.8 GHz, which is consistent with above conclusion. Conversely, the differences between No. 1 and No. 2 configurations are the greatest near the equator (about 0.78 m/s) where the wind speed is lower than 3 m/s. The differences between No. 1 and No. 2 decrease gradually with increasing latitude in the Southern Hemisphere. From 40° S to 40° N, the differences between No. 1 and No. 2 (average about 0.08 psu, 0.1 °C and 0.45 m/s for SSS, SST and WS, respectively) are greater than those between No. 1 and No. 3 (average about 0.005 psu, 0.01 °C and 0.03 m/s for SSS, SST and WS, respectively). Moreover, the WS RMS error with No. 2 shows the obvious reverse change trend compared with No. 1 and No. 3. It further illustrates that it is necessary to correct the sea surface roughness using the scatterometer.

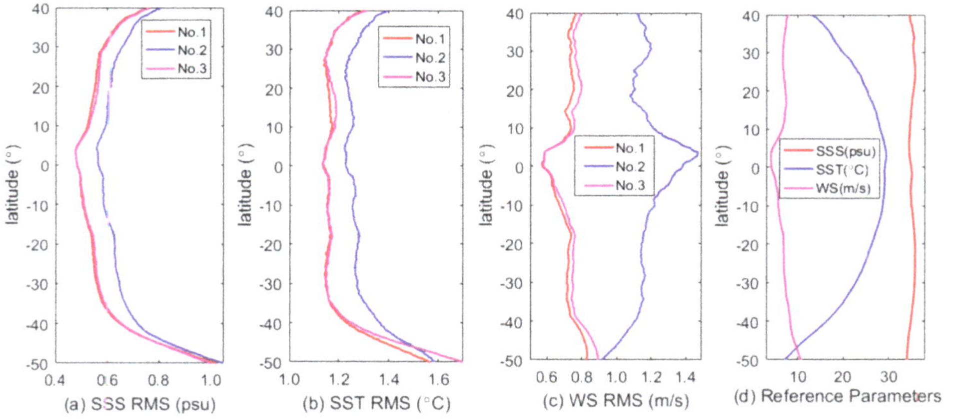

Figure 6. The latitudinal profiles of the errors on the retrieved: SSS (**a**); SST (**b**); and WS (**c**) of MICAP; and the initialized oceanic inputs (**d**). The SSS, SST, and WS errors are denoted by the red, magenta and blue solid line, respectively. The horizontal coordinate of (**d**) represents the range of SSS (red), SST (magenta), and WS (blue).

6. Conclusions

In contrast to the SMOS and Aquarius, MICAP uses \ multi-frequency (1.4 GHz, 6.9 GHz, 18.7 GHz and 23.8 GHz) radiometers combined with the 1.26 GHz scatterometer to measure SSS, SST, and WS. In this paper, to further simply the instrument configuration of MICAP without loss of the SSS, SST, and WS retrieval accuracies, three different configurations (see Table 2) are compared.

From 40° S to 40° N, the average differences of the latitudinal profiles of SSS, SST, and WS RMS errors between No. 1 and No. 3 configurations are about 0.005 psu, 0.01 °C and 0.03 m/s, respectively.

It indicates that No. 3 configuration also could satisfy the accuracy requirements well for global SSS, SST and WS retrievals during only one satellite pass. Thus, the instrument configuration of MICAP can be simplified (without the 23.8 GHz channel) in the future.

By contrast, from 40° S to 40° N, the average differences of the latitudinal profiles of SSS, SST, and WS RMS errors between No. 1 and No. 2 configurations are about 0.08 psu, 0.1 °C and 0.45 m/s, respectively. Moreover, there are two interesting results for WS RMS error. The WS RMS error with No. 2 shows the obvious reverse change trend compared with No. 1 and No. 3. From 10° S to 10° N, the differences of the latitudinal profiles of WS RMS error between No. 1 and No. 2 are up to 0.78 m/s when the WS is lower than 5 m/s. The results indicate that the 1.26 GHz scatterometer of MICAP is necessary for more accurate SSS, SST, and WS retrievals.

In addition, the SSS, SST, and WS RMS errors are also tested for various noise levels on their first guess values. The results indicate that the retrieval accuracies of SSS, SST, and WS are only affected by the noises added to each first guess values, which may be because the MICAP can simultaneously retrieve SSS, SST, and WS.

Acknowledgments: The authors would like to thank Laboratoire de Physique des Oceans for providing the SSS data, as well as Remote Sensing Systems for providing the initialized oceanic and atmospheric data. This work was supported by the National Key Research and Development Program of China under Grant No. 2016YFC1401006.

Author Contributions Xiaobin Yin conceived and established simulation models; Lanjie Zhang performed the simulate and test; Lanjie Zhang and Zhenzhan Wang analyzed test results of performance for MICAP; Lanjie Zhang wrote the manuscripts; and Zhenzhan Wang and Xiaobin Yin edited the article.

Conflicts of Interest: The authors declare no conflict of interest. The founding sponsors had no role in the design of the study; in the collection, analyses, or interpretation of data; in the writing of the manuscript, and in the decision to publish the results.

References

1. Batteen, M.L.; Collins, C.A.; Gunderson, C.R.; Nelson, C.S. The effect of salinity on density in the California Current System. *J. Geophys. Res. Oceans* **1995**, *100*, 8733–8749. [CrossRef]
2. Emery, W.J.; Meincke, J. Global water masses-summary and review. *Oceanol. Acta* **1986**, *9*, 383–391.
3. Lagerloef, G.S.; Swift, C.T.; Le Vine, D.M. Sea surface salinity: The next remote sensing challenge. *Oceanography* **1995**, *8*, 44–50. [CrossRef]
4. Schmitt, R.W. The ocean freshwater cycle. In *JSC Ocean Observing System Development Panel*; Texas A & M University: College Station, TX, USA, 1994.
5. Reul, N.; Fournier, S.; Boutin, J.; Hernandez, O.; Maes, C.; Chapron, B.; Alory, G.; Quilfen, Y.; Tenerelli, J.; Morisset, S.; et al. Sea surface salinity observations from space with the SMOS satellite: A new means to monitor the marine branch of the water cycle. *Surv. Geophys.* **2014**, *35*, 681–722. [CrossRef]
6. Schmitt, R.W. Salinity and the global water cycle. *Oceanography* **2008**, *21*, 12–19. [CrossRef]
7. Font, J.; Boutin, J.; Reul, N.; Spurgeon, P.; Ballabrera-Poy, J.; Chuprin, A.; Gabarró, C.; Gourrion, J.; Guimbard, S.; Hénocq, C.; et al. SMOS first data analysis for sea surface salinity determination. *Int. J. Remote Sens.* **2013**, *34*, 3654–3670. [CrossRef]
8. Lagerloef, G.; Boutin, J.; Chao, Y.; Delcroix, T.; Font, J.; Niiler, P.; Reul, N.; Riser, S.J.; Schmitt, R.; Stammer, D.; et al. Resolving the global surface salinity field and variations by blending satellite and in situ observations. In *OceanObs 09*; European Space Agency: Paris, France, 2010; pp. 587–597.
9. Le Vine, D.M.; Lagerloef, G.S.; Colomb, F.R.; Yueh, S.H.; Pellerano, F.A. Aquarius: An instrument to monitor sea surface salinity from space. *IEEE Trans. Geosci. Remote Sens.* **2007**, *45*, 2040–2050. [CrossRef]
10. Boutin, J.; Martin, N.; Yin, X.B.; Font, J.; Reul, N.; Spurgeon, P. First assessment of SMOS data over Open Ocean: Part II—Sea surface salinity. *IEEE Trans. Geosci. Remote Sens.* **2012**, *50*, 1662–1675. [CrossRef]
11. Kerr, Y.H.; Waldteufel, P.; Wigneron, J.P.; Martinuzzi, J.A.M.J.; Font, J.; Berger, M. Soil moisture retrieval from space: The Soil Moisture and Ocean Salinity (SMOS) mission. *IEEE Trans. Geosci. Remote Sens.* **2001**, *39*, 1729–1735. [CrossRef]
12. Sabia, R. *Sea Surface Salinity Retrieval Error Budget within the Esa Soil Moisture and Ocean Salinity Mission*; Universitat Politècnica de Catalunya: Barcelona, Spain, 2008.

13. Font, J.; Lagerloef, G.S.; Le Vine, D.M.; Camps, A.; Zanife, O.Z. The determination of surface salinity with the European SMOS space mission. *IEEE Trans. Geosci. Remote Sens.* **2004**, *42*, 2196–2205. [CrossRef]
14. Lagerloef, G.; Colomb, F.R.; Le Vine, D.; Wentz, F.; Yueh, S.; Ruf, C.; Lilly, J.; Gunn, J.; Chao, Y.; Decharon, A.; et al. The Aquarius/SAC-D mission: Designed to meet the salinity remote-sensing challenge. *Oceanography* **2008**, *21*, 68–81. [CrossRef]
15. Entekhabi, D.; Njoku, E.G.; O'Neill, P.E.; Kellogg, K.H.; Crow, W.T.; Edelstein, W.N.; Kimball, J. The soil moisture active passive (SMAP) mission. *Proc. IEEE* **2010**, *98*, 704–716. [CrossRef]
16. Le Vine, D.M.; Lagerloef, G.S.; Torrusio, S.E. Aquarius and remote sensing of sea surface salinity from space. *Proc. IEEE* **2010**, *98*, 688–703. [CrossRef]
17. Fore, A.G.; Yueh, S.H.; Tang, W.; Stiles, B.W.; Hayashi, A.K. Combined Active/Passive Retrievals of Ocean Vector Wind and Sea Surface Salinity with SMAP. *IEEE Trans. Geosci. Remote Sens.* **2016**, *54*, 7396–7404. [CrossRef]
18. Akhil, V.P.; Lengaigne, M.; Durand, F.; Vialard, J.; Chaitanya, A.V.S.; Keerthi, M.G.; Gopalakrishna, V.V.; Boutin, J.; de Boyer Montégut, C. Assessment of seasonal and year-to-year surface salinity signals retrieved from SMOS and Aquarius missions in the Bay of Bengal. *Int. J. Remote Sens.* **2016**, *37*, 1089–1114. [CrossRef]
19. Philipps, S.; Boone, C.; Obligis, E. The role of averaging for improving sea surface salinity retrieval from the Soil Moisture and Ocean Salinity (SMOS) satellite and impact of auxiliary data. *J. Atmos. Ocean. Technol.* **2007**, *24*, 255–269. [CrossRef]
20. Yueh, S.H.; West, R.; Wilson, W.J.; Li, F.K.; Njoku, E.G.; Rahmat-Samii, Y. Error sources and feasibility for microwave remote sensing of ocean surface salinity. *IEEE Trans. Geosci. Remote Sens.* **2001**, *39*, 1049–1060. [CrossRef]
21. Zhang, L.J.; Yin, X.; Wang, Z.Z.; Liu, H.; Lin, M.S. Preliminary Analysis of the Potential and Limitations of MICAP for the Retrieval of Sea Surface Salinity. *IEEE J. Sel. Top. Appl. Earth Observ. Remote Sens.* **2018**, submitted.
22. Gaillard, F. ISAS-13 temperature and salinity gridded fields. *SEANOE* **2015**. [CrossRef]
23. Wentz, F.J.; Ricciardulli, L.; Gentemann, C.; Meissner, T.; Hilburn, K.A.; Scott, J. *Remote Sensing Systems GCOM-W1 AMSR2 Monthly Environmental Suite on 0.25 Deg Grid, Version 7.0*; Remote Sensing Systems: Santa Rosa, CA, USA, 2014; Available online: www.remss.com/missions/amsr (accessed on 1 March 2017).
24. Wentz, F.J.; Ricciardulli, L.; Gentemann, C.; Meissner, T.; Hilburn, K.A.; Scott, J. *Remote Sensing Systems Coriolis WindSat Monthly Environmental Suite on 0.25 Deg Grid, Version 7.0.1*; Remote Sensing Systems: Santa Rosa, CA, USA, 2013; Available online: www.remss.com/missions/windsat (accessed on 2 March 2017).
25. Font, J.; Camps, A.; Borges, A.; Martín-Neira, M.; Boutin, J.; Reul, N.; Mecklenburg, S. SMOS: The challenging sea surface salinity measurement from space. *Proc. IEEE* **2010**, *98*, 649–665. [CrossRef]
26. Yin, X.B.; Boutin, J.; Martin, N.; Spurgeon, P.; Vergely, J.L.; Gaillard, F. Errors in SMOS Sea Surface Salinity and their dependency on a priori wind speed. *Remote Sens. Environ.* **2014**, *146*, 159–171. [CrossRef]
27. Revision 13 SMOS L2 OS Algorithm Theoretical Baseline Document, SO-TN-ARG-GS-0007. Available online: http://www.argans.co.uk/smos/docs/deliverables/delivered/ATBD/SO-TN-ARG-GS-007_L2OS-ATBD_v3.7_110622.pdf. (accessed on 13 March 2017).
28. Meissner, T.; Wentz, F.J.; Ricciardulli, L. The emission and scattering of L-band microwave radiation from rough ocean surfaces and wind speed measurements from the Aquarius sensor. *J. Geophys. Res. Oceans* **2014**, *119*, 6499–6522. [CrossRef]
29. Dinnat, E.P.; Boutin, J.; Yin, X.; Le Vine, D.M.; Waldteufel, P.; Vergely, J.L. Comparison of SMOS and Aquarius Sea Surface Salinity and analysis of possible causes for the differences. In Proceedings of the XXXIth URSI IEEE General Assembly and Scientific Symposium (URSI GASS), Beijing, China, 16–23 August 2014.
30. Klein, L.; Swift, C. An improved model for the dielectric constant of sea water at microwave frequencies. *IEEE J. Ocean. Eng.* **1977**, *2*, 104–111. [CrossRef]
31. Meissner, T.; Wentz, F.J. The emissivity of the ocean surface between 6 and 90 GHz over a large range of wind speeds and earth incidence angles. *IEEE Trans. Geosci. Remote Sens.* **2012**, *50*, 3004–3026. [CrossRef]
32. Zhou, Y.; Lang, R.H.; Dinnat, E.P.; Le Vine, D.M. L-Band Model Function of the Dielectric Constant of Seawater. *IEEE Trans. Geosci. Remote Sens.* **2017**, *55*, 6964–6974. [CrossRef]
33. Zine, S.; Boutin, J.; Font, J.; Reul, N.; Waldteufel, P.; Gabarró, C.; Delwart, S. Overview of the SMOS sea surface salinity prototype processor. *IEEE Trans. Geosci. Remote Sens.* **2008**, *46*, 621–645. [CrossRef]

34. Yin, X.B.; Boutin, J.; Martin, N.; Spurgeon, P. Optimization of L-band sea surface emissivity models deduced from SMOS data. *IEEE Trans. Geosci. Remote Sens.* **2012**, *50*, 1414–1426. [CrossRef]
35. Liebe, H.J.; Hufford, G.A.; Cotton, M.G. Propagation modeling of moist air and suspended water/ice particles at frequencies below 1000 GHz. In Proceedings of the AGARD 52nd Specialists' Meeting Electromagnetic Wave Propagation Panel, Palma de Mallorca, Spain, May 1993.
36. Wentz, F.J.; Meissner, T. *Algorithm Theoretical Basis Document (atbd), Version 2*; RSS Tech. Proposal 121599A-1; Remote Sensing Systems: Santa Rosa, CA, USA, 2000.
37. Isoguchi, O.; Shimada, M. An L-band ocean geophysical model function derived from PALSAR. *IEEE Trans. Geosci. Remote Sens.* **2009**, *47*, 1925–1936. [CrossRef]
38. Liu, H.; Zhu, D.; Niu, L.; Wu, L.; Wang, C.; Chen, X.; Wu, J. MICAP (Microwave imager combined active and passive): A new instrument for Chinese ocean salinity satellite. In Proceedings of the 2015 IEEE International Geoscience and Remote Sensing Symposium (IGARSS), Milan, Italy, 26–31 July 2015; pp. 184–187.

remote sensing

MDPI

Letter

End-to-End Simulation of WCOM IMI Sea Surface Salinity Retrieval

Yan Li [1,*], Hao Liu [1] and Aili Zhang [1,2]

1 Key Laboratory of Microwave Remote Sensing, National Space Science Center, Chinese Academy of Sciences, Beijing 100190, China; liuhao@mirslab.cn (H.L.); 18813185108@163.com (A.Z.)

2 University of Chinese Academy of Sciences, Beijing 100094, China

* Correspondence: liyan@mirslab.cn; Tel.: +86-13521484090

Received: 29 November 2018; Accepted: 17 January 2019; Published: 22 January 2019

Abstract: The Water Cycle Observation Mission (WCOM) is an Earth science mission focused on the observation of the water cycle global climate change intensity through three different payloads. WCOM's main payload is an interferometric microwave imager (IMI). IMI is a tri-frequency, one-dimensional aperture synthesis microwave radiometer operating at the L-, S-, and C-bands to perform measurements of soil moisture and ocean salinity. Focusing on sea surface salinity (SSS), an end-to-end simulator of WCOM/IMI has been realized and tested on climatological data. Results indicate a general agreement between original and retrieved SSS, with a single measurement root mean square error of 0.26 psu and with an orbital measurement of 0.17 psu in open sea. In accordance with previous studies, good results are obtained in open sea, while strong contamination is observed in coastal areas.

Keywords: Water Cycle Observation Mission (WCOM); interferometric microwave imager (IMI); one-dimensional (1D) aperture synthesis radiometer; sea surface salinity (SSS); brightness temperature (T_B)

1. Introduction

The Chinese Water Cycle Observation Mission (WCOM), subject to the Strategic Space Science Priority Project of the Chinese Academy of Sciences, is planned to be launched in the near future. The WCOM aims to observe and track the main parameters related to the global water cycle, including soil moisture, ocean salinity, snow water equivalent, soil freeze-thaw, atmospheric water vapor, and precipitation [1].

To achieve the goals above, the WCOM relies on the following three payloads: (1) An interferometric microwave imager (IMI), a tri-frequency one-dimensional (1D) interferometric microwave radiometer operating at the L-, S-, and C-bands to measure soil moisture and sea surface salinity; (2) a polarimetric microwave imager (PMI), a conically scanning polarimetric radiometer operating at multiple frequencies between the C- and the W-band, with full polarimetric capabilities for most of them (it aims to measure land and sea surface temperature, water vapor, and precipitation); and (3) a dual-frequency polarized SCATterometer (DFPSCAT) to measure the snow water equivalent and soil freeze–thaw [2].

L-band microwave radiometry has been widely agreed as the most effective tool to measure soil moisture and ocean salinity from space. In fact, three satellite missions embarking with L-band radiometers have been launched in the past decade, namely ESA's Soil Moisture and Ocean Salinity mission (SMOS) [3], NASA's Aquarius/SAC-D [4], and Soil Moisture Active/Passive mission (SMAP) [5]. However, the retrieval of salinity is still considered challenging due to the low sensitivity of brightness temperature (T_B) to salinity (from 0.8 K down to 0.2 K per psu, depending on ocean surface temperature, radiometer incidence angle, and polarization [6]), making it hard to meet the accuracy and stability requirements for application and research.

This study focuses on the analysis of sea surface salinity (SSS) using the tri-frequency 1D aperture synthesis radiometer IMI onboard the WCOM satellite, and is structured as follows. Section 2 provides an introduction on the instrument and methods applied in the end-to-end simulation. This is followed by the simulation results, including T_B reconstruction and salinity retrieval, which are presented and discussed in Section 3. Finally, the main conclusions are summarized in Section 4.

2. Concepts and Methods

2.1. Interferometric Microwave Imager (IMI)

The system performance characteristics for the WCOM/IMI are listed in Table 1.

Table 1. Interferometric microwave imager (IMI) design characteristics for Water Cycle Observation Mission (WCOM).

	L-band	S-band	C-band
Frequency (GHz)	1.4135	2.695	6.9
Bandwidth (MHz)	25	8	200
Along-track Resolution (km)	35	20	10
Cross-track Resolution (km)	35–75	20–45	15–30
Radiometric Resolution (K)	0.2	1.5	0.6
Field of View (km)	1000	1000	1000

Like SMOS's MIRAS, IMI adopts aperture synthesis technology to overcome the barrier that the antenna size places at the L-band. Specifically, IMI is a tri-frequency (L-, S-, and C-bands) 1D interferometric radiometer consisting of a deployable 6 m parabolic cylinder mesh reflector and a tri-frequency patch feed array. Additionally, as a 1D interferometric radiometer, IMI has a lower system complexity compared with the two-dimensional MIRAS, indicating that it is much easier to control its in-orbit instrument stability and calibration accuracy.

As mentioned above, besides the L-band, the IMI also operates at the S- and C-bands, which can improve its sensitivity to soil moisture on land and assist in taking measurements of SSS and sea surface temperature (SST) on the ocean.

2.2. End-to-End Simulation Model and Method

The end-to-end simulation for ocean salinity in this paper is based on previous work regarding the 1D interferometric radiometer IMI simulation system [7], the sea radiometer transfer model, and a salinity retrieval algorithm. The simulation system is built on the Matlab platform.

The end-to-end simulation flow is shown in Figure 1. It begins with the input salinity data and ends with the retrieved SSS from the simulated "measured" T_B. In the simulation, the original SSS, together with auxiliary data, were processed into the target T_B scenes in different bands. These then formed the visibility function through the radiometer simulation. Afterwards, the visibility function was again reconstructed into T_B. Finally, the retrieved SSS were achieved through the comparison between the reconstructed T_B and the modeled T_B. The modules used in the end-to-end simulation chain are highlighted as grey cases at the top of Figure 1.

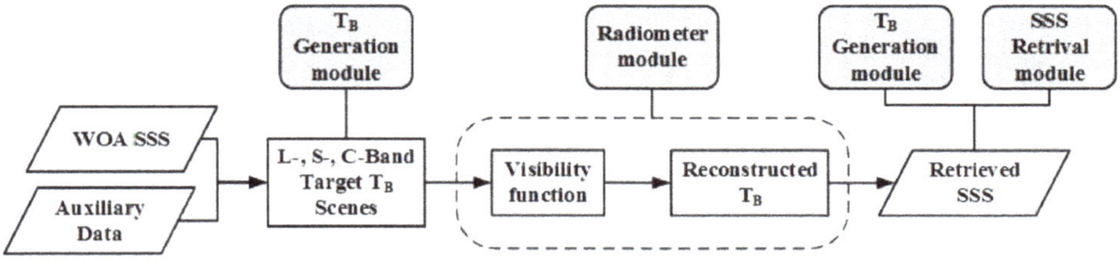

Figure 1. The end-to-end simulation system of sea surface salinity (SSS) for the tri-frequency one-dimensional interferometric microwave imager, which consists of three major modules highlighted as grey cases at the top (i.e., the brightness temperature (T_B) generation module, radiometer module, and sea surface salinity (SSS) retrieval module).

2.2.1. T_B Generation Module

As seen from Figure 1, the T_B generation module was used not only in the measurement scene generation as an input to the radiometer module, but also in the SSS retrieval module to calculate and update the modeled T_B.

According to [8], the T_B at the top of the atmosphere in the Earth reference frame can be calculated using Equation (1).

$$T_B = \left(T_{B,sea} + T_{DN} \cdot \Gamma + T_{GAL} \cdot \Gamma \cdot e^{-\tau_{atm}}\right) \cdot e^{-\tau_{atm}} + T_{UP} \tag{1}$$

where $T_{B,sea}$ is the T_B radiated by the ocean surface and can be further calculated using Equation (2).

$$T_{B,sea} = T_{B,flat} + T_{B,rough} = \left(1 - \Gamma_{fresnel}\right) SST + T_{B,rough} \tag{2}$$

where $T_{B,sea}$ is described as the sum of T_B in the case of a flat sea ($T_{B,flat}$) and additional T_B due to surface roughness ($T_{B,rough}$). $T_{B,flat}$ is the emission of a flat sea surface, which can be described as a function of the SST and the Fresnel reflectivity. $\Gamma_{fresnel}$, which in turns depends on the incident angle (θ) and the complex dielectric constant of sea water (ε). In this paper, ε is calculated using the Klein–Swift model [9] for the L- and S-band and the Meissner–Wentz model [10] for the C-band. For a rough sea surface, $T_{B,rough}$ is computed with a two-scale model [11] for the L- and S-band and is computed with the model from AMSR-E [12] for the C-band.

In Equation (1), T_{DN} and T_{UP} refer to the downward and upward atmospheric radiation, respectively, $e^{-\tau_{atm}}$ is the atmospheric opacity, and τ_{atm} is the optical thickness. The model used for the atmosphere is from the ESA's SMOS mission for the L-band and from AMSR-E [12] for the C-band. Finally, Γ is the sea surface reflection coefficient, which can be expressed as $\Gamma = 1 - \left(\left(T_{B,flat} + T_{B,rough}\right) / SST\right)$, and T_{GAL} is the cosmic and galactic contribution.

After the calculation of T_B in the Earth reference frame, the T_B in the antenna reference frame is obtained through multiplying the polarization rotation matrix as shown below.

$$\begin{bmatrix} T_X \\ 2T_{XY} \\ T_Y \end{bmatrix} = \begin{bmatrix} \cos^2\varphi & \sin^2\varphi \\ -2\cos\varphi \cdot \sin\varphi & 2\cos\varphi \cdot \sin\varphi \\ \sin^2\varphi & \cos^2\varphi \end{bmatrix} \begin{bmatrix} T_{B,H} \\ T_{B,V} \end{bmatrix} \tag{3}$$

where, T_X, T_Y, and T_{XY} are the antenna T_B, φ is the angle of polarization rotation, $T_{B,H}$ and $T_{B,V}$ are the horizontal and vertical components of T_B in the Earth's reference frame, respectively.

2.2.2. Radiometer Module

In Figure 1, the radiometer simulation module processes the "measured" T_B calculated from the T_B generation module into the visibility function using the model of radiometer system. The visibility function calculation is described considering the weighted function of the antenna pattern [13,14], by applying Equation (4).

$$V_{i,j}(u,v) = \iint \frac{T_B(\xi,\eta)}{\sqrt{1-\xi^2-\eta^2}\sqrt{\Omega_i\Omega_j}} F_i(\xi,\eta)F_j^*(\xi,\eta)\widetilde{r}_{ij}(\tau)e^{-j2\pi(u\xi+v\eta)}d\xi d\eta \quad (4)$$

where $F(\xi,\eta)$ is the normalized antenna pattern, which is a complex function including amplitude and phase; $T_B(\xi,\eta)$ is the T_B from the observation scene; $\widetilde{r}_{ij}$ is the fringe-washing function, which accounts for spatial decorrelation effects and depends on the frequency response of the pair of elements collecting the signals being correlated; and ξ and η are the direction cosine coordinates ($\xi = cos\phi sin\theta$, $\eta = sin\phi sin\theta$, with ϕ and θ being the angle in the instrument plane and the angle from the normal to the instrument plane, respectively. For IMI, a 1D radiometer, η = 0 and v = 0 in Equation (4).

Equation (4) is the ideal equation for the visibility function of interferometric synthetic aperture radiometers. To include the image errors introduced by coupling effects between the antennas, $T_B(\xi,\eta)$ in Equation (4) is replaced by $T_B(\xi,\eta) - T_r$, where T_r is the physical temperature of the receivers and assumed to be 300 K in the simulation. This leads to the formulation of the Corbella equation.

The resulting $V_{i,j}(u,v)$ are processed by the T_B reconstruction for generating the "measured" T_B. Due to the difference between antenna patterns described in Section 2.3, it is not proper to directly apply the Fourier-based image reconstruction method. Specifically, the G matrix method [15,16] is used to reconstruct the T_B map as, expressed in Equation (5).

$$\hat{T} = G_i^+ \times V = G_i^H\left(G_iG_i^H\right)^{-1}V \quad (5)$$

where $\hat{T}$ is the reconstructed T_B, $G_i{}^+$ is the pseudoinverse matrix of the G matrix, and subscript i indicates the i-th visibility sample. There are only 18 continuous, non-redundant visibility samples for the eight-element linear feed array, forming a small-sized G matrix, simplifying the calculation of the pseudo-inverse through the Moor-Penrose method, and thus the image reconstruction.

2.2.3. SSS Retrieval Module

The SSS retrieval is based on a nonlinear iterative convergence algorithm where the prior values of the parameters to be retrieved were adjusted in order to minimize a cost function [16]. T_B values were calculated by applying the T_B generation module described in Section 2.2.1 and were compared to the "measured" T_B resulting from the radiometer simulation module.

The L- and S-band-measured T_B data were suitable to retrieve SSS, while the C-band measured T_B was used as auxiliary data to assist in the retrieval of SSS, and to retrieve SST. Thus, the cost function based on all the data of the three bands is expressed as

$$\chi^2{}_2 = \sum_{i=0}^{M1}\frac{\left[Tb_{L,p,meas}-Tb_{L,p,model}\right]^2}{\sigma^2_{Tb,L,p}} + \sum_{i=0}^{M2}\frac{\left[Tb_{S,p,meas}-Tb_{S,p,model}\right]^2}{\sigma^2_{Tb,S,p}} + \sum_{i=0}^{M3}\frac{\left[Tb_{C,p,meas}-Tb_{C,p,model}\right]^2}{\sigma^2_{Tb,C,p}} + \frac{\left[SSS-SSS_{prior}\right]^2}{\sigma^2_{SSS}} + \frac{\left[SST-SST_{prior}\right]^2}{\sigma^2_{SST}} \quad (6)$$

where, $Tb_{L,p,meas}$, $Tb_{S,p,meas}$, and $Tb_{C,p,meas}$ indicate the "measured" T_B at different bands (L-, S-, and C-band, respectively) and polarizations (horizontal and vertical), and $\sigma^2_{Tb,L,p}$, $\sigma^2_{Tb,S,p}$ and $\sigma^2_{Tb,C,p}$, are the corresponding uncertainties. Lastly, SSS_{prior} and SST_{prior} refer to the prior estimated values for SSS and SST associated with the corresponding uncertainties σ^2_{SSS} and σ^2_{SST}. Since the current study focuses on salinity retrieval, the retrieval results of SST are not included.

2.3. Simulation Input

For IMI instrument concept proofing and performance validation, a ground-based demonstrator designed with an eight-element L-band radiometer is developed in 2011 [2] and applied in several ground-based experiments targeting different objects, such as buildings, the sun, the cold sky, a noise point source, etc. A picture of the aforementioned IMI prototype and a schematic of the arrangement of the antenna units in the 1D feed array are shown in Figure 2a and Figure 2b, respectively. The scatterometer units involved in the 1D feed array make the prototype a combined active/passive instrument. Anyway, since the performance of the radiometer instrument is the focus of this paper, only the radiometer feed array is adopted in the simulation.

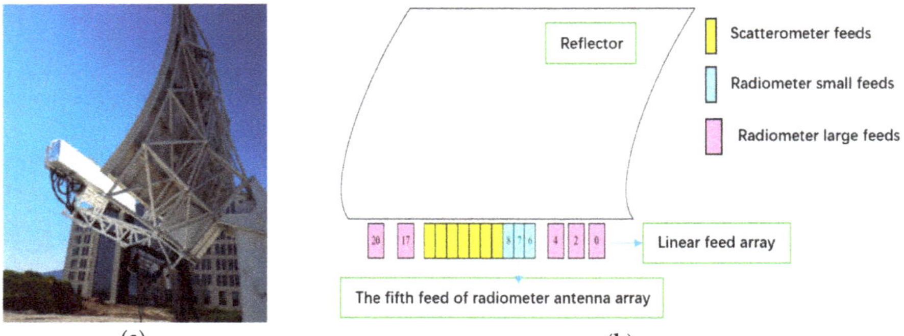

(a) (b)

Figure 2. A ground-based prototype of the one-dimensional interferometric microwave imager (IMI). (**a**) Photo of the ground-based IMI prototype. (**b**) Arrangement of the small and large antenna units in the feed array of the IMI prototype.

Both the actual measured antenna pattern and the distribution of the IMI prototype were applied in the simulation. Specifically, as seen in Figure 2b, the whole eight-element radiometer feed was considered a mixed array constituted by three small-size and five large-size antenna units, in which the minimum antenna spacing of the antenna array Δu was 0.6125 λ. Along-track and cross-track antenna patterns measured from the prototype experiment for each unit are presented in Figure 3a and Figure 3b, respectively.

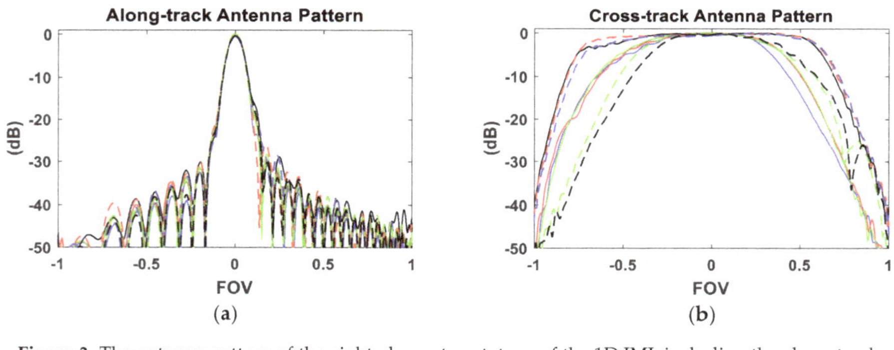

(a) (b)

Figure 3. The antenna pattern of the eight-element prototype of the 1D IMI, including the along-track and cross-track antenna patterns in (**a**) and (**b**), respectively.

The alias-free field of view (AF-FOV) areas were calculated using Δu and applying $\sin\theta_{max} = 1/2\Delta u$, leading to an observation angle included between $-55°$ and $55°$, approximately. All the reconstructed T_B data within the AF-FOV were considered valid for the SSS retrieval.

Other simulation inputs include geophysical data and orbit parameters. The orbit is configured as a sun-synchronous orbit with an altitude of 657 km and inclination of 98°, and the local time of the ascending node is at 6:00 a.m. Geophysical data, including SST, 10-m wind speed, columnar atmospheric water vapor, and cloud liquid water were measured from AMSR-E [17], which are monthly averaged maps that all have spatial resolution of 0.25°. The original salinity map comes from World Ocean Atlas (WOA) 2013 [18], which is an in situ monthly averaged map with a spatial resolution of 1°. In order to be consistent in the simulation, the salinity map is interpolated into a 0.25° × 0.25° dimension.

Besides, the configuration of this simulation included both the measured antenna patterns and the image reconstruction error, and there was no additional noise or drift between the parameters used in the forward models and those used for the retrieval.

3. Results and Discussion

3.1. T_B Reconstruction Results

The specific scene obtained for a sub-satellite point located at 62.68 W, 13.78 N is shown in Figure 4. The results of the T_B generation module at the L-, S-, and C-bands are displayed on the left, center, and right panels, respectively, with horizontal (H-pol) and vertical polarizations (V-pol) on the top and bottom rows, respectively.

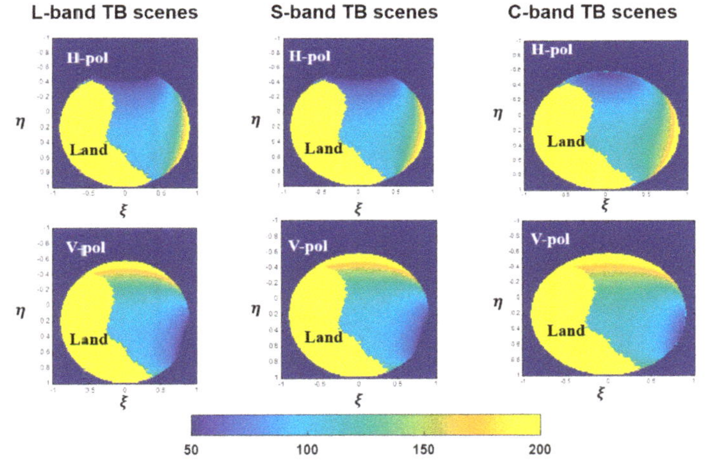

Figure 4. T_B scenes from the forward T_B generation module used as the input of the radiometer simulation. From left to right, the L-, S-, and C-band T_B are presented. The horizontal and vertical T_B are presented on the top and bottom row, respectively.

The T_B scenes in Figure 4 are used as inputs for the successive radiometer simulation to achieve T_B reconstruction results. Due to the lack of measurement data from the S- and C-band antenna array, only T_B reconstruction as a result of the L-band is presented in the following figure, similar results are expected at the other bands.

Since the antenna array of IMI is 1D, the output from the T_B reconstruction of the radiometer module is also 1D. Results of all the measured polarizations are shown in Figure 5, with the reconstructed T_B data represented by the blue solid line and the simulated T_B represented by the black dashed line. Good agreement is found between modeled and retrieved T_B in the AF-FOV areas, with a root mean square error (RMSE) of 0.09 K.

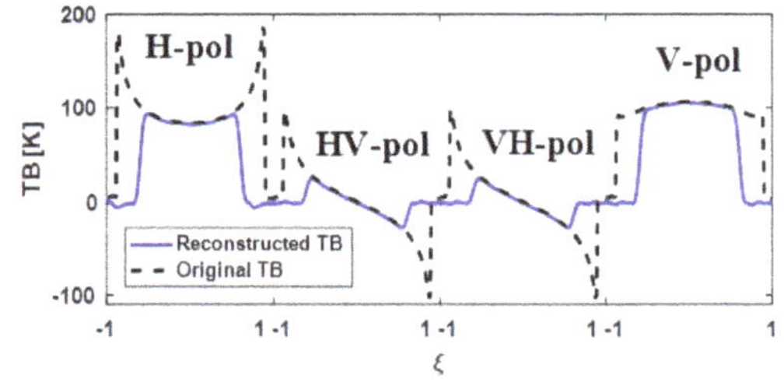

Figure 5. L-band reconstructed T_B output from the radiometer simulation. The blue solid line and the black dashed line present the reconstructed T_B and original simulated T_B, respectively. From left to right are the horizontal polarization (H-pol), cross polarization (HV-pol, VH-pol), and vertical polarization (V-pol) results.

3.2. SSS Retrieval Results

In Figure 6, the results of SSS retrieval from a single measurement are shown. The retrieved SSS from each measurement are mapped into Earth coordinates in order to compare them with the WOA salinity data. The differences between WOA SSS and retrieved SSS are generally within ±0.5 psu, with an RMSE of 0.26 psu.

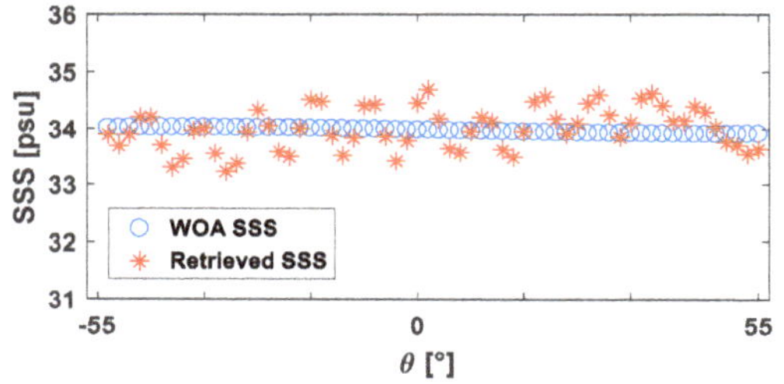

Figure 6. Comparison results of retrieval SSS and World Ocean Atlas (WOA) salinity, which are represented by the red stars and the blue circles, respectively.

Figure 7a presents the half-orbital (about 2930 measurements) results of the SSS retrieval and the original WOA SSS, while their difference is shown in Figure 7b.

As can be seen from Figure 7, although the SSS retrieval results in the center of the orbit and in the open sea (not near land or poles) are proven to be good, there are large errors, especially at the edge of the orbit and when approaching land. Those errors in the results can be attributed to the aliasing effect, instrument imperfection, and the T_B retrieval algorithm. Although T_B measured from observation angles of −55°–55° are treated as AF-FOV and are used to retrieve SSS, the results still presented large errors near −55° and 55° due to the low sensitivity of T_B to SSS. Even small errors in the reconstructed T_B introduced large errors to the retrieved SSS.

To focus on the retrieval results in open sea, orbital SSS retrieval results from the South Pacific Ocean, South Atlantic Ocean, and Indian Ocean—namely Open Sea 1, 2, and 3, respectively—are selected in this study. Figure 8 presents the retrieved SSS of the three open sea areas on the top row and the corresponding error maps compared with the WOA SSS data on the bottom row. To avoid the large errors around the edge of the orbital results, a narrower AF-FOV the from an observation angle of −45–45° are adopted for the retrieval in the three open sea areas.

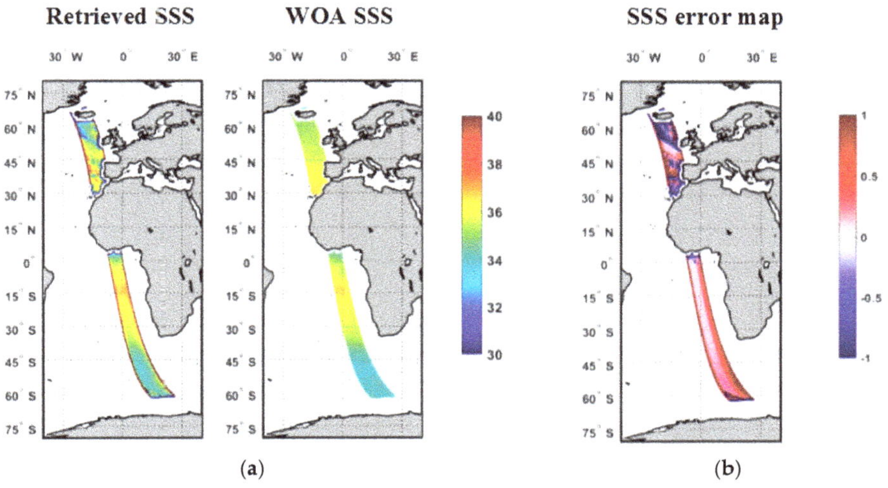

Figure 7. (**a**) Half-orbital SSS retrieval result and the WOA salinity, and (**b**) corresponding error map when compared with WOA salinity.

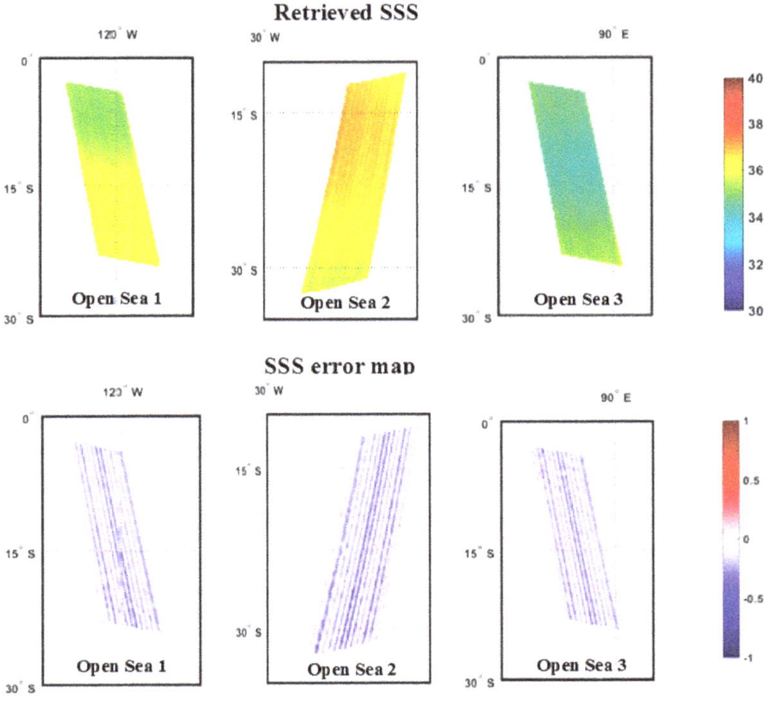

Figure 8. Simulation results of the orbital retrieved SSS in the three selected open sea areas. SSS retrieval results and the corresponding SSS error maps compared with WOA salinity are shown in the top and bottom rows, respectively.

As seen from Figure 8, the retrieval results in open sea are much better than the whole half-orbital result in Figure 7. Table 2 displays quantitative results of the retrieved SSS in the three open sea areas, including their RMSE and standard deviations (std). From the quantitative analysis, the accuracy of SSS retrieval in open sea is about 0.17 psu, which satisfied the objective of an accuracy of 0.2 psu for salinity observation.

Table 2. Root mean square error (RMSE) and standard deviation (std) of the retrieved SSS for the three selected open sea areas.

Open Sea	1	2	3
Area Location	South Pacific Ocean	South Atlantic Ocean	Indian Ocean
RMSE (psu)	0.1680	0.1779	0.1655
std (psu)	0.1209	0.1296	0.1209

Compared with 2D instruments, such as SMOS/MIRAS, the imaging results of the 1D radiometer system manifested larger Gibbs errors [19], even for the observation of open seas. Also, the situation became worse when approaching land. This can be attributed to the number of antenna elements of the 1D radiometer system (i.e., eight) used in this paper being much less than that used in the SMOS/MIRAS 2-D system (i.e., 69). The reduction of the antenna elements leads to the reduction of baselines and, eventually, results with a more severe sampling truncation in the spatial frequency domain. This is challenging for systems with small scale antenna elements, like our IMI prototype. An improved system, which may consist of more antenna elements in the feed array, will be developed in the future based on the IMI prototype. Although larger scale antenna arrays will reduce the Gibbs error, land contamination in the observations cannot be eliminated. Many studies devoted to reducing land contamination in SMOS have been carried out [20]. Future work will also focus on the correction of land contamination for our designed system.

4. Conclusions

This paper introduces an end-to-end simulation system to verify the capacity of measuring salinity for a designed system IMI, which is a tri-frequency (L-, S-, and C-bands), one-dimensional aperture synthesis radiometer.

The end-to-end simulation system consisted of three main modules: a tri-frequency T_B generator, radiometer, and SSS retrieval. First, T_B scenes for the three bands are produced by the T_B generator module. Then, the reconstructed T_B are outputted from the process of the radiometer module. The experimental measurements of the antenna patterns of a ground-based prototype of IMI are used in the radiometer module simulation.

The simulation was based on the actual antenna patterns and distribution of the IMI prototype and included image reconstruction errors. Additionally, it is mentioned that forward model and retrieval apply the same parameters, as well as that no error was considered on the geophysical data. With this configuration, results indicate a RMSE of 0.26 psu for a single measurement in open sea. Large errors near the edge of the orbit and strong land contamination are found in the half orbital SSS retrieval results. The errors in the results can be attributed to the aliasing effect and the low sensitivity of T_B to SSS. Then, three areas are selected in order to focus on the retrieval results in open sea. Simulation results in open sea presented a much better accuracy of 0.17 psu.

Besides the possible underestimation due to the aforementioned simulation configuration, large errors existed in SSS retrieval, which mainly originated from a spatial bias in the radiometer system. An improved system that contains more antenna elements will reduce those errors. To achieve better salinity accuracy, land contamination correction must be done in the future.

Author Contributions: H.L. provided concept and parameters of WOCM/IMI; Y.L. and A.Z. established simulation models; Y.L. performed the simulation and tested the simulation results; Y.L. wrote the manuscripts; and H.L. edited the article.

Funding: This research received no external funding.

Acknowledgments: The authors would like to thank Remote Sensing Systems for providing the geophysical data of AMSR-E, and National Oceanic and Atmospheric Administration for providing the in-situ sea surface salinity data of WOA.

Conflicts of Interest: The authors declare no conflict of interest. The founding sponsors had no role in the design of the study; in the collection, analyses, or interpretation of data; in the writing of the manuscript, and in the decision to publish the results.

References

1. Shi, J.; Zhao, T.; Du, J.; Ji, D.; Xiong, C.; Dong, X.; Liu, H.; Wang, Z.; Jiang, L.; Du, Y. Observing Earth's water cycle from space. In Proceedings of the SPIE International Asia-Pacific Environmental Remote Sensing Symposium, Beijing, China, 13–17 October 2014.
2. Liu, H.; Niu, L.; Wu, L.; Zhang, C.; Zhang, X.; Yin, X.; Wu, J. IMI (Interferometric Microwave Imager): AL/S/C tri-frequency radiometer for Water Cycle Observation Mission (WCOM). In Proceedings of the 2016 IEEE International Geoscience and Remote Sensing Symposium (IGARSS), Beijing, China, 10–15 July 2016.
3. Kerr, Y.H.; Waldteufel, P.; Wigneron, J.P.; Martinuzzi, J.A.M.J.; Font, J.; Berger, M. Soil moisture retrieval from space: The Soil Moisture and Ocean Salinity (SMOS) mission. *IEEE Trans. Geosci. Remote Sens.* **2001**, *39*, 1729–1735. [CrossRef]
4. Lagerloef, G.; Colomb, F.R.; Le Vine, D.; Wentz, F.; Yueh, S.; Ruf, C.; Lilly, J.; Gunn, J.; Chao, Y.; Feldman, G.; et al. The Aquarius/SAC-D mission: Designed to meet the salinity remote-sensing challenge. *Oceanography* **2008**, *21*, 68–81. [CrossRef]
5. Entekhabi, D.; Njoku, E.; O'Neill, P.; Spencer, M.; Jackson, T.; Entin, J.; Im, E.; Kellogg, K. The soil moisture active/passive mission (SMAP). In Proceedings of the IGARSS 2008 IEEE International Geoscience and Remote Sensing Symposium, Boston, MA, USA, 7–11 July 2008; Volume 3.
6. Yueh, S.H.; West, R.; Wilson, W.J.; Li, F.K.; Njoku, E.G.; Rahmat-Samii, Y. Error sources and feasibility for microwave remote sensing of ocean surface salinity. *IEEE Trans. Geosci. Remote Sens.* **2001**, *39*, 1049–1060. [CrossRef]
7. Jin, M.; Liu, H.; Wu, L.; Wu, L.; Wang, R.; Zhang, C.; Yin, X.; Zhao, T.; Sun, W.; Cui, H.; et al. Task Simulation and External Error Sources Analysis for an Ocean Salinity Mission with One-dimensional Synthetic Aperture Microwave Radiometer. *Remote Sens. Technol. Appl.* **2017**, *32*, 346–355.
8. Zine, S.; Boutin, J.; Font, J.; Reul, N.; Waldteufel, P.; Gabarró, C.; Tenerelli, J.; Vergely, J.-L.; Talone, M.; Delwart, S.; et al. Overview of the SMOS Sea Surface Salinity Prototype Processor. *IEEE Trans. Geosci. Remote Sens.* **2008**, *46*, 621–645. [CrossRef]
9. Klein, L.; Swift, C.T. An improved model for the dielectric constant of sea water at microwave frequencies. *IEEE Trans. Antennas Propag.* **1977**, *25*, 104–111. [CrossRef]
10. Meissner, T.; Wentz, F.J. The emissivity of the ocean surface between 6 and 90 GHz over a large range of wind speeds and Earth incidence angles. *IEEE Trans. Geosci. Remote Sens.* **2012**, *50*, 3004–3026. [CrossRef]
11. Wentz, F.J. A two-scale scattering model for foam-free sea microwave brightness temperatures. *J. Geophys. Res.* **1975**, *80*, 3441–3446. [CrossRef]
12. Wentz, F.J.; Meissner, T. *AMSR Ocean Algorithm, Algorithm Theoretical Basis Document*; Remote Sensing System: Santa, Rosa, CA, USA, 2000.
13. Corbella, I.; Duffo, N.; Vall-llossera, M.; Camps, A.; Torres, F. The visibility function in interferometric aperture synthesis radiometry. *IEEE Trans. Geosci. Remote Sens.* **2004**, *42*, 1677–1682. [CrossRef]
14. Vall-Llossera Ferran, M.M.; Corbella Sanahuja, I.; Torres Torres, F.; Camps Carmona, A.J.; Colliander, A.; Martín Neira, M.; Ribó Vedrilla, S.; Rautiainen, K.; Duffo Ubeda, N. MIRAS end-to-end calibration: Application to SMOS L1 processor. *IEEE Trans. Geosci. Remote Sens.* **2005**, *43*, 1126–1134.
15. Corbella, I.; Torres, F.; Camps, A.; Duffo, N.; Vall-llossera, M. Brightness-Temperature Retrieval Methods in Synthetic Aperture Radiometers. *IEEE Trans. Geosci. Remote Sens.* **2009**, *47*, 285–294. [CrossRef]
16. Tanner, A.B.; Swift, C.T. Calibration of a synthetic aperture radiometer. *IEEE Trans. Geosci. Remote Sens.* **1993**, *31*, 257–267. [CrossRef]
17. AMSR2/AMSR—Remote Sensing Systems. Available online: http://www.remss.com/missions/amsr/ (accessed on 1 September 2018).
18. World Ocean Atlas 2013 Version 2. Available online: https://www.nodc.noaa.gov/OC5/woa13/ (accessed on 1 September 2018).

19. Zhang, A.; Liu, H.; Wu, L.; Wu, J.; Niu, L.; Guo, X. Spatial Bias Analysis and Calibration for the L- band 1-D Synthetic Aperture Radiometer. *J. Remote Sens.* under review.
20. Li, Y.; Li, Q.; Lu, H. Land Contamination Analysis of SMOS Brightness Temperature Error Near Coastal Areas. *IEEE Geosci. Remote Sens. Lett.* **2017**, *14*, 587–591. [CrossRef]

MDPI
St. Alban-Anlage 66
4052 Basel
Switzerland
Tel. +41 61 683 77 34
Fax +41 61 302 89 18
www.mdpi.com

Remote Sensing Editorial Office
E-mail: remotesensing@mdpi.com
www.mdpi.com/journal/remotesensing

www.ingramcontent.com/pod-product-compliance
Lightning Source LLC
Chambersburg PA
CBHW041753220926
51847CB00014B/184
9783039210763